The Social Behavior of the Bees

The Social Behavior of the Bees

A Comparative Study *by Charles D. Michener*

The Belknap Press of Harvard University Press, Cambridge, Massachusetts, 1974

Library of Congress Card Catalog Number 73-87379
SBN 674-81175-5
Printed in the United States of America

Preface

Perhaps it is because man is a thoroughly social species that he has long shown an interest in social insects. It is true that some social insects sting, bite, eat wood, can be robbed of honey and wax, pollinate crops, or for other practical reasons are important to us. But quite aside from practical considerations, man has long had emotional or intellectual reactions to social insects (the termites, the ants, some wasps, and some bees) simply because they are social. The colonial life and cooperative activity of other species intrigue us because these are also aspects of our own natural history.

Recent studies of the social behavior of the honeybee, not only those on communication of the distance and direction of food sources but also those on chemical communication of various kinds, on caste determination, and on sensory and nervous equipment, have directly or indirectly stimulated much work on all social insects. Such studies encouraged the establishment of an international journal on these insects (*Insectes Sociaux*), as well as major research journals on bees (*Journal of Apicultural Research, Apidologie*), and ultimately promoted the writing of the present volume.

While this book has been in various stages of consideration or preparation since 1967, I have tried to be aware of pertinent literature through 1972. Many publications appearing during the first half of 1973 also have been considered. The number of recent publications in the field is impressive, as indicated by examination of the quarterly journal *Apicultural Abstracts,* running to some 250 pages annually. Although the hard work of many people over the decades and centuries provides the background on which a review such as this is based, the current activity is such that keeping the manuscript up to date during the process of writing and reviewing was a significant task.

It is a pleasure to acknowledge as many as possible of the numerous persons who have helped in one way or another in the preparation of this book. The illustrations were obtained from many sources. Some have been modified or used unchanged from earlier publications, as is acknowledged in the captions. Others were prepared with skill and care by Mr. Barry Siler of the University of Wyoming, Mrs. Ann Schlager of Lawrence, Kansas, and Mr. J. M. F. de Camargo of Ribeirão Preto, Brazil, as noted in the captions. Sources of photographs are also indicated in captions. I wish to acknowledge deep indebtedness to all the authors and publishers who have freely given permission for the use of illustrations, as well as to artists and photographers whose previously unpublished works are included here.

The preparation of the text was enormously facilitated by the secretarial skill and care of Mrs. Joetta Weaver, who was also responsible for maintenance of the all-important literature files. For careful reading and criticizing of the manuscript, I wish to thank my wife, Mary H. Michener, as well as Thomas P. Snyder. Dr. Karl A. Stockhammer of the University of Kansas read several chapters and made valuable suggestions. Dr. Walter C. Rothenbuhler of Ohio State University kindly read and provided needed help with Chapter 8. Especially important to me has been the thoughtful reading of most of the chapters by Dr. Rudolf Jander, also of the University of Kansas. I have not followed all of his suggestions, but invariably found them useful in one way or another, and many of them are incorporated into the text.

Over a period of many years the National Science Foundation has provided grants that permitted work on bees at the University of Kansas to prosper and that enabled me to visit numerous parts of the world to investigate often strange and intriguing apoid inhabitants. On two occasions the John Simon Guggenheim Memorial Foundation provided grants essential for this travel (Africa, Brazil). At other times opportunity for necessary travel was provided by the Fulbright program of the United States government (Australia), by the Campanha Nacional de Aperfeiçoamento de Pessoal de Nivel Superior and the Conselho Nacional de Pesquisas, Rio de Janeiro (Brazil), and by a variety of other agencies, to all of which I am indebted.

Especially important has been support from the University of Kansas and the National Science Foundation for my students and former students, whose names appear repeatedly among the following pages.

Contents

Introduction

This book, generally speaking, is concerned with the distinctively social attributes of the bees. The features common to solitary relatives as well as the social bees are omitted, except for brief treatments where needed as background. The voluminous literature on the morphology, sensory equipment, and physiology of the honeybee is largely excluded from consideration, except where there is some obvious or probable relation to social behavior. For the most part, solitary bees as well as many other insects have similar equipment and physiology. No attention is given here to practical aspects of the management of bees as pollinators or as honey and wax producers, although a few key references are included.

The origin and evolution of social behavior is a recurrent theme in this book. For such discussion bees are particularly appropriate subjects, because they show a broad spectrum of stages in the evolution of social behavior; in other social insects the intermediate, slightly social forms between the solitary and the fully social are scarce or totally extinct. Moreover, some bees have what is probably the most complex behavior known among invertebrates. And bees make constructs (nests) which result from behavior (often the summation of the behavior of many individuals), but which can be studied and compared more readily than behavior patterns themselves. I have tried to provide a comparative account of the social behavior of the bees, with enough references so that prospective researchers and those desiring more detailed information can find their way easily into the research literature. Such an account is appropriate now because of the recent vast growth of technical literature in the field.

To most people, the word *bee* suggests only the common honeybee, perhaps also bumblebees; but the bees are in fact a superfamily of the order Hymenoptera containing an estimated twenty thousand species. The common honeybee, *Apis mellifera,* is only one among these many species, relatively few of which are social in the usual entomological sense (that is, eusocial; see Chapter 5, G). One objective is to place the honeybee where it belongs within this large group of insects, instead of treating it as something apart.

In the chapters that follow, some of the social bees receive much more attention than others, for two reasons: (1) because much more is known about the lives of some, and (2) because the social systems of some are so much more complex that more pages are required to explain them. For an understanding of social evolution in the bees it would be important to distinguish these two categories. Unfortunately it is not easy to do so. I like to think that most forms receiving little attention in this book exhibit little in the way of social interactions, so that not much space is required for them, while those receiving much space need it for a description of the complexity of their sociality. In a general way this is true; the more highly eusocial forms have been studied more because of their conspicuous sociality. Yet in detail it is not true. For example, there is no doubt of the complexity and diversity of the social life of the stingless honeybees (Meliponini). There should be as much to say about them as about the true honeybees, *Apis.* Yet far less is known about the stingless honeybees; less space is devoted to them simply because they have not been studied in as much detail as *Apis.* They live in areas where there are fewer biologists than in the temperate regions, where most of the work on *Apis* has been done.

Studies of many groups of social bees are still in the early stages—we are learning what kinds exist and the salient facts of their natural history. The first experimental work on halictid social biology is only now being done. At the other extreme, work on *Apis* has long since extended into the physiological and sociological areas; extensive experimental studies have been carried on for many years.

Presumably the next major steps will be into the population biology of social bees, an area scarcely touched upon here, except to some degree in Chapter 20. The principles in this field for bees do not differ from those for other social Hymenoptera. Even if I were able to do so, which I am not, there would be little point in entering deeply into the theory of social and evolutionary biology for bees, for this area has been beautifully treated by Wilson in his book *The Insect Societies* (1971). Much of the theoretical background for the present work should be sought in Wilson's book.

Few authors of books like this imagine that their readers will start at the beginning and read to the end. Even if a reader is not looking for some specific idea or fact, he is likely to have an area of interest somewhat different from the whole book.

As the contents show, the text is divided into three parts. Part I contains background material and could have been an appendix, except for its size and the need for background at the outset. Some bee specialists and other entomologists will want to skip much of Part I, especially Chapter 1. Those interested in behavioral mechanisms or social behavior for its own sake, rather than comparative or evolutionary behavior, would be well advised to bypass the whole of Part I and use the glossary and index to help with any unfamiliar terminology. Chapter 3, on classification, will be best used for reference purposes—to place genera and higher groups mentioned later in the book—except by the few persons interested in a comprehensive classification of the bees.

Part II is the heart of the book, containing the general accounts of bee societies, their functions and evolution.

Part III contains additional material usually not related to obvious principles, but characteristic of particular groups of bees. It is organized according to the main groups of social bees.

A reader seeking information on a particular group of bees, for example the bumblebees, should first open to the appropriate chapter (28) in Part III. He should then check the index for material on bumblebees in Parts I and II. These sections, plus Chapter 28 itself, will give him all the information in the book on these bees.

Part I Melittological Background[1]

This part, consisting of Chapters 1 to 4, is intended to provide selected background information that is needed for a ready understanding of the subsequent chapters. These four chapters make no attempt to serve as comprehensive treatises. Some sections are abbreviated to the point of being scarcely more than outlines. A treatment of these topics in depth and detail would be inappropriate in a book on the *social behavior* of the bees. Thus morphological material is presented not for itself but to indicate something about socially significant structures, for example, structures for carrying food or glands for producing communicative substances (pheromones). Or again, while the classification of bees is outlined in Chapter 3, details are presented only for the social forms and their close relatives.

1. Melittology is the study of any or all groups of bees, that is, the superfamily Apoidea. Apiculture has to do with the management and scientific background for the management of honeybees, usually for honey or wax production or for pollination.

1 *Bees, Their Development, Structure, and Function*

The bees are insects constituting what is usually called a superfamily, the Apoidea. They are a group of flower-visiting wasps that has abandoned the wasp habit of provisioning nests with insect or spider prey and instead feeds its larvae with pollen and nectar collected from flowers or with glandular secretions ultimately derived from the same sources. They are a large and diversified group, here considered to consist of nine families (Chapter 3). They range from slender wasplike forms to the familiar robust shape and from gnat-sized species (2 mm in length) to giants 39 mm long.

In rather technical terms, bees have the basic features of sphecoid wasps (Comstock, 1924), including the characteristic form of the pronotum of that group (short, with a lobe on each side below the base of the forewing, and not reaching the tegula). They differ from such wasps in possessing at least a few plumose (branched) hairs, in having the hind basitarsus without a strigilis and at least slightly broader than the following tarsal segments, and in having the seventh metasomal tergum of the female fully separated into two hemitergites instead of with a continuous, more or less sclerotized band connecting broader lateral areas. Larvae of bees differ from those of sphecoid wasps in usually having only one papilla (the maxillary palpus) on the maxilla instead of two.

The presumably primitive family Colletidae differs from the Sphecoidea, in group features, only in the few characteristics enumerated above, although more specialized bees are far more distinctive. The characters of the plumose hairs and broad hind basitarsus are probably related to the gathering of pollen for food, and this latter function seems far more meaningful than the minor anatomical differences. It seems logical to unite the Sphecoidea and the Apoidea in a single superfamily, as was done long ago, for example by Comstock (1924).

A. Development

The ontogenetic development of bees is similar to that of other holometabolous insects, that is, insects with complete metamorphosis. Each individual passes through egg, larva, pupa, and adult stages. The larva and adult are the feeding stages. The larva is the growing stage, the pupa the stage of differentiation and formation of adult structures. The pupa is sometimes found within a cocoon spun by the larva, and all the immature stages are, in bees, ordinarily found in cells made and provisioned by the adults.

1. Eggs

The eggs are moderate-sized to extremely large, elongate, and usually slightly curved. In many species each female lays only a few eggs (see Chapter 1, B2c). Such forms often have rather large eggs, and the egg size appears to reach a maximum in the almost solitary carpenter bees of the genus *Xylocopa* in which eggs over 15 mm long, the largest insect eggs, are laid by female bees having a total length of less than 30 mm (Iwata, 1964). Some of the subsocial bees also have large eggs while others, for example the South African species of *Allodapula,* have rather small eggs. Finally, in the highly social bees which lay enormous numbers of eggs, they are small.

Most bee eggs hatch within a few days after egg laying. The embryology has been described in detail for *Apis* (Nelson, 1915; Du Praw, 1967), for *Trigona* (Beig, 1971) and for *Chalicodoma muraria* (Carrière and Bürger, 1897). Because the chorion of a bee egg is soft and membranous rather than hard and shell-like, hatching is difficult to discern. In some forms the very delicate chorion splits and is shed, but in most it appears to be dissolved, the larva becoming gradually fully exposed without ever recognizably working its way out of a chorion or pushing the latter back or to the undersurface of its body. *Apis* falls in this latter group, its chorion dissolving (disappearing) from the middle toward both ends of the egg (Du Praw, 1961).

2. Larvae

The larva is a whitish, legless grub (Figures 1.1 and 1.2) which either feeds upon the food stored with it, in mass provisioning forms, or is fed at intervals by adult bees, in progressive feeders. It therefore has little ability to move and its sensory and protective features are greatly reduced as compared with primitive Hymenop-

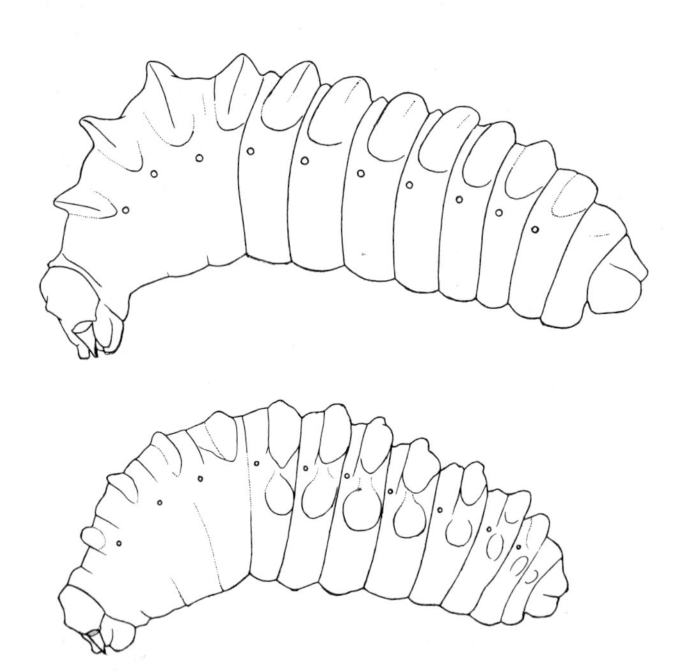

FIGURE 1.1. Larvae of halictine bees. *Above, Lasioglossum (Evylaeus) kincaidii; below, Augochlora pura.* (From Michener, 1953b.)

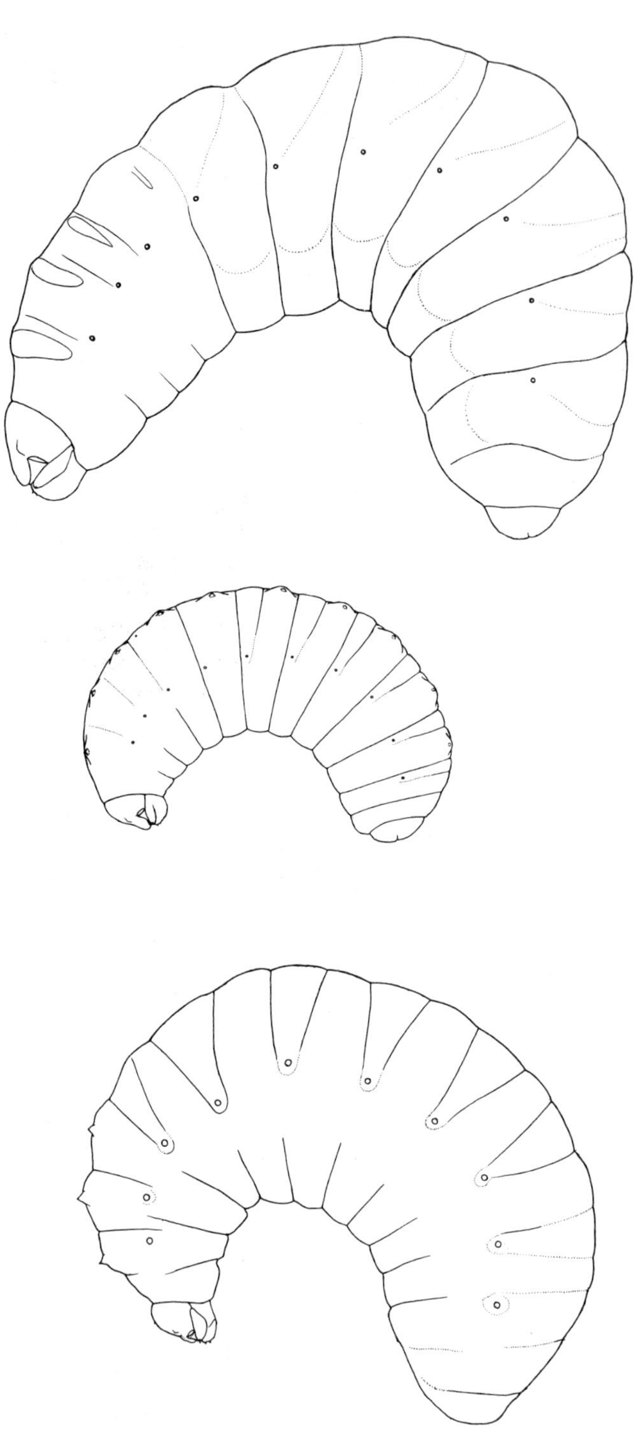

FIGURE 1.2. Larvae of apine bees. *Top to bottom: Apis mellifera, Trigona cupira,* and *Bombus americanorum.* (From Michener, 1953b.)

tera. In most families the larvae are nearly hairless and the antennae are mere cephalic tubercles. Larvae usually grow rapidly; the number of molts is difficult to determine because the cuticle is so delicate, but in various Apidae there are five (Oliveira, 1960). Hackwell and Stephen (1966) report five larval stages also for *Nomia melanderi.* In some other bees there are apparently only three, but the number has not been fully verified for such commonly social groups as the Halictinae and Ceratinini. Batra (1966b) reported five larval stages in *Lasioglossum zephyrum,* and Syed (1963) four for *Braunsapis* and probably *Exoneura.*

The last larval stadium is divided into two parts. The first part is a feeding stage similar to the preceding stadia and, in fact, a great deal of larval growth occurs in this stage. The larva is still very soft, delicate, and usually considerably curled; one can usually see the food or fecal mass in the midgut through the body wall, even in those species in which defecation begins when the larva is only about half grown. In the second part of the last larval stage the cuticle becomes less delicate and often is finely wrinkled; the body straightens and lies lengthwise in the cell instead of being curled. Such a larva is called a prepupa; some authors term it a postdefecating larva or a pharate pupa. In the majority of bee larvae, defecation does not occur until most or all of the food has been consumed; exhaustion of the food supply, cessation of feeding, and emptying of the gut roughly coincide with acquisition of the prepupal body form and appearance. No molt occurs at this time but the larva becomes more opaque. Compared with earlier larval stages as well as with the pupa, the prepupal stage is relatively resistant and it commonly lasts at least as long as the other larval stages taken together. In many bees it is the stage that survives long periods of adversity, sometimes (although not in any social bees) several years.

Figures 1.3 and 1.4 show the gut of the larva and its conversion through metamorphosis to that of the pupa and adult. In parasitic and aculeate Hymenoptera the connection between the midgut and hindgut is typically closed during the feeding stage of the larva and the first defecation occurs with the change into the prepupal condition, as indicated above. The connection closes again during metamorphosis, as does that between the foregut and midgut. Most of the social bees show these closures, as indicated by the defecation pattern described, but the allodapine bees, in addition to megachilids and others,

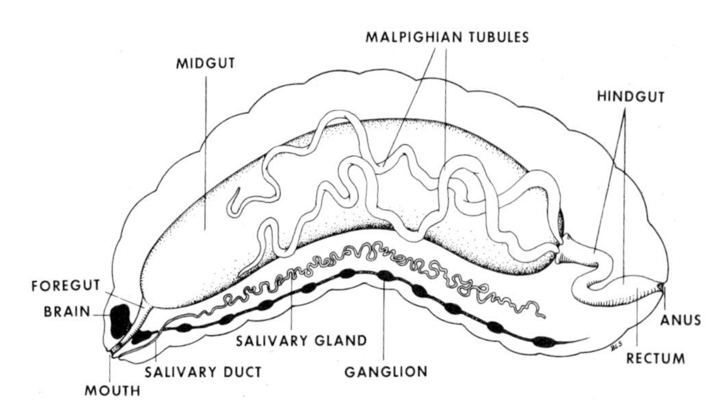

FIGURE 1.3. Diagrammatic view of a bee larva (based upon *Apis*) showing the alimentary tract, Malpighian tubules, and the central nervous system. (Based on Lotmar, 1945.)

start defecation when the larvae are younger, sometimes when they are half grown. Details of larval structure and development of *Apis* are reviewed by Jung-Hoffmann (in Chauvin, 1968).

3. Pupae and teneral adults

The pupa is a relatively delicate whitish stage having the form of the adult insect (Figure 1.4, E) but with the appendages folded and the wings small. Pigmentation of the adult cuticle soon shows through and after a few days the pupal cuticle is shed and the adult bee emerges.

The newly expanded wings of the adult are usually milky and the body integument in some species is paler than in mature bees. In most species full adult pigmentation develops within a day or so, but in some of the allodapine bees, in the Meliponini, and to a lesser extent in the Bombini, full adult pigmentation may not appear for many days. Such callow or teneral (incompletely pigmented, usually with the cuticle somewhat soft) adults are common in the nests and, in some species, even flying about outside.

B. Adults

The adult is the active stage in bees, as in other Hymenoptera. Since the immature stages are nearly immobile and in most species, moreover, are closed in their cells and inaccessible to adults, it is among adults that most of the social interactions have evolved. I com-

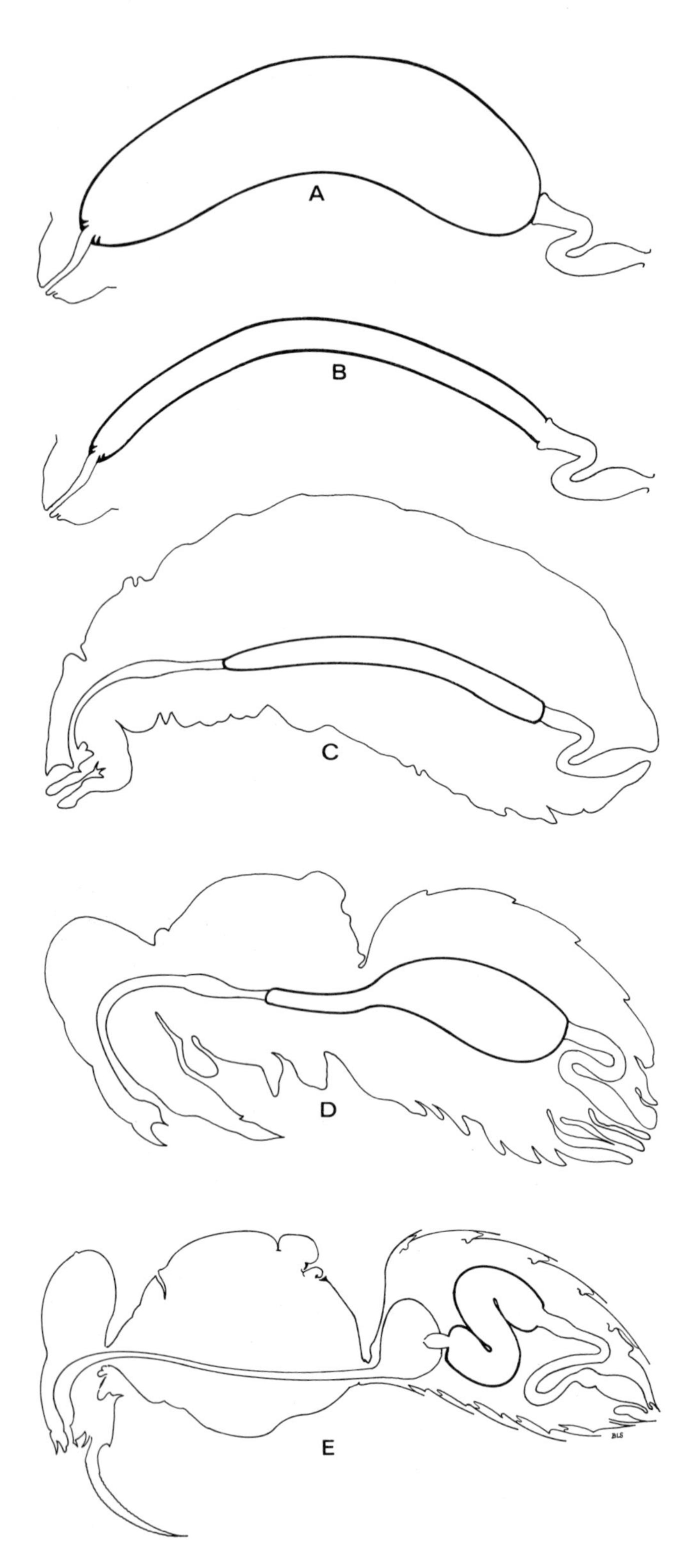

ment below largely on those features of adult structure and function which seem especially involved in social behavior and which therefore need to be understood in order to appreciate succeeding chapters.

Bees partake of all the ordinary morphological and physiological features of insects. Detailed accounts of morphology and physiology of the honeybee have been written (Snodgrass, 1910, 1925, 1956; Zander, 1951). External or skeletal morphology of certain other bees has been treated in some detail by Michener (1944) (for *Anthophora*), Eickwort (1969a) (for *Pseudaugochloropsis*), Camargo, Kerr, and Lopes (1967) (for *Melipona*), and Urban (1967) (for *Thygater*).

1. External structure

a. The body. As in most other Hymenoptera, the division of the body into three major tagmata, head, thorax, and abdomen, does not fully correspond to that of other insects. This is because in the suborder Apocrita (or Clistogastra) to which bees belong, the first abdominal segment is immovably fused with and in fact fully incorporated into the thorax. This segment is called the propodeum. Therefore the division between the second and third tagmata is one segment posterior to the comparable division in other insects. For this reason the terms mesosoma (thorax + first abdominal segment) and metasoma (abdominal segments 2, 3, and so on) have been proposed for morphological usage in the Apocrita (Michener, 1944).

For behavioral and taxonomic studies, no confusion results from using the familiar word thorax instead of mesosoma.

For the abdomen the situation is not so simple, because one constantly needs to refer to the segments by number. The need arises not only in every taxonomic description

FIGURE 1.4. (*A*) Diagrammatic view of the alimentary tract (and head) of a bee larva (based on *Apis,* as in Figure 1.3). The midgut is shown in heavy lines, the foregut and hindgut in light lines. Note that the midgut does not open into the hindgut. The subsequent diagrams show the changes in the gut during metamorphosis. (*B*) represents the gut of a prepupa, with midgut and hindgut connected and the midgut contents voided; (*C*) shows the gut as the pupal body form is being assumed, and (*D*) shows the pupa, both with the midgut closed at each end; (*E*) shows the gut of the adult and its position within the body. (Based on Lotmar, 1945.)

but also in behavioral studies where one has to indicate the locations of glands and the like. To apply numbers one must know where to start. The first abdominal segment is the propodeum, but few students have started counting there. For bees, most students start with the first metasomal segment, that is, the first segment of the tagma that looks like the abdomen, the segment that is actually the second abdominal segment. Other hymenopterists use the word gaster for the swollen part of the abdomen behind the petiole and number the gastral segments. As the petiole can consist of either one or two segments in ants and of only the very base of one segment in bees and various other groups, a given gastral segment may be quite a different segment numerically in another hymenopteron. The numerical chaos is illustrated by Table 1.1.

For this book I have chosen to use the words abdomen and abdominal in all ordinary situations because of their familiarity. When it is necessary to speak of particular segments by number, I use the ordinary system (line 2 of Table 1.1) starting with the first segment of the metasoma as 1, with the adjective metasomal to remind the reader of the starting point.

b. Mouthparts. The mouthparts are a combination of the chewing or biting and the sucking types, the labrum and mandibles being attributable to the former and the maxillae and labium to the latter. The mandibles are the principal tools used in nest making and thus are of major social significance. For example, depending on the kind of bee, they are used to dig soil, cut leaves, gather plant hairs or resin, or to manipulate wax. Their structure varies according to the kind of nest made. But mandibles also have a variety of other functions, such as cutting open corollas to get at nectar in flowers too narrow and deep to be robbed without such mutilation; gripping a resting or sleeping place, in the case of bees that do not rest in

TABLE 1.1. Numbering systems for abdominal segments in Hymenoptera.

1. Abdominal segments	1	2	3	4	5	6
2. Metasomal segments	P[a]	1	2	3	4	5
3. Gastral segments						
a. Bees	P	1	2	3	4	5
b. Formicine ants	P	Pt[a]	1	2	3	4
c. Myrmicine ants	P	Pt	Pt	1	2	3

[a]P = propodeum; Pt = petiole.

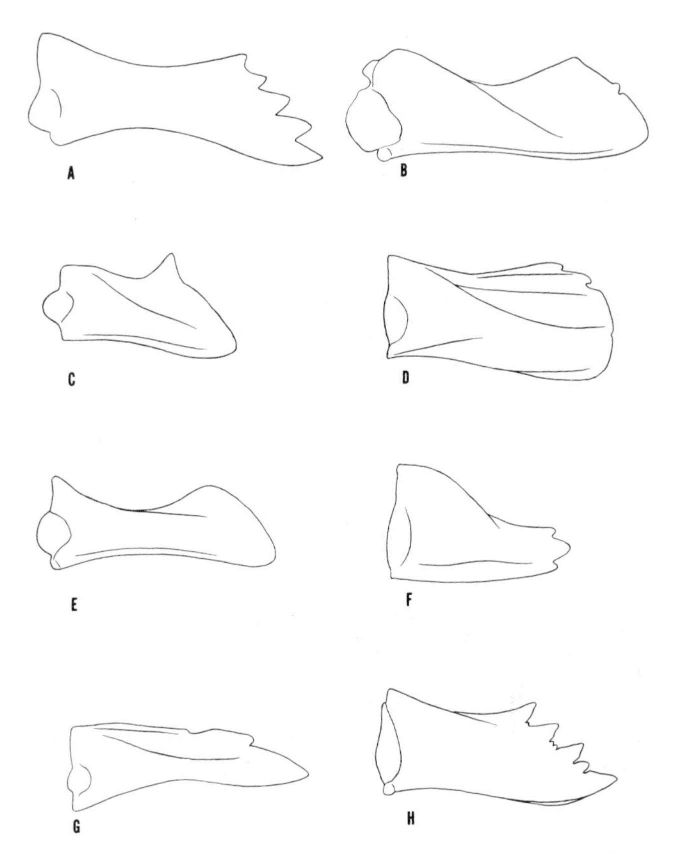

FIGURE 1.5. Mandibles of various female bees, workers where castes exist, seen in outer view: (A) *Trigona corvina,* (B) *Melipona marginata,* (C) *Trigona capitata,* (D) *Bombus americanorum,* (E) *Apis mellifera,* (F) *Allodape mucronata,* (G) *Lasioglossum imitatum,* (H) *Immanthidium repetitum.* (Original drawings by C. D. Michener.)

a nest or other enclosed space; or gripping or pinching an enemy, often to give a firm base for driving in the sting. Figure 1.5 illustrates some different types of mandibles.

The maxillae and labium together form the proboscis, which is used in taking up liquids such as nectar, honey, or water. Finely divided solid matter, such as pollen, can be taken into the mouth, which is at the base of the proboscis. Bees are not able to eat coarse solid materials unless, like sugar, they can be dissolved and actually ingested as liquids. Bees also use the proboscis for a number of activities other than sucking liquids, such as food exchange among workers in highly eusocial bees, "licking" of substances such as pheromones from the

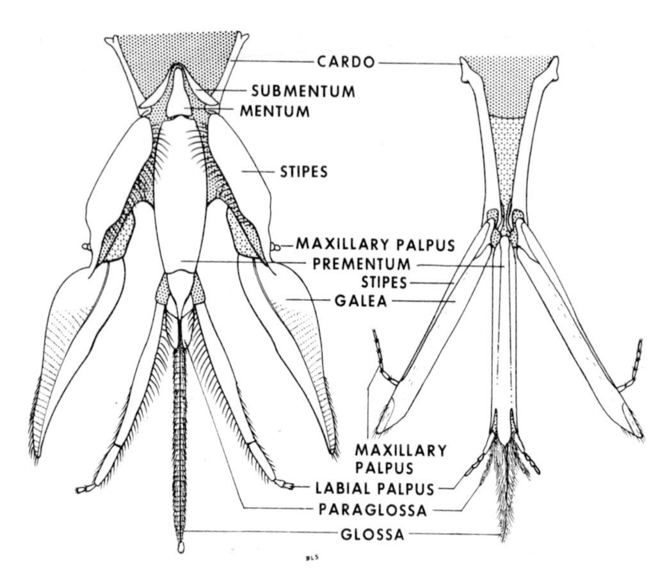

FIGURE 1.6. Diagrams of the proboscides of (*left*) a bee of the family Apidae (based on *Apis*) and (*right*) a bee of the family Halictidae (based on *Pseudaugochloropsis*). (Original drawings by Barry Siler.)

bodies of one another, and spreading and smoothing the cell linings of many ground-nesting bees. In *Xylocopa* the bladelike maxillary galeae form a cutting apparatus for piercing corollas of tubular flowers and a comb on each stipes is used for collecting pollen from the legs (Schremmer, 1972b). The principal features of the proboscis, however, have evidently evolved for its function in taking nectar; its other functions appear to have influenced its structural evolution but little.

The structures of the proboscis for two very different sorts of bees are shown in Figures 1.6 and 2.1. The details of function have been studied for *Apis*, and they are described by Snodgrass (1956). They involve formation of a tube around the glossa by the flat, sheathlike galeae and basal segments of the labial palpi. In this tube liquids rise toward the mouth, as a result of back and forth movements of the glossa, suction through the mouth, and probably capillary action. When not in use the proboscis is folded in a large grove, the proboscidial fossa, on the underside of the head.

c. The scopa and legs. Except in certain Colletidae which carry pollen with nectar in the crop, the structures used for carrying pollen consist of scopal hairs having various locations and arrangements (Figures 1.7 and 1.8), principally on the underside of the abdomen or on the hind legs. Those bees whose scopa forms a corbicula or pollen basket on the hind tibia also use the corbicula for transportation of resins and sometimes other nest making materials. Further details of the scopa and its function are given in Chapter 14 (B and C).

Bees that make their own burrows and cells in the ground or in rotting wood (for example, Halictidae) usually have, on the outer side of the base of the hind tibia, a scale-like structure called the basitibial plate (Figure 1.9), and at the apices of the tibiae, again on the outer surface, there are often small, sharp, outer projections called tibial spines. (These are not the tibial spurs, which are movable structures on the inner surface of the apex of the tibia.) The bees move about in their burrows by pushing out aginst the burrow walls with the tibiae; it is the basal and apical hind tibial structures, basitibial

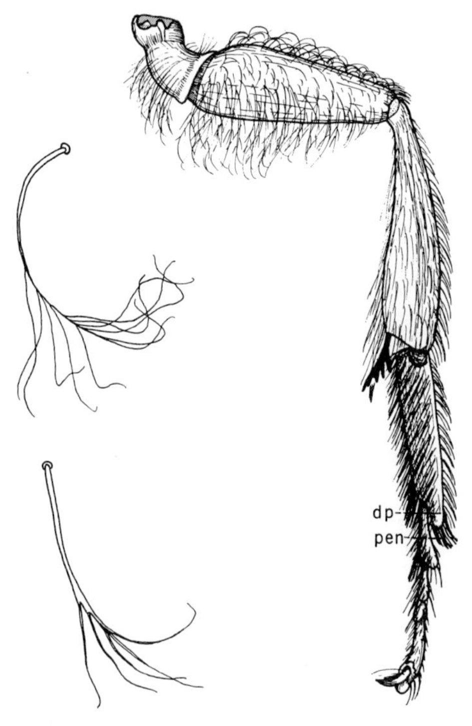

FIGURE 1.7. Hind leg of female of *Pseudaugochloropsis graminea*, a halictid bee, showing the pollen-carrying scopal hairs, especially on the femur. One hair from the femur (*above*) and one from the tibia (*below*) are enlarged. Such a scopa carries dry pollen: *dp*, distal process of basitarsus; *pen*, pencillus. (From Eickwort, 1969a.)

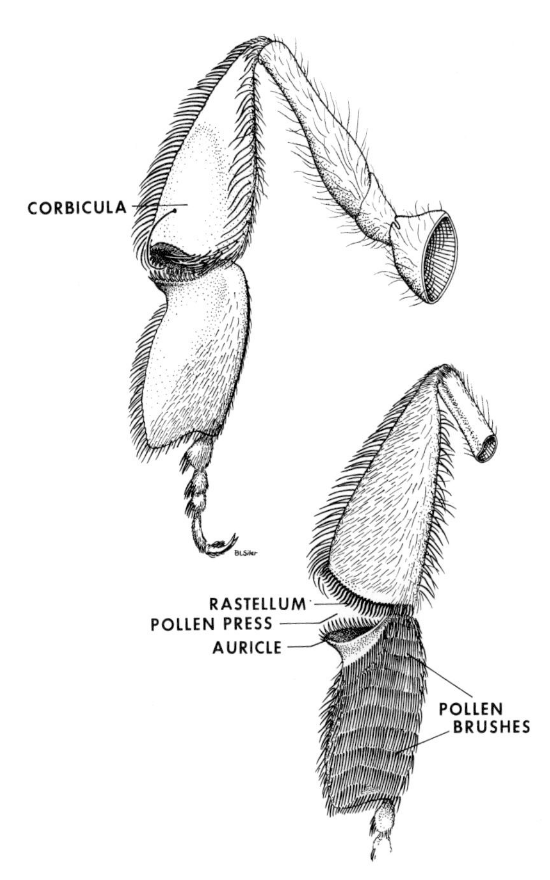

FIGURE 1.8. *Top:* Outer view of hind leg of worker of the common honeybee *Apis mellifera; bottom:* Inner view of tibia and basitarsus. (Redrawn from Ribbands, 1953, after Hodges, 1952.)

plates and tibial spines, that give traction on walls.

A structure that looks like a small paint brush at the apex of the hind basitarsus in females of these same bees is the penicillus (Figure 1.7); it serves, along with the glossa (of the proboscis), to spread the liquid that becomes the waxlike cell lining.

d. The sting. Most female bees have functional stings for individual or social defense. In certain groups, however, stings are reduced and functionless. Among social bees this is true of the Meliponini or stingless honeybees, and among nonsocial groups, of most Andrenidae and certain genera in other families. The sting is a modified ovipositor in which the valvulae of the eighth metasomal segment are fused and expanded basally to form a bulb for containing poison; apically these fused structures form a single sharply pointed stylet. The valvulae of the seventh segment slide lengthwise along the stylet and enclose the poison canal beneath it (Figure 1.10, C). Basally these valvulae bear "valves" which project up into the bulb and, as they move back and forth, pump the poison out of the sting; apically they are the slender lancets. The part of the sting that punctures is thus made up of three sharp parts which enclose the poison canal. The sting and associated glands of the honeybee are shown in Figure 1.10.

Since the ovipositor functions as the sting, it cannot carry eggs, which are laid through an opening at its base. For the same reason, male Hymenoptera do not sting.

2. Internal organs

a. The gut. The intestinal tract is shown in Figure 1.11.

The cibarium is the pump which draws food into the mouth. It can be dilated by the contraction of muscles extending from its anterior wall to the facial region (clypeus); circular muscles cause its contraction.

The crop, often called the honey stomach, is enormously stretchable. When filled with nectar, the crop is stretched until its walls are delicate and transparent and it presses the rest of the viscera to the rear and lower side of the abdomen. The contents are regurgitated when pressure is applied on the distended crop by the viscera because of telescoping of the abdominal segments.

The mandibles do not break up pollen grains but only help in their ingestion. The pollen therefore remains in

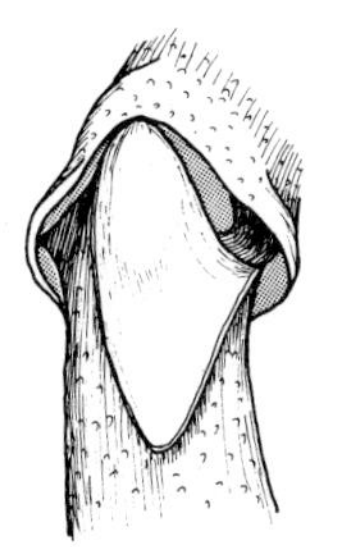

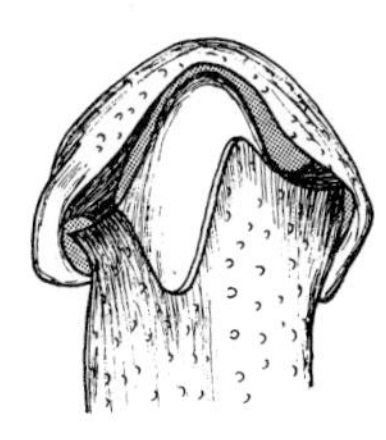

FIGURE 1.9. Outer surface of the base of the hind tibia, with the apex of the femur, of the females of the following halictids (*left to right*): *Pseudaugochloropsis graminea, Andinaugochlora micheneri,* and *Augochloropsis metallica.* The basitibial plate, shown in these drawings, is found in most ground-nesting bees. (From Eickwort, 1969a.)

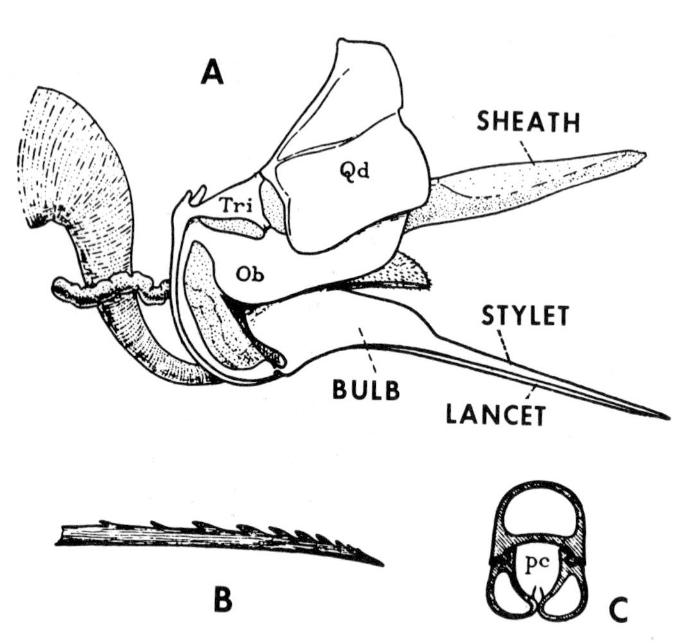

FIGURE 1.10. (*A*) Side view of sting and associated structures of worker of *Apis mellifera.* The large poison sac is partially shown at the left; Dufour's gland is the smaller, more or less horizontal structure at the left: *Qd,* quadrate plate or eighth metasomal tergum; *Ob,* oblong plate or second valvifer; *Tri,* triangular plate or first valvifer. (*B*) The apex of a barbed lancet (first valvula) of the sting. (*C*) Cross section of the sting shaft: *pc,* poison channel. The stylet is above, the two lancets below. (From Snodgrass, 1956, © Cornell University; used by permission of Cornell University Press.)

the crop intact. Solid particles (pollen grains) are removed from the crop by the proventricular valve and passed back, along with some of the liquid from the crop, into the midgut or ventriculus, where most of the digestion and absorption occurs. The intact walls of the pollen grains persist but in the hind gut, two or three hours after ingestion, most of the pollen grains are seen to be empty (Witcomb and Wilson, 1929). Digestive enzymes as well as the digested content of the pollen grains must therefore be able to pass through the pollen grain walls.

The rectum of most bees is of moderate size, but it becomes enormously distended in winter as a reservoir for water and feces in honeybees, which are active in the hive but do not defecate during cold weather.

b. Flight muscles. The large indirect muscles of flight (Figure 1.12) are so called because they are not attached to the wings directly but move the skeletal parts (notum and pleuron) that move the wings. Flight mechanisms do not differ meaningfully between social and nonsocial bees; bees of both types must fly from nest to flowers and return. Nonetheless, in the course of evolution among bees, the muscles have acquired special social significance. (1) They are largely responsible for both heat production and for the fanning which makes possible nest cooling. Social temperature control (Chapter 17) is therefore dependent upon them. (2) They produce vibrations used in various ways. Communicative use of vibrations (with or without wing movements) is well established in highly social bees—communications with respect to food sources and between rival gynes of *Apis,* for example. The vibrations are perceived through the substrate, probably not as sound through the air. Vibrations are also used to help modify the environment. Thus they may be used in collecting pollen or to help in digging or in pushing through a narrow space. For example, energy transmitted from wing muscle vibrations through the jaws to a pebble perhaps helps to loosen it so that it can be removed. Such use is by individual bees; collective use for such purposes is not known. Collective use of buzzing (often with wing movements) as though to frighten intruders is common in social bees, and the buzzing of an individual bee when threatened or captured must sometimes result in its escape.

c. Ovaries. Ovaries of egg-laying females are of special interest in a consideration of sociality because of differences between queens and workers. In solitary bees each ovary consists of three or four ovarioles, closely appressed together and opening posteriorly in a short lateral oviduct. The two lateral oviducts, one for each ovary, open into an exceedingly short common oviduct (Figure 1.12). In each ovary the germ cells start in the slender anterior ends of the ovarioles. As they move backward, ova become interspersed with clusters of nurse cells. Thus the middle part of each ovariole contains an alternating series of growing oocytes and clusters of nurse cells. The nurse cells disintegrate as the oocytes reach full egg size, so that the posterior parts of some ovarioles contain large eggs without nurse cells. Total egg production in solitary bees probably ranges from about eight (for example, in *Andrena erythronii,* Michener and Rettenmeyer, 1956) to forty (for example, *Megachile rotundata,* Gerber and Klostermeyer, 1970).

Most social bees do not have more ovarioles than do their solitary relatives. Thus in social Halictidae there are

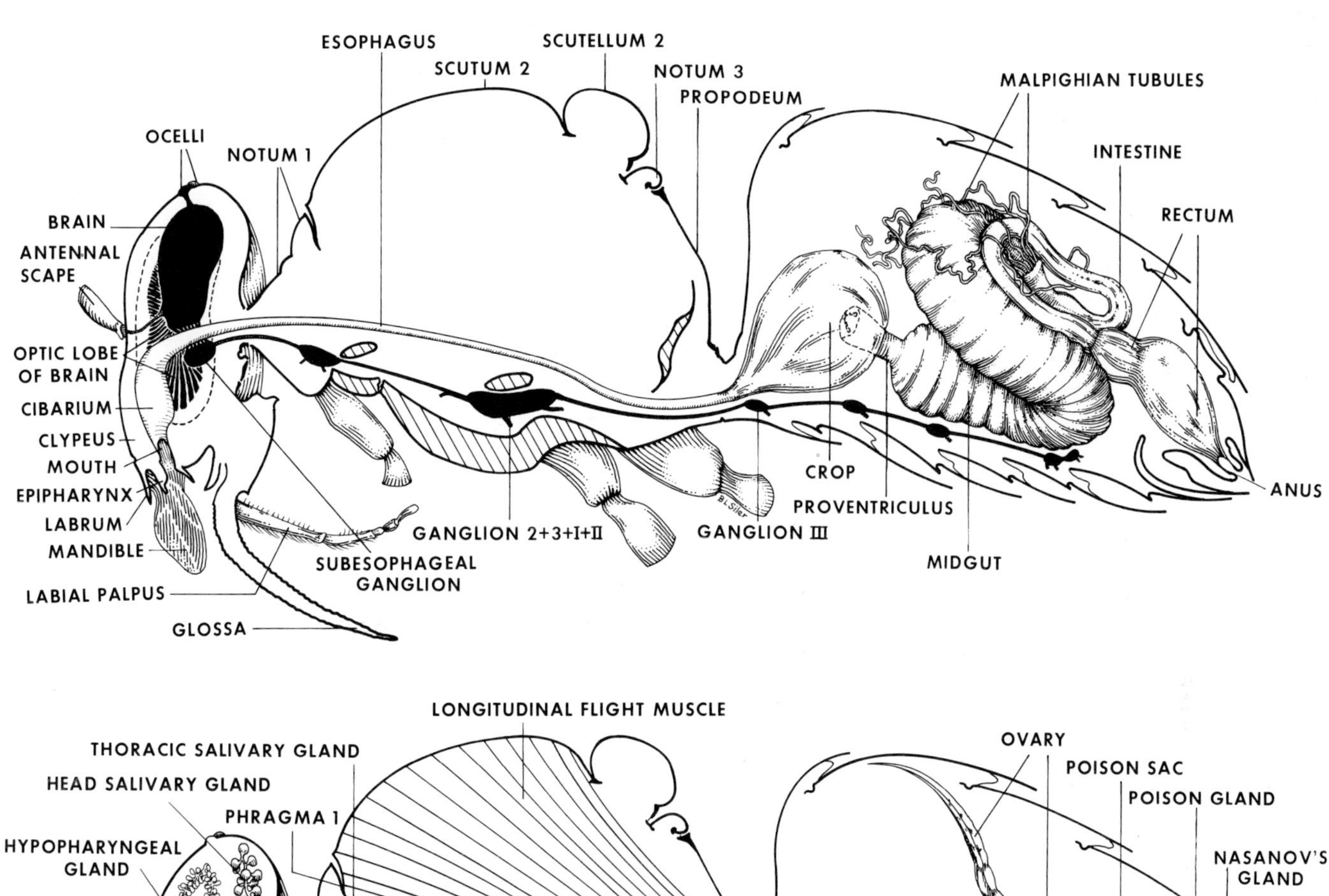

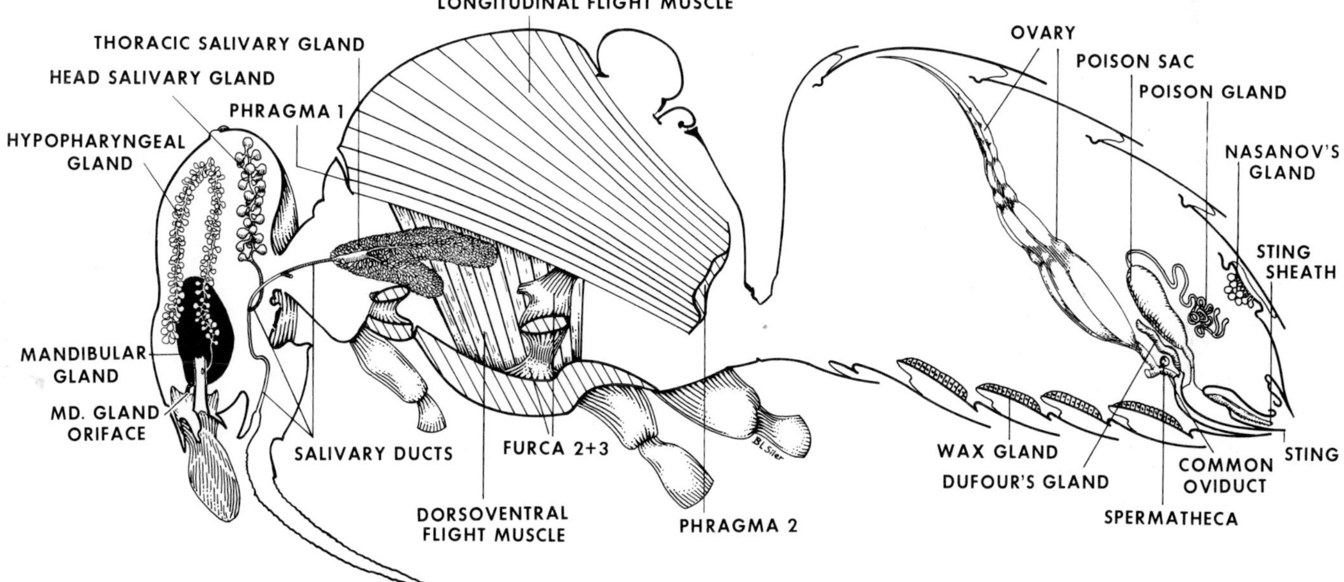

FIGURES 1.11 and 1.12. Diagrams showing certain organ systems of an adult female bee. The objective is to show as many of the major organs as possible for any bee, but some of the glands are specifically those of honeybees of the genus *Apis. Arabic numerals* indicate structures associated with thoracic segments; *roman numerals* indicate abdominal segments. (Original drawings by Barry Siler.)

three ovarioles per ovary as in solitary Halictidae, and in social Anthophoridae and most social Apidae there are four per ovary as in solitary members of these families. In queens the ovarioles become longer, sometimes fold to permit additional length, and more enlarging oocytes or eggs become recognizable in each than in workers or solitary relatives. The eggs of social bees may also be relatively small, so that many of them can be produced simultaneously. Some halictid queens, as well as bumblebee queens, produce hundreds of eggs or even a very few thousand. Most surprising is the fact that meliponine queens may produce hundreds of thousands of eggs with such ovaries. Only *Apis* differs in ovariolar number from its relatives: queens have numerous ovarioles (150 to 180 per ovary); workers have a small but variable number (Figure 9.18). Nonetheless the productivity of an *Apis* queen is probably no greater than that of some queens of *Trigona* in the Meliponini.

d. Glands. Nest building, communication, and defense among bees often involve glands whose secretions are either building materials, food, communicative substances (pheromones), or agents of defense. The following account explains the positions (Figure 1.12) and lists the major functions of the more obvious exocrine glands. Others exist but their functions are unknown or in no known way related to social behavior. Major works on glands are by Cruz Landim (1967) and papers listed by her (see also Altenkirch, 1962).

The mandibular glands lie one in each side of the head; each includes a storage sac from which the secretion escapes through an opening in the articulatory membrane mesally at the base of the mandible. They are widespread in Hymenoptera but their functions are well known only in highly social bees, where they are often unusually large (Chapter 20, A6). There is remarkable diversity in the functions of these glands in such bees (Kerr and Costa Cruz, 1961). As explained in detail in subsequent chapters, they are known to produce food for larvae, defense substances, social pheromones, sex pheromones or attractants, and the like.

The hypopharyngeal (often formerly called pharyngeal) glands are paired structures that discharge through openings on either side of the hypopharyngeal plate. The primitive condition in bees seems to be a small group of secretory cells, each with its own duct to the exterior. Their function in such cases is unknown; possibly they produce digestive enzymes, such as invertase. Only in the Apidae do the cells discharge into a single long duct on each side. In the highly social bees the duct of each gland is enormously elongated, often longer than the body of the bee, with individual cell ducts opening into the main duct throughout its length (Figures 1.13 and 1.14). The glands lie convoluted in the facial region of the head. In workers of *Apis,* the Meliponini, and probably in fe-

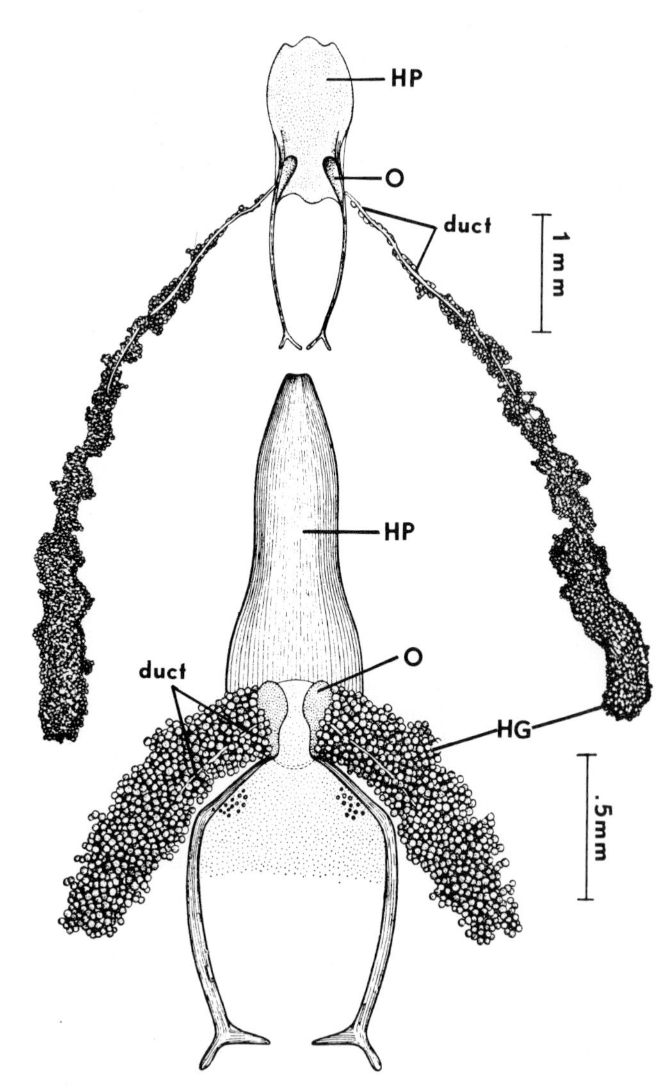

FIGURE 1.13. Hypopharyngeal glands of (*above*) a worker bumblebee, *Bombus atratus,* and (*below*) a female nonsocial apid, *Euglossa cordata: HP,* hypopharyngeal plate; *HG,* hypopharyngeal glands; *O,* oriface area of the duct. (From Cruz Landim, 1967.)

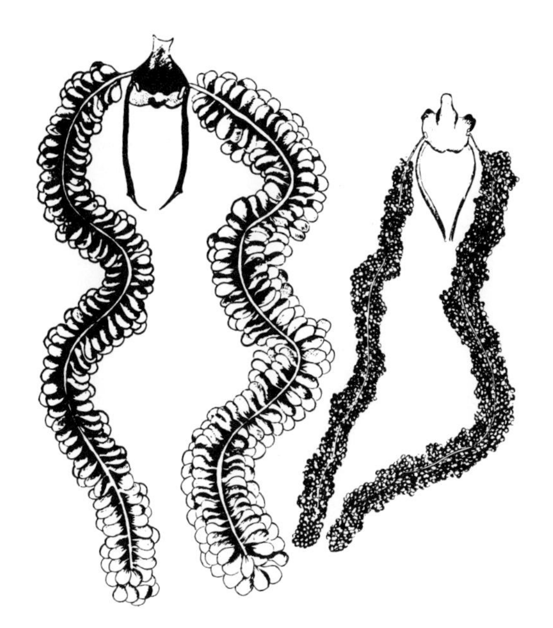

FIGURE 1.14. Hypopharyngeal glands of workers of (*left*) *Apis dorsata,* (*right*) *Melipona quadrifasciata.* (From Cruz Landim, 1967.)

males of *Bombus,* hypopharyngeal glands secrete part of the larval food. At least in *Apis* workers they also secrete the invertase which breaks down sucrose in nectar—part of the honey making process. They are absent in males and rudimentary in queens of *Apis.*

The salivary glands proper, or labial glands, include one or two pairs of glands. One, which is small or absent except in the Apidae, lies in the posterior part of the head. The other is in the thorax. All open through a single median duct that discharges on the anterior side of the proboscis near the base of the glossa (Figure 1.15). Simpson (1960) shows that in *Apis* these glands produce a watery material that is used to dissolve solid sugar, clean the queen's body, and perhaps to soften materials being chewed. This may be the liquid seen added to wax when it is manipulated. An oily secretion of unknown function is produced in the same glands.

Wax is produced by epidermal glands on those parts of certain abdominal terga and sterna that are overlapped and hidden by the posterior parts of preceding sclerites. The wax is secreted as a scale through thin, shining areas of the cuticle (Figure 12.3). Scattered minute glands or groups of glands are found in this location in many bees. Only in the Apidae do they become dense, continuous over extensive areas, and clearly associated with wax production, and even in this family the function is not certain for the tribe Euglossini. Locations of the wax glands in the four tribes of Apidae are shown in Figure 1.16. Wax is pushed out from between the terga or sterna by hind leg movements (Figure 1.17) during which bristles of the basitarsus stick into the wax scale. Once free, the scale is chewed until it is malleable.

Nasanov's gland, an odor- or pheromone-producing gland in *Apis,* occupies approximately the position of a wax gland on the apical abdominal tergum of workers (Figures 1.12 and 15.6), but the openings are through partially eversible membrane basal to the tergum proper. Perhaps it is homologous to wax glands, to the more widespread abdominal glands found in various other groups of bees, or to both. It has often been called the scent gland, but as bees have various glands that produce scents, it seems best to use the more specific term.

The sting gland or poison gland is a slender, bifurcate structure that opens into a large poison sac and thence into the base of the sting (Figures 1.10 and 1.12). The venom, contrary to popular fancy, does not contain formic acid. It is a complex mixture with peptides being the principal active agents, at least in *Apis.* The gland is reduced or absent in Meliponini, in some of which, however, the sac persists without known function (Kerr and Lello, 1962). Opening at nearly the same place, but discharging into the sting chamber beside the sting instead of into the sting, is Dufour's gland or the basic gland. It was formerly thought to produce part of the venom but does not do so. In bees such as the Halictidae, it is very large and secretes at least part of the waxlike material used to line the cells. In the Apidae it is much smaller and of unknown function although its secretion may form a waxy covering for the egg and may also attach the egg to the bottom of the cell in which it is laid (Kerr and Lello, 1962). This gland is absent in Meliponini, whose eggs are laid on food and the cells closed, so that such a coating and attaching material would be superfluous.

A setose membrane at the base of the sting is the source of the major alarm pheromone of *Apis* workers.

The Koshevnikov glands, located laterally on the first valvifer or between the first valvifer and the seventh metasomal hemitergite (stigmal plate) of the sting apparatus, are the source of an attracting scent pheromone

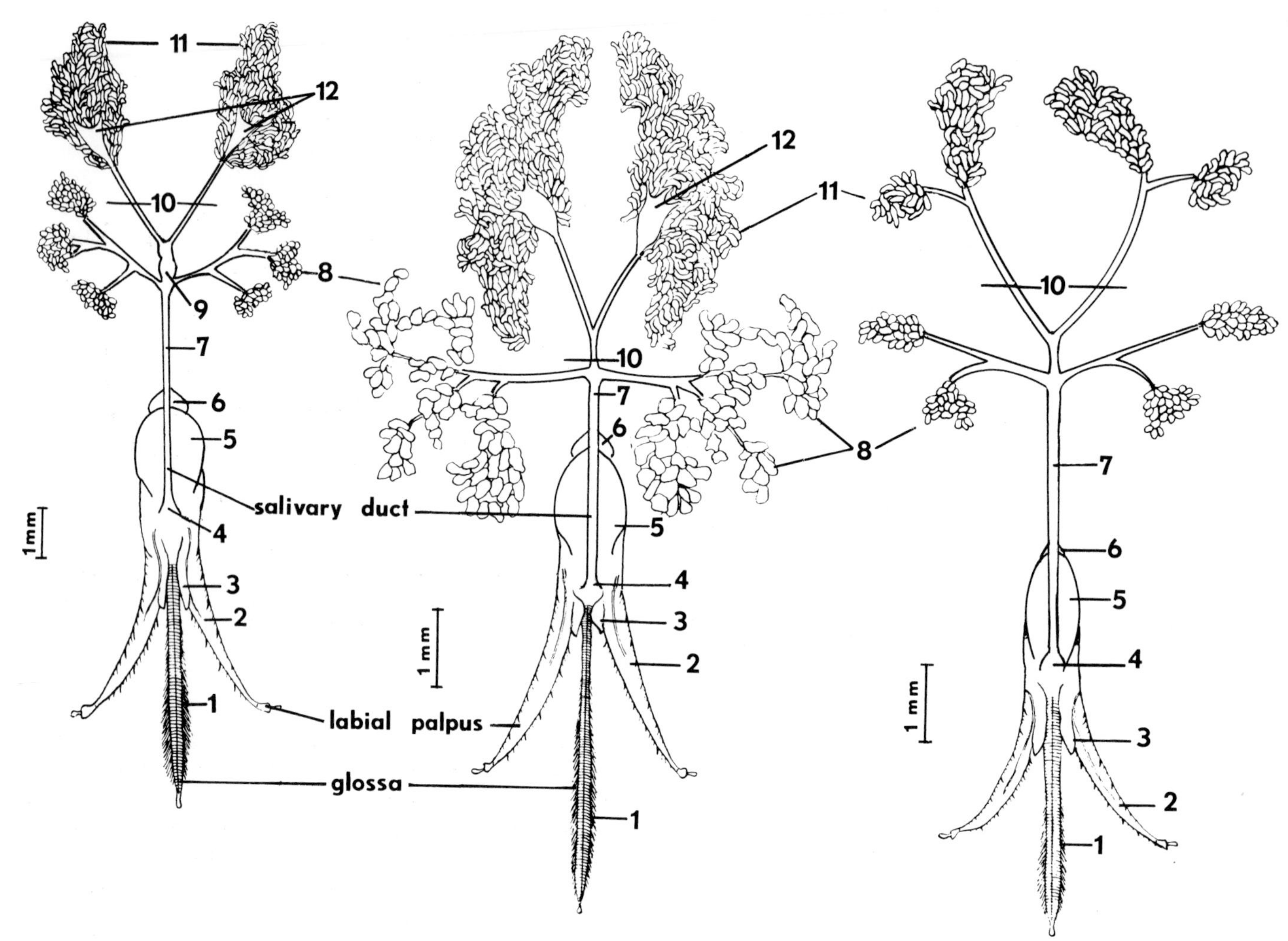

FIGURE 1.15. Salivary systems of apid bees. *Left, Bombus atratus; center, Apis mellifera; right, Melipona scutellaris. 1,* glossa; *2,* labial palpus; *3,* paraglossa; *4,* opening of salivary duct; *5,* prementum; *6,* submentum; *7,* salivary duct; *8,* salivary glands of head; *9,* salivary reservoir; *10,* level of foramen magnum; *11,* thoracic salivary glands; *12,* salivary gland reservoirs. (From Cruz Landim, 1967.)

of the *Apis* queen. The gland occurs also in workers and, less well developed, in many other female bees (Altenkirch, 1962).

e. The brain. Diagrams of the nervous system of bees appear in works on bee anatomy and do not need repetition here. However, the complexity of the brain is interesting in connection with behavior. The number of cells in most bee brains probably approaches 10^6, although in small species it may not greatly exceed 10^5. Recent figures for the honeybee (Witthöft, 1967) indicate about 851,000 for workers and 1,210,000 for males. As compared to most bees, the latter figure is doubtless large because of the enormous eyes and optic lobes of male *Apis.* If optic lobes are ignored, the worker *Apis* has some 30,000 more brain cells than the male.

Compared to higher vertebrates, bees exhibit remarkably elaborate behavior activated by mechanisms involving relatively few cells. The number of nerve cells in the brains of higher vertebrates is in the range of 10^9 to 10^{12}.

The number of cells in a bee's brain, however, is not

constant. Just as in higher animals and man, the number decreases with age, and they become increasingly vacuolated. Rockstein (1950) determined the average number of cells in medial transects of brains of newly emerged *Apis* workers as 522. In workers of two, four, and six weeks of age this mean number decreased to 445, 434, and 369, respectively, and at 51 days of age reached 325. Whether such reduction is in any way related to behavioral changes that occur with age, and particularly to the peculiar senescent (meaningless and nonadaptive) behavior sometimes seen in old bees, is unknown.

3. Sensory organs

The sensory equipment of honeybees has been studied in great detail. There is no reason to believe that this equipment or its functioning is very different in other bees, and such studies as have been made with *Bombus* (Brian, 1954; Jacobs-Jessen, 1959; Free and Butler, 1959), for example, indicate no major differences. The following account, therefore, while based on *Apis,* is presumably widely applicable to solitary as well as social bees. The information provided here by way of background is summarized in far greater detail in various works, for instance, Ribbands (1953), Frisch (1967a) and chapters by Neese, Chauvin, Markl, and Heran in Chauvin (1968). The book by Frisch is especially valuable; many references found in it are omitted in this chapter for the sake of simplicity and brevity.

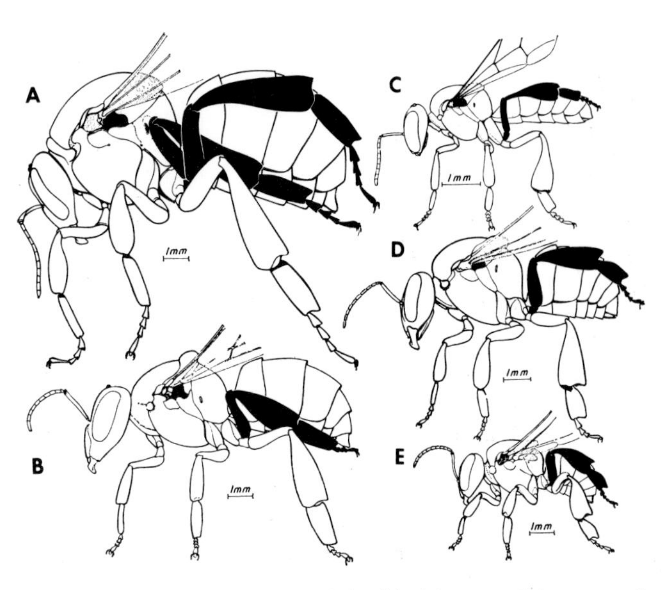

FIGURE 1.17. Movements of the hind legs used in removal of the wax scales from between the abdominal sclerites of workers. (*A*) *Bombus,* (*B*) *Apis,* (*C*) *Trigona silvestrii,* (*D*) *Melipona,* (*E*) *Trigona postica.* (From Cruz Landim, 1963.)

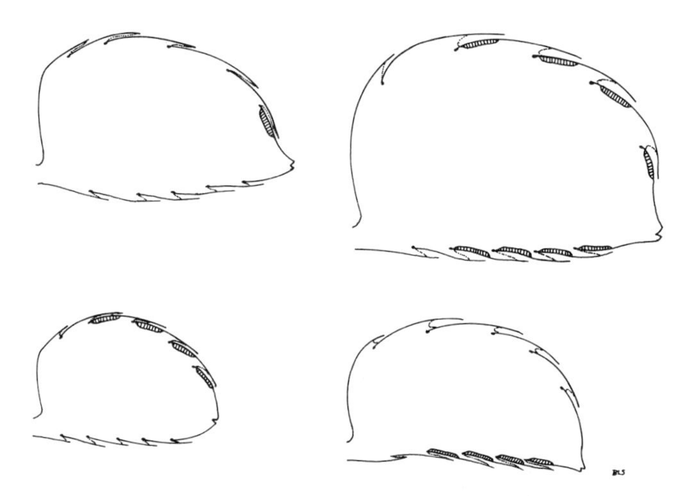

FIGURE 1.16. Diagrams of abdomens of the main apid groups showing locations of wax glands. *Above: Euglossa cordata,* female, and *Bombus atratus,* worker; *below: Melipona quadrifasciata* and *Apis mellifera,* workers. (From Cruz Landim, 1963.)

a. Eyes. There are two compound eyes, occupying large parts of the sides of the head, and three simple eyes (ocelli) on the upper part of the face or on the vertex, between the summits of the compound eyes.

At the surface the compound eyes consist of thousands of hexagonal facets which are the surface indications of the ommatidia. Each ommatidium contains lenses (the outermost one is the corneal lens, which forms the surface facet) which concentrate light on the single rodlike light-sensitive element, the rhabdom. Thus the individual ommatidium has no broad, light-sensitive retinal surface on which an image shape can be perceived. Each ommatidium can respond only to intensity, color, and polarization of light; photographs through the cornea of an insect eye showing an image in each facet are deceptive since they suggest a type of vision that does not exist.

The ommatidia diverge, as suggested by the convexity of the ocular surface, so that each covers a different visual field. The bee must perceive its environment by integration in the nervous system of information received by the thousands of ommatidia. Visual acuity depends on the number of ommatidia and varies inversely with the angle of the visual field subtended by each. There are about 13,800 ommatidia for the two eyes of a worker *Apis* (Dade, 1962), and if one considers the eight sensory cells per ommatidium, this makes about 25,000 sensory cells per mm^2, compared to nearly 80 million per mm^2 in the retina of man. The subtended angle of each ommatidium is smallest on the vertical axis near the middle of the eye, and increases toward the periphery. The angle covered by a single facet is nearly three times as great in a horizontal plane as in a vertical plane, so the bee is very astigmatic. Visual acuity is best in a narrow vertical field anterolaterally directed. Even in this area the ommatidia of *Apis* workers are at angles of about 1° to one another in the vertical plane; this fact plus studies of bee responses indicates an acuity of 1/100 to 1/60 that of man, depending on the method.

The low-resolution vision indicated above is not nearsightedness; it applies at all distances and there is no focusing mechanism. In ability to discriminate dark and light shades the bee's eye also is far inferior to our own, but bees can be trained to discriminate both relative and absolute light intensities: for example, shades of gray (Daumer, 1963). Dark adaption is not greatly different from ours (Seibt, 1967).

In perception of movement, the bee's eye is at least as good as ours. This has been determined by the responses of bees to moving patterns, for example black and white stripes. By moving such a pattern of stripes in front of a bee and recording its responses, it can be shown that bees can still detect the stripes when they pass the eye at rates up to about 300 per second. That is the flicker fusion frequency for *Apis* (Autrum and Stoeker, 1950). For man the equivalent is only 15 or 20 per second, depending on light intensity. Thus to us fluorescent lights seem steady, but for bees they must flicker at 60 cycles per second. Since the bee often flies close to vegetation, patterns are constantly moving rapidly across the visual field. It is the flickering of such patterns across the visual field which the eyes of flying insects are particularly well adapted to perceive and which seem basic to their ability to respond to form.

Hertz, some forty years ago, discovered that bees' perception of form is based on different principles than ours. Bees appear to recognize objects not on the basis of what we call their shapes, but on the frequency with which edges cross their field of vision, that is, on the degree of "disruption." Thus it was difficult for Hertz to train bees to distinguish figures in the upper row of Figure 1.18 or those in the lower row, but she could readily train them to distinguish any figure in the upper row from any in the lower. Bees easily learn to discriminate different degrees of disruption, but the more highly dissected the figure, the easier it is to train bees to it as a food place. Obviously, the more highly dissected it is, the more lines will flicker across the field of vision as the bee flies. The relation of this to the often highly dissected floral shapes is obvious. On the other hand, a bee returning to its nest responds most easily to simple shapes, a not surprising bit of behavioral evolution since nest entrances are usually simple holes. Among the secondary sources cited above, Ribbands (1953) in particular gives a good review of the studies of Hertz and subsequent authors, and Schnetter (1968) provides a source of newer data.

It is not possible, however, to explain all pattern discrimination of bees in terms of relative disruptiveness of patterns. For example, bees easily distinguish different orientations (vertical, slanting, horizontal) of straight edges (Wehner and Lindauer, 1966).

Color vision of bees is remarkably similar to our own in spite of its independent origin. Both bees and man have trichromatic vision, that is, color vision based on three types of cells, each maximally sensitive to wave lengths representing one of three primary colors. Any color discriminated by bees can be made by mixtures of the three wavelengths: 530 mμ (green), 430 mμ (blue) and 340 mμ (ultraviolet). The principle of complementary colors applies to both man and bees. Colors can be ar-

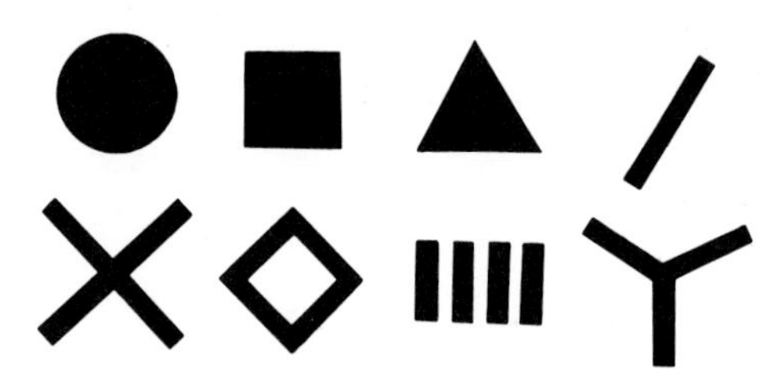

FIGURE 1.18. Figures to which honeybees (*Apis mellifera*) have been trained. Bees could not easily distinguish figures in the upper row or in the lower row, but they could be trained to distinguish figures in the upper row from those in the lower, and the latter were preferred to the former. (From Frisch, 1967a, after Mathilde Hertz.)

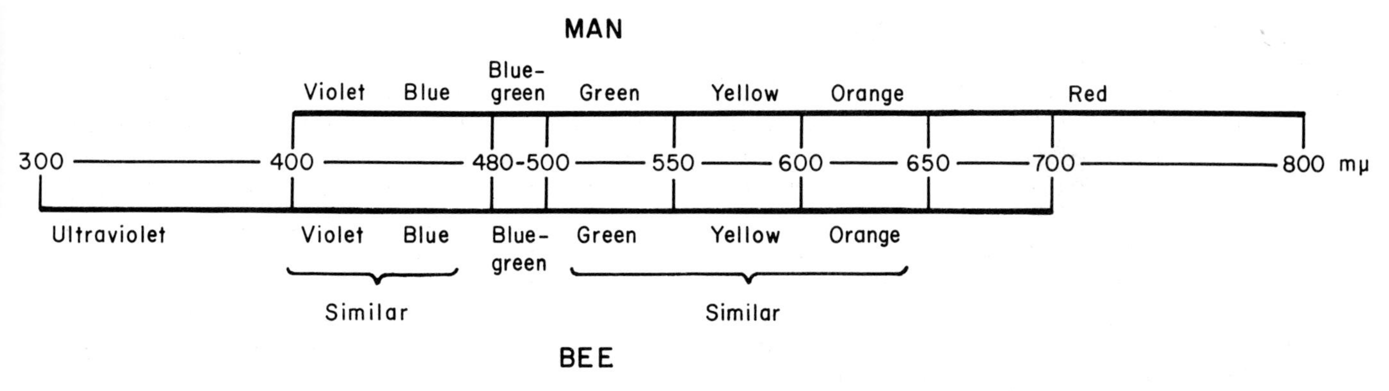

FIGURE 1.19. Colors of the spectrum for the human eye (*above*), and for the honeybee *Apis mellifera* (*below*). (From Frisch, 1967a.)

ranged in a circle (Figure 1.20) closed by purple (bee purple for bees). Any pair of colors on opposite sides of the circle complement each other to give white (bee white) (Daumer, 1956).

In spite of these remarkable similarities, color vision of bees differs from that of man. To judge by training experiments, bees can distinguish far fewer hues than we do, although contrary to early studies they do distinguish some subdivisons of their major color areas, for example, orange, yellow, and green. Also, their visible spectrum differs from ours, for they are highly sensitive to ultraviolet radiation and insensitive to red. (Sensitivity extends to slightly longer wavelengths in *Bombus* than is *Apis.*) Visible spectrums for man and *Apis* and color circles for man and *Apis* are shown in Figures 1.19 and 1.20. Purple is the mixture of light of the two extremes of the visible spectrum, red and violet for man, yellow and ultraviolet for the bee, hence "bee purple." "Bee violet" refers to a spectral region around 400 mμ or to a mixture of blue and ultraviolet, a color that training experiments indicate

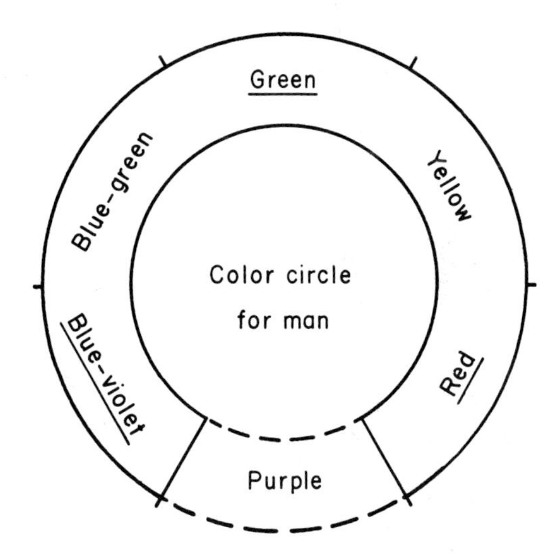

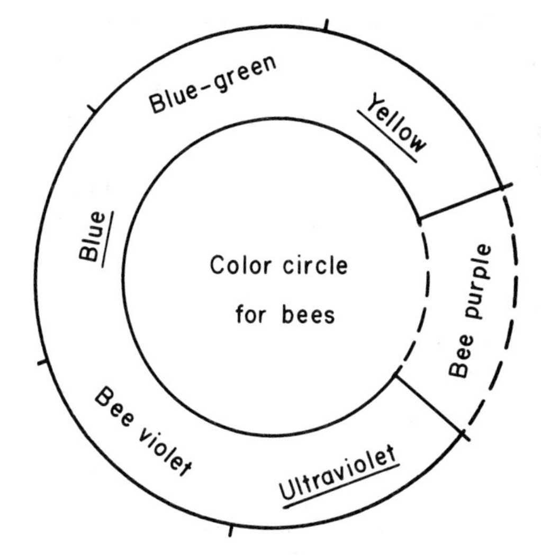

FIGURE 1.20. Color circles for man and the honeybee. The three primary colors are underlined. Intermediate colors are produced by mixing the primary colors. Complementary colors are opposite each other in the circles. (From Frisch, 1967a, modified from Daumer, 1956.)

is well distinguished by bees. A mixture, in the proportions found in sunlight, of all wave lengths that can be perceived, we call white. Bee white, of course, is similarly explained, but differs from our white because of the bees' different visible spectrum (Figure 1.19).

Bees' reactions to flowers, the form, colors, and odors of flowers that determine the bees' responses, and the coevolution of flowers and bees constitute an enormous field that cannot be treated in detail here. Daumer (1958) has made especially important contributions with regard to the colors of flowers as seen by bees. Many flowers reflect varying amounts of untraviolet, so that flowers that we see as similar in color may look very different to bees. Many flowers also have nectar guides to which bees respond by extending the proboscis and probing. In some flowers such markers are visible to us—dark bases of petals or other marks indicating the locations of nectaries as well as of stamens and pistils. In other flowers the guides are invisible to us but recognizable to the bee because they do not reflect ultraviolet, like the rest of the flower. Responses by bees to nectar guides are sometimes reported even for individuals that have never seen a flower before. But Free (1970c) reports that initially they have no effect, although bees easily learn to associate them with the food source. They facilitate the learning necessary for a bee to efficiently manipulate a particular kind of flower.

Besides form and color, bees, like other arthropods and unlike vertebrates, can perceive the direction of the plane of vibration of polarized light. (Light vibrates transversely, that is, at right angles to the direction of travel. Ordinarily such vibration is in all transverse directions, but in fully polarized light the vibrations are all in one plane.) As explained in Chapter 14 (D), light from the blue sky is partially polarized, maximally so at right angles to the line from the observer to the sun, so that the pattern of polarization in the sky varies with the apparent movement of the sun. Orientation and communication involving the sun's position can therefore often continue even when the sun is invisible, providing that some blue sky is visible. Frisch (1967a) gives a detailed summary of current knowledge of the mechanism of perception of the plane of polarization.

In summary, the basic elements perceived by the compound eyes of bees appear to be (1) light intensity, (2) degree of disruption of patterns (hence movement), (3) different orientations of straight edges, (4) color, and (5) direction of the plane of polarized light. Moreover, bees learn to recognize different arrangements and combinations of such basic elements. Thus they can discriminate a right-left sequence of two colored areas from a left-right sequence of the same two areas. This conclusion was reached using the sort of technique illustrated in Figure 14.5. The relative positions of objects can be used in homing (Gasbichler, 1968). Such abilities have nothing to do with recognition of disruptiveness of patterns; the mechanisms and details are little understood.

The ocelli probably have nothing to do with sight in our sense but serve to alert the bee to activity at certain light intensities, and help to maintain nervous activity and muscular tonus appropriate for given light conditions.

b. Chemoreception. The sense of smell in bees is located on the antennae, usually on segments 5-12 in females and 5-13 in males. (Males have one more antennal segment than females.) These segments are covered with sensilla of various sorts; Frisch (1967a) lists seven kinds whose structure suggests that they serve for odor perception. One kind, the pore plates, of which there are about 3000 on a worker *Apis* antenna, is known to have an olfactory function, as established electrophysiologically. The functions of the others remain incompletely known. Some presumably respond to atmospheric humidity, others to pheromones produced by the queen *Apis*. There is a special type for CO_2 perception (Lacher, 1967). Temperature perception also is apparently most acute in the antennae, and some of these sensilla are probably involved.

A bee can get a much more minute picture of the origins of odors than we can because its organs of smell are located on external projections (antennae) which it can direct over the surfaces of objects, even into crevices, etc. Thus, as it moves about, a bee can make a sort of topographical olfactory analysis and can learn and respond to a sequence of odors on a substrate.

Extensive studies of odor perception have been made with *Apis*, using discrimination in training experiments as the principal experimental tool. For many substances, especially floral odors and bee pheromones, bees are more sensitive, sometimes far more sensitive than man. Thresholds vary from $\frac{1}{10}$ to $\frac{1}{100}$ of those for humans, and bees are often attracted by odors imperceptible to us. To certain other materials, however, especially the (to us) unpleasant odors associated with filth and decompositon, bees are less sensitive than is man. Vareschi (1971) has made the most recent study of odor discrimination in *Apis,* using electrophysiological methods to study responses of olfactory cells as well as behavioral methods.

Other classes of chemoreceptors are those of taste (that is, contact chemoreception), present not only on the mouthparts but also on antennae and legs. Most studies of these sensilla have had to do with perception of sweetness. Bees respond to most, but not all, of the sugars that taste sweet to man. The nonsugar sweets (such as artificial sweetners) that are sweet to man are not to bees. The sensitivity of bees to sugar is not especially high; starved bees will respond to a $\frac{1}{16}$ M sucrose solution, but this is near the threshold level.

c. Mechanoreception. Tactile hairs are probably scattered over the body and appendages. The tactile sense seems especially well developed on the antennae and mouthparts. Other mechanoreceptors are the chordotonal and campaniform sensilla, which respond to positions of the body and deformations of the cuticle, respectively.

Mechanoreceptors can have a variety of functions often different from the presumably primitive ones and sometimes of interest in connection with social behavior. Probably no bee could construct a proper nest without knowing the direction of the pull of gravity. There are small groups of hairs at the neck and the base of the abdomen. They are like minute tactile hairs, but they normally touch only nearby parts of the body, being deflected by any tendency of the head or abdomen to hang downward. They thus serve to perceive the direction of the gravitational force; the bee knows which way is down and can build, dance, or otherwise behave appropriately.

Similar clumps of hair at the bases of the appendages, along with chordotonal organs in them, doubtless serve as proprioceptors to indicate the positions of the appendages. These sensilla also serve as a part of the gravity-detecting apparatus and perhaps serve to detect vibrations in the substrate or of objects being touched by the antennae. Since bees do not appear to respond to airborne sound, but only to vibrations in the substrate and surfaces being touched, these sensilla presumably provide the basis for responses to vibrations.

Tactile hairs, for example on the eyes of *Apis,* as well as antennal positions perceived by the chordotonal organs (Johnston's organ) in the second antennal segment probably of all bees, provide information on air resistance in flight. When integrated with visual cues, such information enables a bee to respond to wind velocity and direction. This is of course important for orientation, for example, in return to the nest (Chapter 14, D).

C. Conclusions

Although some social bees show the most complex behavior of any invertebrate animals, their structure, development and physiology are not correspondingly elaborate. Indeed in these features social bees are essentially like nonsocial bees and like other groups of aculeate Hymenoptera. It is difficult to point to any feature of structure, development, or physiology, as described in this chapter, that is not found in both nonsocial and social bees. The apparatus already possessed by the nonsocial bees was adequate to permit social evolution. Of course the more highly social bees like *Apis* have special glands of social significance, not to mention caste dimorphism, to be discussed in detail in later chapters. Nonetheless, most of what has been described above could equally have been said for nonsocial bees and even for a wide range of other Hymenoptera.

2 *The Origin and History of Bees*

Evolution is the major unifying concept of biology. Most chapters, particularly in Part II of this book, relate to the evolutionary origin and elaboration of social behavior among the bees, regardless of which lines of descent illustrate particular evolutionary attainments. The present chapter, however, attempts to sketch the lines of descent—first, the ancestry of the bees as a whole, and second, the divergence and history of the various groups of bees, a topic to be detailed in Figure 2.2.

A. Ancestry

No doubt the primitive Hymenoptera resembled modern sawflies (suborder Symphyta or Chalastogastra) in major features, such as plant-feeding habits, caterpillar-like larvae that defecate throughout their lives, relatively complete wing venation, thoracic structure similar to that of Neuroptera and allied orders believed to be ancestral, and the sawlike or piercing ovipositor for placing eggs in plant tissues.

From such insects must have arisen the main parasitic groups of Hymenoptera, such as the ichneumonoids, chalcidoids, and proctotrupoids, from an already parasitic ancestor of which came the wasps (Vespoidea, Sphecoidea), and thence the bees. All these groups are members of the suborder Apocrita (or Clistogastra), characterized by transfer of the first abdominal segment into the thoracic tagma and development of a constricted, flexible joint between the resulting "thorax" or mesosoma and the metasoma (second and following abdominal segments, Chapter 1, B1). Probably this development is connected with flexibility useful in bending the abdomen to insert eggs deeply by means of a piercing ovipositor into the tissues of hosts, probably originally insects but possibly plants. The larvae of the parasitic groups are mostly internal parasites in insect hosts. They are modified by reduction of sensory organs (loss of eyes, great reduction of antennae, palpi, setae), and loss of legs and of nearly all pigmentation and sclerotization of the body cuticle. Strictly speaking, such insects are not true parasites, for the presence of one or more in the body of a host normally results in the death of the latter. The parasite consumes most of its host like a predator, yet one individual host suffices. The term parasitoid has been coined for such forms, but since it seems to cause no confusion, the simpler word parasite is employed here.

Presumably because early death of the host would result from release of foreign wastes into its blood and tissues, the parasitic Hymenoptera developed a mechanism (failure of the midgut to join the hindgut, mentioned for bees in Chapter 1, A2) that prevents defecation of the larva until its growth is complete. At that time death of the host is no longer of concern to the parasite and in any event would usually occur soon as a result of destruction of tissues.

Although the stinging or aculeate Hymenoptera (wasps, ants, and bees) show no signs of having arisen from any existing parasitic groups, they must have come from ancestors having the general characteristics of the parasites. The aculeate Hymenoptera have the same major body divisions as the parasites and are for this and other reasons placed in the same suborder. The larvae, either sealed in cells with all the necessary food or fed and cared for continually by their mothers or sisters, are usually almost as reduced in external structure as are larvae of the parasitic groups. One of the evidences of the origin of aculeate Hymenoptera from parasitic ancestors is the delay in most aculeates of the midgut-hindgut connection until larval maturity. This is a feature which makes little sense for most aculeates unless it is an ancestral relic. In some specialized aculeates (for example, many xylocopine and megachilid bees) the beginning of defecation is advanced, so that the first feces are voided when the larva is only one-third or one-half grown. This reversion shows that sanitation of the larval cell is not a necessity for larval survival.

It is easy to put together series showing progressive loss of the main features of the parasites and acquisition of those of the aculeate Hymenoptera. Such series can be seen in various morphological features and are reasonably well correlated with behavioral series. Many ichneumonids and braconids have poison glands associated with their ovipositors. The poison is injected by the ovipositor to partially, and usually temporarily, paralyze hosts, especially those that are difficult to cope with. The ovipositor and associated poison may, however, be used as a sting in defense, as can easily be discovered by carelessly picking up many moderate-sized braconids or ichneumonids, such as *Ophion*. Some ichneumonids are external parasites, placing their eggs upon, rather than within, the bodies of their hosts. It is no great step from the behavior of such parasites to that of many scoliid wasps and their relatives, such as tiphiids. These insects are aculeates, in that the ovipositor has lost its egg-laying function and serves only as a sting for defense and for paralyzing hosts. They are similar to the main parasitic groups in that the female locates a host and parasitizes it without moving it to a prepared site; an egg is laid on the paralyzed host, which is then abandoned by the wasp. Most pompilid wasps differ in that the host or prey (a paralyzed spider) is placed in a cell constructed of mud or excavated in the soil and, after the egg is laid, the cell is closed by the mother wasp. Among many other wasps (for instance, most eumenines and sphecids) the insect or spider prey is small, so that several or many paralyzed individuals are required to provision a single cell. Still others (vespines) bring chewed insect tissue to feed their larvae progressively. Such sphecid and vespid wasps have fully abandoned the parasitoid mode of existence and have become predators.

Although most aculeate and parasitic Hymenoptera use other arthropods as larval food, adults feed on sweets as well as on blood of their hosts or prey. In some and perhaps most groups, sweets are necessary for survival of adults. Since the most generally available sweet material in nature is nectar from flowers, contact between flowers and adult parasitic and aculeate Hymenoptera has been maintained through much of the evolution of the group, and the labiomaxillary complex of the mouthparts is well adapted for sucking or lapping nectar from shallow flowers.

Since flowers contain a rich protein source, pollen, in addition to nectar, it is not surprising that certain wasps that had reached the predatory stage somehow abandoned predation as a means of provisioning cells for their larvae and instead store pollen as larval food. One such group is the Masaridae, a family of wasps closely related to the Vespidae and in the superfamily Vespoidea. The masarids include some genera with greatly elaborated mouthparts for taking nectar and pollen from deeply tubular flowers. The degree of morphological and presumably behavioral adaptation of such masarids (for example, *Pseudomasaris*) to flowers as sources of protein as well as of carbohydrates seems fully as great as that of many bees and, indeed, Malyshev (1968) calls them vespoid bees. However, since the masaridae did not experience a great adaptive radiation but remain relatively scarce, they are regarded as a family among the vespoid wasps.

The other group of wasps which abandoned predation as a source of larval food was from an entirely different source than the masarids, namely the wasps of the superfamily Sphecoidea. From this group arose the bees. Although bees did not arise from wasps that would fall in any existing sphecoid group, the similarity of primitive bees to sphecoid wasps (usually put in the family Sphecidae) as a group is very close indeed, much closer than to any other Hymenoptera. In view of their extensive adaptive radiation and the enormous number of genera and species, it is customary to give the bees the status of a superfamily, the Apoidea, although the degree of morphological difference from the Sphecidae does not justify this. In any event the bees are a probably monophyletic offshoot of the sphecoid wasps, an offshoot that happened upon a new adaptive situation (angiosperm pollen for larval proteinaceous food) and thrived.

Bees, entirely dependent on flowers for food, could not have arisen before the appearance of the angiosperm plants. Angiosperms became the dominant vegetation in middle Cretaceous times, although they existed earlier, perhaps mostly in arid uplands where opportunities for fossilization were poor. How early bees arose from the sphecoid wasps is unknown; it might have been as late as the middle Cretaceous. It should be noted that although angiosperms became very numerous at that time, most of them did not belong to the families which are now regularly visited and primarily pollinated by bees. The majority of angiosperms living at that time were probably pollinated for the most part by beetles (Baker and Hurd, 1968). The first known fossil bees are of Eocene age, but at that time they were already quite specialized and unlike any of the groups of wasps from which they must have arisen.

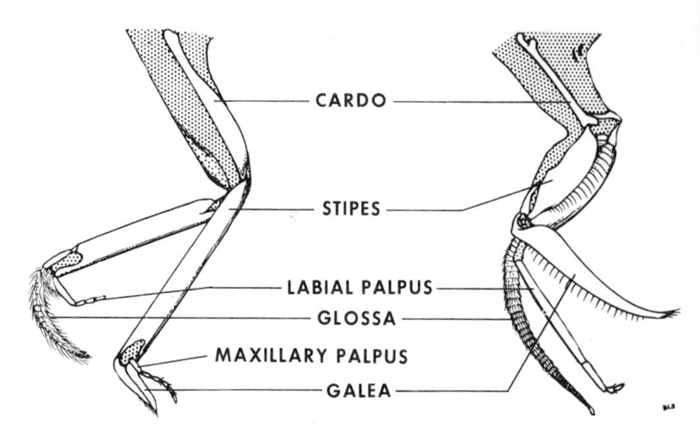

FIGURE 2.1. *Left:* Proboscis of a "short-tongued" bee, a halictid, *Pseudaugochloropsis graminea,* in side view. *Right:* Proboscis of a "long-tongued" bee, *Apis mellifera,* in side view. (Original drawings by Barry Siler.)

B. Evolutionary history among bees

Evolution among the bees is a theoretical subject, there being little in the way of fossil material on which to base a knowledge of the history of the group. The few fossils, often not well preserved, have mostly been studied only by persons with little knowledge of bee classification as the subject is now understood. Except for fossils in amber, the only parts reasonably well preserved are the wings. Although the wings of bees exhibit many important characters, useful in recognizing subfamilies, tribes, genera, and even species, there is enough variation within many of the families that it is often difficult to be certain of the family of a bee from the wings alone. In spite of these difficulties and our resultant ignorance concerning the evolutionary history of bees, some points seem worth making here.

Primitive angiosperms had relatively shallow flowers, such as can be used as pollen and nectar sources by short-tongued insects, including many beetles, wasps, and the short-tongued bees. The first five families listed in Chapter 3 are characterized by usually short mouthparts (Figures 1.6 and 2.1) and are often grouped as the short-tongued bees. Some, in fact, have long tongues (the halictine genus *Thrinchostoma,* for example) and others have moderately elongated mouthparts due to elongation of the basal parts of the proboscis (Halictinae). Such elongation of the mouthparts as occurs among the short-tongued families has been achieved independently by several quite different means and must be an adaptation to the narrow tubular flowers that have arisen among the more specialized angiosperms. Be this as it may, the short-tongued bees constitute a diversified group, probably the remnants of an apoid radiation that occurred at a time when most of the angiosperms had shallow flowers. The abundance of the most primitive of the families, the Colletidae, in the southern continents and especially in Australia, where many other archaic organisms also survive, fits in with the idea of the primitiveness of this family and implies a lesser degree of primitiveness for the other short-tongued families. In the Colletidae the glossa is nearly always broad and bilobed, as in wasps.

In contrast to the short-tongued bees, those with long tongues, that is, the last four families in the classification (Chapter 3), all have similar equipment for taking advantage of nectar sources within deep tubular flowers. This involves the elongation of the glossa, maxillary galeae and basal segments of the labial palpi, forming the sucking apparatus described in Chapter 1 (B1b). Plants with deep flowers probably arose in co-evolution with their principal pollinators, among which must have been these bees which seem to represent a second major radiation of the Apoidea. That this has been a more recent radiation than that of the short-tongued families is suggested by the close similarity of the long-tongued families to one another, particularly the Anthophoridae to the Apidae, so that major gaps between them are difficult to identify. Indeed, in an earlier classification (Michener, 1944), the Anthophoridae and Apidae were placed together in a single family.

Although the radiation of the long-tongued bees presumably postdated that of the short-tongued, it must have been well along during the early periods of the Tertiary,

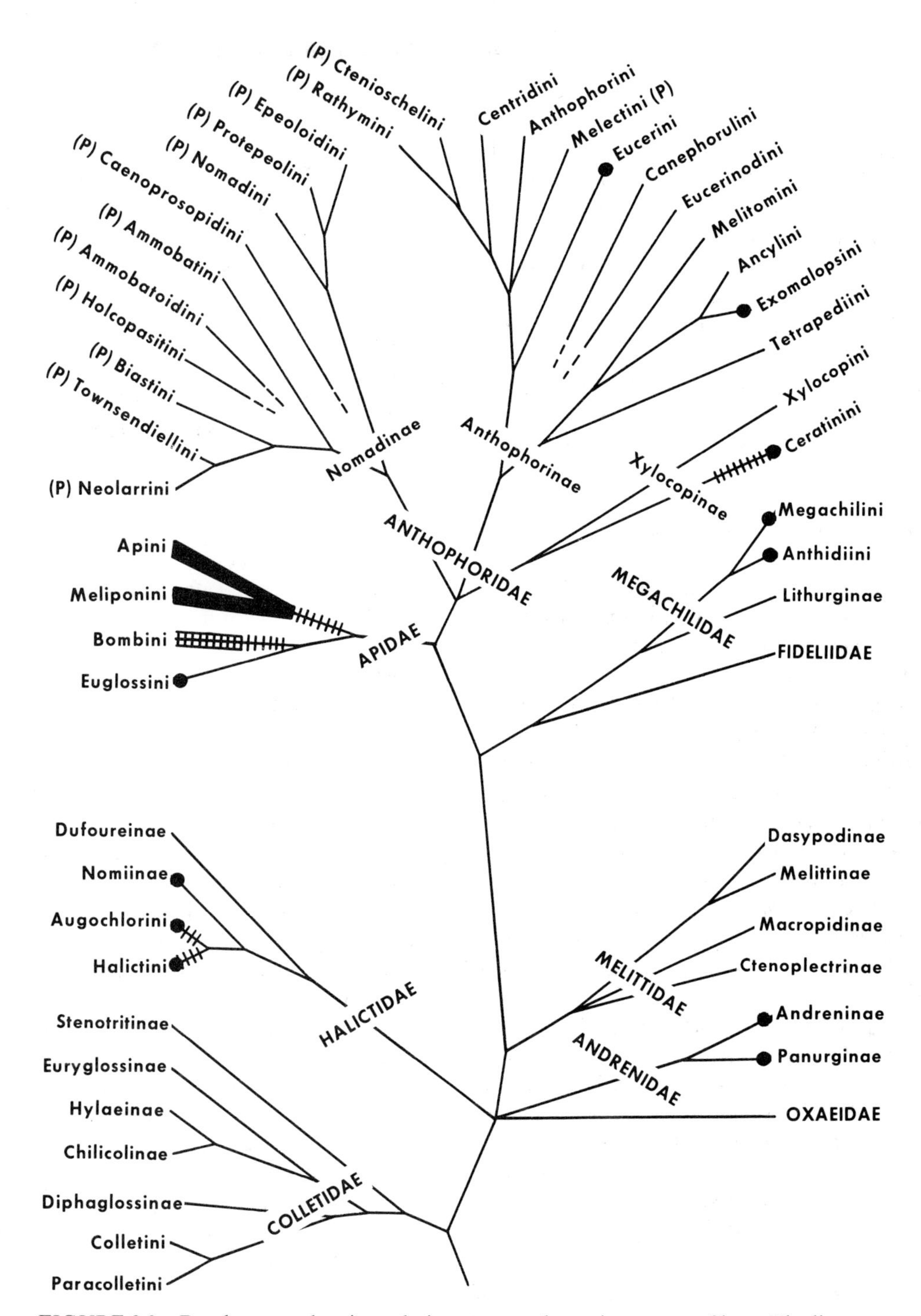

FIGURE 2.2. Dendrogram showing relations among the major groups of bees. The lines are subjectively determined lines of descent. Lines ending in *black dots* represent taxa in which some species live in parasocial colonies, at least part of the time. Lines with *crossbars* represent taxa containing primitively eusocial species. When the crossbars are connected at the sides (Bombini), all of the nonparasitic species are primitively eusocial. *Heavy bars* represent taxa all species of which are highly eusocial. Tribes which are entirely parasitic and therefore could not contain colonial species are marked (P). (Original drawing by Barry Siler and C. D. Michener.)

for there are excellent fossils of highly specialized, long-tongued bees of late Eocene age.

There is no fossil evidence of social behavior except in cases where one can recognize the worker caste among forms closely related to existing ones. Kelner-Pillault (1969) has described a minute *Trigona* of the subgenus *Hypotrigona* from the Baltic Amber (late Eocene). Its immediate relatives are unknown, but in amber from the state of Chiapas, Mexico, of Oligomiocene age, there are fossil bees of the genus *Trigona* so similar to existing species, not found in the vicinity but occurring from Costa Rica to Brazil, that they clearly belong in the same subgenus (*Nogueirapis*) and are scarcely different at the specific level (Wille, 1959). Moreover, in probably Oligocene amber from the Dominican Republic there are fossils of the African group of *Trigona,* subgenus *Hypotrigona* (Wille and Chandler, 1964). These fossils are all of worker bees of a highly social group. It is therefore clear that social behavior, of the type described below as highly or complexly eusocial, existed at least as early as the late Eocene in Europe and the Oligomiocene in the Western Hemisphere. Fossil apids perhaps related to *Apis* also occur in the Baltic Amber but belong in a distinctive and extinct genus, *Electrapis.* Nothing is known of its sociality or lack of it. *Apis* is unknown until early Miocene; it seems likely that specimens of the extinct subgenus (or genus?) *Synapis* of that age are workers of a social species. In summary, social behavior among bees probably first arose early in the Tertiary. In view of the numerous groups in which incipient social behavior is known, sociality has probably been arising from time to time down until the present period.

Figure 2.2, which should be examined with the caution suggested in the text of Chapter 3, gives an idea of the two radiations mentioned and the forms contained in the resultant families.

3 *Systematic Positions of Social Bees*

In order that the place of social bees can be properly understood in relation to the bees as a whole, the classification of the Apoidea is outlined below. It may seem that an unnecessarily large number of families, subfamilies, and tribes is recognized, but when one must organize twenty thousand species into meaningful groupings, a considerable hierarchy of categories constituting a higher classification is needed for convenience of expression, information storage and retrieval, indication of relative similarities, and for such help as it can give in predicting unknown characteristics. For an example of the last, most Xylocopinae, unlike most other bees, nest in wood or stems. Nests of many species of this subfamily have never been found, and their location is therefore an unknown characteristic. With the aid of the classification, however, one can predict that they are in wood or stems. (Predictions are not infallible: the small genus *Proxylocopa* nests in the ground.)

The classification of bees has been subject to many differences of opinion. Some students of the group prefer to place all bees in one family, the Apidae, while others recognize variable numbers of separate families. The greatest differences of opinion involve the wholly subjective ranking of the various groups in the classificatory hierarchy and not the limits of the groups themselves. The classification utilized here is a middle-of-the-road one, recognizing a moderate number of families. Nearly all of what follows, however, could readily be restated in terms either of a more conservative or a more finely split classification without losing validity. In other words, there is much better agreement among bee specialists on the relationships among groups than on the levels which the groups should occupy in the classification.

Figure 2.2 is a tentative phylogenetic diagram for the bees, showing the groups considered as tribes, subfamilies, and families. No modern phenetic (degree of difference) or cladistic (lines of descent) studies of the whole group have been made. Both would be desirable. This illustration is useful, however, in showing probable cladistic lines with some phenetic adjustments, all arrived at without the rigor of formalized methods. The levels of social attainments of the various groups are shown by symbols, as indicated in the caption.

Like most diagrams of this sort, Figure 2.2 is as much a classification as a justifiable statement of the historical sequence of the branching of the lines of descent.

The classification used below is that of Michener (1944) as modified by the same author (1965a) and by suggestions of Rozen (1965) and others. A detailed defense of it is inappropriate and unnecessary here.

As shown in Figure 2.2 and in the classification below, social behavior of some level exists, and has doubtless arisen independently, in various families of bees. Aggregations of nests are made by some species of every family and of every nonparasitic subfamily and tribe, excepting perhaps the Anthidiini, Bombini, Canephorulini, Ceratinini, Chilicolinae, Ctenoplectrinae, Eucerinodini, Hylaeinae, Macropidinae, and Tetrapediini. Of course the various parasitic bees make no nests, hence have no nest aggregations. Nonetheless, genuine colonies in which two or more adult bees occupy a single nest are known in relatively few groups and eusocial organization in still fewer.

Brief statements follow the comments on each group of bees to show the levels of social development so far known in the group. The levels indicated are in terms of the categories defined in Chapter 5 and refer to the maximal social development, not ontogenetic stages. Thus bumblebees are listed as primitively eusocial, although each overwintered female is solitary when she establishes her nest and then subsocial until the first progeny mature. Because it is so widespread, nest aggregation is not indicated in this classification. The genera mentioned elsewhere in the book are listed to show their systematic positions.

1. Family Colletidae

This is regarded as the most primitive family of bees, largely because of the short and usually bilobed or bifid glossa, which resembles that of other Hymenoptera, including the sphecoid wasps. There are several subfamilies, in general divisible into two groups, the moderate-sized to large, hairy forms (Diphaglossinae—including the Caupolicanini of Michener, 1944—and Colletinae) and the usually small forms with short, sparse hair (Hylaeinae, Chilicolinae, and Euryglossinae). The Hylaeinae and Euryglossinae have the habit, not found elsewhere among the bees except in *Lestrimelitta* (Apidae), of carrying all their pollen for larval food to the nests in the crop, mixed with nectar, instead of on scopal hairs of the body or legs. The Stenotritinae are possibly not even colletids. They are large, hairy forms with a blunt rather than a bilobed glossa. Nothing is known of their nesting habits.

Colletids nest in the soil, in pithy stems or holes in wood, or (rarely) in rotting logs or stumps. The nests consist of burrows with cells either constructed in the burrows or diverging into the substrate from the burrows. The distinctive feature of the cells of the whole family is the lining, which consists of a cellophanelike secreted material, painted onto the inner wall of the cell by the paintbrushlike action of the short, broad glossa.

All major groups of colletids are either found only in the Southern Hemisphere or are best represented there, especially in Australia, although the family as a whole is worldwide.

Some included genera are *Colletes* (Colletini), *Leioproctus* and *Paracolletes* (Paracolletini), *Chilicola* (Chilicolinae), *Hylaeus* (and its subgenus *Prosopis*) and *Meroglossa* (Hylaeinae), *Euryglossa* and *Pachyprosopis* (Euryglossinae), and *Stenotritis* (Stenotritinae).

No species of the family is known to be parasocial or eusocial.

2. Family Oxaeidae

The separation of this small group from the Andrenidae by Rozen (1965) is apparently justified. It consists of large, fast-flying bees found in the American tropical and subtropical regions.

The nests are deep burrows in the soil.

The included genera are *Oxaea* and *Protoxaea.*

No species is known to be parasocial or eusocial.

3. Family Halictidae

This enormous family contains some of the most abundant small and middle-sized bees in the world; indeed, in most temperate and in some tropical areas halictids are seen more often than any other bees except the honeybees. Halictids are often called "sweat bees" because in hot weather they alight on the skin and lap up perspiration; then, if caught under clothing or between parts of the body, the females sting. This family can be recognized by the reduced mentum and submentum of the labium as well as by other details of structure (see Michener, 1944).

The nests are burrows in the soil or in rotting wood.

a. Subfamily Dufoureinae. A small subfamily found in the Holarctic, African, and oriental regions, with one Chilean genus.

Some included genera are *Dufourea* (and its subgenus *Halictoides*) and *Systropha.*

No species is known to be parasocial or eusocial.

b. Subfamily Nomiinae. A large group in the Old World tropical and South Temperate regions, with a few species in the Holarctic region.

The principal included genus is *Nomia.*

While most species of *Nomia* are probably solitary, some of those of the Eastern Hemisphere are parasocial. Of these some are communal, some perhaps quasisocial or possibly semisocial.

c. Subfamily Halictinae. Because of their habits, members of this worldwide subfamily are more appropriately called sweat bees than are other halictid groups. Many species are colonial; colonial behavior appears to have arisen repeatedly and a broad spectrum exists from solitary to primitively eusocial behavior.

(1) Tribe Augochlorini. This tribe consists of usually brilliant green—although sometimes brassy, blue, purple, or black—sweat bees found mostly in South and Central America, with a few species ranging as far north as Canada.

Some included genera are *Andinaugochlora, Augochlora, Augochlorella, Augochloropsis, Neocorynura, Paroxystoglossa,* and *Pseudaugochloropsis.*

Presumably most species of Augochlorini are solitary, but some are communal, semisocial, or primitively eusocial.

(2) Tribe Halictini. This is a worldwide tribe of enormous size, less abundant in the Neotropical region than in any other, probably because of partial replacement by the Augochlorini. Species of the Halictini are sometimes brilliantly metallic green or blue, but usually show only weak greenish or bluish coloration, or the integument is black.

Some included genera are *Agapostemon, Halictus, Lasioglossum, Paralictus, Pseudagapostemon, Ruizantheda, Sphecodes,* and *Thrinchostoma.*

Many species of Halictini are solitary, but some are communal and very large numbers are primitively eusocial. Colonial species occur in all the genera listed except for *Thrinchostoma,* which is little known, and *Paralictus* and *Sphecodes,* which are parasitic in the nests of other bees (see Chapter 19).

4. Family Andrenidae

This large group is the principal family with two subantennal sutures on each side of the face. It is found in all continents except Australia. A few species in each subfamily live in colonies. Nests are burrows into the soil.

a. Subfamily Andreninae. This subfamily consists of very many moderate-sized species found principally in the Holarctic region, although with a few species in Africa and South America. They are among the commonest spring bees in most North Temperate areas.

The enormous but homogeneous genus *Andrena* contains nearly all the species of the subfamily.

Most species are solitary but a few are parasocial, presumably communal.

b. Subfamily Panurginae. This widespread subfamily contains moderate-sized to minute forms, many of them noteworthy for their ornate yellow markings, others wholly white, yellow, or black. A high percentage of the species are oligolectic, that is, restricted in their pollen collecting to particular species or groups of flowers.

Some included genera are *Calliopsis, Meliturga, Meliturgula, Nomadopsis, Panurginus, Panurgus, and Perdita.*

While most species are probably solitary, many *Perdita* and *Panurgus,* at least two species of the large genus *Panurginus,* and at least one species each in the small genera *Meliturgula* and *Meliturga,* are communal.

5. Family Melittidae

This small but diverse family, consisting of four subfamilies, is absent in Australia (except for one rare species) and South America. Nests are burrows into the soil or, for *Ctenoplectra,* burrows in wood.

The Ctenoplectrinae (genus *Ctenoplectra*) is found in the Old World tropics, the Macropidinae (genus *Macropis*) in the Holarctic area. The remaining subfamilies are Holarctic and African, being especially well represented in southern Africa. The Melittinae contains the genus *Melitta,* the Dasypodinae includes *Hesperapis* and *Dasypoda* among others.

No species of the family is known to be parasocial or eusocial.

6. Family Fideliidae

This family is known only in South Africa and Chile. It is similar and possibly ancestral to the Megachilidae.

Nests of the genera *Fidelia* and *Neofidelia* have been studied; they are burrows into the soil.

There is no evidence of communal or eusocial behavior.

7. Family Megachilidae

This is a very large, worldwide family, noteworthy for the fact that females of nonparasitic species (like those of Fideliidae) carry pollen by means of a scopa on the venter of the abdomen instead of on the hind legs. The family is also interesting because, unlike most other bees except some Apidae, material for cell walls and linings, or at least for partitions between cells, is brought into the nests from outside. Such material may be pieces of leaves, chewed leaf material, resin, mud, pebbles, or hairs removed from plants. In other groups except the Apidae, cells are excavated in or built from the substrate material and usually lined with secreted material. Secreted cell linings are absent in the Megachilidae.

a. Subfamily Lithurginae. This small subfamily is found in the tropical and warm temperate parts of the whole world.

The principal genus is *Lithurge,* which nests in wood.

No species is known to be communal or eusocial.

b. Subfamily Megachilinae. This large subfamily is divisible into two tribes as follows:

(1) Tribe Megachilini. This tribe is large and abundant in every continent and includes the familiar leaf-cutter bees as well as others which use the various materials listed above for cell construction (except that none use plant hairs).

Included genera are *Chalicodoma, Chelostoma, Heriades, Hoplitis, Megachile, Osmia,* and others.

A few species live in colonies which are communal or perhaps quasisocial but the vast majority are solitary. The colonial species are in *Chalicodoma* and *Osmia.*

(2) Tribe Anthidiini. This tribe, much smaller and less abundant than the Megachilini, nonetheless occurs in all continents although only one species reaches Australia. Its species use the various materials listed above for nest making, except for mud. As in the Megachilini, a few species are colonial.

Some included genera are *Anthidium, Trachusa,* and others having names using *Anthidium* as the root.

Most species are solitary but occasional species in the genera *Dianthidium, Heteranthidium,* and *Immanthidium* are communal.

8. Family Anthophoridae

This is another very large, widely distributed family. It is highly diversified and its subfamilies are sometimes considered to warrant family rank.

a. Subfamily Nomadinae. This subfamily includes numerous tribes of bees, all of which live as parasites in the nests of other bees. The interrelationships of these tribes are little understood and have nothing to do with the subject of this book; their arrangement in Figure 2.2 means little. Parasites in this and other groups of bees have larvae which feed on food provided by the host; they are thus cuckoos or social parasites, not true parasites or parasitoids.

The names of the major included genera can be recognized from the roots of the tribal names, except for the well-known genera *Epeolus* and *Triepeolus,* here placed in the Nomadini.

b. Subfamily Anthophorinae. This subfamily includes both parasitic tribes (Ctenioschelini—the Ericrocini of Michener, 1944—Melectini, Rathymini) and nonparasitic ones. Nests of the latter are usually in the soil, sometimes in wood; the cells, excavated into the substrate, are lined with waxlike material as is common in families 2 to 5 above. The nonparasitic tribes are listed as follows:

(1) Tribe Exomalopsini. A tribe of rather small and commonly robust bees containing American, mostly Neotropical genera.

Among included genera are *Exomalopsis* and *Paratetrapedia.*

So far as known all *Exomalopsis* are colonial, probably communal but perhaps with somewhat more elaborate relationships.

(2) Tribe Ancylini. A small group of bees found in arid areas around the Mediterranean and eastward into Asia; probably should be united with the Exomalopsini. Included genera are *Ancyla* and *Tarsalia.* Their nesting behavior is unknown.

(3) Tribe Tetrapediini. A small tribe from the American tropics. It is related to the Exomalopsini.

The included genus is *Tetrapedia.*

So far as known the species are solitary and nest in wood.

(4) Tribe Melitomini. A tribe (the Emphorini of Michener, 1944) found in the Western Hemisphere. It contains relatives of the Exomalopsini, from which most differ by their large, robust form. Most species are restricted to particular groups of flowers for their pollen collecting.

Included genera are *Diadasia, Melitoma,* and *Ptilothrix.*

So far as known there are no parasocial or eusocial species.

(5 and 6) Tribes Canephorulini and Eucerinodini. These tribes are not closely related. Each is recognized on the basis of a single South American species whose nesting behavior is unknown. The genera are *Canephorula* and *Eucerinoda.*

(7) Tribe Eucerini. This is a large tribe, common in all continents except Australia. The species are robust and fast flying. The males of nearly all species are noted for their long antennae.

Among the included genera are *Alloscirtetica, Eucera, Melissodes, Peponapis, Svastra, Synhalonia, Tetralonia,* and *Thygater.*

Most species are solitary but at least a few species of *Eucera, Melissodes,* and *Svastra* often live in colonies, presumably communally.

(8) Tribe Anthophorini. A worldwide tribe of moderate-sized to large, robust, fast-flying, hairy bees. There are few genera but many species.

Most of the species fall in the two genera *Amegilla* and *Anthophora.*

No species is known to be parasocial or eusocial.

(9) Tribe Centridini. A tribe similar to the Anthophorini but limited to the tropical and warm temperate parts of the Americas. It was called the Hemisiini by Michener (1944) and the principal genus was then known as *Hemisia* rather than *Centris.*

Included genera are *Centris* and *Epicharis.*

No species is known to be parasocial or eusocial.

c. Subfamily Xylocopinae. This subfamily contains anthophorids which nest for the most part in wood or in pithy stems, unlike most of the Anthophorinae. The cells are unlined. The tribes are as follows:

(1) Tribe Ceratinini. These are small, relatively hairless, slender bees which usually nest in hollow or pithy stems or in rotted woody vines or stems. They found throughout the world. In the tropical and southern parts of the Old World several genera constitute a unit informally called the allodapine bees, repeatedly referred to in subsequent chapters.

Among the included genera are *Allodape, Allodapula, Braunsapis, Ceratina, Compsomelissa, Eucondylops, Exoneura, Exoneurella, Halterapis, Inquilina, Macrogalea, Manuelia,* and *Nasutapis.* Of these all but *Ceratina* and the Chilean *Manuelia* are in the allodapine group. Three of the genera, *Eucondylops, Inquilina,* and *Nasutapis,* are parasitic in nests of other allodapines, and some of the other genera contain one or more parasitic species.

Ceratina, Halterapis, and probably *Manuelia* are solitary. The other nonparasitic genera contain subsocial and often primitively eusocial species.

(2) Tribe Xylocopini. This tribe consists of large, often very large, robust bees that nest in coarse, pithy or hollow stems or usually in solid wood, except for the Palearctic genus *Proxylocopa,* which nests in the soil. The tribe is found in tropical and warm temperate parts of the world.

Included genera are *Lestis, Proxylocopa,* and *Xylocopa.*

Except for reports of what are here called quasisocial relations, as occasional events, perhaps abnormalities, colonies are unknown in the Xylocopini.

9. Family Apidae

Although of only moderate size, this family is of special interest because it includes all of the highly eusocial bees as well as the primitively eusocial bumblebees and the solitary and parasocial Euglossinae. Unlike all other bees, females (except for parasitic and robber species and queens of the highly eusocial forms) carry pollen in a corbicula or pollen basket on each hind tibia.

a. Subfamily Bombinae. These are middle-sized to large, heavy-bodied bees found principally in the Holarctic and Neotropical regions.

(1) Tribe Euglossini. This tribe, found only in the American tropics, contains beautiful, large, metallic green, blue, red, and purple bees, as well as yellow-and-black or white-and-black hairy forms resembling bumblebees. They have very long tongues and the males of most or all species are involved in orchid pollination. There are four nest-making genera and two parasitic ones.

Among included genera are *Euglossa, Eulaema* (the *Centris* of Michener, 1944), and *Euplusia.*

Some Euglossini are solitary, others communal or quasisocial.

(2) Tribe Bombini. The familiar hairy, most often yellow-and-black bumblebees are primarily Holarctic, although they range south to Tierra del Fuego and the high mountains of Java.

The tribe contains two genera, *Bombus* and *Psithyrus.*

The species of *Bombus* are nearly all primitively eusocial, although at least one is a social parasite, as is *Psithyrus.*

b. Subfamily Apinae. These are minute to moderate-sized bees. All species live in perennial, highly eusocial colonies.

(1) Tribe Meliponini. This tribe consists of the stingless honeybees, found in the tropics of the world and especially abundant in the American tropics.

The genera are *Dactylurina, Lestrimelitta, Melipona, Meliponula,* and *Trigona.*

(2) Tribe Apini. This tribe, the true or stinging honeybees, was originally restricted to Eurasia and Africa, but one species, the common honeybee, has been introduced into all parts of the world.

The only genus is *Apis.*

4 *Some Terminology for Bees' Nests and Social Life*

Some terms like nest entrance or burrow are self-explanatory, and many other aspects of nest structure that are special to each group of social bees are described in Part III. There remain, however, certain terms applied to nest structures and functions that require explanation here. Moreover, although kinds of colonies are differentiated in the next chapter, a number of terms relating to colony life and interactions of individuals within colonies are treated here. They merit more detailed attention than mere glossary definitions.

A. Nests

A *nest* is a construct made, in our case, by bees. It is the place where the eggs are laid and the young reared. It is also the place where adults, at least females, spend most of their time not used for foraging. It may be a mere burrow in a pithy stem, a cluster of mud or resin cells cemented to a twig or a wall, a burrow system with cells radiating from it, a cluster of cells in the soil, or an elaborate construct like that of the honeybee. It may consist entirely of excavation into the substrate (for example, some species of the genus *Perdita,* in which the cells are unlined cavities in the sand), or it may consist entirely of materials brought to a site for construction of the cells (some Megachilidae). More often it consists of an excavation in the substrate to which materials are added to smooth, waterproof, or otherwise modify the cell walls and sometimes parts of the burrow system. It may consist largely of secreted wax, often supplemented and mixed with foreign materials, for example pollen (*Bombus*) or resin (Meliponini). A detailed account of nests of bees was given by Malyshev (1935); briefer or more specialized accounts are given by Michener (1964b), Sakagami and Michener (1962), and Wille and Michener (in press). Nest burrows are sometimes divided into cells by means of cross partitions; in other cases the cells are especially excavated in the substrate or constructed in a cavity.

A *cell* is typically the compartment in which a single immature bee is reared. Sometimes it is only a space excavated in the substrate, but almost always there are especially constructed or modified cell walls, or at least partitions between cells. A cell usually is approximately the size of one fully developed bee of the species concerned. In the genus *Bombus* there are regularly several young together in each cell and the cells are enlarged as the larvae grow, a unique feature. In *Megachile policaris* and *Lithurge fuscipennis* several larvae may grow up together in a single enormous cell (Krombein, 1967; Gutbier, 1915) but in all other known species of these genera each cell is inhabited by a single young bee. In the allodapine bees the partitions between cells have been lost and the young are therefore reared, not in cells, but in a common space which is the nest burrow; the nest is typically nothing but a simple burrow in a pithy stem or a burrow already made by a boring insect in a woody stem.

Cocoons, not being constructed by adult bees, might not be considered as parts of nests by some persons. But for some nests, especially those of *Bombus,* they are important structural elements quite aside from their prime function of protecting pupae. Many bee larvae, after finishing feeding, spin a protective cocoon of silk which is secreted by the salivary glands of the larva—homologues of the thoracic salivary glands of adults. Most Apidae spin cocoons before defecation and the larval feces are therefore inside the cocoons; other bees that make cocoons spin after defecation. The cocoon is usually impregnated by substances which harden among the silk fibers and form a firm matrix for them. Often, for example in many Megachilidae, there are several layers of silk and matrix material. Cocoon formation has been studied in detail only in *Apis,* where a colorless material, probably from the Malpighian tubules, forms part of the matrix and yellow and brown material from the hindgut forms the rest (Jay, 1964).

The cocoon protects the young bee during its metamorphosis—the delicate stages when it changes from larva to pupa and from pupa to adult—as well as the pupa itself, which is also easily injured. Some groups of bees, however, spin no cocoons. Examples are the subfamilies Halictinae and Xylocopinae, and the tribe Anthophorini. Perhaps their pupae are less easily damaged than those of cocoon makers. The larva of species whose pupae are naked has a reduced salivary opening without the strongly projecting lips that form the spinneret of cocoon-spinning species.

B. Colonies

A *colony* is made up of all of the mature individuals occurring in a nest, irrespective of the type of organization among them, plus immature stages being actively cared for. Male bees generally leave the nests soon after maturity and before mating and do not return, or return only at night and during inclement weather. Adult bees of a colony are, therefore, ordinarily all females; males when present are best thought of as visitors rather than as members of the colony. Only in highly social bees such as honeybees do males remain in nests with colonies; they are fed by workers or use available food stores in the nest and in other ways also qualify as at least temporary members of the colonies.

Eggs, larvae, and pupae are not considered parts of a colony in mass-provisioning species. This is not merely to justify the long-established expression "solitary bees," for the cells are closed and there is no contact between adults and young in such nests. Social interactions, if they occur, have to be primarily among the adults. Two or more mature females are therefore necessary to form a colony in mass-provisioning species. In progressive feeders, however, the young are obviously a significant part of the social organization and are considered to be part of the colony. In such forms a minimal colony may consist of one female and her immature progeny.

Colonies of bees range in size from those with a total population of only a female and her progressively fed offspring (even one bee and one larva) or a few adult females (sometimes only two similar females or one queen and one worker) to the honeybee (*Apis*), in which strong colonies have 60,000 adult individuals, and some of the tropical stingless honeybees (*Trigona*), whose colonies are reported to attain populations of 180,000 or more. A group or aggregation of nests, for example of solitary bees, does not constitute a colony in the present sense.

Biologists use the word *social* in a variety of ways. Ecologists may speak of a society consisting of all the organisms of a certain area; others speak of the social relations among birds in a flock or animals in a herd, or of the social relations between members of a sexual pair. Special and narrower definitions are applied to colonial insects. All these meanings, however, indicate not only interactions among individuals of a group, but interactions that produce effects qualitatively different from the mere summation of the independent activities of the individuals. If a number of organisms are close together, yet do not influence one another (in a detailed sense, a most unlikely occurrence), one may speak of the group as an aggregation but not as a society.

The word social is used in Chapter 5 and thereafter only in an informal way, and other words or prefixed modifications are used to indicate more specifically what is meant. Throughout the book the word social refers to colonies, not to sleeping clusters or sexual pairs or other such groupings, even though its use for such groups is elsewhere perfectly legitimate, even for the same insect species.

Social organisms in the narrow entomological sense, those called eusocial in subsequent chapters, live as groups of adults of different generations, with cooperative activity and with different individuals performing different functions necessary for the group. If one ignores physically connected individuals such as polyps of colonial hydroids, only ants, termites, some bees, some wasps, and man (and to a lesser degree certain other vertebrates) are social in this sense. But this book covers a broader spectrum of behavior than merely the eusocial, for one of its objectives is to describe the many incipient social groups found among bees. Colonies of bees, irrespective of their level of social organization, are our subject.

Various types of individuals may be found in bee colonies. In Hymenoptera the *males*, as indicated above, are usually short-lived, play no significant role in the social organization, and usually leave the colony early. Their only known function in most cases is fertilization of young females. The caste system is therefore limited to the female sex. Males of social Hymenoptera are no more different from the females than are those of solitary species, and the designation of the male (called the drone in honeybees) as a caste is only confusing. We should therefore speak of males and females, the latter divided into castes, in social Hymenoptera.

Castes in Hymenoptera, then, are the forms of females that perform different functions in a colony. They differ from one another at least behaviorally and physiologically, often also in structure, and some of the differences are permanent, not merely due to age. The separation of females into castes which have different functions is evidence of *division of labor*.

In bees the only castes are *queen* and *worker*. The differences in function and structure between these castes are detailed in Chapter 9, but very briefly, queens are specialists at laying eggs and workers at foraging and various nesting activities. In kinds of bees that lack external caste differences, the internal differences are usually not developed in freshly emerged young adult females. In such species the caste differences become evident when the bee establishes a nest, mates, and embarks on queenlike behavior, or alternatively fails to do so and becomes a worker. It is a convenience to restrict the word *queens* to females that are actually functioning as queens in the context of a colony—the main feature is that they lay eggs in the presence of other individuals. *Workers*, on the contrary, do not lay or do so less frequently. In bee colonies containing two or more layers, the principal layer is normally called the queen. The others are laying workers.

The word *gyne* is used for both potential and actual queens. Expressions like functional or reproductive queen, hibernating queen, nonreproductive queen, queen larva, and the like all refer to individuals to which the general term gyne can be applied. I would use queen only for the first. Gyne is especially applied to individuals not functioning as queens but presumed to be potential queens. In many species all young adult females are gynes, castes becoming recognizable by physiology and behavior later in life.

A *monogynous colony* contains only a single gyne; a *polygynous* one contains two or more gynes. Monogynous and polygynous are here used synonymously with haplometrotic and pleometrotic, respectively, because the first pair of words is simpler and more likely to be remembered. The gynes may or may not all be queens. In semisocial colonies some gynes never develop functional ovaries, but become workerlike and are called auxiliaries.

In various kinds of bees, but especially some species of Andrenidae and Halictidae, *male polymorphs* have been noted. The largest males have extraordinarily large heads and jaws and broad, often angulate, genal areas. This is evidently merely a matter of allometry, and no one has suggested a division of labor among males on the basis of this variation alone. Recently, however, Rozen (unpublished) in North America and Houston (1970) in Australia have discovered species of bees in which some of the males are large-headed, short-winged, and presumably flightless. Rozen's is a species of *Perdita* (Andrenidae); Houston's is a *Lasioglossum* of the Australian subgenus *Chilalictus* (Halictidae). Houston suggests that such males may constitute a soldier caste which spends its entire time in the nest, a view supported not only by the shortened wings but by small size of the eyes, enormous mandibles, enlarged fore femora, etc. These males have normal genitalia and may mate, but probably only with their sisters. Unfortunately no behavioral observations have been made. If the large-headed males indeed function as soldiers, the case is well worthy of intensive study. Nothing comparable is known among males of other social insects, and some reasons for lack of castes among males are enumerated in Chapter 20. If the normal and large-headed male forms are indeed behavioral castes, they ought to be discussed in many of the succeeding chapters. As nothing more is known about them, I have chosen not to do so; paragraphs about our ignorance of them in so many chapters seem unnecessary.

The word communication is understood in different

ways by different persons. In the past (for example, Michener, 1969a), I have considered any activity of an individual bee that influences another member of its colony to be social communication. Hence I viewed the coordination of members of a colony as a product of communication. However, on considering the effects of inaction as well as action, and noting that almost anything that an organism does or does not do may influence another organism, I have concluded that my earlier use of the term encompasses virtually all behavior and therefore should be narrowed. An excellent discussion of the problem is that of MacKay (1972).

In any colony there are interrelations among individuals, such that behavior of one influences behavior or development of another. All these interrelations are termed *social interactions.* Feeding of a larva by a bee is an example of a social interaction. The food influences growth of the larva, affecting size and perhaps caste, or lack food results in death. And the adult bee is also influenced at least behaviorally. Some social interactions, unlike feeding of a bee larva, are *signals,* that is, their function is primarily the transmission of information. One might argue that the food in the above example should be looked at as a signal that stimulates larval growth, but the informational content is not the main function of the food. It is therefore not called a signal. Transmitting of information is called signaling, whether or not the activity is goal-directed and whether or not there are recipients. If there are no recipients, signals do not result in social interactions. Thus there can be signaling without social interaction. To return to the above example, the hungry larva may signal its need, even in the absence of an adult that can respond.

There are many activities in bee colonies that result in nest structures (or conditions of brood or stored food) to which other bees respond. The results are *indirect social interactions*—indirect because information transfer is not between two bees but between a construct made by bees and another bee. The construct (or brood or food) is made or cared for with other primary objectives, not for signaling, although the information content may be essential for colony integration. One could speak of an informational component of the construct as an indirect signal.

As an aside, note that through similar constructs, brood, or stored food, an individual bee may influence its own subsequent behavior. This is simply feedback, although previously I have spoken of such a bee as communicating with itself.

The word *communication* is here limited to the sending or transmitting of signals whose primary function is to influence the behavior or development of other individuals and which actually are perceived and elicit responses. Of course this explanation is for social communication among individuals. For most purposes among bees, the communication is between members of a single colony, although sex attractants and the like function among individuals from different colonies. Communicative signals can be either physical or chemical in nature, and may or may not involve contact between signaler and receiver. The energy output in communicative signaling is generally small, often less than that involved in a response.

Part II Comparative Social Behavior

This part, consisting of Chapters 5 to 20, is a general account of social behavior as it occurs among the bees. Because of differences in kinds and amounts of information available for different groups of bees, it is often not possible to be rigorously comparative, nor would such an approach always be desirable. Wherever practical, however, each aspect of social behavior is considered, first for the least social bees to which it pertains and then for groups that are progressively more specialized with respect to that behavioral feature.

The organization of Part II is largely in terms of functions—defense, caste determination, foraging, and the like—but it has been convenient to introduce certain chapters organized around social levels (5, 6, and 20) and one (7) centered on cues in the nest structure mediating certain behavior. Chapter 5 concerns kinds of social behavior among bees. Chapters 6 to 11 concern in general the dynamics of development (ontogeny), maintenance (homeostasis, foraging, communication), and reproduction of colonies. Chapter 20 is about evolution of social behavior in bees.

It seems desirable to cite here a number of important reviews of social behavior in insects and especially in bees. So far as appropriate in the body of the text, references are to original sources or to an author's own summation or to major works on particular kinds of bees. Yet the reviews serve important purposes in assembling information, listing references, and sometimes in providing new insights and integration of ideas. The following list will guide readers to useful review papers and books which contain citations to many papers excluded from the references at the end of this book. Reviews of social behavior in bees: Buttel-Reepen, 1903, 1915; Kerr and Laidlaw, 1956; and Michener, 1969a. Reviews of social behavior in Hymenoptera: Bischoff, 1927; and Michener, 1961b. Reviews of social behavior in insects: Allen, 1965a; Brian, 1965a; Goetsch, 1940; Maidl, 1934; Michener and Michener, 1951; Richards, 1953; Wheeler, 1923, 1928; and Wilson, 1966, 1971.

5 *Kinds of Societies among Bees*

As indicated previously, colonies of bees range from those that seem almost insignificant—two or three bees in a burrow in the ground or in a hollow stem—to the large colonies of the honeybees. The kinds and amount of division of labor and communication among bees in colonies vary greatly. Man being a classifying animal, a classification of the social relationships has been developed. This chapter explains the various levels of colony organization found among bees—that is, a classification of kinds of insect societies as it is applied to bees. This classification is useful but, as demonstrated in Chapter 20 (C), it stresses only a few attributes and by itself does not present a good picture of the social variation found among the bees.

For the less complicated levels there is not a great deal to say. They are intrinsically simple and less remarkable than highly social levels. Because the less highly social forms often nest hidden in the ground, they have received less attention from researchers than highly social species. For the lower social levels this and the next chapter contain all the information that is included in this book. Thus on such topics as aggregations, communal colonies, and quasisocial colonies, almost nothing is found after Chapter 6. For semisocial and eusocial colonies, however, the treatment in Chapters 5 and 6 is a mere outline, as all the subsequent chapters deal with them.

A. Sleeping clusters

Sleeping clusters are not colonies, for the bees (or wasps) in a cluster are not inhabiting a nest, rearing young, and the like. These clusters differ in this respect from all subsequent categories. They consist of groups of males, rarely with a few females, which assemble in certain places to pass the night. There may be up to two or three hundred bees involved, usually of one species but occasionally mixed. Bees of all families except the Apidae sometimes form such clusters. (Probably because the families are small and rare, I know of no records of such clusters for Fideliidae and Oxaeidae.) Species whose males make overnight burrows or sleep in curled leaves or such protected places probably do not form sleeping clusters in the open, although sometimes several males will crowd into one burrow or curled leaf. The clusters are commonly on exposed stems, the bees clinging sometimes with their legs, often also with their mandibles. Some species adopt a bizarre position, projecting rigidly at right angles from the twig or stem which they grasp only with the mandibles.

These sleeping aggregations are not without interest for persons concerned with social behavior. There are clearly interactions among the bees, although it is not known what attracts large numbers of males to a single place to sleep or what selective advantage such behavior might have. The same males often return night after night to the same stem, showing that the males have topographic orientation abilities similar to those of females and that inability to find the nest after leaving it is probably not the reason that most male bees never return to their natal nests.

The literature is full of descriptions of sleeping aggregations of male bees. Most authors made no serious effort to relate their findings to others; they simply described an unusual clump of bees. Le Masne (1952) has given a general account of sleeping aggregations.

B. Levels of social organization

Species are often called solitary, communal, social, and so forth. Such terms are generally applied to the most complex type of organization attained during life cycle of the species, as was done in Chapter 3. Such usage is appropriate and useful, but when precision of reference to a particular colony or interaction is needed, it often does not suffice to allude to the organizational level attained by the species. Only the highly eusocial bees remain permanently at that level. Nearly all other forms change during their life cycles, usually from solitary to some social pattern. The names for the organizational levels are therefore applied to the various ontogenetic stages through which a colony may pass. To use the example cited in Chapter 3, a bumblebee nest is established by a young gyne; she is solitary. Soon she becomes the mother in a subsocial colony, and later the queen of a primitively eusocial colony. But in a different context the same bumblebee species may legitimately be referred to as a primitively eusocial species, since that is the most elaborate social level which the species attains.

Table 5.1 lists the various types of bee organizations here recognized and summarizes some of their principal features. The propriety of this particular classification is further examined in Chapter 20.

C. Solitary bees

In the majority of species of bees each female makes her own nest, or sometimes several of them, without regard to the locations of other nests of the same species. Such bees mass-provision the cells by placing enough pollen and nectar in each to provide for the entire growth of a larva. After oviposition, the mother seals the cell and goes on to construct and provision another. Ordinarily she dies before her progeny mature and emerge from their cells; therefore there is no contact between generations.

Contact between generations does occur, however, even in some bees that are best called solitary. For example, in the European *Halictus quadricinctus* and in *Augochlora pura* of the eastern United States, the mother maintains the nest and is frequently still present when her progeny mature (Verhoeff, 1897; Stockhammer, 1966). The same is true of many *Xylocopa,* for example, *X. tranquebarorum* (Iwata, 1938, 1964), in which, after the reproductive season, a nest may contain a mother and several adult daughters. (Some new nests may contain two bees of about the same age but such associations are apparently temporary.) So far as known in such bees, however, cell construction and provisioning is the work of a single bee. In *Ceratina* the last cell is often protected by the mother

TABLE 5.1. Levels of social organization among bees.

Level	Castes and division of labor	Colonies with adults of two-generations (matrifilial)	Cooperative work on cells	♀♀ structurally similar; gynes (if any) survive alone	Progressive feeding
Solitary	—	no colonies	—	+	—
Subsocial	—	—	—	+	+
Parasocial					
Communal	—	—	—	+	—
Quasisocial	—	±	+	+	±
Semisocial	+	—	+	+	±
Eusocial					
Primitively social	+	+	+	+	±
Highly social	+	+	+	—	±

Note: + indicates that column heading applies, — indicates that it does not; ± indicates variability.

bee instead of by a partition; death of the mother frequently leads to destruction of the cell contents (Chapter 7, D2).

I also do not consider as communal or even as colonies, as the terms are here used, groups of several bees (sisters?) that sometimes live temporarily together in old nests before dispersing to start new nests individually. Sometimes (as in *Meroglossa torrida* from Queensland, Michener, 1960a, and *Neocorynura fumipennis* from Costa Rica, Michener, Kerfoot, and Ramírez, 1966) one bee may begin to provision new cells in the old nest before the others leave. Such units, with minimal interactions among bees, may sometimes have been evolutionary antecedents to parasocial colonies, but evidence for this hypothesis is minimal.

Probably the majority of species of solitary bees have only a single generation per year, the adults emerging and flying about during a relatively brief season, sometimes only two or three weeks. Such species pass the rest of the year in the nest. The feeding stage of the larva is ordinarily brief, often only a few days, and most of the year is passed in the prepupal stage or as young adults either still in their natal cells or in special hibernating or estivating places.

Some solitary bees, however, regularly have two generations per year, for example one in spring and another in the autumn, while others go through a succession of overlapping generations so that, except in the spring, all stages can be found at any time during the warmer months of the year. Examples of North American bees having such life histories are *Megachile brevis* (Michener, 1953) and *Augochlora pura* (Stockhammer, 1966). In the tropics, where there is no winter, some species of solitary bees breed continuously throughout the year (for example, *Neocorynura fumipennis* in Costa Rica; Michener, Kerfoot, and Ramírez, 1966). Others, however, are as seasonal as bees of temperate areas (like *Andrena vidalesi* and *Calliopsis hondurasicus,* active only in the early dry season in lowland Panama, Michener, 1954).

D. Aggregations

Among all families of bees there are species in which each female makes her own nest, as described for the solitary forms, but in which the nests are grouped. Aggregating of nests may always be facultative; some nests are usually scattered. But for many species aggregations are very characteristic. Similarly, there are often groups of nests of communal, semisocial, primitively eusocial, and even highly eusocial bees.

Nest aggregations occur most commonly among bees that burrow in the soil. Aggregations of such burrows may vary from a few nests so scattered that one wonders if they constitute an aggregation at all, to small, dense clusters of nests like those of *Lasioglossum versatum* in Kansas (Michener, 1966b), or to enormous numbers of nests occupying rather large areas. In *Lasioglossum versatum* groups of up to 60 or 100 nests occupy areas of soil often only about 30 to 60 cm on a side in the midst of acres of apparently identical and unoccupied terrain. I have observed a clay bank 1 to 4 m high along the side of a small canyon in eastern Colorado that was densely packed for a distance of 500 m with nests of *Anthophora occidentalis.* Figure 5.1 shows a part of a bank inhabited by an aggregation of another anthophorid, *Paratetrapedia oligotricha.* By far the largest aggregation of nests of solitary bees that has been recorded was described by Blagoveshchenskaya (1963) in the Soviet Union. It was 10 to 150 m wide, over 7 km long, and occupied an area estimated at 360,000 m^2 along one bank of the River Barysh. The principal bees were *Dasypoda plumipes,* averaging 21 nests per m^2, and two species of *Halictus,* averaging 13 per m^2, giving a total of 12,240,000 nests of these three species in the entire aggregation.

Some bees that make burrows in stumps or logs instead of soil also form aggregations. For example, one may find numerous nests of carpenter bees (*Xylocopa*) in a single post or building while nearby none are found in similar posts or buildings. Iwata (1938) records four nests of *Xylocopa tranquebarorum* in adjacent internodes of a single bamboo stem, and in 1964 the same author further describes the departure of young adult females to establish their own nests nearby, and inheritance of the parental nest by the youngest daughter. Bees nesting in stems, however, usually have no opportunity to aggregate, as they are forced to use whatever sites are available.

Bees, mostly megachilids, that construct exposed cells of mud and other materials brought to the site sometimes also form aggregations of nests; for example, *Chalicodoma muraria* sometimes covers large portions of walls in southern Europe and North Africa with masses of its

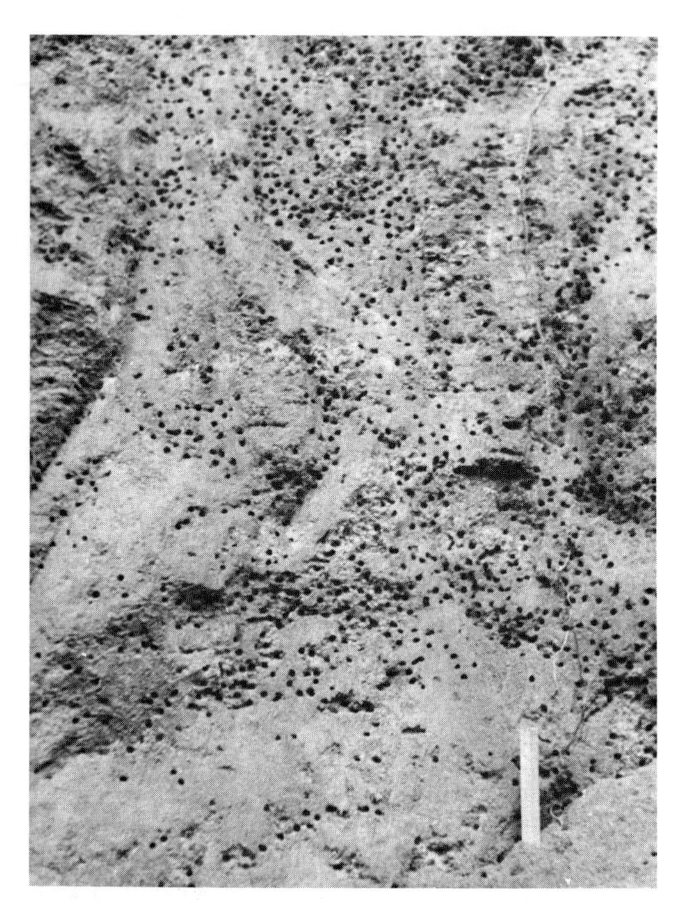

FIGURE 5.1. Part of a vertical earthen bank in Rio de Janeiro densely perforated by nest burrows of an anthophorid bee, *Paratetrapedia* (*Trigonopedia*) *oligotricha.* The white bar at the lower right is a ruler about 11 cm in length. (Photograph by C. D. Michener.)

cells. There are places in Egypt where the almost stone-hard cells of this bee have covered the designs and hieroglyphs of the ancient civilization.

A few Apidae are known to form aggregations, as follows: certain species of *Euplusia,* many individuals of which may make their cells near together in a given crevice or in a thatched roof; certain species of *Trigona;* and *Apis dorsata,* which sometimes builds numerous large nests near together. There is a record of 156 nests of this *Apis* in a single tree in India (Singh, 1962).

The question frequently arises as to whether aggregations result from limitations of suitable habitat or from other factors including social interaction. Some aggregations occupy all of certain limited suitable habitats (for instance, a patch of sandy soil or a rotten log) and therefore probably result from limitations of the habitats. However, it is clear that other factors often contribute to the formation of aggregations. A study (Michener, Lange, Bigarella, and Salamuni, 1958) of nests of solitary bees in burrows in roadside banks in the southern Brazilian plateau showed that certain soil layers and exposures were the ones principally occupied by the bees, indicating the importance of edaphic and light factors. The more detailed groupings of nests, however, could not be explained by known chemical or physical attributes of the soil or by its exposure. These groupings were presumably due either to a tendency of bees to return to the sites where they were born or to interactions among bees or between bees and existing nests (see also Chapter 6, A).

E. Parasocial colonies

Communal, quasisocial, and semisocial groups are so similar superficially that the convenient collective term parasocial has been proposed for them. Parasocial colonies are simply any colonies in which the adult bees consist of a single generation, unlike the eusocial forms in which two generations of adults are ordinarily present. They are almost always small colonies and the relationships among the individual bees appear to be at a simple level.

As described elsewhere, the most important advantage these minimal colonies afford their inhabitants is probably improved defense. One or more bees usually are present, and can eject some kinds of intruders which can easily damage the cells or progeny of a solitary bee while the mother is away, or which can establish themselves inside the nest of a solitary bee.

1. Communal colonies

A communal colony consists of a group of females of the same generation using a single nest, each making, provisioning, and ovipositing in her own cells. Hence dissections of nest populations ordinarily show that all mature females have enlarged ovaries and have mated. This definition is designed for bees that make nest burrows in addition to cells; such a nest occupied by a communal colony is sometimes called a composite nest. Some bees, however, construct cells in the open or in large cavities not made by the bees, often with cells in contact with one another, sometimes on top of one another, different cells being made simultaneously by different females. Such females may be regarded as constituting a communal group. In the absence of nest structures other than cells (burrows, for example), however, the distinction between aggregations and communal groups becomes trivial, or at least there is no practical way to draw the distinction.

Various unrelated megachilids are the principal makers of such clumps of cells, either exposed or in large cavities—the work of several or many females (Malyshev, 1935; Michener, 1968b). Grouping is well developed; for example, one mass of cells of *Immanthidium repetitum* of South Africa consisted of about 1750 cells (Michener, 1968b), and Fabre (1914) describes solid masses of cells of *Chalicodoma sicula* in France covering several square meters of surface and with "hundreds and thousands" of bees flying about. The cell masses of *C. muraria* described in section D above are similar.

Economy of labor cannot be recognized in such work and there is little evidence of improved defense, although one can imagine that the numerous bees might improve the defenses against certain enemies. Unfortunately, no one has made the critical observations on rates of predation and parasitism in nests of lone bees vs groups, either for these bees or for the plainly communal bees described below.

There is not much to say about the social behavior in communal groups, partly because there is not much social behavior. Mutual tolerance and perhaps attraction to nest entrances are the most evident behavioral features of the communal forms by which they differ from solitary relatives. As with aggregating bees, the communal grouping is usually and possibly always facultative; lone individuals of the same species nest successfully. In burrowing bees economy of labor results from communal grouping in that only one main burrow has to be made; defense of the burrow is presumably also improved. For most purposes, however, each female seems quite independent of the others.

Subsurface union of nests of halictines is known for various aggregating species (Michener, 1966c; Sakagami and Michener, 1962) but unfortunately little is known of the behavior of the bees in such situations. I suppose that the two bees in a pair of joined nests might be thought of as constituting a communal colony if they are mutually tolerant and allow the connection to persist. Most known cases of subsurface joining of nests, however, concern primitively eusocial groups, so that relationships of the individuals are presumably modified by already established eusociality within colonies. Nest cavities of the solitary halictines *Augochlora pura* and *Neocorynura fumipennis* occasionally meet, but in both species the bees make a new wall of loose particles of the substrate, wood (*Augochlora*) or soil (*Neocorynura*), reseparating their nests. This may be no more than the response of the nesting bees to any excess space—they close off their own nest areas from other cavities.

It is not uncommon for a nest of bees which are ordinarily solitary to contain two or more adult females, especially if the nests are in an aggregation. In most such cases that have been studied, the number of branch burrows or groups of cells corresponds to the number of adult females utilizing the nest. It is therefore assumed that each bee has its own branch or cell group. Such cohabitation by two females of ordinarily solitary species occurs occasionally in species such as *Andrena erythronii* (Michener and Rettenmeyer, 1956), *Nomia punctulata* (Hirashima, 1961), *Eucera longicornis* (Nielsen, 1902) and various other Anthophorinae (Michener, 1966a). Sometimes two or more sisters of *Xylocopa sauteri* use their dead mother's nest entrance but have separate branch burrows (Iwata, 1964). Nests like these can almost be thought of as solitary nests that happen to have their entrances inside of a common burrow.

Likewise various halictine[1] bees which are normally solitary occasionally exist in communal colonies. Thus for *Halictus quadricinctus,* Vasić (1967) seems to have shown that there are sometimes composite nests, two or four females using a common entrance burrow, each building her own cluster of cells. Sometimes, however, this European species, famous for its large size and beautiful cell clusters, makes a nest occupied by two or three females working presumably communally on a single giant cell cluster of over 30 cells (Grozdanić, 1966). In certain species of bright green *Augochloropsis* of the Americas (such as *A. diversipennis* or *humeralis*) two or more females may

1. I have previously listed *Lasioglossum esoense* of Japan as occasionally communal (Michener, 1969a). However, this resulted from a misreading; coexistence of more than one working female has not been found in the species. *L. esoense* is solitary, although young adults sometimes coexist with an old mother.

work in a single burrow, each seemingly making and provisioning her own cell cluster. In both species nests occupied by single females are far more common (Michener and Lange, 1959; Smith, 1901).

In various other species of bees, communal colonies are of such frequent occurrence as to be typical or at least common, although it is likely that lone females can also succeed in all cases. In the enormous genus *Andrena,* most species are solitary, some nesting in aggregations. However, *Andrena bucephala* (Perkins, 1917) and *A. ferox* (Yarrow and Guichard, 1941) live in colonies that are probably communal. Observations of both were made in Britain. No detailed studies or dissections have been made; the observations that have been published are only to the effect that numerous females are seen going in and out of a single burrow entrance in the ground. Perkins estimated several hundred to a thousand bees (total; both sexes and parasites) from three nearby burrows of *A. bucephala.*

Certain other andrenids are often or usually communal. While many species of the North American genus *Perdita* are solitary, nests of *Perdita opuntiae* and *P. lingualis,* unrelated species in different subgenera, are found with up to ten or eleven females each. Such nests are burrow systems; those of the former were made near Boulder, Colorado, in sandstone rock (Custer, 1928a), and the latter were studied in sand banks of the Kansas River (Michener, 1963a). Rozen (1971) has reported two nests of *Perdita utahensis* that contained four and eight females. Dissections were made of bees from nest populations of *P. lingualis* and *P. utahensis.* In the latter all had functional ovaries. As to the former, while some young, often unfertilized females with not much enlarged ovaries were present, they were not carrying pollen and had evidently recently emerged from pupae; all bees carrying pollen were fertilized and with enlarged ovaries. Hence the colonies of this species also must be communal.

In the Palearctic andrenid genus *Panurgus* apparently most species are usually communal, although some are solitary. Several European and North African species have been recorded as nesting several bees to the burrow. The first such records are at least 130 years old. Recently Rozen (1971) has dissected the bees from nests of two species (nest populations of one to twenty females) and found that all had well-developed ovaries.[2] This is an attribute of communal groups that is unlikely to be found in any other type of colony except the quasisocial.

Other panurgine andrenids with communal colonies are the European *Panurginus labiatus* (Malyshev, 1924), *Panurginus albopilosus* from Morocco (Rozen, 1971; two and four bees in nests, dissections made), *Meliturgula braunsi* from South Africa (Rozen, 1968; from one to thirty-six bees in nests), and the European *Meliturga clavicornis* (Stephen in Rozen, 1965; one to two bees in nests).

So far as known all species of the Western Hemisphere anthophorid genus *Exomalopsis* regularly nest in burrows in the soil inhabited by colonies that may be communal. Michener (1966a), however, suggested that the young females possibly cooperate in cell provisioning. This suggestion is based on the observed foraging activities of females whose ovaries were not fully developed. These females were collecting pollen and doubtless taking it into nests. Females with fully developed ovaries were also collecting pollen at the same time. The suggestion is that the females with underdeveloped ovaries were provisioning cells or helping to do so and that the eggs for such cells were provided by other females whose ovaries were fully enlarged. If this is so, colonies are semisocial, at least until ovarian development of all the females has occurred.

An example, treated in greater detail in Chapter 6, of a burrowing halictine regularly showing communal behavior is *Pseudagapostemon divaricatus.* This bee burrows into earthen banks on the southern Brazilian plateau. Most of the nests contain several (or occasionally up to 40) females in a colony; all have about the same degree of wear of wings and mandibles and are of about the same age, being members of the same generation. All are fertilized, with enlarged ovaries. The number of cells being provisioned at a given time in a nest equals the number of female occupants, leading to the assumption that each female makes her own cells (Michener and Lange, 1958c). Probably each bee makes one or more lateral burrows, each with several cells. Similar communal behavior occurs in certain other halictine bees: for example, in *Pseudagapostemon perzonatus* (Michener and Lange, 1958c) and *Ruizantheda mutabilis* (Claude-Joseph, 1926) in South America, and in *Lasioglossum* (*Ctenonomia*) *albescens* (Sakagami, 1968) and *L. ohei* (Sakagami, Hirashima, and Ohé, 1966) in Malaya and Japan, respectively. Other halictids that commonly live with several females in each nest are certain Asiatic and African species of the genus *Nomia* (Batra, 1966a; Michener, 1969b).

2. The publication merely mentions the number of ovarioles, which is of no special significance, but a letter from Dr. Rozen specifies that they were enlarged.

Communal colonies sometimes involve two different species. For example, eight females of *Svastra obliqua* and two of *Melissodes* sp. were seen in Colorado going in and out of one burrow in the soil (Custer, 1928b). Cells of both species were found when the nest was excavated. Several authors (Sakagami and Michener, 1962) have reported smaller numbers of bees of two species, usually only one individual of each, sharing the same burrows simultaneously.

2. Quasisocial colonies

These are small colonies in which from two to several females, usually of about the same age and of the same generation, cooperatively construct and provision cells, more than one bee working on a given cell (Table 5.1). As in communal groups, each female has enlarged ovaries and is mated, indicating that each is an egg layer. The term quasisocial was suggested to me for such groups by Dr. Suzanne W. T. Batra in 1968. Most accounts of quasisocial colonies are equivocal; the condition seems to be rare and may usually exist as an ontogenetic stage or occasional situation rather than as the terminal stage in normal colonial development.

Although some species of *Euglossa* are solitary, several species of that genus and of the related genus *Eulaema* live in small to rather large colonies about which very little is known (Bennett, 1965; Dodson, 1966; Sakagami, Laroca, and Moure, 1967). In *Euglossa imperialis,* a large and magnificently green species studied by Radclyffe B. Roberts in Costa Rica, the number of cells being constructed and provisioned is sometimes less than the number of foragers, suggesting cooperative activity by the several (up to eight) females, all of which have enlarged ovaries (Roberts and Dodson, 1967). There is no evidence of division of labor except that, of course, for any one cell only one female can lay the egg.

A single nest of the large, black *Eulaema nigrita* was observed in Brazil by Zucchi, Sakagami, and Camargo (1969). It contained three bees which built a clump of cells in a cavity provided for them. The authors noted that while a given female is primarily responsible for building her own cells and provisioning them, the other females do limited work on them and also provide limited additions to the provisions. These assistants are chased away by the "owner" of the cell when she is present. Presumably each bee is the owner of certain cells but contributes somewhat to others. Such contributions could be simply errors—responses of a bee to cues normally provided by one of her own cells but sometimes produced by a neighbor's cell when in the proper condition. Whether errors or not, they occurred rather frequently. The situation is weakly quasisocial, or intermediate between communal and quasisocial.

Although the numerous species of the genus *Osmia* are ordinarily solitary bees, *O. emarginata* has several times been recorded as working in some sort of group. Sometimes a lone female constructs her nest, which consists of a group of cells made of chewed leaf material and attached to a protected place on a rock or (usually) in a crevice between rocks. However, two or more females may work together on a single clump of cells. For example, Deleurance in France (1949) describes a group of 23 newly made cells, 21 of them finished, closed, and containing eggs, but two being worked on by three females of *O. emarginata,* so that a given cell was seemingly not the work of one bee alone, but the joint effort of two or three. Working females were able to tolerate the presence of one another. The ovaries and spermathecae were not examined but it seems likely that such a group is quasisocial. Grandi (1964), in a more extensive account of the same species in Italy, mentions as many as five females working on a single cell cluster, but says that each female works alone on a given cell, which she defends against intrusions of the others but which may be built taking advantage of prior construction by other individuals. This is a communal or aggregative group of bees like those of *Chalicodoma;* it would be desirable to verify Deleurance's observations.

Friese (1891) records three females of *Osmia vulpecula* working on one group of cells but gives no details.

Occasional colonies of allodapine bees fall temporarily into the quasisocial category when two or more egg layers are present without workers. There are no cells, hence there is no cooperation among egg layers in cell construction and provisioning; but the young are fed progressively, and probably each female is involved in the process.

Some species of *Nomia* are possibly quasisocial, although knowledge of their colonies is inadequate (Batra, 1966a; Michener, 1969b). The best known of such species is from India, *N. capitata,* for which only three nest populations (four to six bees each) have been dissected. In each nest there were two to four females with enlarged ovaries, apparently egg layers, all inseminated. All foraging was done by such bees. One of the bees with enlarged ovaries in each nest was much more worn and presumably older than the others; probably she was the mother but perhaps an older sister. Such a bee was not found collecting pollen, a finding suggestive of eusocial behavior. In at least one nest two pollen collectors were putting pollen into a single cell.

Some polygynous halictine colonies pass through a quasisocial stage before worker production begins, although the equivalent stage is usually semisocial. Michener and Wille (1961) indicate that the occasional polygynous young spring colonies of *Lasioglossum imitatum* in Kansas contain several fertilized gynes, sometimes all with well-formed ovaries, suggesting quasisociality. In other similar colonies of the same species, however, one gyne was laying eggs while the others became auxiliaries, a semisocial arrangement.

Many of the interactions in which two or a few bees use the same nest or even apparently cooperatively provision the same cell may be more accidental than socially meaningful. This view is especially supported when bees of two species provision cells in the same nest (Chapter 19, A) or even contribute to the provisioning of the same cell, as in the following case. Bohart (1955) describes a nest made by two species of *Osmia.* The species, *O. lignaria* and *californica,* are solitary species of different subgenera that commonly nest in burrows in wood. They differ in the material used for closing cells, shape of the pollen mass, placement of the egg, and kinds of pollen used. Yet in one nest two females, one of each species, worked simultaneously for several days. Bohart says, "The first four cells contained eggs and were provisioned largely by *lignaria,* although there was some pollen collected by *californica* in the fourth cell. The fifth, sixth, and seventh cells were provided with eggs and provisioned chiefly by *californica.* However, about half the pollen in the fifth cell was collected by *lignaria.* Most of the first partition was made by *lignaria,* but a small hanging fragment (perhaps torn away by *lignaria*) was made by *californica.* The next partition was made by *lignaria* but surfaced above by *californica.* All succeeding partitions, including the massive terminal plug, were made by *californica.*"

This is a case of gradual supersedure, one female taking over the nest of the other. But if the bees had been of the same species, and if they had been observed when both were provisioning the same cell, the relationship would have been called quasisocial. Actually, the phenomenon is probably evidence of competition for nest sites, and its implications for social behavior are hardly significant.

3. Semisocial colonies

These are small groups which show cooperative activity and division of labor among adult females as in eusocial groups. Dissection of a nest population therefore shows some mature females with slender ovaries and often unmated, in contrast to one or more with enlarged ovaries. The existence of division of labor, with both egg layers and workerlike individuals among the adult females, distinguishes semisocial groups from the communal and quasisocial ones. All the females are of the same generation, unlike those in the matrifilial colonies of eusocial forms. When a nest of a semisocial colony is opened, only one cell may be found in the process of being built or provisioned, even though several bees may be collecting provisions. This is the best indirect evidence of cooperative activity. Semisocial and quasisocial colonies are little known and may grade insensibly one into the other. Their principal differences from other types of colonies are shown in Table 5.1.

Semisocial relationships can arise even in bees that do not ordinarily live in colonies at all. In the Brazilian *Augochloropsis diversipennis* I found an occasional pollen collector with slender ovaries; one of them was unfertilized. This is in spite of the fact that in this species most nests are occupied by only one bee, only an occasional one containing two or more females. These findings suggest rare semisocial group relationships, such as are common in another species from the same area, *A. sparsilis.* Similarly in *A. brachycephala,* which only occasionally has two bees to a nest, a few workerlike individuals are known (Michener and Seabra, 1959).

Semisocial colony formation and cooperative behavior are regular ways of life of some species, perhaps necessary for reproductive success, in contrast to the usually or always facultative communal and quasisocial colonies. *Augochloropsis sparsilis* (Michener and Lange, 1958d, 1959) and *Pseudaugochloropsis* spp. (Michener and Kerfoot, 1967) are large, usually brightly metallic green or brassy halictine bees of the American tropics which regularly make semisocial colonies. They are the only known insects in which the semisocial stage seems to be the normal terminal level of colony organization rather than an unusual condition or a stage in colony development. For further information see Chapters 6 (B2) and 22.

Polygynous young colonies of other halictines (Vleugel, 1961) are often temporarily semisocial, in that division of labor develops among gynes, one becoming the egg layer or queen, the others auxiliaries or, in effect, workers. This was inferred from ovarian development in some, but not all, such colonies of *Lasioglossum imitatum* (Michener and Wille, 1961); was indicated at least temporarily for polygynous spring nests of *Halictus maculatus* (Bonelli, 1964); and was recognized for several other species by Knerer and Plateaux-Quénu (1966a), who state that, on the average, larger females have better-developed ovaries than smaller ones (but see Chapter 9, A).

In *L. versatum* (Michener, 1966b), which has many gynes in its spring nests, up to two thirds of them show ovarian enlargement and presumably lay eggs, but the remainder, although fertilized, have slender ovaries. Both those with enlarged and those with slender ovaries collect pollen, there being no evident inhibition of foraging on the part of egg layers by the auxiliaries. We have here something intermediate between a communal and a semisocial colony. It is possible that those gynes with slender ovaries will later lay eggs (or have done so earlier); data are not adequate to show whether auxiliary status is permanent in this species, but it probably is.

Spring nests of *Augochlorella* (Ordway, 1965, 1966) show an interesting diversity. In May (in Kansas) some of them contain lone individuals. The others, however, contain two to five gynes, so that the mean number per nest for all nests exceeds two. More than one egg layer was found in some 30 percent of the nests. A colony consisting of two or three such egg layers and no others is communal or quasisocial. But some 40 percent of all the gynes retain slender ovaries, so that in many nests, in addition to one or more queens, there exist one or more auxiliaries. When there is only one queen, the colony is a semisocial group; when there are two or more it is the same kind of intermediate noted for *Lasioglossum versatum,* that is, communal with respect to relations among queens but semisocial with respect to the whole colony. Both queens and auxiliaries forage and carry pollen, although later in the year, after the colonies have become monogynous and primitively eusocial, only the workers carry pollen.

In bumblebees groups of gynes can be made artificially after the spring aggressive phase (Röseler, 1965), which normally results in monogyny. Such groups are semisocial in that there is division of labor among the gynes. Also, Lehmensick and Stein (1958) observed division of labor among gynes in an unusual colony consisting largely of gynes.

In at least a technical sense, any eusocial colony which loses its queen and in which one or more workers begin laying is also semisocial, since bees of a single generation constitute the colony. Laying by workers can occur in colonies of nearly all eusocial species, if orphaned. A colony in which a replacement queen is produced, for example, an *Apis* colony after supersedure, is also semisocial until the old workers have died and daughters of the new queen have taken their places. As the social organization of such colonies is that typical of ordinary colonies of the species, no special discussion of it is required.

F. Subsocial colonies

These are family groups each consisting of one adult female and a number of immature offspring which are protected and fed by the adult. The mother leaves or dies before or at about the time that the young reach maturity. There is no division of labor among adults, as is found in semisocial and eusocial groups. Many common small birds and mammals are subsocial by the definition given above. Perhaps by happenstance, the only subsocial bees are those in which the one cell–one larva arrangement has at least partly broken down.

Young colonies of *Bombus,* before workers are produced, are subsocial; the queen progressively feeds the growing larvae in a more or less subdivided common cell.

The allodapine bees are more consistently subsocial, although many species are subsocial only during a portion of the life of a colony. The young are reared together in a common burrow and in almost all species are fed progressively (see Chapter 27). Each nest is started by a lone female. When such a female is feeding her young, protecting them, and cleaning the nest, the group is a typical subsocial one. Later, in many species, the same colony may become eusocial, semisocial, or even quasisocial. However, many nests of probably all species do not progress beyond, or else return to, the subsocial stage. For example, if only one mature daughter stays in the natal nest, the colony is again subsocial after the mother dies.

Some species of allodapines, however, are regularly

subsocial, only rarely with two mature females working in the same nest. Examples are *Exoneurella lawsoni* from Australia (Michener, 1964c) and *Allodape mucronata* from South Africa (Michener, 1971).

G. Eusocial colonies

In the preceding types of colonies, interactions among adults, where they occur, ordinarily involve individuals of the same generation. In subsocial groups, two generations are involved, but one of them consists of immatures. However, the true social Hymenoptera, for which the word eusocial was coined by Batra (1966b), live in colonies which are family groups[3] consisting of individuals of two generations, mothers and daughters. Temporary exceptions exist, as when an *Apis* colony has just replaced its old queen with a new one who is a sister of the workers, so that the colony is technically semisocial. Usually in bees a eusocial colony contains only one queen, and the bulk of the females are workers (daughters). Division of labor, with some individuals functioning as egg layers or queens and others as workers, that is, with more or less recognizable castes, occurs in both the semisocial and eusocial colonies but not in the other types of colonies (see Table 5.1).

1. Primitively eusocial colonies

In primitively eusocial bees female castes are externally indistinguishable in structure, except for allometric macrocephaly (often intergrading) of queens of a few species (Quénu, 1957; Sakagami and Fukushima, 1961; Sakagami and Hayashida, 1968; Sakagami and Moure, 1965; Sakagami and Wain, 1966). The castes commonly, but not always, differ in mean size, and sometimes all or nearly all of the females are recognizable as to caste by size alone. The gynes of primitively eusocial bees are equipped morphologically and behaviorally to live alone and to establish colonies as lone individuals. Swarming does not occur.

Castes do differ, of course, in ovarian development and in behavioral attributes, as detailed in Chapter 9 (B). In some species, such as *Lasioglossum zephyrum*, there are not two isolated castes but there is continuous variation in female behavior and physiology. It is convenient to call the one that lays the most eggs the queen. In closely related species of the same subgenus (*Dialictus*), the castes are often rather distinct.

In the *Allodape* group and most eusocial species of halictinae, most or all young adult females are unrecognizable as to caste. They become recognizable upon dissection when they are older. If they are fertilized and have enlarged ovaries, they are queens. Old individuals (for practical purposes recognized by worn mandibles and wings) that are unfertilized and have slender ovaries are workers. Intermediates between the castes occur in many species.

The bumblebees differ from most other primitively eusocial bees in that the great majority of females—in some species all of them—are readily assignable to caste by size; typical gynes are far larger and in some species differently colored than typical workers. There are also a few halictine species in which all individuals are assignable to caste by size alone.

The only food stored in many primitively eusocial bee colonies is in the brood cells, where it is used cheifly by growing larvae. An exception occurs in many allodapine nests in which food is briefly placed in the nest burrow, there being no cells. The principal exception, however, is in the bumblebees (*Bombus*), where food is stored outside the brood cells, as in highly eusocial bees.

Integration within colonies of primitively eusocial bees probably depends largely on actions and substances found also in nests of solitary and parasocial relatives. In eusocial bees a single cell may be made and provisioned cooperatively by several workers. This statement applies even to primitively eusocial halictines (Batra, 1964; Michener and Wille, 1961; Plateaux-Quénu, 1963), in which the coordination is apparently just like that in semisocial colonies of the same and related species. At least in *Bombus* and probably in halictines, dominance-subordinance patterns exist, the queen being dominant and certain workers dominant to all but the queen. The aggressive behavior which establishes these patterns is presumably important in maintaining the social organization. Stealing and eating of eggs is part of this picture. Only in *Bombus* is other presumably uniquely social behavior often recognizable; as noted in Chapter 10 (B), for one group of this genus a pheromone influencing caste determination is reported (Röseler, 1970). Communi-

3. Deegener (1918) calls such a group a gynopoedium, and I suppose that the adjective gynopoedial has priority over the recently much used word matrifilial (= mother, daughter). I have nonetheless used the latter, since it seems easier in English.

cation concerning locations of food sources and nest sites in primitively eusocial bees almost surely occurs, if at all, only through floral odors on returning bees and in their nectar or pollen loads. Larvae are fed largely or wholly on nectar and pollen. Although the populations of colonies of primitively eusocial bees range from two females to several hundred, most colonies contain under twenty adult females.

There are many more kinds of primitively eusocial bees than of highly eusocial ones. Most primitively eusocial bees, perhaps one thousand species or more, are in the tribe Halictini; but there are some in the Augochlorini, in the allodapine group of Anthophoridae, and in the Apidae there are a few hundred species of *Bombus*. Only in *Bombus* are all nonparasitic species eusocial; the other groups all contain solitary, subsocial, or parasocial species as well.

2. Highly eusocial colonies

The well-known, highly or complexly eusocial bees differ more in complexity of social integration from other bees—solitary to semisocial, subsocial, and primitively eusocial—than do these other forms among themselves. In one sense, therefore, looking at social organization of bees phenetically, a classification of behavior should first separate the highly eusocial bees from all the rest. However, because of relationships often thought to be basic and of evolutionary interest, it is customary to heavily weight certain attributes, especially the coexistence of adults of two generations, and to place the primitively and the highly eusocial groups together. This matter is treated in detail in Chapter 20 (C).

In highly eusocial bees the female castes are strikingly different from one another, not only behaviorally, physiologically, and usually in size, but also in external structure. Gynes lack the structures for collecting and manipulating pollen and doubtless also lack behavior patterns needed for foraging. They are unable to survive long outside of their colonies; new colonies are therefore established by swarming, that is, with the aid of workers from the parent colony.

Colonies are long-lived and sustained through periods of adversity by food for adult as well as larval consumption, stored in the nests but not in brood cells. Integration within colonies is complex and involves a variety of behavior patterns, pheromones, and physiological adaptations that would have no obvious function in solitary forms. Aggressive behavior among individuals of the same colony is rarely evident, nor is the egg eating that is often associated with such behavior. Communication concerning food sources and, at swarming time, concerning nest sites, is well developed in many of these bees. Larvae are fed at least in large part on glandular secretions of workers. Populations of colonies are commonly in the thousands (up to over 60,000 for *Apis*, 180,000 for some species of *Trigona*), although some species often have colonies of only one or two hundred.

The highly social bees are all in the apid subfamily Apinae, and constitute all species of this subfamily. There are two groups, the tropical stingless honeybees (tribe Meliponini) and the true honeybees (tribe Apini, genus *Apis*).

H. Progression and reversion

It is tempting to look at the sequence of increasingly complex societies described in sections C, D, E, and G of this chapter as an evolutionary sequence. However, as indicated in Chapter 20, the next to the last social level described, primitively eusocial, has probably arisen not only from semisocial and from subsocial ancestors, but also in some cases directly from solitary ancestors. No doubt any of the lower social levels could do likewise.

Like other characters, those that relate to social interactions undergo selection and recombination, and the social organization of each species or population is an adjustment to its particular environment. Up to and including the primitively eusocial level, the social developments are probably all reversible. Species with primitively eusocial stages in colonial history nearly always also have solitary stages, and selection might favor reduction or elimination of the former. This has clearly happened in various parasitic bees (Chapter 19, B). Variations in social levels in different populations of the same species (Chapters 20, B2, and 24) suggest evolution of decreasing as well as increasing interactions within certain nonparasitic species within the primitively eusocial group. Thus oscillation from one social level to another has probably occurred, and reversion may be almost as common as progression to more elaborate social levels.

Only the highly eusocial bees have reached a point where, because of the interdependence of the very different castes and the inviability of a lone bee, reversion to any other social level would be very improbable. No species exists which suggests such reversion.

6 *The Origin and Growth of Aggregations and Colonies*

The origins and histories of colonies include some of their most interesting aspects. This chapter deals with the origin and ontogeny of groups and colonies of bees, not in an evolutionary sense, but during the lives of the individual bees and colonies concerned.

It is easy to feel that the stages in colonial development give insight into the stages of evolutionary history. This notion (recapitulation) as applied to morphology played an important role in the history of evolutionary thought. In detail, ontogeny commonly does not recapitulate phylogeny; probably for no animal do all structures pass through the same stages and in the same sequence in the growth of the individual as in the evolution of the species. But the idea should not be merely ignored. If a postulated evolutionary sequence flies in the face of an ontogenetic sequence, it is well to re-examine the former, although it may well be correct.

Colonial development or ontogeny ordinarily involves progressively increasing population size and complexity of interactions. Therefore, in a very general way, colonial ontogeny must be parallel to the evolution of successively more elaborate levels of social behavior. But the recapitulation idea cannot be used constructively to help answer many of the interesting problems of social evolution in bees. Thus the fact that bumblebees pass through solitary and subsocial stages before becoming eusocial cannot be used as evidence that a proto-bumblebee's mature colonies were subsocial. In this case part of the problem is semantic. A mother bee only needs to feed her young progressively in order to be called subsocial. There is no evidence as to whether progressive feeding or eusocial behavior appeared first in ancestors of our bumblebees. The same confusion does not arise with halictid bees because all of them are mass provisioners.

The fact that many halictid (and wasp) colonies pass through a semisocial stage before becoming eusocial suggests that evolutionarily, semisocial behavior antedated eusocial behavior in at least some phyletic lines of these insects. I happen to agree with this view. But in other species eusocial colonies arise each year without semisocial antecedents. If evolutionarily speaking, mutual tolerance among females first arose in such forms when daughters remained with their mothers, forming eusocial colonies, then the advantages of colonial life early in the active season (spring) might later have led to semisocial associations at that season. It is quite possible that in some groups evolution of the semisocial stage antedates the eusocial stage, as in ontogeny, but in others the semisocial stage may be a later development. These matters are more fully elaborated in Chapter 20.

Among the more important studies of social insects that ought to be made are those of ontogeny of colonies, including population growth curves and life tables for workers that mature at different stages of colonial growth. Almost nothing is known of the influence of colony age on worker survival and function. Further understanding of the adaptive mechanisms of social insects requires such investigation.

A. Aggregations

Individual nests of solitary bees do not usually survive long. Occasionally such a nest may be occupied repeatedly so that a single burrow in the ground, for example, may be used by succeeding generations over a period of months or years. Generally, however, burrows of solitary bees in more or less flat ground become closed soon after the death or departure of the mother bees, and this usually occurs a few weeks after the establishment of the nests. Burrows in banks commonly stay open longer because rain or wind are less likely to fill them with soil, and burrows in wood last until the wood decomposes. Nonetheless, even nests in banks or wood usually are occupied for only short periods. The average solitary bee nest is used by the bee that makes it for only a few weeks, although in some forms, such as certain species of *Xylocopa,* the mother bee may occupy her burrow for about a year. Young adults emerge from their pupae and typically leave the nests a few weeks to nearly a year after nest construction and provisioning by the mother.

In contrast, aggregations of bee nests are sometimes very long-lived. One can sometimes return year after year to the same place and see the same groups of nests. Stoeckhert (see Knerer and Atwood, 1967) in Germany knew of an aggregation of nests of *Lasioglossum malachurum,* a primitively social species, that was at least 37 years old.

Little is known about the establishment and ontogeny of aggregations. Some bees seemingly tend to return to the vicinity where they were reared to burrow into the soil, so that an aggregation develops near a successful nest of earlier years (Malyshev, 1935). Iwata (1938, 1964) reached similar conclusions as to the origin of adjacent *Xylocopa* nests in bamboo stems. Such a response has selective value since, of all the possible nesting sites, one suitable enough to have produced adult bees one year has a good chance of being a good site again the following year. Nothing is known of the ways in which the bees, in such cases, are able to recognize and return to the site of their birth. No doubt physical landmarks, such as are used by all bees in finding their way back to nests, are important, but chemical cues may also play a part.

Dr. G. E. Bohart writes (in a personal communication) that both the alkali bee *Nomia melanderi* and a gregarious leafcutting bee, *Megachile rotundata,* "seem to receive and act upon information they acquire upon emergence from their nests. Leafcutting bees will tend to return to nest in the same kinds of materials from which they emerged (for example, straws, or blocks of wood, or sheets of plastic). Emerging alkali bees tend to renest close to the area from which they emerged, but if they are emerging from a block of soil brought from another area, they will occupy unused blocks of the same kind of soil before they will go into the native soil, which may be as close or even closer."

Aggregations formed by return to the natal site are likely to grow, if at all, only gradually, as more and more progeny are produced and nest in the same place. If the population level is stable, such an aggregation can grow only at the expense of individuals nesting at other sites.

Dense aggregations, however, are probably often established by a number of bees simultaneously making nests at a new site. Evidence of such behavior comes from work on *Nomia melanderi* by Dr. Bohart. He noted (in a personal communication) that a suitable site, artificially constituted, may go unoccupied for years. Then suddenly, within a day or two, thousands of bees burrow into it; so many may be attracted that nests are made in unsuitable marginal soil outside the nesting site. He writes that small numbers of *Nomia* did arrive at a site near Riverton, Wyoming, for several days prior to the nesting surge. They would scratch around and sometimes dig under the surface. But only after the first full tumulus was seen did other nests quickly appear. "The first nests were established around 4:30 or 5:00 p.m., and by about 7 o'clock, there were already several hundred, and by the following evening, at least 10,000." The bees established themselves near one corner of one site. Nests quickly spread in all directions from the starting point, nearly all within a few meters. This resulted in many of the nests being in rather dry soil outside the prepared site, even though suitable soil was present for a long distance in the opposite direction. Within a few days, the nests in the dry soil were abandoned and presumably the bees re-established in the better soil which, by this time, was largely occupied.

The importance of odor of cells, brood, or bees is suggested by observations on *Amegilla salteri* (Michener, 1960b; Cardale, 1968) and *Andrena flavipes* (Butler, 1965). Females of the *Amegilla,* an Australian species, were attracted to soil and cells from a nesting aggregation and entered windows of a laboratory, buzzing in the vicinity of cells and earth from the nesting site even though these were completely hidden by cloth. Young females of the *Andrena,* a European species that ordinarily nests in dense aggregations in banks, were taken from certain aggregations, marked, and released near the same or other aggregations. They joined the aggregation nearest to the point of release even though some had been born elsewhere and had never seen the site near which they were liberated. Obviously, selection of the place to make a burrow was influenced in this case not by memory of the place of birth but by some factor, probably odor, coming from the existing nest aggregations. Apparently suitable unoccupied banks were present, so that the selection of an occupied bank by the transplanted individuals was probably not merely the result of soil characteristics, exposure, or other physical factors.

Lasioglossum versatum, probably semisocial in spring, clearly primitively eusocial in summer, is of interest because of the very small areas of the aggregations in which its nests are found, as mentioned in Chapter 5 (D) (Michener, 1966b, c). New nests of this species have not been seen established in the usual fashion, by bees burrowing down from the surface. It is therefore possible that most new burrows are established in the spring when the bees make new tunnels to the surface from their hibernating quarters in the soil. If bees occupying different locations in the ramifying depths of the nest form independent new tunnels to the surface, several potential colonies will exist where there was one the previous year. If most new nests arise in this way, the small size and density of the aggregations would be explained. No other bee is known in which this sort of nest establishment seems probable.

If a given aggregation site is watched, year after year, it is found that the number of bees fluctuates from season to season. Extinction of aggregations is common. Natural enemies of bees do well in aggregations. I have seen aggregations of *Lasioglossum versatum* in Kansas reduced by mutillid wasps to only a few burrows still occupied by bees, with almost no cells containing larvae or pupae of bees. Aggregations of *L. zephyrum* near Lawrence, Kansas, were decimated in the summer of 1970, one by mutillids, another by myrmosids. Parasitic bees of the genus *Paralictus* also take a heavy toll of *L. versatum.* Bohart, Stephen, and Eppley (1960) have reported near destruction of aggregations of *Nomia melanderi* in Utah, largely by the depredations of bee flies (Bombyliidae).

Successional changes of vegetation are also important in eliminating aggregations of nests. Many species of bees nest in more or less bare ground; when vegetation becomes dense in such places, these species of bees disappear. I have seen aggregations of *Lasioglossum rohweri* very large and dense in areas of soil having scattered vegetation only a few centimeters tall. Invasion by dense grasses growing from 50 to 80 cm in height seemed related to the disappearance of the bees.

Bees nesting in banks are in general peculiarly favored by human activity. Whereas under primeval circumstances, their nesting sites are largely limited to the sides of rivers and ravines, road building and other such activities often produce more numerous and more widely spread nesting sites than ever exist naturally. On the other hand, further construction often results in destruction of such sites. Nesting aggregations on level ground, in wood, or elsewhere are also frequently destroyed by man.

B. Parasocial colonies

These colonies are little known and their origins and development less so; the only sorts about which such information is available, however fragmentary, are a few of the communal and semisocial species. All of the observations concern species that burrow into banks or into more or less flat ground. In all cases the crucial problem for this chapter is how the group of adult females is established or assembled. Thereafter production of young and death of the adults should follow essentially the pattern for solitary bees.

There are at least two ways in which parasocial colonies are formed: (1) by two or more young females staying in the nest in which they were reared, and (2) by young females joining a nest already initiated by a lone bee. Both methods of colony formation occur in many species that have parasocial colonies, although in different relative frequencies.

Field observations leading to the recognition of the two methods of colony formation are of the following sorts. It is sometimes noticed, in facultatively parasocial halictines, that solitary females are commonly in new nests containing no old cell remnants, while the communal or semisocial colonies of the same species are all in old nests having such cell remnants. This is interpreted to mean that when there are two or more bees working in a nest, they matured there in the cells whose remnants are found in it. In such cases the members of the colony are probably sisters or at least close relatives. On the other hand, when new nests lacking old cell remnants are inhabited by two or more females, bees must have joined a nest already started by one of them. In this case the probabilities are that they are not close relatives. The distinction is especially important for the semisocial colonies, where the theory of origin of a worker caste is involved (see Chapter 20). An investigator must open a nest during a rather short period in its history in order to distinguish between the two kinds of colonies (probable sisters vs unrelated females). There are therefore no adequate data to show the relative frequencies of the two. However, my impression is that colonies in old nests (probable sisters) are the more common type.

Formation of colonies of unrelated females involves discovery of existing nests by females looking for a nest site. Visual and perhaps olfactory components are relevant to the finding of such a nest. A visual component is demonstrated by the attraction of female bees to holes or dark spots in an insectary or bee room. Here the bees, in unfamiliar circumstances, fly about investigating each nail head or dark spot on the wall as well as each genuine hole. The effectiveness of such searching in *Lasioglossum zephyrum* and *Augochlorella striata* is shown by the examples cited in Chapter 20. In the case of *L. zephyrum* and also the nest of *Pseudagapostemon divaricatus* described below, bees newly entering a burrow were often attacked at the nest entrance and temporarily prevented from joining the colony. They usually managed to get in after a short time, however, and then seemed peaceably accepted.

There is little to be said, because little is known, about growth of parasocial colonies after they are established. Nests of some probably last for several years, at least several bee generations. Large populations, as are reported for *Andrena* in Chapter 5 (E1), probably indicate several years of occupancy and the build-up of a larger colony each year. The case of *Augochloropsis diversipennis* cited below also indicates growth of the colony over a few generations, although in this case colonies of over four bees are not known.

As shown in Chapter 22, a nest of *Pseudaugochloropsis* is likely to remain small unless a colony develops, either from the progeny of the bee that established it or by additional individuals entering the nest from outside. The large majority of the nests are occupied by lone females and contain up to six cells. Those occupied by semisocial groups of two to seven females contain up to 60 occupied cells. Once a *Pseudaugochloropsis* nest is inhabited by a colony of two or more adult females instead of the single founder, its likelihood of survival and growth seems greatly improved (Michener and Kerfoot, 1967).

1. Communal colonies

A little is known about the origins of communal colonies of several species of bees, as indicated in the following paragraphs. Occasional nests of the ordinarily solitary but somewhat gregarious North American bee *Andrena erythronii* are occupied by two bees. Such nests appear at times when the soil surface is dry and hard, because females have difficulty in starting their own burrows and sometimes enter existing burrows. The result is an occasional composite nest with the entrance burrow used communally by two bees, but with two major branches and groups of cells, presumably one made by each bee. Nothing is known of the interactions between the bees should they meet (Michener and Rettenmeyer, 1956).

Another instructive species from the standpoint of ontogeny of communal colonies is *Augochloropsis diversipennis.* This bee nests in vertical earthen banks on the southern Brazilian plateau. The bees overwinter in the nests; in spring many disperse and start new nests. Many nests are isolated but one aggregation was found where the bees seemed to have crowded into an area having particularly soft soil (Michener and Lange, 1959). Here nest burrows inhabited in summer by two or three females were common. Such groups result from females remaining together in old nests. Each female seemingly excavates her own new branch burrow, starting among the old cells and terminating in a new and deeper cluster of cells. The adults constitute a communal group by virtue of using an old burrow for the entrance to their individual nests. This is not without its social significance, however; the cell clusters are placed far deeper, hence are presumably better protected, than in the case of nests made by lone bees starting at the surface, and the entrance is more commonly guarded than is the entrance of a nest occupied by a single bee. The deepest nests seemed to be the result of three or four generations of use of a single entrance.

Colonies of *Pseudagapostemon divaricatus* studied by Michener and Lange (1958c) in Brazil are perhaps the best studied of any communal halictine. This species also nests in earth banks. Some of the nests are occupied by single females behaving like those of other solitary halictines, but the majority are inhabited by communal groups (Figure 6.1). Some groups are in old nests and are presumed to consist largely of sisters, whereas others are in burrows not occupied by the previous generation and are presumably composed of unrelated females. A particularly instructive nest of the latter sort came to contain the largest colony of the species that has been recorded. Like many burrowing bees, the females tend to start burrows in depressions and cavities, apparently being attracted by shadows or dark spots. A burrow was started by a single female in an unusually large cavity that we had dug to study another nest. During the ensuing weeks, females flying about the bank were attracted to the large cavity, and once in it and examining its surface, they often found and attempted to enter the burrow there. In the course of a few weeks forty females established themselves in this burrow. The populations of other nests ranged from one to eighteen adult females.

The only well-studied Old World, usually communal halictine is *Lasioglossum* (*Evylaeus*) *ohei*, from Japan (Sakagami, Hirashima, and Ohé, 1966). Its social arrangements are like those of *Pseudagapostemon divaricatus*, with about one female per branch burrow, but the nest architecture is different (Figure 6.2). There is no evidence that any colonies of *L. ohei* result from invasion of females; all might have resulted from re-use of old nests by bees reared in them the preceding year.

Legewie's papers (1925a, b) suggest that *Halictus sexcinctus* is another species that occurs in communal groups. He says that in spring the bees are solitary, but nests are enlarged by the several sisters of the second generation, each of which makes and provisions her own group of cells. The work needs verification, partly because Legewie's mistaken ideas about parthenogenetic reproduction in halictids may have led him to misinterpret his observations.

2. Semisocial colonies

The commonest semisocial colonies arise early in the history of halictine nests. A queen and one or more auxiliaries commonly are found in a single nest before the season of production of ordinary workers. In the majority of cases such groups of bees occur in the old nests in which the gynes hibernated, while lone females of the same species are in new nests. This indicates that the groups are made up predominantly of sisters or close relatives that never abandoned their parental abode. This is not always true, however.

Krunić (1959), for example, found in a population studied in Yugoslavia that spring polygynous (presumably semisocial) colonies of *Lasioglossum malachurum* arise when single females that have started new burrows are subsequently joined by other females. Observations made in Germany, however, indicate that such colonies result from two or more females remaining together in their natal and overwintering nests (Stöckhert, 1923). It is likely that the difference is not regional but local and annual, depending upon the density of the bees, the characteristics of the soil surface (recent moisture, and so on) that determine ease of starting burrows, and especially the hibernation place (in natal nests or elsewhere), which is apparently determined for this species by autumn weather. If the gynes do not hibernate in the nests, they are unlikely to find old nests in the spring and establish themselves there, but if they do remain in their nests through the winter, two or more may stay there also in the spring.

In the semisocial *Augochloropsis sparsilis*, the females maturing in a nest seem regularly to remain there, to clean out the old cells and re-use them, rearing a new brood in the same place and even in the same cells (Michener and Lange, 1959). The same is apparently true of various species of *Pseudaugochloropsis* (Michener and Kerfoot, 1967). However, some females of both *A. sparsilis* and *Pseudaugochloropsis* leave, instead of remaining in the old nests, and as solitary individuals excavate new nests, where they are sometimes joined later by other individuals.

C. Allodapine bees

These insects (Michener, 1962, 1965b, 1971; Sakagami, 1960a) present a particularly confusing picture with respect to the types of associations listed in Chapter 5. Seemingly, any adult females in a nest cooperatively feed the young that are there, defend them, and clean the nest. Understanding of the development of the colonies is not difficult, but the terminology presented in Chapter 5 is not always helpful. Usually in any allodapine colony, one or occasionally two bees are the principal egg layers while any other mature females present are workers.

A new nest is established by a lone female which lays

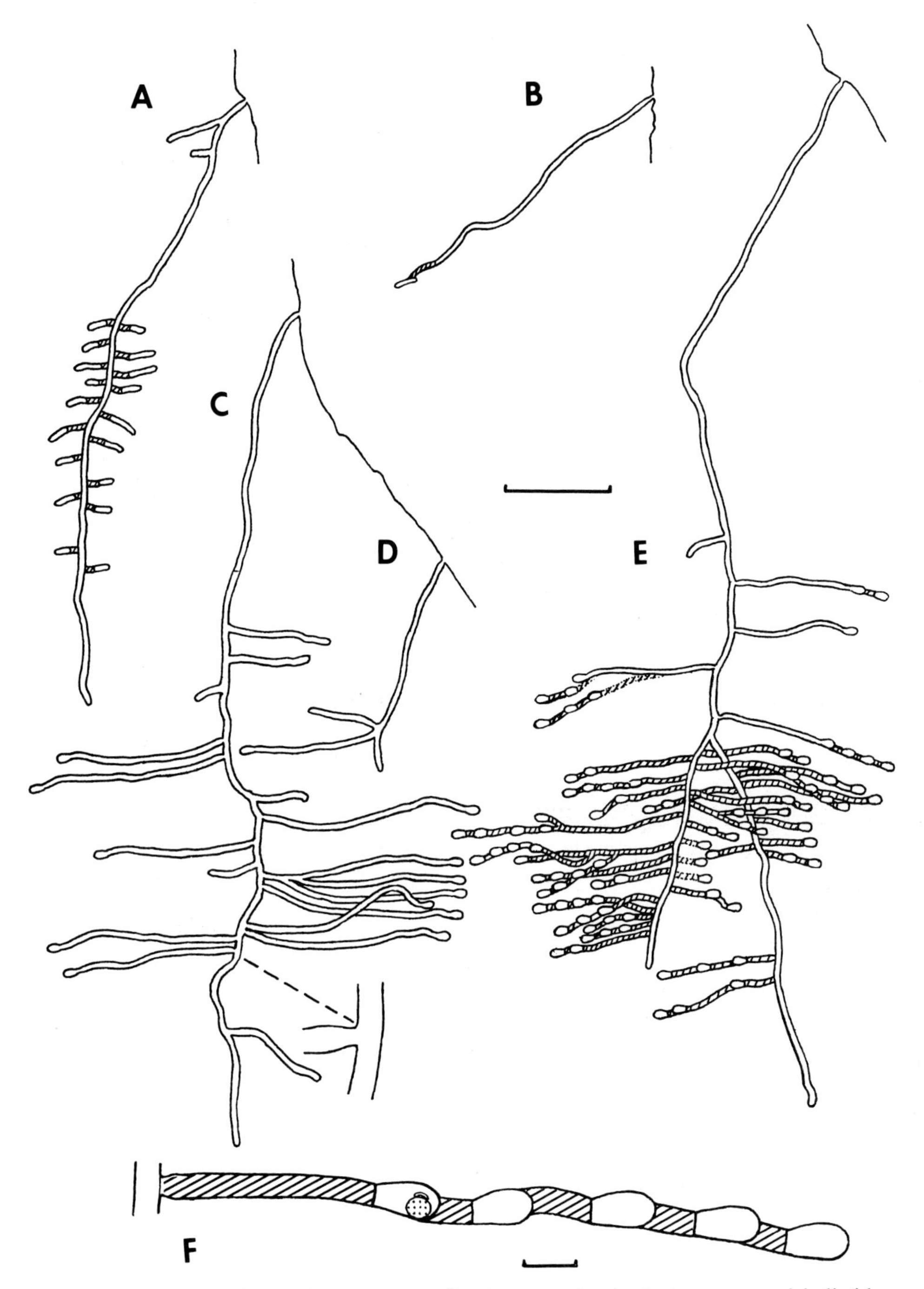

FIGURE 6.1. Nests of *Pseudagapostemon divaricatus,* a facultatively communal halictid. (*A*) and (*B*) Nests with 18 and one hibernation chambers respectively. (*C*) and (*D*) Nests with several and one individual lateral burrows, respectively, each ending in a cell. *D* was inhabited by only one bee. An enlarged section shows the constricted mouth of a lateral burrow. (*E*) A nearly completed nest, constructed by an estimated 18 bees; lateral burrows filled with earth except for the cells. (*F*) A completed lateral burrow, showing the contents (pollen mass and egg) only of the last cell to be made. Shaded parts of burrows were earth-filled. The scale line for Figures *A* to *E* represents 8 cm, that for Figure *F* represents 1 cm. (From Michener and Lange, 1958c.)

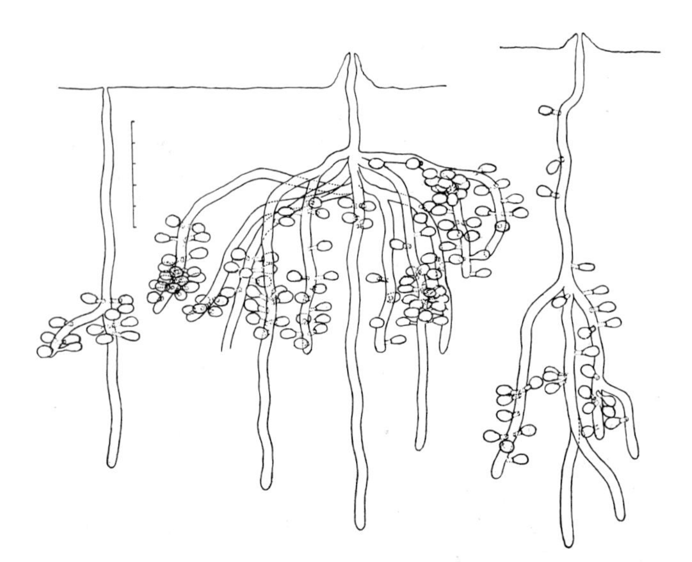

FIGURE 6.2. Nests of *Lasioglossum ohei,* a facultatively communal halictid. The nest at the right was unusual in having a few cells along the main burrow above the level of the branches. (From Sakagami, Hirashima, and Ohé, 1966.)

her eggs and rears her young together, usually in a hollow stem, with no separate cells. The young in most species are fed progressively. The group therefore is subsocial at this stage, and in a few species further colonial development seldom occurs (Chapter 5, F). In most species, however, some of the young females that mature remain in certain nests, so that tiny colonies of adults exist in some of the stems. The original female usually lays several eggs and then, while rearing the resulting larvae, lays no more, or only an occasional egg, until the first brood is nearly mature, after which she commonly lays a few more eggs. Often the nest then is inhabited by one or two or even more young adult daughters, functioning at least for the time as workers, and the original mother, now a queen. The group is therefore primitively eusocial. When the mother leaves or dies, which occurs long before the young from the second group of eggs reach maturity, a group of females of the same generation is left in the nest; they care for larvae that are their own brothers and sisters. At this stage the colonies are best described as semisocial, since usually one of the sisters has enlarged ovaries and is therefore queenlike. Rarely, two or more adult females in the nest have enlarged ovaries and presumably lay eggs, and the group can be described as quasisocial; but this is an unusual and probably temporary state of affairs. If the egg layer in a semisocial group lives long enough, she will be joined by her own adult female offspring and a primitively eusocial group will be re-established.

Allodapine colonies never contain more than a very few (one to six) recognizably fully mature adult female members. When other females are present in a nest, as is frequently the case, they are newly emerged, young adults that have not yet dispersed and that perhaps play no active role in care of the young. In temperate climates activities are terminated by cool autumnal weather, and in the spring all nests are inhabited by lone females. In the tropics dispersal of lone females probably occurs continuously and at such a rate that no colony can become large; how long a colony survives in the tropics is unknown.

D. Primitively eusocial colonies

A colony of primitively eusocial bees normally has a distinct origin, a period of growth, a period of maturation when reproductives are produced in greatest numbers, and then, due to the life history of the insect, the colony dies, only the fertilized gynes surviving to establish new colonies. The life cycle, while variable in certain respects, is markedly different from that of the highly eusocial bees, in which such a colonial life history is not detectable.

Most colonies of primitively eusocial bees are started by individual gynes acting like solitary bees—constructing and provisioning cells, laying eggs, and rearing a first brood of young. Most of these young become workers. A detailed account of the solitary phase of a eusocial halictine is given by Sakagami and Hayashida (1961). In all known cases, the ovarian activity of the gyne ceases or is greatly reduced while the first brood of young is growing, commencing again when this brood matures. Thereafter, the gyne, or now queen, ordinarily spends most or all of her time in the nest, the foraging being done by her daughters. Thus the primitively eusocial colony is established.

Some colonies of some species of primitively eusocial halictine bees are started by semisocial or communal groups instead of by lone gynes. In most cases they are probably semisocial; the gynes with small ovaries become auxiliaries or in effect workers, and die early so that a single queen survives to be associated with workers that are her progeny. Enlarged ovaries of only a single individual in such a group, implying reproductive division of labor, have been described for various halictines. In other species (*Lasioglossum versatum, Augochlorella,* Chapter 5, E3) there is often more than one egg layer in the spring nest, but some other gynes are auxiliaries. Sometimes the auxiliaries do not remain workers. Thus in *Halictus scabiosae* (Knerer and Plateaux-Quénu, 1967a) the queen in the semisocial spring nest is the guard and becomes increasingly aggressive, denying entrance to her auxiliaries even though they may be collecting pollen for her nest. They are ultimately driven away and establish their own individual nests, often by usurping unguarded burrows of their own or other species, before maturation of the worker progeny of the semisocial colony. On the other hand, auxiliaries may continue foraging for a time after the appearance of new, genuine young workers, as in one nest of *Halictus ligatus* mentioned by Knerer and Plateaux-Quénu (1966a), providing for continuous presence of more than one bee in the burrow. Whatever the details, the colony by summer becomes primitively eusocial, matrifilial, and monogynous, as if it had been started by a single gyne.

In most of the primitively eusocial bees, reproduction goes on more or less continuously after the first workers have matured. In the halictines this is true, for example, of *Augochlorella* and of *Lasioglossum* of the subgenus *Dialictus* (Chapters 23 and 24). In some halictines (*Lasioglossum,* subgenus *Evylaeus,* some species), however, young are produced in separate batches even after the first brood, with no egg-laying activity by the queen between these episodes (see Chapter 25). In *Lasioglossum (Evylaeus) malachurum,* there are three broods in central Europe and seemingly four in Italy, the last brood in the autumn being bisexual, as in most temperate climate halictines, and producing, synchronously with death of the old queens and destruction of the social organization, the young gynes which mate and then overwinter. In *L.* (*Evylaeus*) *marginatum* there are similar successive broods, but they are annual; the colony does not break up until the death of the queen and production of a usually bisexual brood at an age of five or six years (Plateaux-Quénu, 1959, 1962).

Although the queen of *Lasioglossum marginatum* is long-lived and the colony growth is correspondingly protracted, in most halictines the colony survives only during the favorable season of one year. In temperate climates colonies are generally established in spring or early summer and break up with the production of new, young gynes and males in autumn. In tropical areas, so far as known, survival of colonies is commonly equally short, there usually being a particular season of the year when most of the colonies of a given species are established and another season when activities of the species are at a minimum, although not necessarily stopped altogether as in the temperate winter. Other species are active throughout the year, and in such cases the use of the nest may be continuous; one might argue that if this is true the colony is long-lived, as in highly eusocial bees. No such species has been properly studied. Eickwort and Eickwort (1971) believe that the Turrialba, Costa Rica, population of *Lasioglossum* (*Dialictus*) *umbripenne* uses its nests for long periods, but their observations were carried out during only two months.

Even in some temperate species nests may be re-used, year after year, but new colonies establish themselves in old nests each spring. At least I believe that when young gynes stay in their natal nest through the winter and start a colony there in the spring, it is properly called a new colony.

In some primitively eusocial bees the queen lives through more or less the whole period of colonial activity. However, in other forms she does not. For example, she dies during the summer in *Augochlorella* spp. (Ordway, 1965), *Halictus confusus* (Dolphin, 1966), *Lasioglossum* (*Dialictus*) *versatum* (Michener, 1966b), *L.* (*D.*) *zephyrum* (Batra, 1966b) and *L.* (*D.*) *rhytidophorum* (Michener and Lange, 1958b). In these species the queen is replaced by one of her daughters, often before midsummer. Males are present in early as well as late broods of such species, a necessity for fertilization of replacement queens. Even in such species, however, males are produced in greatest numbers in late summer or autumn, and at the same season all the females produced are gynes which overwinter after mating.

Bombus is the primitively eusocial group that most closely approaches the highly eusocial bees, partly because food is stored in the nest for adult consumption, thus reducing the influence of bad weather. Even in these bees colonies are always established by lone gynes and ordinarily last for less than a year.

As in other primitively eusocial bees, the establishment,

development, and termination of a colony of *Bombus* are closely tied to seasonal changes and to the aging of the queen. As in many primitively eusocial Halictini, the young of the first brood, and usually those produced for some time thereafter, are females which become workers. Only toward the termination of activities, typically in autumn but for some species as early as late spring, are males and young gynes produced. At about the same time the queen dies, and with the death of the workers the colony is extinct. After mating, the young gynes estivate, or hibernate, or in some tropical places perhaps promptly start new nests.

A partial exception to the automatic death of colonies of primitively eusocial bees occurs in *Bombus atratus,* studied by Zucchi in Brazil. Colonies containing groups of daughter gynes may carry on after the death of the original queen. Ultimately these gynes fight, one survives, and a monogynous condition is re-established. The same nest may be used for several years, the colony being alternately monogynous and polygynous. It is not especially important in this case whether one speaks of a perennial colony or a succession of colonies in the same nest.

E. Highly eusocial colonies

The highly eusocial bees differ from the others in that their colonies are perennial. Barring accidents, predation, parasites, and disease, a colony of highly eusocial bees is, so far as known, immortal in the sense that, given suitable conditions, an *Amoeba* is immortal. It can go on dividing, growing, and dividing again, indefinitely, without being halted by natural events of the life cycle of the species. There is no portion of the life cycle that does not involve eusocial relationships.

New colonies of highly eusocial bees are established by swarms, a swarm being a group of workers and a queen that leave the parental colony and its nest to establish another colony and nest elsewhere. Once the new colony is established, its size increases through rearing of large numbers of workers. At the same time, quantities of food, both pollen and honey, are stored so that the growth becomes relatively independent of periods of inclement weather. At a certain stage young reproductives are produced; then a gyne (either the old queen or a young one, depending on the kind of bee) and some of the workers leave as a swarm to establish another new colony.

F. Colony size

Colonies of social insects attain populations which, although highly variable, are more or less characteristic of the species and indicate something about colonial ontogeny. From the standpoint of its social affairs, the meaningful population is best measured as the number of adult workers plus queens present. Sometimes there are temporarily large numbers of young gynes which play only a minor part in the social organization and all or most of which will disperse. They are not considered in this section.

Maximal worker populations are ordinarily attained just before the maximal production of reproductives. Ontogenetically this time is the culmination of the colony growth period and, except in the highly eusocial forms, is only shortly before the breakup of the colony.

Table 6.1 gives data on sizes of colonies of numerous social bees. The rather moderate populations shown for Meliponini (*Melipona* and *Trigona*) perhaps result from the understandable tendency of students to count bees only in the smaller colonies. The figures shown in that tabulation are reasonably firm, being based on counts or seemingly rather careful estimates. In Table 6.2 are presented estimates of colony sizes of some Meliponini which reach much higher totals. The methods of making these estimates are not known to me and the figures are probably more impressions than true estimates. Yet they presumably do show that the figures for Meliponini in Table 6.1 are far from maxima.

Allodapine bees can frequently be found in substantial groups, sometimes up to 15 or 20 adult females in a single dry hollow stem. As noted earlier, in such cases most of the individuals are young, commonly still pale in coloration, that is, they are tenerals or callows. And even among those that are fully colored, some are unworn and show no evidence of working as yet. Every intergradation exists between such individuals and the working members of the colony, so that counts of the numbers of such members become quite arbitrary. Nonetheless the counts are clear indicators of the small size of the colonies. In some

TABLE 6.1. Colony size (number of adult workers and queens) in various social bees at the season when the maximum number of workers is active.

Species	Population size	Reference
Augochloropsis brachycephala	2–4	Michener and Seabra, 1959
A. sparsilis	2–5	Michener and Lange, 1959
Pseudaugochloropsis spp.	2–6	Michener and Kerfoot, 1967
Augochlorella striata and persimilis	2–5	Ordway, 1966
A. michaelis	2–5	Michener and Lange, 1958a
Lasioglossum (Dialictus) versatum	5–60	Michener, 1966b
L. (D.) imitatum	2–25	Michener and Wille, 1961
L. (D.) rhytidophorum	2–10	Michener and Lange, 1958b
L. (D.) zephyrum	4–45	Batra, 1966b
L. (D.) umbripenne	60–84	Wille and Orozco, 1970
L. (Evylaeus) duplex	2–9	Sakagami and Hayashida, 1968
L. (E.) malachurum	15–26	Chapter 25, B
L. (E.) marginatum	54–486	Plateaux-Quénu, 1962
Bombus medius	2184	Michener and LaBerge, 1954
B. lucorum	> 200	Sladen, 1912
B. ruderatus	> 400	Hoffer, 1882
B. hortorum	100	Sladen, 1912
B. muscorum	> 50	Sladen, 1912
B. affinis	176	Plath, 1934
B. terricola	50–150	Plath, 1934
B. bimaculatus	> 60	Plath, 1934
B. impatiens	> 450	Plath, 1934
B. ternarius	> 100	Plath, 1934
B. vagans	> 70	Plath, 1934
B. fervidus	50–125	Plath, 1934
Trigona iridipennis[a]	2,250	
T. spinipes	5,500	
T. julianii	963	Juliani, 1967
T. mosquito	1,175	
T. braunsi	100–750	
T. araujoi	2,500	
T. nebulata komiensis	195–2,000	
T. schrottkyi	300	
T. cupira	2,900	
T. corvina	7,200	
Melipona marginata	160–243	
M. anthidioides	894	
Apis mellifera	4,000–60,000+	

[a]Except as otherwise indicated, sources of information for the data on *Trigona* and *Melipona* are given by Wille and Michener (in press).

species, such as *Exoneurella lawsoni* and *Allodape mucronata,* there is ordinarily only one adult working bee in a nest. In all species, nests of this type are sufficiently numerous that the mean number of working or mature females per nest does not reach two, although in *Exoneura variabilis* it does reach 1.9 (Table 6.3). The maximum number of working females, queens plus workers, known from a single nest is six. Rarely are as many as one third of all the nests of a population of allodapine

TABLE 6.2. Colony size (number of adult workers and queens) in colonies of Meliponini from São Paulo, as estimated by Lindauer and Kerr (1960).

Species	Colony size
Melipona quadrifasciata	300–400
Melipona scutellaris	400–600
Trigona silvestrii	400–600
Trigona capitata	1,000–1,500
Trigona mombuca	2,000–3,000
Trigona testaceicornis	2,000–3,000
Trigona droryana	2,000–3,000
Trigona jaty	2,000–5,000
Trigona postica	2,000–50,000
Trigona spinipes	5,000–180,000

bees inhabited by two or more mature working females, and in the nearly solitary species such nests may constitute only 3 to 10 percent of the total.

Why is it that the colonies of some social bees never contain more than a very few adult females (workers and queen) while those of other kinds consist of perhaps 180,000 individuals? Of course this question has two aspects: what are the control mechanisms, and what is the adaptive significance of colony size? The latter will be dealt with in section G below and in Chapter 20.

TABLE 6.3. Numbers of adult females per nest for various allodapine bees.

Species	Mean total no. of adult ♀♀	Mean no. of mature adult ♀♀	Maximum no. of mature adult ♀♀
Exoneurella lawsoni	1.7	1.1	2
Allodape mucronata	1.1	1.1	2
A. panurgoides and ceratinoides	1.1	1.1	3
A. panurgoides and ceratinoides (Feb.)	1.8	1.1	3
Allodapula dichroa	1.2	1.0	2
A. variegata	2.3	1.1	3
A. acutigera	1.2	1.1	3
A. melanopus and turneri	2.6	1.4	6
Braunsapis facialis	1.5	1.4	3
B. leptozonia (Natal)	1.8	1.5	3
B. foveata and leptozonia (Cameroon)	1.4	1.4	4
Exoneura variabilis	2.1	1.9	6

SOURCE: Michener, 1971.

There are, of course, numerous answers to the question of mechanisms, each telling part of the story.

The population growth of a colony depends on the relations between the egg-laying rate, the individual maturation rate, the death rate, the departure rate of females (if some of those produced leave to nest elsewhere) and the sex ratio. In some forms like the allodapines all these are important; many of the young adult females do disperse and a significant percentage of the progeny at all seasons are males, all of which leave the nest. More highly social bees limit both male and gyne production seasonally, so that during the growth phase of the colony the laying rate, the maturation rate, and the death rate are the determinants of population growth.

As to the laying rate, the queen has an innate capacity which varies enormously among species. Some of the halictine and allodapine queens average only one egg every several days, attaining rates of one egg per day for short periods. In contrast, queens of *Apis* attain rates of 3000 eggs per day, although 1000 to 2000 is more usual, and perhaps 80 percent of the eggs laid reach maturity. Egg production may be greatly augmented by laying workers, or in some species there may be two or more queens.

The egg-laying potentials of queens (and workers) may not be reached for various reasons. For example, workers may not construct enough cells, or they may not bring enough food to the queen. Failure to bring enough food to allow attainment of the reproductive potential may be due to outside influences, such as bad weather or insufficient forage, or may result from colonial influences, such as an inadequate number of workers or inappropriate age structure of the worker population. Females of bees (and other Hymenoptera) regularly resorb a part of their egg production instead of laying all that the ovaries manufacture, the percentage resorbed presumably depending at least in part upon the number that can be placed in appropriate cells. Potential and actual egg production, as they vary during the life of a queen, are obviously of much social importance and deserve intensive study. One would like to know, for example, whether the increase in productivity with age that occurs in queens of *Lasioglossum marginatum* (Figure 9.8) results from hormonal or other changes in the queen, from an increase in the number of cells to be filled, from an increase in the queen's food, from interactions among these, or from other factors. Clearly such matters play a major role in determining the growth and development of a colony, but they remain largely unstudied.

Pollen supplies influence laying and brood rearing by *Apis*. Although laying queens eat the glandular secretions of workers, presumably these secretions are reduced in quantity when food supplies are scarce. Thus the relation of the queen's laying to the food available to the colony is not surprising.

Worker population size in *Apis* also influences productivity of the queen, partly irrespective of the amount of food brought in by the workers. The interactions are complex. The queen tends to lay in areas of comb where workers are dense, but the workers probably tend to be dense in areas where the queen has laid (Fukuda and Sakagami, 1968). Moreover, workers influence laying by the number of cells prepared. Perhaps when there are too few cells, workers retain nutrients, therefore soon increasing wax production for new cell construction (Free, 1966). Obviously there is an intricate web of relationships influencing population growth in the *Apis* colony.

In addition to the factors mentioned above which limit egg production, there are of course the death producing factors operating throughout the lives of the individual bees and thereby serving to limit colony size. Disease, parasites, and predators take their toll at all stages. Cannibalism of eggs and larvae by adults in the colony is well known. Moreover, the longevity of adult workers varies greatly among different kinds of bees. In some of those with small colonies (Halictinae), workers of some species probably survive as adults for only about three weeks. In other species of the same subfamily and genus (*Lasioglossum marginatum*) the workers live for over a year. In *Apis* in summer, workers live for a maximum of about six weeks; in *Bombus* some live for more or less the whole summer. Given the same reproductive activity, it is clear that the longer workers live, the larger the population of the colony will be.

One result of social behavior is improved survival of young under increasingly homoeostatic conditions. The mortality of young is unknown but considerable in halictine bees. Brian (1965b) gives survival from 100 *Bombus agrorum* eggs as 71 larvae, 52 pupae, 34 adults. Fukuda and Sakagami (1968) give equivalent figures for *Apis* as 94 larvae, 86 sealed cells, and 85 adults under normal conditions where the brood area was covered with bees and temperature therefore well controlled.

The age structure in colonies of bees has been studied

seriously only in *Apis* (Sakagami and Fukuda, 1968). In this elaborately social form the survivorship curve has essentially the same form as in man, a result of protection of eggs and young by the society, up to the adult stage, and even of adults until they become foragers. The destruction of whole colonies by catastrophes is not considered, and could not be considered, in Sakagami and Fukuda's work. Ordinary nonsocial insects usually have high mortality periods at about the times of egg hatching and of metamorphosis, as well as among old adults; for honeybees the first two such periods show relatively minor mortality and the survivorship curve is truncated near the point of maximum age. Because of the protection afforded by provisioned cells, it is likely that compared to most insects all bees and wasps show relatively small preadult mortality, although it is probably higher in forms with mass provisioning than in progressive feeders where repeated examination doubtless often permits removal of mold, parasites, and the like.

G. The importance of numbers

Social behavior involves living together by two or more conspecific individuals. The importance of such associations is obvious in many cases, as is made evident repeatedly throughout this book. I wish only to cite here certain cases where, aside from division of labor, numbers of individuals living together seem to influence the activity of the group.

In *Lasioglossum zephyrum* lone bees are considerably slower to start cell provisioning than bees living in small groups (Michener, Brothers, and Kamm, 1971a). Probably this is a result of mutual stimulation; a lone bee may spend long periods doing nothing while several bees in a burrow seemingly disturb and thus activate one another.

To turn to highly eusocial bees, *Apis* cannot swarm unless at least 200 bees are involved. Comb construction requires a certain minimum of bees, which varies according to the egg layer present. Darchen (1957; in Chauvin, 1968; see also Townley, 1970) found that a group of only 50 workers with a fertilized queen could construct comb; 75 was the minimum population that built with a virgin queen. If the queen were freshly dead, the group needed was about 200 workers; otherwise comb was not built. Without a queen Darchen obtained no comb building from small groups of workers. Even with 5,000 there was no building for 14 days, but it then began at the same time that laying workers appeared. Darchen postulates a pheromone originating in the queen or laying workers that stimulates comb construction but when the population is large enough, the pheromone, if it exists, is presumably not needed. It is said that a group of 10,000 workers can construct comb without either queen or laying workers.

H. Concluding remarks

A major weakness of the preceding account is that except for some data on mature colony sizes, it is nearly all descriptive. There are almost no hard data on the ontogeny of colonies based on adequate numbers of nests observed throughout their periods of activity.

The study of the origin, growth, maturity, and demise of colonies of bees will be full of important material for those who would understand the ecology or even the life history of these insects. Nearly all studies of parasocial and primitively eusocial bees, except bumblebees, have been made by destroying the nests and killing and dissecting the bees. Any knowledge of the ontogeny of the colonies therefore has had to be obtained by comparison and by indirection, not by the continuing observation of colonies through their histories. The situation is somewhat better for bumblebees and highly eusocial bees because their colonies are larger and can more readily be observed. Even here, however, few statistics such as population growth rates are available; yet clearly these must be among the most important attributes of a colony. Not only should such information be gathered, but the factors that cause differences in population growth rates (both among colonies of a species and among species) also must be studied if the social phenomena involved are to be adequately understood.

7 *The Social Significance of the Nest and Its Contents*

We ordinarily think of bees in the sunlit world of flowers in which they fly, generally isolated from one another and their nests, and in which vision is a major guide for orientation. The contrast of this picture with the conditions inside the nest could hardly be more striking. Except for the few bees like *Chalicodoma* that make exposed cells, the nest, where bees spend most of their lives, is a dark place in which they move by walking. In the highly eusocial species each individual is in frequent if not almost constant contact with its nest mates. Activities are tactually, chemically, gravitationally, or sonically stimulated, for eyes are useless in the dark.

The whole of social organization requires social interactions. The question of the early stages of evolution of such interactions often arises. Organisms respond to various environmental stimuli, including the condition of their nest; they eat, hide, fly, and so on, according to interactions between these stimuli and their own internal conditions. One of the evolutionarily simplest forms of social interaction for an incipient social insect to use for integrating its colonies would be modifications of environmental features like nest structures, to which the species responds anyway. Solitary bees must respond to holes in their nesting substrate, to their own partly made nests, to their own partly provisioned cells, and the like. A species evolving social behavior can doubtless take advantage of this fact for indirect social interactions between individuals. The bees themselves would not need to confront one another or evolve suitable responses to conspecific adult females. The following sections give various examples of this sort of interaction, some of it even in the most highly eusocial bees, which have evolved, in addition, elaborate direct communication among adults. Direct interactions among adult members of a colony are, however, considered only in subsequent chapters.

The structure of the nest and the provisions and brood that it contains have various profound influences on the behavior of bees, sometimes also on their structure, including even sex and caste. Some of these influences will be taken up in this chapter; others, such as the effect of nest structure and contents on defense, will be dealt with later. The feedback influence of the nest structure on subsequent construction is a type of indirect social interaction that Grassé (1959) has called stigmergy. A major part of the following material concerns stigmergic responses, although we will not ignore effects of the construct on nonconstruction activities. Finally, interactions between young (that is, brood) and adults will also be considered.

A. Social significance of nest structure

The primary functions of the nest construct are for protection, brood rearing, and food storage. The nest's structure and content, however, do provide for indirect social interactions among bees and are therefore of major importance in the integration of colonies. There is feedback from such factors as construct, provisions, and brood, through the bees to subsequent construction, to colony growth and maturation, and to nearly all other activities.

The construct, at least the nest entrance and main

burrow, is often influential even in a parasocial colony. Members of such a colony, as indicated previously, often are unrelated individuals. A female seeking a nesting site can be seen to be attracted by burrow entrances, which she investigates and may enter and utilize. The existence of a nest is an invitation to other bees to enter and live colonially.

In the Halictinae, nests of related species are usually similar irrespective of social organization. That is, a solitary species will make a nest similar in basic structure to that of a related primitively eusocial relative, although different in the number of cells. Also, within a species, nests are basically similar whether made at a solitary, parasocial, or primitively eusocial stage in the life history. Thus a construct made by a lone bee can be approximately duplicated by a group working cooperatively.

Probably individuals of the group respond to the same cues that stimulate and modify the activities of lone individuals (Michener, 1964b). There is not yet good experimental evidence on this subject, but there are observations supporting this viewpoint. A lone bee, whether of a solitary species or a solitary stage in the life cycle of a social one, does not work continuously at cell construction, provisioning, or any other single activity. As it moves around in excavating, carrying excavated material away from the cell or bringing material to the cell to form the lining, it appears that the cell itself is often the cue that stimulates further work or controls the type of such work. In *Lasioglossum* (*Dialictus*) the bee often stops work for considerable lengths of time and wanders about or rests, appearing to resume work when it happens to come to the cell under construction. In many (but by no means all) halictine bees, also, two or three cells are often in different stages of construction at the same time, so that the appropriate reaction for a given place differs according to the particular cell or construction site.

The following are examples of cooperative activities in halictine colonies that must be mediated by physical or chemical cues from the nest structure: In *Lasioglossum* (*Dialictus*) *zephyrum,* more than one female can be active in the construction of one cell (Batra, 1964) and up to six in provisioning a cell. In the rather unrelated *Lasioglossum* (*Evylaeus*) *malachurum,* Legewie (1925a) reported three or four foragers bringing pollen into one or two cells, or six to ten provisioners bringing pollen to three cells in a single nest; and for *L.* (*E.*) *calceatum* Plateaux-Quénu (1963) showed that four workers provisioned a single cell. Batra (1968) suggested that a chemical cue (pheromone), perhaps the distinctly sweet, pungent odor of the freshly completed cell of *Lasioglossum* (*Dialictus*), serves to stimulate the provisioning of the cell. She noted that if a cell lies unused for a day or two, until this odor disappears, it is never used but is filled with earth. Such abandonment is not inevitable, even in the species with which she worked, but when it does not occur, the cell may have been relined.

Both Stockhammer (1966) and Batra (1968) agree that completed, occupied cells are never cut into from the outside by excavating Halictinae, whether solitary or colonial. Cells of *L. imitatum* seemed as well protected from burrowing *L. versatum,* as did cells of *versatum* itself. The avoidance is possibly related to a secretion which soaks into the substrate around the cell;[1] it is applied from the apex of the abdomen, probably originates in Dufour's glands, and it or perhaps a product of the developing young bee or both are effective throughout the growth of the latter. On the other hand, abandoned cells from which bees have emerged may be destroyed by subsequent construction activities.

From these observations it seems probable that cooperative work of mutually tolerant halictines is possible because of responses to physical and chemical attributes of the construct not differing from comparable responses of solitary individuals. Thus the indirect social interactions should be exactly the same as the feedback responses of solitary individuals. This similarity could explain the repeated independent origins of sociality in this group of bees.

Interestingly, it is postulated that in *Apis,* also, a larval secretion inhibits workers from cutting into capped cells containing prepupae or pupae (Morse and McDonald, 1965). These authors believe that a substance secreted by the larva when the cocoon is spun prevents bees from manipulating the wax around the cocoon and that the

1. Stockhammer's work was on the solitary halictine *Augochlora pura,* and a more extensive study of this and related matters, using the same species, has recently been made by May (1970a). She concluded that a complex of factors prevent nesting bees from cutting into cells containing brood: texture and shape of cell walls; odor of fresh pollen, perhaps adsorbed on the lipid cell lining or the larval cuticle; and, for older cells, a species-specific pupal odor. The odors alone, as shown by extracts and other experimental methods, do not inhibit excavation in the absence of the characteristic texture and shape, but these physical features do not inhibit excavation after the occupant of the cell has emerged, when the odors are presumably gone.

frequent removal of the cap of a queen cell at any time after its construction results from the fact that queen cocoons, unlike those of workers or males, are not spun in contact with the wax cell cap.

In most bumblebee (*Bombus*) species, after the colony is established and growing, new cells are constructed on the tops of clumps of cocoons. The number of the cocoon clumps is therefore one of the factors limiting cell construction and oviposition rate. This is an important control mechanism, since it tends to assure that the number of workers emerging from the cocoons is large enough to feed the larvae hatching from the eggs (Free and Butler, 1959). It is actually a more finely tuned mechanism. Brian and Brian (1948) and Brian (1951) report for *B. agrorum* that the number of eggs laid on a group of pupal cocoons is proportional to the size of the cluster of cocoons, there being about three eggs per cocoon. This seems to mean that there is a fairly sensitive regulation of productivity during the growth phase of the colony according to the number of workers being matured. The influence of the nest on the cell-making and egg-laying activities of the queen is apparent.

There are other obvious ways in which the content of an individual nest can influence bumblebee behavior. For example, some species of the genus normally place pollen in wax pockets beside the cells; the larvae feed directly from such pollen supplies. Other species store pollen away from the larvae in wax cells and old cocoons which may be extended with wax to form tall, pollen-filled cylinders. However, occasionally there is a time in the development of a colony of a pocket-making species when there are no growing larvae and hence no pockets. When this occurs pocket makers adopt the habits of pollen storers and put pollen into old cocoons or wax cells (Sladen, 1912).

Construction in the highly social bees involves large numbers of individuals which secrete or collect the construction material. It seems almost certain that the construct itself or its contents (young bees or stored food) provide the coordinating cues. Nonexperimental but suggestive evidence for this is provided by observations on workers of *Trigona carbonaria* constructing numerous resinous pillars in the nest (Michener, 1961a). Bees carrying bits of the cerumen construction material seemed to wander about until they happened upon a high point of some sort, to which they attached the material. It seemed that the higher the point, the more likely that a bee finding it climbed to its summit and deposited a load of building material there. The result was that the high points grew to become the pillars.

Probably most of the nest construction activities of the stingless bees are similarly stimulated by existing features of the nest. The following details concerning nest construction by *Trigona braunsi,* a minute African species studied by Bassindale (1955) (called *T. gribodoi* by him), are representative of construction activities in the genus and are very suggestive of the influence of the structure on the behavior. For constructing cells, cerumen is brought to the site by bees of a certain age group (described as the second stage for this species in Chapter 12, B3). The cerumen may be brought from various locations in the nest, as follows: empty storage pots, old brood cells, the cerumen dump, and in Bassindale's artificial hives, wax from other sources provided by him.

The sources of the materials are further described as follows. The empty storage pots may be cut down and the cerumen removed directly to the new brood cells, or it may be deposited first in the dump and later taken for brood cells. Cerumen is removed from the brood cells themselves, as in all melipone bees, as soon as the larvae within construct cocoons, the walls of the old cell being thinned by removal of cerumen from all exposed surfaces. The dump is a place in the nest where reserve building materials are accumulated, either wax secreted by the glands of bees or cerumen removed from cells or storage pots or other nest structures. The wax scales appear between the dorsal abdominal segments of young worker bees. A bee showing such scales stands on the dump and scrapes the back of the abdomen from front to rear with a posterior leg (Figure 1.17). The distal end of the basitarsus catches the translucent white wax scale and pulls it away from the abdomen; it then falls to the dump. Usually the bee which produced it or another bee picks up the scale and manipulates the wax, incorporating it into the material of the dump.

Cerumen is removed from the dump by bees working along the surface with their mandibles. Each bee cuts off a strip of material perhaps half as long as the body of the insect; the strip is held at one end by the jaws, the rest of it extending beneath the body of the bee. To start a new cell a strip is applied to the side of an existing cell. Additional strips are added, forming, with appropriate mandibular smoothing and shaping, the cup or base of the cell, which is then extended upward by strips added to the edge. Several bees contribute to the building of any one cell, there being no evident communication among them except that provided by their own constructing activities. The impression that one gets from this and other studies is that bees of the right age group respond to the nest construct in various ways, now removing cerumen, now adding it, depending on what they encounter as they move about.

Data on *Apis* summarized by Lindauer (1961) appear also to show numerous responses of great integrative importance on the part of honeybee workers reacting to conditions in the hive, not only to the brood but also to the construct and stored food materials. The source of the information is supposed to be a sort of summation of the previous activities of the colony; the recipient is the individual worker which, in wandering about the hive (patrolling is Lindauer's term), gathers its own information. Lindauer's belief is based upon observations such as the following: a worker bee was found to spend 56 hours in patrolling out of a total of 177 hours that she was observed (Figure 12.4). She does various things as she wanders—tending brood, building comb, capping comb, cleaning cells, etc. All these activities appear to be responses to the conditions that she finds at the moment as she moves about. Figure 12.5 shows the activities of a single bee watched nearly continuously for 10 hours during her eighth day of adult life. Note that patrolling often terminates in some activity presumably useful to the colony. Thus the situation is similar to, and better documented than, that postulated for stingless bees.

The following paragraphs concern more specific relationships of *Apis* to nest structure or content. Some activities are probably parts of the complex of behavior described by Lindauer while others are different.

Apis brood cells are re-used after emergence of a bee, and the refurbishing of such a cell for subsequent use is a complicated process. Perepelova (see Ribbands, 1953) arranged for continuous observation of 30 cells from the time of the workers' emergence until new eggs were laid in them. Each cell was visited by 16 to 29 different bees, one at a time. The average of 40.8 minutes per cell spent in refurbishing is divided as follows: removing remains of cap, 13 minutes; smoothing cell edges, 6.5 minutes; cleaning walls, 8 minutes; cleaning the bottoms of cells, 13.3 minutes. It must be the construct that guides all this activity.

In *Apis,* whose combs are made exclusively of wax, chewed and manipulated along with an oral secretion, wax secretion is stimulated by an interesting construct-related mechanism. House bees that accept nectar from foragers normally soon process it and regurgitate it into cells for storage or subsequent use. But if there are not enough such cells, and if the quantity of nectar being brought in exceeds the immediate needs of the colony, these bees are forced to retain nectar in their crops for long periods. In the absence of storage cells, the house bees themselves become storage containers. The sugar in such nectar is not indefinitely stored, however, but is assimilated, with the result that wax is secreted by the wax glands. Thus the raw material for more storage space is produced as a result of the lack of such space—the greater the nectar surplus the greater the wax secretion (Butler, 1954a).

Lau (1959) and Darchen (in Chauvin, 1968; see also Townley, 1970) have given extensive treatments of comb construction in *Apis,* and there is good evidence of stigmergy. A comb starts on the roof or side of the nest cavity, at first as little tubercles of wax. Increasingly, new wax is added to existing tubercles, much as with the pillars described above for *Trigona carbonaria.* Eventually such masses unite to form the beginning of a comb of cells.

Each of the multitude of building workers performs her tasks largely independently of her fellows. She often moves about, feeling nearby parts of the comb. Her work is not performed in any regular order—she will add wax, smooth it, or remove it, in any sequence, often moving from building site to building site, so that she must be guided by what she and her associates are building. There is no interchange of building material among bees. Thus from a vast number of seemingly independent individual actions guided by the growing comb and by the sensory and nervous limitations of bees, the beautiful wax comb of hexagonal cells is produced.

Constructing *Apis* are surrounded and overlain by chains of bees not immediately involved in the construction. The shape of the comb determines the location of the chains that in turn control the building of the comb (Darchen, 1962).

A comb grows with rather thick and irregular wax added at the lower edge; this is soon formed into round cell cups, the margins of which are gradually extended. Each new cell is placed in the angle between two preceding cells. As they are drawn out, the walls are thinned, except that strength is maintained by thickened edges, and the hexagonal form is attained as a result of close proximity to other cells. The thickenings at the edges of the walls seem to move outward from the septum of the comb as the cells grow, wax being scraped away from the inner part of the thickening to leave only the thin finished wall, and added to the outer surface. Figure 7.1 shows the growth of cells at the edge of a comb.

The sensory apparatus that enables *Apis* workers to do the elegant job of building a wax comb was investigated by Martin and Lindauer (1966). They found that the hair plates at the neck of the worker, which are also essential for orienting communicative dances with respect to gravity (Chapter 15, D2d), are necessary for construction of cells and comb; without them bees cannot even begin to build. They consider sense organs at the tips of the antennae important for controlling the thickness of walls and their smoothness. But surprisingly, the diameters of the cells and the correct orientation of their walls (angles of 120° to adjacent walls) are not dependent upon antennae. Bees without antennae can still make cells of the normal shape and orientation. What is used for measurement of the regular diameters and angles is still unknown. Perhaps it is not surprising that it has been found difficult to learn the mechanism for this ability; it is likely to involve complex integration, since the bees can make cells of two sizes (producing workers and males) but of the same shape. Indeed, Taber and Owens (1970) report that queenless colonies build comb that is neither of worker nor of male cells, but intermediate. They further emphasize that there is substantial irregularity in size and shape of cells; somewhat too much emphasis has been placed on their extreme regularity. The sensory and motor mechanisms for cell building must be complex and closely controlled by feedback.

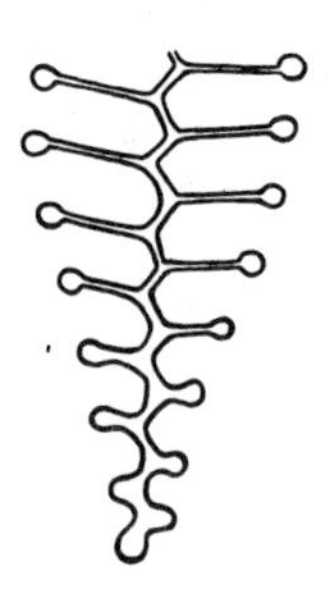

FIGURE 7.1. Vertical section through the growing edge of a comb of *Apis mellifera.* (From Ribbands, 1953, after B. Lineburg.)

It is assumed that the cell size and shape influence the queen *Apis* in her egg laying, so that she ordinarily lays unfertilized eggs in the cells of the right size to produce males and fertilized eggs in the other cells (see Chapter 8). The difference in size between worker and drone cells is shown in Figure 7.2. It is also assumed that cell size and shape influence the feeding of the developing larvae by the workers of *Apis* (see Chapter 10, C2). As is well known, the feeding of the female larvae determines the caste into which they will develop. If the nature of the feeding as well as the sex of the eggs that are laid are influenced by the cell size and shape, the great importance of the construct for this indirect social interaction is clearly demonstrated.

Artificially large worker cells in *Apis* cause larvae in them to be fed more and to produce larger workers than normal, hence increasing the honey production from the colony (Glushkov, 1964). In a wide variety of solitary bees also, unusually large individuals can be produced by feeding more than the bees ordinarily provide in a cell; this has been demonstrated by casual observations on *Megachile* and *Augochlora.*

In drone cell construction by *Apis,* also, the existing structure and brood influence subsequent building (Free, 1967a). The number of drone cells constructed is inversely related to the number already present. How the bees "know" when enough have been built, or sense the number already present, is obscure. One can only suggest that it has to do with the constant patrolling of the colony by the worker bees.

B. Responses of larvae to the construct

Larvae are so intimately surrounded by the brood cell walls that there must be many responses of larvae to contacts with these walls. Such responses have been studied only in the context of orientation of the fully fed larva. In many bees and wasps, whether solitary or social, it is important that the larvae, before pupation, orient themselves with their heads toward the cell entrances so that newly emerged adults will start their escape in the correct direction. Those that head the wrong way often wear themselves out burrowing and die. Or they might cut through other cells and destroy their siblings. Some species seem to orient by gravity. The gynes of *Apis* do so, always heading downward; gyne cells open downward. Allodapine bee larvae do likewise, heading upward

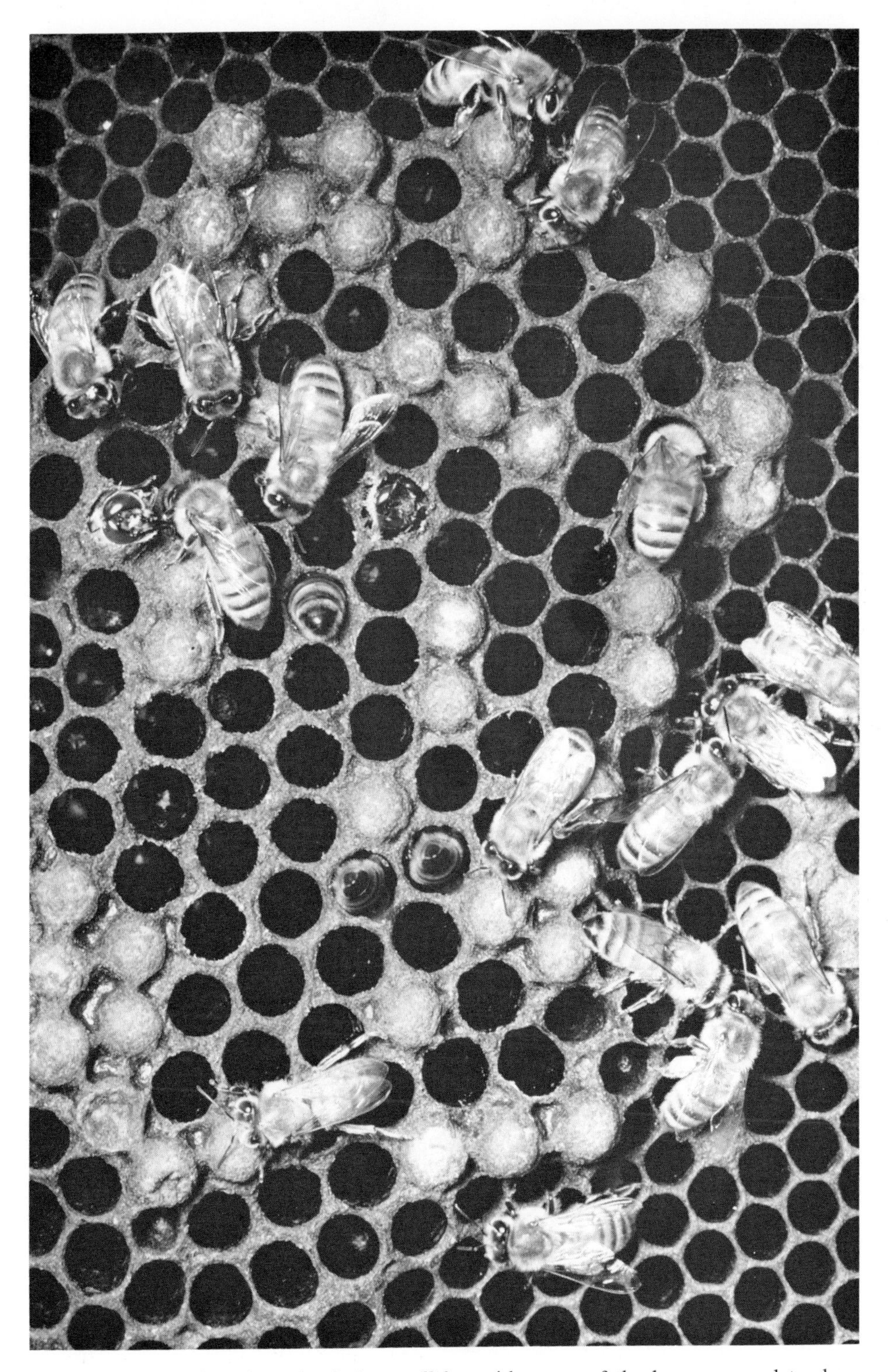

FIGURE 7.2. A brood comb of *Apis mellifera* with many of the bees removed to show the cells clearly. The large central area consists of drone cells; young males, recognized by the enormous eyes meeting on top of the head, can be seen emerging. The marginal areas (*above, right, below*) consist of smaller worker brood cells. In the drone area the cocoons and cell caps are domed, providing more space inside; in the worker areas the cell caps are nearly flat (*extreme right*). (Photograph by E. S. Ross.)

regardless of the orientation of the nest; in this case there is probably no influence on survival. Forms with more or less horizontal cells or with cells that may slope either up or down cannot use gravity as a cue for the direction in which the larva should head. In known cases such a larva uses, instead, the form or texture of the interior of the cell in which it is being reared, built by its mother or sisters. There is thus an indirect interaction, via the construct, from adults to larvae and often from one generation to the next.

Such matters have been investigated in social bees only in *Apis*. As demonstrated by Jay (1963a), using artificial cells of various sorts, mature worker and male larvae, in horizontal cells, respond differentially to the smooth inner ends of the cells vs the rough inner surfaces of the cell caps. They head toward the caps and defecate on the inner ends of the cells. The resultant adults therefore head in the right direction.

C. Responses of adults to stored food

Only *Bombus* and the highly eusocial bees store food in quantity outside of brood cells for use of adults and for transfer to larvae or brood cells as needed. Allodapines may store small amounts of pollen on the walls of the nest, but nothing is known of the cues that cause them to obtain more or less pollen.

In *Bombus* as well as the highly social bees, information presumably obtained from the food already stored influences further food collecting. Free (1955b) has shown for *Bombus* that addition of syrup to the honey storage pots causes a decrease in nectar-gathering activities by foragers. In one experiment he added syrup to half of the pots in five nests; storage space therefore was still available, but nectar-collecting trips dropped from 87 to 6 during equal periods of time, while pollen-collecting trips did not change greatly in frequency (Table 7.1). In a second experiment, addition of pollen caused a significant but less marked decrease in pollen-collecting trips. A given worker is usually temporarily constant to a given activity; nectar foragers therefore became nearly inactive in the first experiment while pollen foragers continued working. In the second experiment pollen foragers reduced their activity while nectar foragers kept on working at the prior rate. In the brief duration of the experiments there was no tendency for individual bees to change

TABLE 7.1. Influence of syrup added to stores on numbers of foraging trips from nests of *Bombus pratorum*.

	Five experimental colonies		Five control colonies	
	Before syrup	Syrup added	Before syrup	Syrup not added
Pollen loads	12	14	3	13
Nectar loads	87	6	94	74

SOURCE: Free, 1955b.
Note: Trips were counted for equal periods before and after addition of syrup.

activities, for example, for nectar gatherers to start gathering pollen instead. However, since bees' tasks change in both directions during their lives, such changes would probably have been noted had the observations been continued.

Inhibition of foraging by adequate supplies is not complete in the higher eusocial bees, as shown by the accumulation of vast quantities of honey, for example in attics or between walls, by *Apis*, and by the fact that removal of honey from hives is not necessary to keep the bees collecting nectar. But lack of stored provisions does strongly stimulate foraging in *Apis* colonies.

The influence of food stores on nest recognition by colony odor is dealt with in Chapter 18 (C) and on communication with respect to food sources in Chapter 15 (A and D1). Of course the flow of food within the nest (Chapter 16) also involves various responses to stored food.

D. Interactions between adults and young

In most groups of social insects interactions between adults and young (that is, brood—eggs, larvae, pupae) are universal and important parts of the social organization. Exchange of food between larvae and adults is well known in ants and vespid wasps, and it has often been supposed that larval activity or secretions are of great importance in maintaining the social group. In most kinds of bees, however, there are no contacts between adults and young because the cells in which the young are reared are closed before the eggs hatch, each cell being mass provisioned with enough food to provide for the entire growth of the larva. Progressive feeding, which of course involves adult–larva contact from time to time during the growth of the larvae, occurs among bees only in *Apis, Bombus,* and most allodapines. Even the highly eusocial meliponines have mass-provisioned cells which, together with the cocoons spun by the mature larvae, completely enclose the immature stages for the whole developmental period. This section is a survey of the kinds of contacts between young and adults that do exist among bees.

In spite of the lack of contact between larvae and adults, there is some evidence that adults are somehow aware of certain developments in closed cells. The best information on the subject is provided by recent work of Batra and Bohart (1969) on *Nomia melanderi,* the alkali bee, a solitary (although aggregating) halictid. The bees were induced to construct nests in a layer of soil between sheets of glass so that their activities could be observed. It was found that cells were left closed if normal larval development was occurring. On the other hand, fungal growth on pollen mass or larva (involving death of the latter) somehow stimulated the mother in many cases to re-open the cell and pack earth into it. The result was a barrier to the spread of the mold and reduced sporulation, possibly due to relatively anaerobic conditions.

1. Halictidae

Although the above case of *Nomia melanderi* does not suggest contact between larvae and adults of bees which mass provision their cells, there are observations of such contact in certain primitively eusocial halictine bees. In some species of *Lasioglossum* (*Dialictus*) the bees open the cells from time to time and "inspect" the eggs or developing larvae or pupae (Batra, 1968); there is no evidence of progressive feeding. The cells are opened as though the inspecting bee knows where the entrance is, although the point is often not recognizable to the human observer. Similar observations were made for *Lasioglossum* (*Dialictus*) *zephyrum,* except that the glossa of the inspecting bee was seen applied to the pollen ball in a cell (Batra, 1964). After inspecting, the bees reclose the cells.

A few open cells are occasionally seen in nests of various halictines as excavated in the field, but there is often the possibility that the cell closure, frequently made of loose earth, collapsed due to the excavating by the observer. However, Batra (1968) noted that cells of *Halictus rubicundus* reared in an insectary where the activities in the nests could be observed through glass were left continually open and the brood periodically inspected. Moreover, *Lasioglossum* (*Evylaeus*) *cinctipes, linearis, malachurum, marginatum,* and *pauxillum* regularly leave the cells open and the brood therefore exposed (Knerer and Plateaux-Quénu, 1966c). In these species the females not only inspect the brood but remove larval feces, exuviae, dead larvae, and the like.

Knerer (1969b) argues that in such forms nectar is added to the food during larval growth, so that progressive feeding occurs; the bases of his statement are observations of inspections, such as those mentioned above, and his finding that the weight of food mass plus larva increases about 60 percent in *Lasioglossum malachurum* during larval growth, an increase that seemed too large to be caused solely by absorption of moisture from the cell atmosphere and surrounding substrate by the pollen mass or the larva itself. Unfortunately for the theory, the same or even greater weight increases occur in solitary halictines (*Augochlora pura*) even in the complete absence of adult bees and clearly have nothing to do with progressive feeding (May, 1970b, 1972). In this case the larva rather than the provisions apparently takes up water vapor from the atmosphere. There is therefore no evidence for the intimate and progressive trophic relations which Knerer suggests between generations of social halictine bees.

In some species there may be local and temporal variation in cell closure. For example, in *L. malachurum,* cells, instead of being left open, may be temporarily closed, then opened when the eggs hatch, and closed again when larval feeding is complete. In other areas cells of the same species are reported to be either continuously open or continuously closed (see Chapter 25, B1).

2. Ceratina

Although the above comments on halictines relate in all cases to primitively eusocial species, there is no obvious reason that this type of behavior might not occur also in solitary species (which would therefore border on subsocial). No solitary bee that regularly opens its cells or keeps them open is known, however. In the genus *Ceratina* there is sometimes contact between the mother bee and one of her progeny. The last cell in a series of cells arranged end to end in a burrow in a stem is often unclosed, and the developing young in that cell is therefore available to the mother bee who rests in the burrow above. The bee ordinarily remains alive until maturation of her progeny; her disappearance often results in death of the uppermost larva in the nest (Malyshev, 1935). The other larvae are closed into cells formed by partitions across the burrow. Shiokawa (1966) records the rather common omission or removal of cell walls between cells in *Ceratina.* He feels that this is most likely the result of a bee removing the last cell wall, exposing a larva, and then provisioning a space above the level of that removal; if so, it probably means that the bee that removed the cell wall came in contact with the young below it. These contacts would seem to mean little, but may suggest how regular larva-adult interactions could arise, especially in view of the close relationship of *Ceratina* to the allodapine bees.

3. Allodapine bees

One allodapine genus from southern Africa, *Halterapis,* mass provisions its young after the manner of *Ceratina,* although the burrow is not divided by partitions into cells. Each larva feeds on a large mass of provisions stored for it. Only the uppermost one in the series of larvae in a hollow stem has any potential contact with the mother bee, for the food masses and growing larvae prevent her from reaching the larvae below.

In all the other genera of allodapines the immature stages are all in regular contact with the adults in the nest. This is the only group of bees in which the young are moved about within the nest, like the young of ants. The nest is a simple burrow, usually in a stem, with no cells. In genera such as *Allodape, Braunsapis,* and *Exoneurella* the immature stages are generally kept in order from youngest (eggs) at the bottom of the burrow to oldest (pupae) at the top, that is, nearest the entrance. If the young are artificially mixed, the bees correct the sequence. In the genus *Exoneura* the sequence is not so carefully maintained, but in general it is as indicated above.

Larvae have tubercles and hairs and projections (least developed in *Exoneura*) which help to support them in the proper positions in the nest, once placed there (see Chapter 27, F). The pupae also, unlike those of all related bees, have long hairs, which probably serve not only to hold the pupae in the proper position in a frequently vertical burrow but also for protection against the rough walls of the burrow when being moved about.

In allodapine genera such as *Allodape, Braunsapis, Exoneura,* and *Exoneurella* the adults make a paste of pollen and presumably nectar which they place on the ventral sides of the larvae, one food mass to each larva. In another group of allodapine bees (the genus *Allodapula*), the larvae in each nest are of nearly the same age and support themselves in a cluster in the burrow, the larvae and the chorions of the eggs from which they hatched supporting the common food mass. Additional food is added from time to time by the adult bees. Details are to be found in Chapter 27 (E).

Rayment (1951) reports trophallaxis, that is, interchange of substances, between adults and larvae of *Exoneura* and says that the projections on the bodies of larvae of *Exoneura* are mouthed by both the adult bees and larvae for an exudate. Sections show no gland cells and there is no evidence that he is correct.

Allodapine bees remove larval feces regularly from their nests. In genera such as *Allodape, Braunsapis, Exoneura, Exoneurella,* the feces are never allowed to accumulate, for the adults have access to the whole nest at all times. In *Allodapula,* however, the adult bees cannot readily get below the clump of larvae which snugly plugs the nest burrow. Therefore feces collect below such a clump and are removed only after the larvae enter the prepupal stage. The prepupae and pupae are arranged end to end as in other genera, and do not plug the nest burrow as do feeding larvae.

4. Bumblebees

In *Bombus* several eggs are laid together in a distendible cell, and after hatching the larvae are fed progressively. In one group of species, called the pocket makers, the larvae are for the most part fed through wax pockets at the sides of the cells. There may be one, two, or even three such pockets at the sides of each cell. Pollen, mixed with a certain amount of nectar, is deposited in the pockets by foragers and pushed into the pockets, usually by house bees. Such pollen forms a considerable food store and, as it is pressed inward, forms the floor on which the larvae in the cell rest and feed (Figures 10.2 and 28.4). This is a type of progressive feeding, but because there is always a mass of pollen between the adult bees and the larvae, there is no contact between the two so long as this method of feeding is used. However, the adults occasionally provide food to the growing larvae by opening the top of the cell and injecting liquid food with the proboscis (Katayama, 1966; Sakagami and Zucchi, 1965); indeed this is the regular way of feeding larvae that are to become gynes (Free and Butler, 1959). The pollen pockets are useful only for relatively dry, semisolid material, so that if captive colonies are given pollen only in a liquid mixture with honey, the bees inject food through the tops of the cells regularly, whether the larvae are to become workers or gynes. In the other large groups of species of *Bombus,* food for larvae is also regularly inserted in liquid form by opening the cell at the top. To complete the statement of contacts or possible contacts between larval and adult bumblebees, it must be noted that *Bombus* cells containing large larvae are often irregularly open at the tops, seemingly because of the inadequate quantity of building material. In any case, the backs of large larvae are often partly exposed.

The presence of larvae of *Bombus* has an influence on the pollen collecting by foragers. Free (1955b) removed larval cells from nests and found a more or less immediate decrease in the frequency of bees returning with pollen loads relative to that of bees returning with nectar only.

5. Apis

The only other bees in which there is progressive feeding and regular contact between larvae and adults are those of the genus *Apis.* There is only one larva per cell. The cell is left wide open until the larva is mature. Feeding is progressive, by large numbers of small meals. A far higher percentage of the food is of glandular origin than in any other bees, except possibly Meliponini (see Chapter 10, C).

Larvae of *Apis* must enormously influence activities of the workers that secrete larval food. As explained in Chapter 12 (B2), it is mostly young workers that secrete this food, but in case of need, older ones (foragers) do so. Free (1961) gave evidence suggesting that the presence of larvae provides the stimulus for maximal hypopharyngeal gland development in foragers and perhaps also young bees. A rich protein diet, necessary for such glandular development, does not by itself produce maximal development of the glands. Unsealed brood has another influence on worker physiology (Kropáčová and Haslbachova, 1971); it inhibits ovarian development, probably being more potent in this effect than are the pheromones of the queen.

As Ribbands (1953) says, Lineburg reported that nurse bees made over 10,000 visits to each larva during its growth. This is presumably exceptional, but Lindauer (1952) found, after egg laying, 59 bee visits concerned with cell cleaning, 1926 visits of larval inspection, and 143 feeding visits. These activities required 54, 72, and 109 minutes, respectively. Capping the cell was fantastically expensive of time—657 visits totaling 6 hours and 20 minutes. Considering the number of bees involved in these sorts of work and the large number of cells, it is likely that most of the 2785 visits involved were made by different bees, influenced in their behavior by the cues of the construct and the larva in the cells. Jung-Hoffmann (1966) indicates that different workers bring the different foods that together form the brood food; those bringing the clear component (see Chapter 10, C2) are on the average older than those bringing the white and yellow components.

Free (1967b) verified that workers of *Apis* can quickly change from nectar collecting to pollen collecting as needs of the colony shift, although as noted in Chapter 14 (A) the reverse change is easier. Free found that both odor of brood and contact through screen with bees tending larvae stimulated pollen collecting, suggesting possible pheromonal communication from brood to foragers. Access of foragers to the brood itself, however, was the most important stimulant to pollen collecting. Jaycox (1970), using extracts of larvae, obtained some evidence of an olfactory cue or pheromone from larvae that apparently stimulates pollen collecting by workers. But this does not rule out the effect of physical contact or a contact pheromone as a stimulant to the forager.

The presence of queen larvae appears sometimes to stimulate production of drone comb in *Apis* (Free, 1967a). The mechanism is unknown, although a pheromone from the adult queen's mandibular glands is probably important for comb building (Darchen, in Chauvin, 1968).

Interesting social interactions between larvae and adults of *Apis* have been noted in connection with studies of the resistance of colonies to American foulbrood, a bacterial disease of larval *Apis* caused by *Bacillus larvae.* There are differences in resistance to the disease among *Apis* strains, and genetic components of the differences are demonstrable. Thompson and Rothenbuhler (1957) showed that larvae nursed by adults of a resistant line were more likely to survive feeding of *Bacillus larvae* to the colony than like larvae nursed by adults of a susceptible line. At least part of the resistance in this case was related to the adults. Those of resistant lines may secrete an antifoulbrood factor into the food given the larvae or may remove spores of foulbrood from the infected syrup in the crop by action of the proventriculus, or the mechanism may be something else.

Another kind of resistance to foulbrood has been shown to be behavioral. The bees of the resistant lines remove dead larvae and pupae from the hive (whether killed by the disease or not) much more quickly than do those of susceptible lines. Thus the source of infection is reduced. They take about the same time to remove dead young whether there are 100 or 2000 of them in the hive, suggesting that some of the idle bees normally seen in a colony are stimulated to activity by large numbers of dead (Jones and Rothenbuhler, 1964; Rothenbuhler, 1964). A more recent summary of these studies is by Rothenbuhler (1967).

The possible subtle influences of feeding or perhaps other features of hive environment on larvae of *Apis* have been reported in the studies of Avdeeva (1965). She introduced Caucasian queens into colonies of central Russian bees and vice versa. The resultant workers deviated morphometrically (measurements of tongue, terga, wax plates, and basitarsi) in the direction of the race of the nurse or host colony.

E. Concluding remarks

Nest structure, chemistry, and content are important parts of a bee's environment and in various ways influence not only its behavioral but sometimes its other attributes. This must be true of solitary as well as social bees. From the social viewpoint, a special importance of the nest is that it mediates activities that are indirectly responses to actions of other bees. One bee constructs, provisions, or oviposits in a nest; another responds to what it finds there, thus reacting indirectly but not directly to the actions of the first bee. The coordination of activities of members of a colony, especially in the case of small and primitive colonies, is largely dependent upon such indirect social interactions. Nests and their contents also have similar functions in highly eusocial colonies, even though direct bee-to-bee communication through contact, vision or pheromones has evolved in these forms for use in various contexts. In short, for both solitary and social bees, nest-building behavior and nursing behavior are sequences of movements that are chain reactions or series of feedback loops. Solitary bees, therefore, are predisposed toward social cooperation.

8 *Male Production and Sex Ratio*

In most animals more or less equal numbers of the two sexes are assured by the chromosomal control of sex. Social evolution in the Hymenoptera, however, involves females only. Escape from the roughly 1:1 sex ratio of most animals was necessary if nonreproductive female workers were to be produced in large numbers without equivalent male production and wastage. The sex determination mechanism found in Hymenoptera is doubtless one of the prerequisites to the development of hymenopteran sociality and, as explained in Chapter 20, may have been a major element in that development.

A. Sex determination

1. Genetic aspects

To an important degree sex in Hymenoptera is controlled behaviorally through fertilization of eggs or lack of it. Fertilized (diploid) eggs of Hymenoptera ordinarily produce females, unfertilized (haploid) eggs produce males parthenogenetically. This basis for sex determination is widely known as Dzierzon's rule, named after a Polish apiculturalist of the last century who first discerned it in work with the honeybee.

Except where modified in various tissues by somatic polyploidy, somatic cells of female Hymenoptera are ordinarily diploid, of males, haploid. Thus the outstanding fact of hymenopteran genetics is the haplodiploid sex-determining mechanism, males being of haploid origin, females of diploid origin. Among bees haploidy of males and diploidy of females have been verified for various eusocial halictids (*Lasioglossum* spp., *Augochlorella,* and so on) and for *Bombus, Trigona, Melipona,* and *Apis,* as well as for certain solitary forms.

A working hypothesis of sex determination in Hymenoptera, placing it within the purview of the dosage or balance theory of sex determination, was originally put forward by Cunha and Kerr and has been more recently explained by Rothenbuhler, Kulinčević, and Kerr (1968) and Kerr (1969). According to this hypothesis, sex determination in ancestral Hymenoptera and perhaps many contemporary ones depends upon almost completely additive female-determining genes (x) at one or more loci and nonadditive or scarcely additive male-determining genes (y) at different loci. The effects we will call FD (female determination) and MD (male determination), respectively. As to diploids, the effect of $2(x)$ is 2FD, while the effect of either y or $2(y)$ is merely MD. The sex-determining values stand in the following relationship: 2FD > MD > FD. A genic equivalent would be $xx > yy$, $yy = y$, and $y > x$. These expressions show that diploids become females and that haploids become males.

In *Apis* (and also the braconid wasp, *Bracon*) there are multiple x alleles at a single major female-determining locus (X); they have lost their additivity unless heterozygous. They are designated x^1, x^2, etc. Different genic combinations and their phenotypic results are shown below:

$$x^1x^1yy \text{ gives FD} < \text{MD} = ♂$$
$$x^2x^2yy \text{ gives FD} < \text{MD} = ♂$$
$$x^1y \text{ gives FD} < \text{MD} = ♂$$
$$x^1x^2yy \text{ gives 2FD} > \text{MD} = ♀$$

Diploid maleness in *Apis* (first two lines above) is lethal by an interesting social mechanism described in section A2 below. Kerr and Nielsen (1967) found that young homozygous diploid *Apis* larvae obtained by inbreeding, although males, had some female-like characters and considered that this supported the genic balance theory summarized above.

Evidence for the existence of only one major X locus in *Apis* is that, in a cross between a queen (for example, x^1x^2) and her brother (say, x^1), 50 percent mortality of the larvae occurs, presumably because of homozygosity (x^1x^1). In ordinary *Apis* populations such deleterious results of inbreeding are reduced by behavior that favors outbreeding, by multiple matings of queens, and by multiple alleles at the X locus. There were about 12 such alleles (x^1 to x^{12}) in a population studied by Laidlaw, Gomes, and Kerr (1956).

In *Melipona, Meliponula, Trigona,* and perhaps most bees, brother-sister mating results in no such high mortality. Kerr (1969) says for *Meliponula bocandei* and *T. xanthotricha* that "in no case was mortality in inbreds greater than 11.4 percent" (quotation corrected in personal communication from W. E. Kerr). There are believed to be at least two femaleness loci, X_a and X_b, in such forms. In other words, instead of multiple alleles at one locus as in *Apis,* these bees have more than one major femaleness locus. The sex determination is otherwise as in the general hymenopteran case described above and not as in *Apis.* In *Melipona* each locus has two alleles. Unlike *Apis,* the homozygotes in *Melipona,* even double homozygotes (for instance, $x_a^1\ x_a^1,\ x_b^1\ x_b^1$), do not become diploid males but develop into females. As explained in Chapter 10 (C3), this system is coordinated with caste determination in *Melipona,* only the double heterozygotes (such as $x_a^1\ x_a^2,\ x_b^1\ x_b^2$) being potential gynes.

Occasionally (usually one percent or less) in most races of *Apis* and regularly in *A. m. capensis,* females instead of males develop parthenogenetically from unfertilized eggs (Anderson, 1963). Thus in the subspecies *capensis* from southernmost Africa, if the queen of a colony is lost, and if the workers fail to make a new queen from one of the old queen's eggs or young larvae, the colony nonetheless need not die out, for female-producing eggs can be laid by unfertilized workers. This thelytokous parthenogenesis is automictic (Tucker, 1958), that is, meiosis and recombinations occur. The union of two products of meiosis must result in heterozygosity of the *x* alleles to form such females.

In *Apis,* and presumably in all bees, the fertilized eggs are different from the unfertilized ones in numerous ways from early in development. Reinhardt (1960) reported differences in the thickness of the chorion, in the yolk material, and other variables at an age of one hour, but Du Praw (1967) indicates that some of these supposed differences are within the range found among female eggs. Sex differentiation, however, does involve various parts of the egg and begins at a very early stage. Beig (1972) reports that male eggs of *Trigona postica* are larger than female eggs when laid. Perhaps large eggs do not get fertilized. In this case sex determination precedes laying, but in *Apis* sex determination must occur at about the time of laying, when female eggs are fertilized.

2. Behavioral control of sex

At mating, the spermatheca of the female is filled with sperm cells which can survive as long as the female bee herself lives. She releases these cells, a few at a time, as the eggs are laid if they are to be fertilized; she has a muscular mechanism (for details see Snodgrass, 1956) which controls egress of the sperm cells from the spermatheca. Fertilization of the eggs and hence sex of the progeny therefore is under nervous control by the mother bee unless she is unmated.

In view of this arrangement, in these insects control of the sex of the offspring and of the sex ratio are subject to evolution to a degree ordinarily impossible, and many aspects of hymenopteran evolution have been dependent upon such control. For example, among some parasitic Hymenoptera, fertilized female-producing eggs are laid in large individuals of the host and unfertilized, male-producing eggs are laid in small individuals. Since females are typically larger than males and presumably require more nutrition, it is reasonable that female-producing eggs should be placed in large hosts. The sex ratio depends on the relative frequency of hosts of different sizes.

In bees, the factors that determine whether a given egg is to be fertilized or not are less clear. Great variation in frequency of males may occur in different parts of the range of a single species, suggesting a rather subtle sex control mechanism subject to selection according to local environmental conditions. An example of such variation is found in the halictine bee *Lasioglossum (Dialictus) imitatum,* which produces almost no males among its early season progeny in Kansas but about 45 percent males in Ontario (Michener and Wille, 1961; Knerer and Plateaux-Quénu, 1967c).

It seems clear that conditions external to the queen herself commonly influence the sex of the progeny which she produces. Such an influence was illustrated by Knerer and Plateaux-Quénu (1967d) for early stages in the reproductive lives of halictine gynes under artificial conditions of short day length and high temperature. Thus *Lasioglossum cinctipes,* which in the field produces a first brood consisting of workers only, produced under these laboratory conditions four times as many males as females. In *Halictus ligatus* and *scabiosae,* whose first broods would also normally be workers, there was an influence not only from the unusual light cycle and temperature relations but from the number of bees in a nest. First broods from lone gynes in captivity consisted mostly of males while polygynous colonies produced mostly females. One lone female of *H. scabiosae* made five cells with small food stores, all of which produced males. Then visits to the nest by another female, which never carried pollen, began; the next seven cells made and provisioned by the first female had larger food stores and produced females. Many more data are needed, but the suggestion is that presence of a second female influenced both cell provisioning and egg fertilization by the first female. Such data indicate the possible importance of both physical and social conditions in sex determination of bees.

Presumably it is true that workers of *Apis* ordinarily determine the sex of their queen's offspring by the size of cells that they make (see Chapter 7, A and Figure 7.2). It has been supposed that either a small (worker) cell snugly fitting the queen's abdomen or a large gyne cell cup will stimulate the queen, nervously, to release sperm and produce a female egg. On the contrary, the large but ordinary shaped (hexagonal) male or drone cells do not have this influence, and the queen therefore lays unfertilized eggs in them. Koeniger (1969, 1970), however, reports that it is with her front legs that the queen "inspects" cells, an activity shown in Figure 11.1, and that interfering with both of them by amputation or by attachment of pieces of paper to them results in deposition of fertilized eggs in most male cells. One free, intact front leg is enough to almost eliminate such errors.

As in many other matters, however, bees behave flexibly. For example, Jordan reported that in an *Apis* hive artificially provided only with male cells, and thus without worker cells, the queen laid fertilized eggs in male cells (see Ribbands, 1953). All that can be said is that just as we do not know all the stimuli that cause workers to make male instead of worker comb, so also we do not know all of the factors that cause the queen to lay fertilized or unfertilized eggs.

Butler (1954a) reports that as a queen moves about in the brood area of her colony, if she traverses patches of worker and drone comb, laying eggs apparently indiscriminately in cells of both sizes, the progeny developing in the larger cells will become males, those in the smaller ones, workers. He watched a queen passing from worker cells to male cells and back again as she moved over a comb containing patches of male cells; she did not hesitate in any way when she moved from one cell type to the other. Although she was laying eggs in quick succession, she was able to control accurately the fertilization and hence the sex of the offspring for about 50 workers and 50 males that were reared from the mixed comb.

Reinhardt (1960), in contrast to Butler, reports that the queen pauses from 15 to 30 minutes in changing from laying fertilized to unfertilized eggs or vice versa. It is not obvious how to relate the two different findings.

Queens of *Apis* sometimes attain laying rates of six eggs per minute; averages of between one and two per minute over long periods are common. Since several sperm cells must presumably be released (or not released) from the spermatheca for each egg laid, it is difficult to see how the associated structures could be promptly cleaned of live sperm cells so that a subsequent egg would pass unfertilized. Possibly all sperm cells cling to the egg or some agent promptly kills those that do not.

In *Apis* there are supplementary mechanisms for reinforcing the usual sort of sex determination. In artificially inbred colonies about 50 percent of the fertilized eggs (laid in worker cells) become diploid males because of homozygosity (section A1 above). In a colony of bees, however, these male larvae never mature. This is not because of any innate failure, for such larvae can be reared artificially, nor is it because of male larvae being in worker cells. It is because of social action. The workers somehow recognize and eat the larvae of diploid males when they are about six hours old (Woyke, 1963, 1965; Woyke and Knytel, 1966). Woyke (1967) gives evidence for a substance secreted by such larvae that causes the cannibalism. Thus there appears to be a behavioral mechanism which insures early elimination of diploid males and thus prevents wastage of food and space in the colony. Diploid males are nearly sterile; they produce little semen, the average volume of their testes being only

10 percent of the volume of haploid males' testes (Woyke, 1973). (Genetic study does show, however, that the few sperm cells produced are viable and contain not only genes of maternal origin but also genes of paternal origin, such as are never found in sperm of normal haploid males—Woyke and Adamska, 1972.)

3. Sources of males

Unfertilized eggs can have two origins. They can be laid by unfertilized females; in social bees this frequently means by workers. Alternatively, they can be laid by fertilized females that withhold sperm cells; this usually means they are laid by queens.

There is no evidence that the production of males (along with young gynes) late in the life of a colony of primitively eusocial bees is a result of near exhaustion of the supply of sperm cells in the spermatheca of the queen. In fact, dissections show that old queens ordinarily still have an abundant supply of sperm cells even though they presumably have not mated since early in adult life.

This observation seems to be supported by findings such as the following. In *Bombus,* when worker production ceases, males are usually produced, and they are followed by a return to female production, these becoming gynes. If the male eggs are laid by the queen, this laying sequence is further evidence that she does not lay unfertilized eggs merely because she has run out of sperm cells. In primitively eusocial halictines, support for the same view is found in the production of males at the same season and in the same parts of the same nests as females.

The sources of the unfertilized eggs, however, are not fully known, and in many cases they may not come from the queens. This is a matter of great interest, but probably does not contradict the view that old queens ordinarily still carry an adequate supply of sperm cells. In many semisocial and primitively eusocial species, certain unfertilized workers can be found with some ovarian enlargement and with one or more ovarioles each apparently producing an egg that is large enough to be laid. Batra (1968) verified egg laying by such individuals of the subgenus *Dialictus,* even though they had been foraging as workers on the same day. Noll (1931) supposed the origin of most males of *Lasioglossum malachurum* to be from unfertilized workers. His judgment was based on few data, mainly on the observation that in a dequeened colony the production of males (the only progeny) was approximately the same as the male production of a normal colony containing a queen. He therefore assumed that the normal source of male-producing eggs was the workers. Ovarian development is rare in the first brood of workers of this species, but occurs in some 25 percent of the second brood (Bonelli, 1948), to judge by a sample of only 32 dissected. In *Lasioglossum marginatum,* also, much of the male production appears to be by unfertilized workers (Plateaux-Quénu, 1959, 1962). In *Bombus atratus* eggs laid by workers produce about 90 percent of the males (Zucchi, 1966, and personal communication).

Workers of highly eusocial bees do not mate. They are physically unable to do so and do not attract males. (In most other bees all females or at least all young females probably can mate. This is true even for *Bombus,* where mated workers or extraordinarily small queens are known.)

Males of the Meliponini are produced in cells that do not differ from those in which workers are produced and are intermingled among those in which workers develop (Figure 10.6). Egg laying by meliponine workers is a widespread phenomenon. Some of the eggs laid by workers are trophic, that is, they serve as food for the queen. In certain groups of *Trigona* trophic eggs are large, subspherical, and incapable of development. They are laid before the queen's egg for any given cell (Chapter 16, D2). Other eggs laid by workers are of the usual elongate form and are laid after the queen's egg, so that the cell is closed up with two or occasionally more eggs in it (Beig, 1972). More than one egg per cell has been reported in *Melipona* and in *Trigona* of the diverse subgenera *Plebeia, Hypotrigona,* and *Scaptotrigona.* For *T. postica* the worker-laid eggs average 1.31 mm in length, compared to 1.19 for female-producing eggs laid by queens, the presumably haploid, male-producing eggs from workers being significantly larger than the diploid, female-producing eggs. Within three days after hatching, the larger and more active male larva bites and kills the female larva, so that, at least in *T. postica* studied by Beig, a cell with two (or more) eggs always gives rise to one adult male.

Sometimes the queen may lay a large, male-producing egg, either by itself in a cell or immediately after laying a smaller, female-producing egg in the cell. In either case a male bee is produced. Dr. W. E. Kerr has stated (personal communication in connection with Beig's work) that 95 percent of the males of *Trigona postica* are progeny of workers and that the same is true of certain species of *Melipona*. Beig himself is less specific, but clearly most males do originate from workers' eggs. The percentage of cells with two or more eggs varied both seasonally and among colonies of *T. postica*—4.5 to 45.5 percent in different colonies in the same month. The factors that lead to the appropriate numbers of males being produced are unknown.

In *Apis*, as noted in section 2 above, the situation is very different, in that most males arise from unfertilized eggs laid by the queen.

B. Sex ratios

Among bees, sex ratio is subject to much interspecific variability even in the nonsocial forms (Table 8.1). Evolution of sex ratio is, of course, bound up with the social evolution in Hymenopteran societies where workers are females and produced in greater numbers and at different seasons than males. Various allodapine bees, however, range from solitary to primitively eusocial in small steps. Sex ratios of many species have been recorded (Table 8.2). Unfortunately, the ratios merely seem erratic, not related in a recognizable way to level of social attainment.

One may assume evolution of both increasing and decreasing sociality. A decreasing percentage of males should accompany increasing sociality; if the sociality then decreases or vanishes, the scarcity of males may continue for a time, leading to some of the erratic relations between sociality and sex ratio that can be noted in Tables 8.1 and 8.2.

In the highly eusocial bees, which have large numbers of workers at all times and produce them more or less continually, the percentage of males maturing is always small (except in the case of progeny of unmated queens or laying workers). In *Apis*, males are produced in comb having cells larger than worker cells. An excess of such comb results in an increased male production but not in an indefinite increase, some other factors serving to limit excessive increase in male production (Allen, 1965a). Weiss (1962) says that in undisturbed colonies of *Apis mellifera* about 13 percent of the brood cells are for males, although only under favorable circumstances are so many

TABLE 8.1. Percentage of males in samples of pupae of some solitary, communal, and semisocial bees.[a]

Species	Percent ♂	Number of adults emerged or pupae examined (and reference)
Solitary:		
Anthophora flexipes	34	200 (Torchio and Youssef, 1968)
Euplusia surinamensis	59	297 (D. H. Janzen, in preparation)
Augochlora pura	27	550 (Stockhammer, 1966)
Neocorynura fumipennis	38	141 (Michener, Kerfoot, and Ramírez, 1966)
Ceratina smaragdula (spring)	25	65 (Kapil and Kumar, 1969)
Ceratina smaragdula (summer and autumn)	47	250 (Kapil and Kumar, 1969)
Osmia excavata	63	2820 (Hirashima, 1958)
Communal:		
Pseudagapostemon divaricatus	62	222 (Michener and Lange, 1958c)
Semisocial:		
Augochloropsis sparsilis	43	106 (Michener and Lange, 1959)
Pseudaugochloropsis graminea	20	31 (Michener and Kerfoot, 1967)
Pseudaugochloropsis nigerrima	43	37 (Michener and Kerfoot, 1967)
Social parasite:		
Psithyrus variabilis	47	72 (Webb, 1961)

[a]Because of differing habits of the sexes, field samples of adults do not accurately reflect sex ratio of production. Even samples of pupae may be deceptive unless proterandry is considered.

TABLE 8.2. Percentage of males in samples of pupae of allodapine bees.

Species	Percent ♂	No. of pupae	Percent of nests with multiple ♀♀
Halterapis nigrinervis group	11.1	27	0
Exoneurella lawsoni (summer)	36.0	114	< 10
E. lawsoni (autumn)	65.5	29	< 10
Allodape mucronata	54.8	73	< 10
A. panurgoides and *ceratinoides*	29.0	107	6–15
A. friesei	41.4	29	? 10
A. rufogastra	29.1	79	? 10
A. exoloma	12.5	64	? 10
Allodapula acutigera	37.0	165	> 10
A. dichroa	42.3	137	< 5
A. variegata	30.5	59	10
A. melanopus and *turneri*	37.8	172	< 25
Braunsapis facialis	30.0	180	> 40
B. bouyssoui	24.4	45	? 40
B. leptozonia (Natal)	14.1	71	> 40
B. foveata (Natal)	6.4	47	? 40
B. foveata and *leptozonia* (Cameroon)	23.1	65	40
B. luapulana	41.0	39	? 40
B. draconis	55.9	34	? 40
B. stuckenbergorum	16.0	25	? 40
B. simplicipes	40.0	60	? 40
Exoneura variabilis	40.1	207	30
E. hamulata	42.0	93	> 35

SOURCE: Michener, 1971.
Note: Genera are arranged approximately in the order of increasing sociality, as indicated by the frequency of workers. The last column, designed to give some idea of the social level of each species, represents the percentage of nests that contain brood and are inhabited by two or more mature, working females. Young adults that might soon disperse were ignored in preparing these figures. Question marks precede figures that are dubious because of insufficient data.

males reared. For *Apis florea* Thakar and Tonapi (1962) say that in a good season as much as 25 percent of the whole brood area may be for male cells.

Various authors (see Hartl and Brown, 1970) have shown that the sex ratio of animals which like the bees have haplodiploid genetic systems should approach 1 : 1. This refers, of course, to reproductive females, that is, queens, in relation to the number of males. Tables 8.1, 8.2, and 8.3 are based on complete sex ratios, both workers and gynes (for the social species) being counted as females. However, if for the social species, all females that are to become workers could be excluded, a preponderance of males would often become evident.

In the highly eusocial bees, a male-biased sex ratio is even more evident when workers are excluded. Except in the genus *Melipona* the number of males produced per young gyne is enormous in highly eusocial bees. Possibly this serves as insurance that the large amount of energy utilized to produce a gyne will not be wasted for lack of mating. This suggestion is possibly convincing for *Trigona* and its relatives, in which males are small, like workers. It is less impressive for *Apis,* in which males are nearly as large as gynes, but in this genus a gyne mates several times, males only once each.

The real explanation for the discrepancy between the predicted 1 : 1 sex ratios and the actual male-biased ratios of the reproductives may be that it is not legitimate to exclude workers as nonreproductives. As explained elsewhere (section A3 above and Chapter 20), workers in many Hymenoptera are responsible for much male production. For most social bees, if each worker were considered as having a realistic fraction of the reproductivity of a gyne, it is likely that a sex ratio of 1 : 1 would be approached.

C. Seasonal variation in sex ratio

In most primitively eusocial bees (and perhaps solitary ones as well, if there are several generations per year, see Stockhammer, 1966) the sex ratio varies with the season (Table 8.3 and Figure 8.1). During early development of the colony, which in temperate regions ordinarily means during spring and the first half or more of summer, a high percentage of the progeny are females which become workers. Late in the summer males are produced in large numbers along with young gynes. For the Halictini listed in Table 8.3, as well as for several other Halictini on which sparse data have been published (Knerer and Plateaux-Quénu, 1967d; Plateaux-Quénu, 1967), two groups are evident for the early season (spring or early summer) progeny, that is, for the first brood. In some of the species 40 to 53 percent of the individuals are males, while in the others 0 to 10 percent are males. Species in the first group are those in which queens are short-lived and must be replaced during the summer by new gynes, which of course must mate; hence the substantial male production. Queens in the second group usually live about as long as the colony, and produce almost entirely worker bees in spring and the first half of summer. Another way to make this statement is to say that some of

TABLE 8.3. Percentage of males in samples of pupae of some primitively eusocial bees.

	Percent (N)			
Species	First brood	Midsummer	Autumn	Reference
Lasioglossum rhytidophorum	41(17)	12(90)	38(83)	Michener and Lange, 1958b
L. imitatum	0[a](53)	15(296)	48(40)	Michener and Wille, 1961
L. versatum	5(20)	27(200)	11[b](95)	Michener, 1966b
L. zephyrum	4(56)	25(491)	30(119)	Batra, 1966b
L. marginatum[c]	0	0	70(3834)	Table 25.2
L. duplex	10	None[d]	56(368)	Sakagami and Hayashida, 1961, 1968
Augochlorella spp.	19(73)	31(26)	44(120)	Ordway, 1966
Exoneura variabilis	None[d]	40(207)	Few	Michener, 1965b
Exoneurella lawsoni	None[d]	33(70)	50(73)	Michener, 1964c
Bombus americanorum	0	0	60(1780)[e]	Webb, 1961

Note: Sample sizes are indicated in parentheses.

[a] Knerer and Plateaux-Quénu (1967c) report that about 45 percent are males in the first brood in Ontario, but do not indicate the number of cells examined. Data above are from Kansas.

[b] Data probably taken too late, after emergence of many males.

[c] Unlike other species, which have annual colonies, the colonies last for five or six years. Figures in the "autumn" column are for the last year of colonial life, in the "midsummer" column, for intermediate years.

[d] No middle brood produced by *L. duplex;* few or no young produced until summer by *Exoneura* and *Exoneurella.*

[e] These figures are based on total queen and male production, worker production arbitrarily being considered as occurring in summer. For four other species of *Bombus* comparable figures ranged from 53 to 62 percent. Free and Butler (1959) support Sladen's estimate of about twice as many males as queens, or 66.6 percent males, for *Bombus* in general.

the females produced in spring and early summer by species in the first group become queens, while they all become workers in species of the second group. It is by no means clear why there should be two discrete groups of this sort without intermediates. Indeed it is probable that further study will uncover species which are intermediate.

For some other groups of bees, such as *Bombus* (Table 8.3), Meliponini, and *Apis,* males are produced seasonally, but discussion of the matter under this heading does not seem particularly useful.

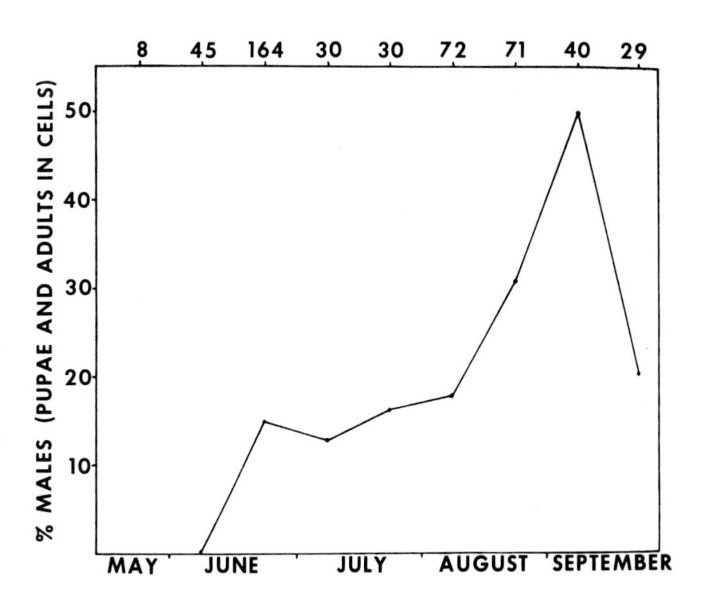

FIGURE 8.1. Graph showing the mean percentage of males of *Lasioglossum* (*Dialictus*) *imitatum* produced at various times of year at Lawrence, Kansas. The number of pupae (and adults not yet emerged from their cells) examined for each point on the graph is indicated at the top of the graph. Great variations in the sex ratio among pupae often occur from place to place or nest to nest. The drop in percentage of males in late September is probably a result of early emergence of males, leaving largely females behind. (From Michener and Wille, 1961.)

9 *Caste Differences*

As explained in Chapter 4 (B), castes in bees are physiologically, behaviorally, and sometimes morphologically different forms (not age groups) of females occurring together in one colony. Solitary, communal, quasisocial, and subsocial bees have no castes. Semisocial and eusocial bees by definition have at least incipient castes. They are queen and worker. This chapter deals with differences in size, structure, physiology, and behavior between these castes in the various groups of bees.

Degree of caste difference varies with the completeness of division of labor and is therefore the easiest measure of social level among eusocial bees. Morphological caste difference is rather closely related to behavioral and physiological differences, so that when the castes differ in size or structure, the amount of such difference is a reasonably good although not infallible measure of social level.

In most semisocial and primitively eusocial bees the castes are approximately equally modified from the all-purpose female of the solitary bees. In highly eusocial forms, however, the worker is more like the ancestral form than is the queen in caste characteristics. The greater differentiation of the queen in highly eusocial forms is not so much because she does all or most of the egg laying and has an augmented laying capacity, as because she has lost the foraging behavior of her ancestors and the associated structures for collecting, carrying, and manipulating food and other materials. The worker, on the other hand, has partly or largely lost reproductive capacity and makes less direct genetic contribution to the next generation than the queen; in some forms like *Apis* she commonly makes none. The queen lives much longer than do females of comparable solitary bees, while the workers live for relatively brief periods, perhaps often shorter than the adult lives of related solitary forms.

As indicated above, the queen may or may not have lost her capacity, both morphologically and behaviorally, to forage and to construct a nest and to do other such work. In primitively eusocial forms in which she engages in such activities during part of her life, she may, while alone, make as large a nest and provision as many cells as do female solitary bees. Workers, however, devote themselves largely to foraging and nest building; more worker attention goes to such work than in comparable female solitary bees. These and some other widespread caste differences are indicated in Table 9.1. Many of the differences are taken up below, in more detail, for various groups of bees.

A. Caste differences within semisocial colonies

In *Augochloropsis sparsilis* there is division of labor among fertilized females, usually only one per colony having enlarged ovaries at a given time. This species lives in small colonies which contain up to four active females in summer. Most of the foraging is done by fertilized females with slender ovaries, most of which are probably permanently workers, although some might later experience ovarian development. However, occasional females, also with slender ovaries, are unfertilized. The number of such individuals ranges from one out of 86 examined in the first generation to between 15 and 20 percent in the second annual generation. Thus there are both fertilized and unfertilized workers or workerlike individuals

TABLE 9.1. Some common differences between queens and workers in bees.

Queen	Worker
Augmented laying capacity	Reduced laying capacity
Mates	Commonly does not mate
Long life (often with hibernation*)	Short life, no hibernation
Foraging activities sometimes reduced or lacking, at least after maturation of first young. Morphological correlates of lack of foraging sometimes evident.	Foraging activities strong
Nest building activities often reduced or lacking	Nest building activities well developed
Defense of colony and nest often little evident	Defense of colony and nest well developed
Size commonly larger	Size commonly smaller
In seasonal climates, commonly develops large fat bodies, but ovaries remain slender in autumn*	Develops no extensive fat bodies
Ovarian development often occurs only after diapause*	Ovarian development (if it occurs at all) can occur without diapause
Consumes more proteinaceous food	Consumes less proteinaceous food; crop rarely full of pollen
Socially dominant, either by means of pheromones or behavior	Socially not dominant

*Item not applicable to highly eusocial bees.

in the population of the species. In view of the small colony size, however, it is obvious that many colonies altogether lack unfertilized individuals. The unfertilized bees wear themselves out, their mandibles becoming worn and their wings tattered, at a season when their fertilized sisters (whether workers or egg layers) are still quite fresh looking (Michener and Lange, 1958d, 1959). A suggestive (but for the limited series studied not statistically significant) mean size difference exists among the three groups of females: unfertilized workers $<$ fertilized workers $<$ bees with enlarged ovaries.

In the semisocial colonies of *Pseudaugochloropsis,* which contain from two to seven females, there may be no workers or as many as half the females may be unfertilized workers. For example, one colony of *P. costaricensis* contained three egg layers or queens, three workers, and one additional unfertilized but presumably young adult female (Michener and Kerfoot, 1967). There are also occasional unfertilized bees with enlarged ovaries. There is a mean size difference between egg layers and workers, the latter averaging smaller, but only slightly so. Workers are commonly found on flowers and carrying pollen into the nests. The extent of division of foraging labor in these bees is not known, although there are certainly differences in egg laying and mating behavior between the castes. Egg-laying bees of this genus are commonly found collecting food on flowers with workers, but this would not be surprising even if the egg layers in a colony never forage, since a majority of the nests are inhabited by lone females and such solitary bees have to forage as well as lay eggs.

Differences in ovarian development, average size, foraging behavior, and length of life are shown by Knerer and Plateaux-Quénu (1966a) among gynes in temporarily polygynous halictine nests. These are temporarily semisocial spring colonies, in which some of the overwintered founding gynes, normally all but one in a colony, become workers (auxiliaries) and show the expected workerlike features of not enlarged or but little enlarged ovaries, small average size,[1] short lives (unless the auxiliaries later leave to establish their own nests), and foraging behavior. In species where much of the guarding is done by queens, the auxiliaries do not guard much but forage instead. This was noted, for example, in *Lasioglossum calceatum* (Bonelli, 1968), and appears to be true of certain species studied by Knerer and Plateaux-Quénu.

Spring colonies of *Augochlorella* studied in Kansas, when semisocial, show the interesting features of no rec-

1. That the largest gyne has the most enlarged ovaries, stays in the nest, and becomes the queen is indicated for at least three species by Knerer and Plateaux-Quénu (1966a). Statistical data (histograms) are given, unfortunately, for only one of them, *Lasioglossum pauxillum.* These data actually show that the average size of queens is about the same as (indeed slightly less than) that of auxilliaries, and that the largest females found were auxiliaries. The connection between the size of a gyne and its fate in a semisocial group therefore remains to be established. Data that are probably pertinent, based on a laboratory study of orphaned groups of workers of *Lasioglossum zephyrum,* are presented in Chapter 11 (C1).

ognized differences in foraging behavior among the bees, no caste-correlated size differences, and sometimes more than one egg layer (Ordway, 1965). Yet there appears to be division of labor as indicated by variation in ovarian development. There is evidence, although equivocal, that auxiliaries die earlier than do egg layers. The same statements apply to *Lasioglossum versatum* (Michener, 1966b).

B. Caste differences in primitively eusocial colonies

Like those of semisocial bees, castes of primitively eusocial bees commonly are difficult to distinguish. In most species the queens average larger than workers, and because of different growth rates of the parts of the body (allometric growth), the proportions may differ also. The average size, however, is not always different, and in halictine and allodapine bees the difference is typically slight, the sizes of the castes broadly overlapping. In bumblebees and a few halictines the castes differ more strikingly in size, although not always discontinuously. All females emerge from the pupa in a similar condition, with slender ovaries, and of course they are unfertilized and unworn. Those that become queens are distinguishable after a time by the enlarged ovaries and by the presence of sperm cells in the spermatheca. those that have become workers can be recognized ultimately by their still slender ovaries and usually by a lack of sperm cells in the spermatheca, combined with evidences of age such as worn mandibles (Figure 9.1) and wing margins (Figure 9.2). An unworn bee that is unmated and has slender ovaries cannot be placed as to caste unless it happens to fall in the size range that is diagnostic for one caste or the other.

Ovaries that have been enlarged can later, at a time when reproductive activity has temporarily or permanently ceased, shrink and become again quite slender. This is shown for *Lasioglossum marginatum* in Figure 9.8. Such ovaries can be recognized by their irregular shape and the stretched calyces or lateral oviducts. They do not have the slender, uniformly tapering form found in young bees. Thus a worn bee with sperm cells in the spermatheca can confidently be called a queen even when her ovaries are slender if they show these signs of regression.

Differences between castes must be looked at in two ways, not wholly independent from one another. One may be concerned with (1) the size of the morphological or behavioral discontinuity (if any) between castes, or (2) the average degree of difference between them, that is, the difference between mean sizes of workers and of queens. That these ideas are not the same is shown by considering *Bombus,* in which the degree of difference is great in size, behavior, and physiology (longevity and the like), yet intermediates exist in many species. On the other hand, in *Halictus sexcinctus,* the degree of difference is minimal, yet there are no intermediates (to judge by a rather small study discussed below). Data are often not available to show whether behavioral or physiological discontinuities occur between castes. Size discontinuities or their absence can more often be demonstrated.

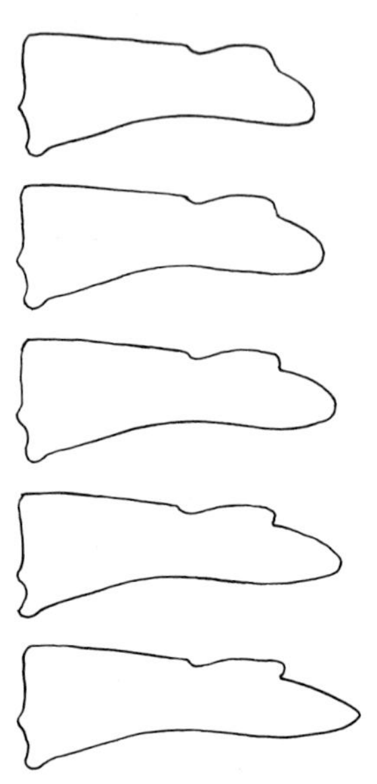

FIGURE 9.1. Mandibles of females of *Lasioglossum imitatum,* from badly worn (*above*) to fresh and unworn (*below*). (From Michener and Wille, 1961.)

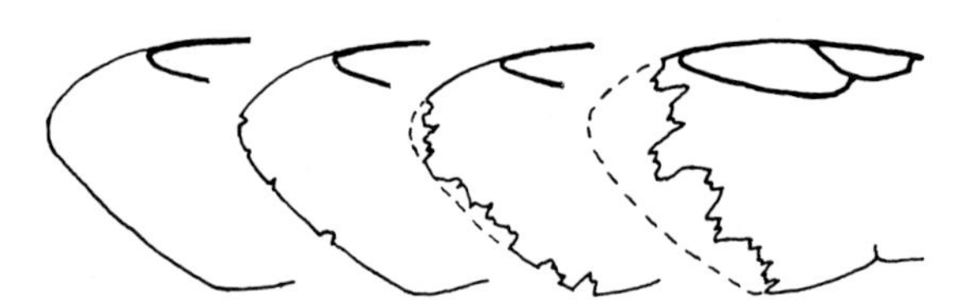

FIGURE 9.2. Apices of forewings of females of *Augochlorella* from unworn *at left* to very tattered *at right.* (From Ordway, 1965.)

In the many eusocial halictines and in *Bombus*, once a colony of adults is established, the queen leaves the nest infrequently or not at all. When she does leave, as noted sometimes in laboratory colonies of *Lasioglossum zephyrum*, she visits flowers for her own food but brings back no pollen load. In *Lasioglossum marginatum* the queen even loses her power of flight after her first brood of workers is produced.

Gynes are long-lived in comparison to workers. For example, gynes of *Lasioglossum nigripes* live 12 to 15 months, workers 3 to 5 weeks (Plateaux-Quénu, 1965a); gynes of *L. marginatum* live 5 to 6 years, workers 12 to 15 months (Plateaux-Quénu, 1960b, 1962); gynes of *L. malachurum, duplex,* and others live perhaps 11 to 13 months, workers 3 weeks or in midsummer even less (Bonelli, 1948; Sakagami and Hayashida, 1968). Surmises for *L. umbripenne* are gynes about a year, workers about a month; and for *L. imitatum,* gynes 11 months, workers 3 weeks. Gynes of *Bombus* in temperate climates live for a little over a year, while workers may survive for over 2 months. Brian (1952) found that half of the marked workers of *B. agrorum* were dead at an age of 3 weeks, although some lived over 60 days.

In *Bombus* and halictines of temperate and cold climates, overwintering is usually by fertilized young gynes only. Young gynes in autumn develop fat, and their ovaries do not enlarge, being seemingly in a reproductive diapause. In contrast, at the same season workers are nearly free of fat body and some ovarian enlargement occurs in many individuals, although often not reaching the point where eggs could be laid. Fertilization is not necessary for fat storage by young gynes. Falling temperature does not stimulate fat storage, at least not for all species, since gynes of some, such as *Bombus pratorum* and *bimaculatus,* enter winter quarters in midsummer.

The following sections provide fuller documentation for certain kinds of caste differences.

1. Halictinae

The number of workers per queen is usually higher than in semisocial halictines. While eusocial colonies containing one queen and one worker occur, such a condition is probably always a short-term one.

a. Relative sizes of castes. Among the best-studied caste differences are those of size. In most species allometry has not been studied, but at least it can be said that the size differences usually do not lead to noticeable differences in proportions. In the following material, size is indicated by a number of different measurements —head width, wing length, abdominal width—no one of which can give a truly adequate representation of general size. However, in view of the lack of striking allometry (except in *Halictus aerarius* and a few other species as discussed below), any one of the measurements used presumably reflects size differences between castes in roughly the same way.

Figures 9.3 to 9.5 show size relationships between workers and queens of species of *Lasioglossum* (and related genera in 9.5). In *L. marginatum* there is no significant caste size difference. In some species, like *L. rhytidophorum,* the mean sizes of the castes differ but there is such overlap that frequency curves for all females show no bimodality. In *L. pauxillum* a curve for all females is strongly bimodal and the castes are largely distinguishable by size alone; the same is true of *L. cinctipes* (see Figure 10.1). In one population of *L. umbripenne* the castes are completely separable by size, although in another population of the same species this is not the case. Finally, in *L. malachurum* the castes are completely separable by size. As can be seen from the captions of these figures, the sequence outlined above is in no sense a taxonomic one; the two subgenera *Dialictus* and *Evylaeus* are fully intermingled. Figure 9.3 shows frequency distributions of the size classes for four species. Differences in relative numbers of the two castes should be ignored in studying this figure, for the numbers of queens and workers collected may be heavily dependent upon the season, method of collecting, or other irrelevant factors. In order to provide additional details, data for *L. imitatum* have been provided in another style, as histograms showing the percentage of each caste or seasonal group in each size class (Figure 9.4). Unfortunately comparable data have never been published for one of the most interesting species, *L. malachurum.*

Knerer and Atwood (1966) and Knerer and Plateaux-Quénu (1966b) have attempted to improve size discrimination for the castes of halictines by using two measurements, wing length and abdominal width, and plotting size-frequency polygons for each caste (Figure 9.5). For most species this method added little to understanding or discrimination of castes, and unfortunately does not present data on the relative numbers of individuals of the various size classes within each caste. However, for a few species, such as *Halictus scabiosae* and *ligatus,* on

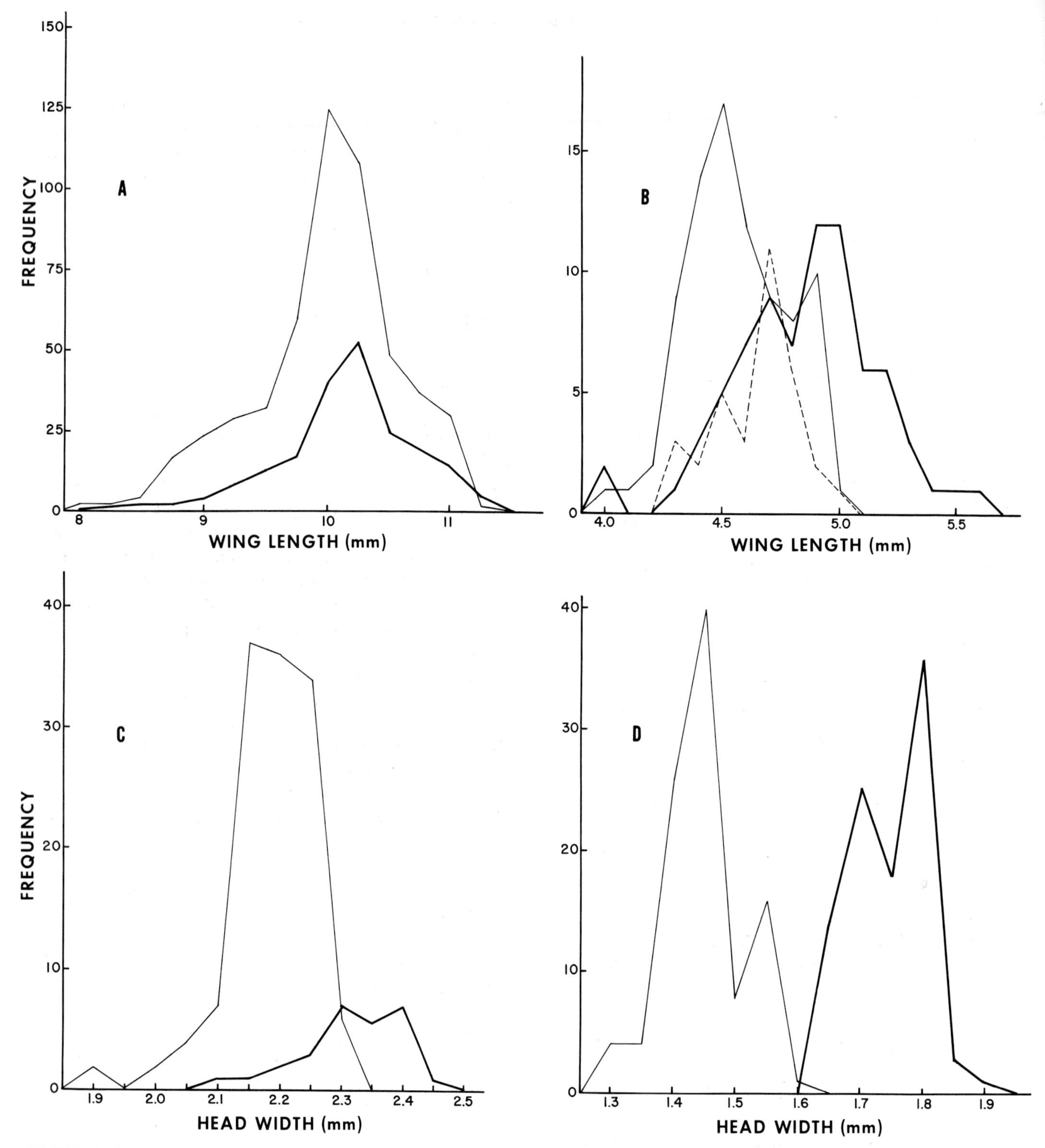

FIGURE 9.3. Frequency distributions showing sizes of workers and of gynes of four species of *Lasioglossum* (Halictidae). (*A*) *L.* (*Evylaeus*) *marginatum* (data from Plateaux-Quénu, 1959). (*B*) *L.* (*Dialictus*) *rhytidophorum* (data from Michener and Lange, 1958b). (*C*) *L.* (*Evylaeus*) *duplex* (data from Sakagami and Hayashida, 1968), (*D*) *L.* (*Dialictus*) *umbripenne,* population from Damitas, Costa Rica (data through the courtesy of A. Wille; see Wille and Orozco, 1970). *Heavy lines* represent gynes, *light lines,* workers. *Broken line* for *L. rhytidophorum* represents laying workers. (Original drawings by Barry Siler.)

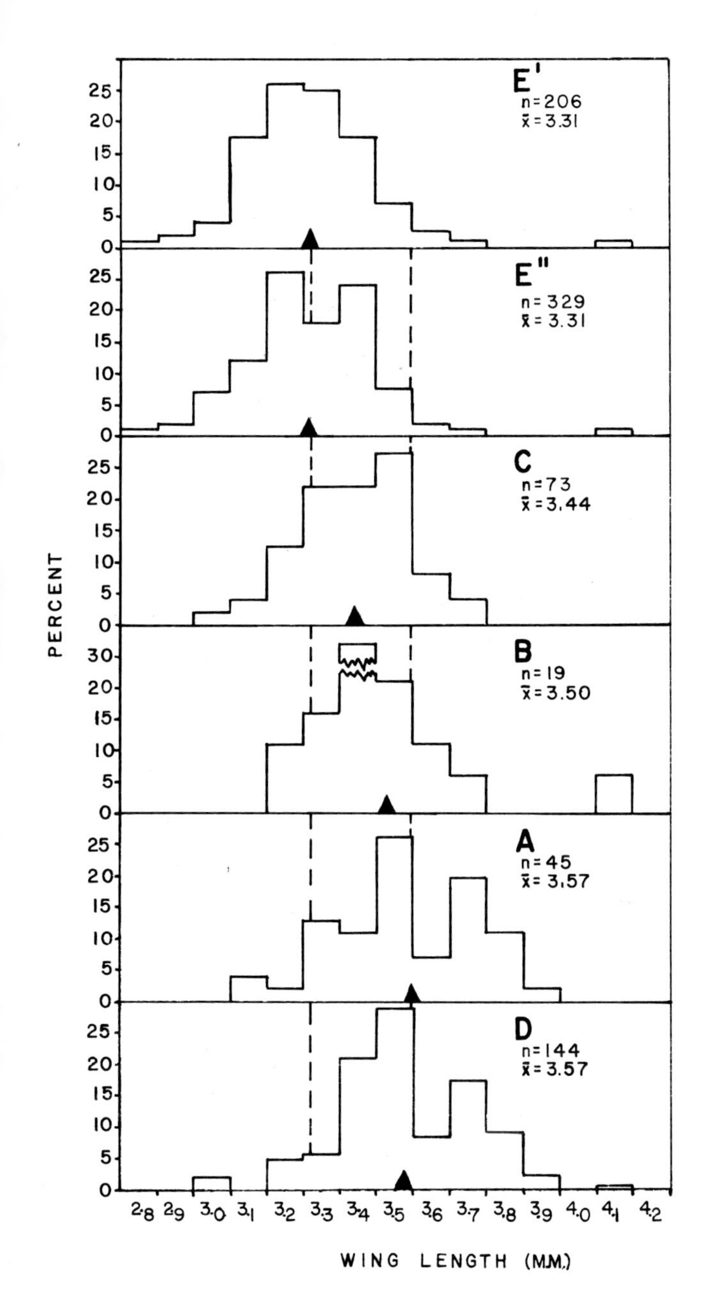

FIGURE 9.4. Histograms showing wing lengths of females of *Lasioglossum* (*Dialictus*) *imitatum*. *Black triangles* show means; *vertical broken lines* represent minimum and maximum means, and serve only to facilitate comparison of the histograms. *E'* and *E''*, workers with slender and very slender ovaries; *C*, workers with one or two enlarged oocytes, that is, laying workers; *B* and *A*, queens, *B* with ovaries less well developed than *A*; *D*, overwintering gynes. (From Michener and Wille, 1961.)

the basis of the data presented, caste discrimination is improved by the use of the two measurements. Obviously, morphometric studies using numerous measurements and multivariate techniques such as principal components and canonical analyses should lead to better insight into the morphometric caste differences and to better caste discrimination, but such studies have never been carried out on halictine bees.

In *Lasioglossum* (*Evylaeus*) *malachurum* (central Europe) workers are produced in two separated broods (early summer and midsummer) and gynes as well as males in a third brood (late summer). In southern Europe the number of worker broods is three (Bonelli, 1948). Each brood is separated by an interval from the next. Likewise, in some other species of the subgenus *Evylaeus*, such as *calceatum*, *cinctipes*, *duplex*, *nigripes*, and *pauxillum*, there is one brood of workers followed after an interval by a brood containing the young gynes, along with males. The difference between mean sizes of the castes in such forms is often considerable, as are behavioral differences, but in *L. duplex* it is only 6.7 percent, in the same range as for *nigripes*, compared to 13 to 18 percent for the other species listed (Sakagami and Hayashida, 1968).

Intermediates in size between workers and queens are numerous in *L. duplex*, from Japan, in which 12 to 15 percent of the workers are larger than their own mothers (Sakagami and Hayashida, 1968), and in this species a frequency distribution of sizes is not bimodal (Figure 9.3), in contrast to that of its close relatives in the subgenus *Evylaeus*. *L.* (*Dialictus*) *rohweri* in Ontario, where it is said to produce only one, temporally isolated, brood of workers, shows few size intermediates between castes, according to Knerer and Atwood. In Kansas, where the same species produces workers continuously over a longer season, merging into that of queen production, there is a broad overlap of sizes of the castes so that, as in many other eusocial *Dialictus*, measurements of all females show no bimodality. Likewise *L.* (*D.*) *umbripenne*, at a Costa Rican locality where the castes are produced at different seasons, shows discontinuous sizes with no caste intermediates (Figure 9.3). At another nearby but climatically different Costa Rican locality where queens and workers are produced at the same season, sizes of the castes are more similar and overlap broadly (Eickwort and Eickwort, 1971), more or less as in Figure 9.3 (C) for *Lasioglossum duplex*.

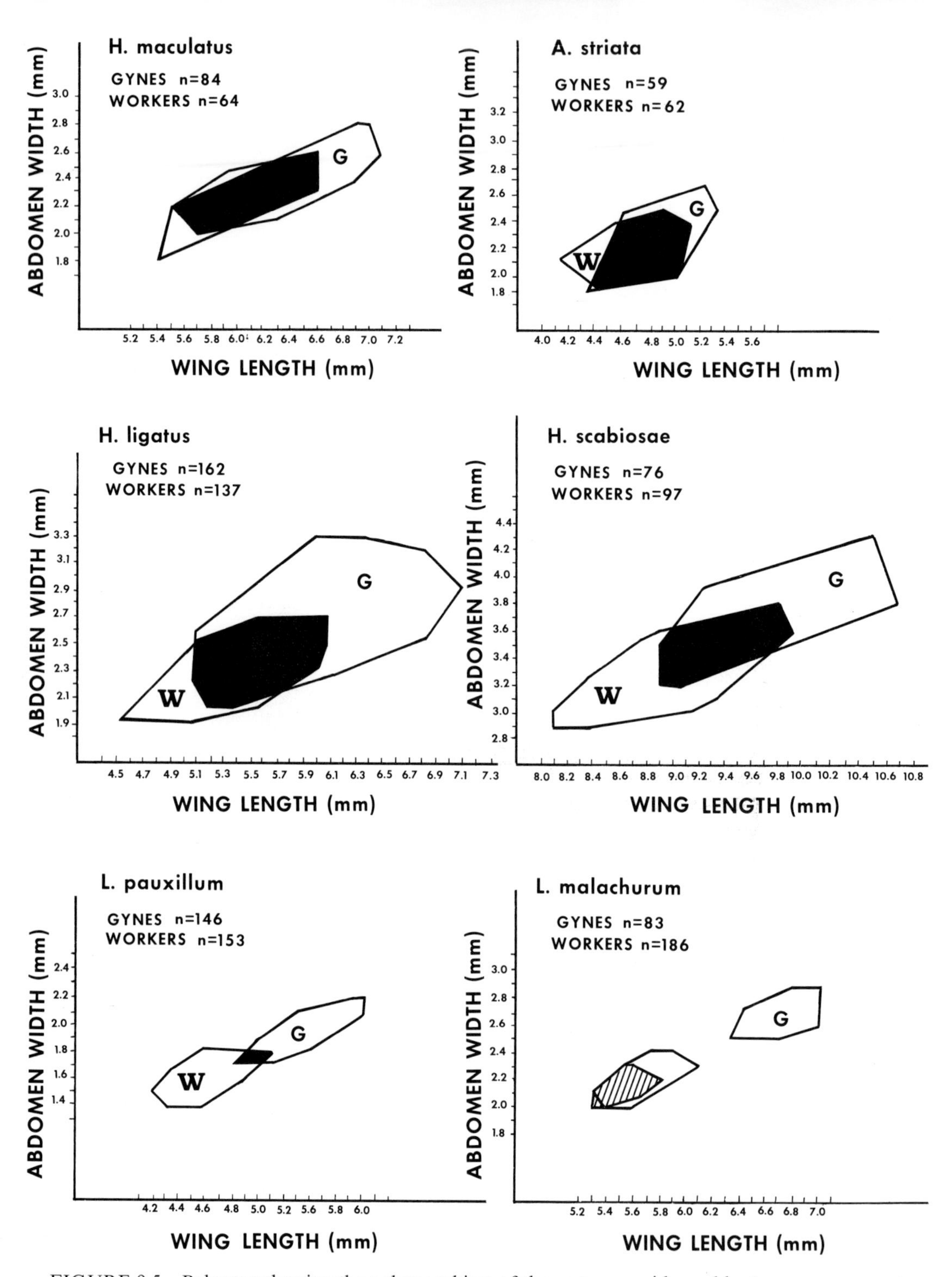

FIGURE 9.5. Polygons showing the polymorphism of the castes as evidenced by two measurements, wing length and abdominal width, in six halictid species, *Halictus maculatus, Augochlorella striata, Halictus ligatus* and *scabiosae,* and *Lasioglossum pauxillum* and *malachurum.* G = gynes; W = workers. *Overlapping polygons* represent workers of the last species, one for each of the two worker broods found in Central Europe and France. *Shaded polygon* is for the first worker brood; *clear polygon* is for the second. (From Knerer and Atwood, 1966, © American Association for the Advancement of Science, and Knerer and Plateaux-Quénu, 1966b.)

In summary, with some exceptions like *L. duplex,* castes intergrade less or not at all in species that produce gynes and workers in seasonally isolated broods. Castes intergrade in size in species and populations that produce gynes and workers in contiguous or overlapping seasons.

In certain halictids enormous differences in size among females have been found, with heads and jaws of the largest female relatively gigantic, as a result of allometry (Sakagami and Moure, 1965). No such species have been studied biologically in adequate detail; the best data are on *Halictus aerarius* in Japan (Sakagami and Fukushima, 1961), a species in which the larger individuals have darker mouthparts, legs and sterna as well as relatively large and distinctively shaped heads. Extreme differences in head size are shown in Figure 9.6, but all intermediates exist at least in some populations; most individuals are nearer the small extreme. Additional studies are needed, but dissections strongly suggest that the large, large-headed females are gynes, the small females usually workers, with the two castes connected by intermediates in spite of the great difference between them. However, some small females are also gynes, being able to mate, overwinter, and establish nests in the spring. Thus, while caste is on the average related to size in this species, as in many others, the relation is not invariable.

An inadequately known but perhaps similar case is that of *Augochlora* subgenus *Oxystoglosella* (Chapter 23, B). The large-headed females may be nest founders, all gynes, and almost discontinuous in size from the smaller headed bees. These gynes are short-lived and are replaced by their largest progeny, which, however, are smaller and have heads of ordinary size, broadly overlapping that of workers.

b. Behavioral and physiological differences between castes. We turn now to the physiological differences between castes, largely reflected by longevity and ovarian development, and behavioral differences noted by direct observation and by recording whether or not the bees have mated (spermathecal contents). When one dissects halictines, one usually finds, as indicated above, that queens have enlarged ovaries and have been fertilized. Workers usually show the opposite conditions, slender ovaries and a high frequency of unfertilized individuals. Ovarian differences are shown in Figure 9.7. In species that produce gynes along with workers, or immediately after workers with no interruption, much intermediacy in physiological caste attributes is usual, just as for size.

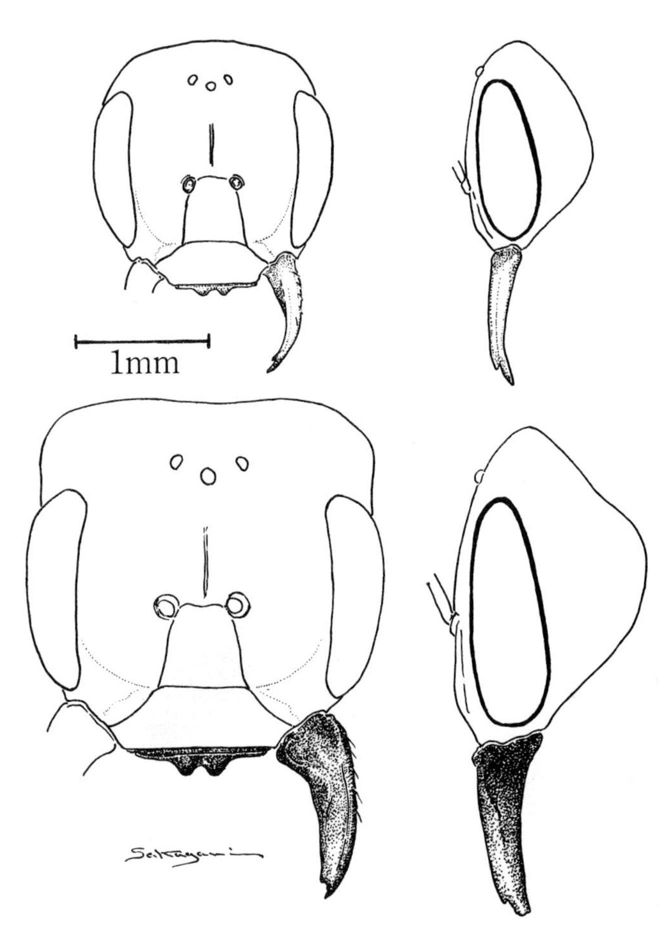

FIGURE 9.6. Front and lateral views of heads of two females of *Halictus aerarius* from the same nest. (From Sakagami and Fukushima, 1961; original drawing provided by S. F. Sakagami.)

In *Augochlorella striata* and *persimilis* (not *A. michaelis*), workers as well as queens are usually fertilized (Michener and Lange, 1958a; Ordway, 1965). In *Halictus confusus* most workers, except for the first brood, are fertilized (Dolphin, 1966). Various combinations of the caste attributes are especially evident in *Lasioglossum rhytidophorum* and *zephyrum,* in which unfertilized bees with enlarged ovaries, fertilized workers, and bees with ovaries of every intermediate size are common (Batra, 1966b). Even in species whose castes are somewhat more different, or at least between which there are fewer intermediates, like *L. imitatum* (Figure 9.4), there are moderate numbers (6 to 12 percent in a summer) of interme-

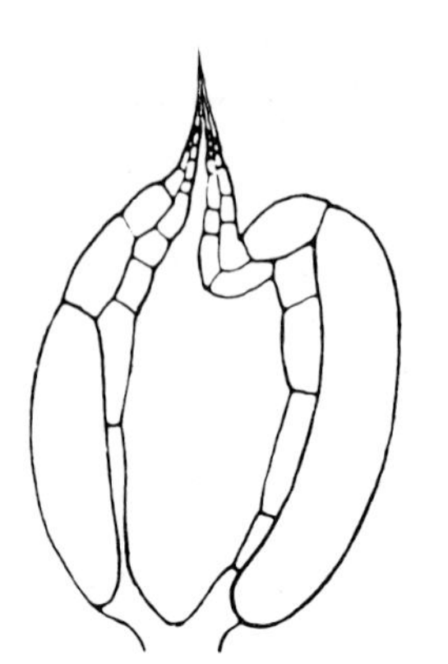
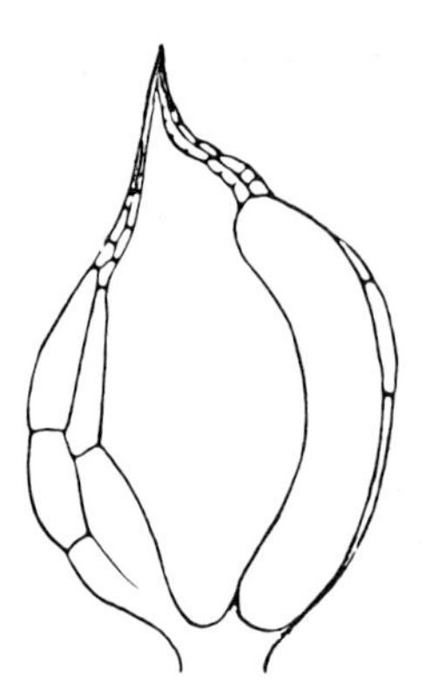
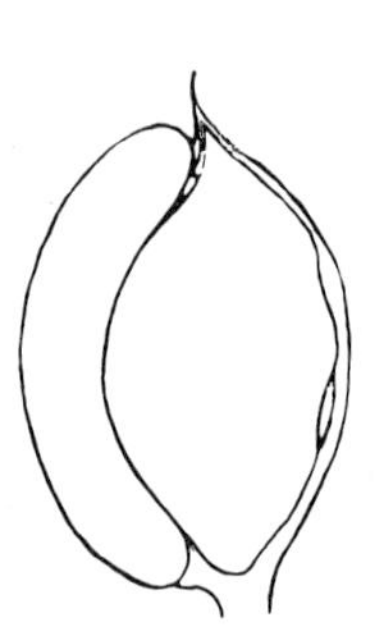
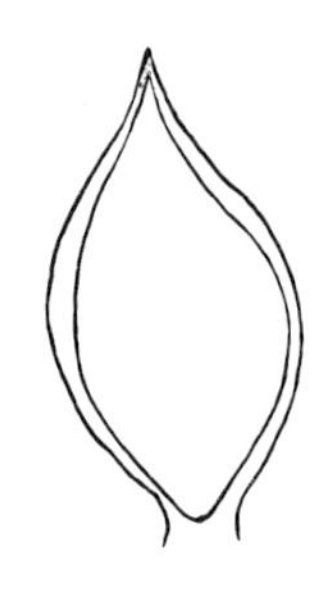
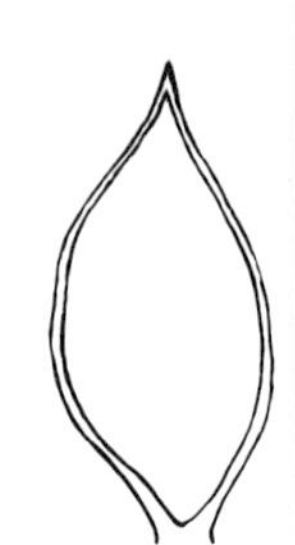

FIGURE 9.7. Sketches of ovaries of females of *Lasioglossum* (*Dialictus*) *imitatum. At left,* two queens; *center,* a laying worker with one egg about ready to lay; *right,* two workers. (From Michener and Wille, 1961.)

diate workers which are workerlike except for one or occasionally two enlarged oocytes (middle drawing, Figure 9.7) (Michener and Wille, 1961). In *L. versatum* 0 to 40 percent of the workers, depending on place and date, showed some ovarian enlargement (Michener, 1966b). At least some eggs from such workers are laid rather than resorbed; in *L. zephyrum* and *versatum* females that forage for pollen are known to lay (Batra, 1964, 1968). Since the genuine queens in the presence of workers do not forage for pollen, these bees must be among the intermediates. The mean size of intermediates is between that of queens and that of workers (Figures 9.3 and 9.4). They are probably also intermediate in a whole range of physiological and behavioral attributes.

Turning now to species (or populations) which produce separate gyne and worker broods, we find, in spite of usually greater behavioral and physiological differentiation, a certain degree of intermediacy in ovarian development in some individuals of all species. *Lasioglossum duplex* is a complete exception to the rule of distinct caste differences with separate caste broods, for although workers and queens are produced at different seasons, they intergrade in every feature (Sakagami and Hayashida, 1968), just as do the castes in *L. zephyrum.*

In *Halictus sexcinctus,* according to Bonelli (1965a), workers and the sexual forms are produced in two broods well separated in time, but the workers are the same size as the gynes (no statistics given) and although unfertilized, have enlarged ovaries early in life and probably lay some eggs. The amount of difference between queens and workers is thus slight, but there are no intermediates, the workers being unfertilized (of necessity, because of the season when they are produced) and much shorter-lived than queens.

Although there is no difference in size between castes of *Lasioglossum* (*Evylaeus*) *marginatum* (Figure 9.3A), the castes are markedly different in behavior and physiology. The species deserves special consideration as the halictine which produces larger colonies than any other, apparently monogynously, and which not surprisingly has queens with the most enlarged ovaries to be found in any halictid. No other halictid is known that has more than one fully formed egg at a time; occasionally in ordinary halictids there is one in each ovary approaching full size. In her first year the ovaries of a queen of *L. marginatum* are somewhat similar to those of other halictids, as shown by a sketch from Plateaux-Quénu (1959) (Figure 9.8A). However, as the years pass the queen's ovaries enlarge more at each laying season. Sketches of ovaries of a queen from a three-year-old and a five-year-old nest show about 10 and 30 full-sized eggs, respectively (Figure 9.8). Each of the six ovarioles thus contains several full-sized eggs in sequence. Except for several laying workers in the last year of the colonial cycle, and with the further exception of laying workers in queenless nests, workers of *L. marginatum* always have slender ovaries.

Egg laying by some workers is very probable even in *L. malachurum,* the halictine whose castes are most different as judged by a size gap. As indicated in Chapter 8 (A3), workers are likely to be the sources of most of the male-producing eggs in this species.

Direct observations of behavioral differences between castes of halictines largely concern foraging; these differences have already been explained. Observations of behavior within the nest are few. For *Lasioglossum versatum,* Batra (1968) remarks that after her first progeny mature, the queen primarily inspects cells, makes pollen balls (from pollen brought in by other bees), and oviposits, without foraging or constructing cells or making burrows. None of these activities is exclusive to the queen, although in this species relatively few workers lay eggs. In this and other species of the subgenus *Dialictus* queens do not guard at nest entrances in the field or rarely do so, although this is a special activity of queens in various other halictine bees, for example, *Lasioglossum* (*Evylaeus*) *calceatum* (Bonelli, 1965b) and *Halictus scabiosae* (Quénu, 1957). Moreover, in artificial semisocial laboratory colonies of *L.* (*Dialictus*) *zephyrum* the queen commonly guards (Michener, Brothers, and Kamm, 1971a).

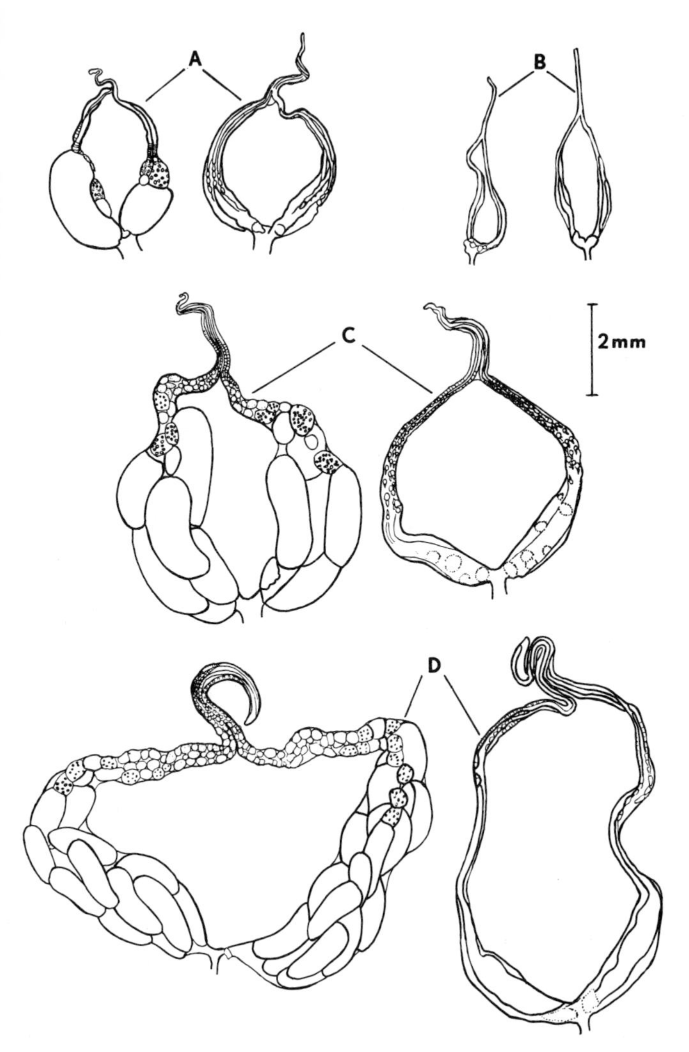

FIGURE 9.8. Ovaries of females of *Lasioglossum* (*Evylaeus*) *marginatum.* (*A*) A gyne at time of first laying (*left*) and after first laying (*right*). (*B*) A gyne (*left*) and a worker (*right*) soon after emergence from the pupa. (*C*) A queen in a third-year nest at time of laying (*left*) and after laying (*right*). (*D*) A queen in a fifth-year nest at time of laying (*left*) and after laying (*right*). *Dotted areas* represent groups of nurse cells. (From Plateaux-Quénu, 1959.)

Current studies of differential behavior of the bees in laboratory colonies of *L. zephyrum* by D. J. Brothers and me seem to indicate that even in this species with seemingly much intergradation between castes, the castes are quite well differentiated by behavior within the nest when relative frequencies of various activities are considered. Thus nudging of other bees in the burrow is more frequent by queens than by others. Not surprisingly, the queen is less often nudged than others. The most characteristic bit of queen behavior often follows nudging; if the nudged bee turns to face the nudger, the latter (usually the queen) quickly backs down the burrow, followed by the former moving down head first. Workers' behavior varies; those that guard most, rarely turn and follow the queen. They persistently remain near the entrance, whereas other workers, especially those most active in work on cells and in foraging, readily turn and are frequently led to deeper parts of the nest near the cells by following the queen's backing.

2. Allodapine bees

In the allodapine bees also there is a variety of intermediates between the egg layers and the workers; some colonies lack the latter and some colonies temporarily lack the former. Mean size and behavioral differences exist between the castes, comparable to those found in most primitively eusocial halictines (Michener, 1962, 1963b, 1965b, 1971; see also Table 9.2 and Figure 9.9). Ovaries of workers of *Exoneura,* as compared to those of egg layers and young adults, are shown in Figures 9.10 and 9.11.

TABLE 9.2. Polymorphism among allodapine bees.

Species	1. Percentage of fertilization among supernumerary ♀♀ (and N)	2. Mean size of supernumerary ♀♀ < size of layer	3. More young in nests with 2+ ♀♀ than in those with one
Exoneurella lawsoni	75 (4)	?	?
Allodape mucronata	0 (5)	?	Yes (NS)
A. panurgoides and *ceratinoides*	56 (9)	?	? Yes
Allodapula dichroa	100 (4)	?	?
A. variegata	40 (5)	?	?
A. acutigera	83 (24)	Yes	Yes
A. melanopus and *turneri*	70 (52)	No	Yes
Braunsapis facialis	20 (30)	No	Yes
B. leptozonia (Natal)	10 (19)	Yes	Yes
B. foveata & *leptozonia* (Cameroon)	5 (40)	Yes (NS)	Yes
Exoneura variabilis	25 (36)	Yes	Yes

SOURCE: Michener, 1971.
Note: "Supernumerary females" are workers and workerlike individuals. (NS) means not statistically significant by Kruskal-Wallis test. ? means too few data to form a useful judgment.

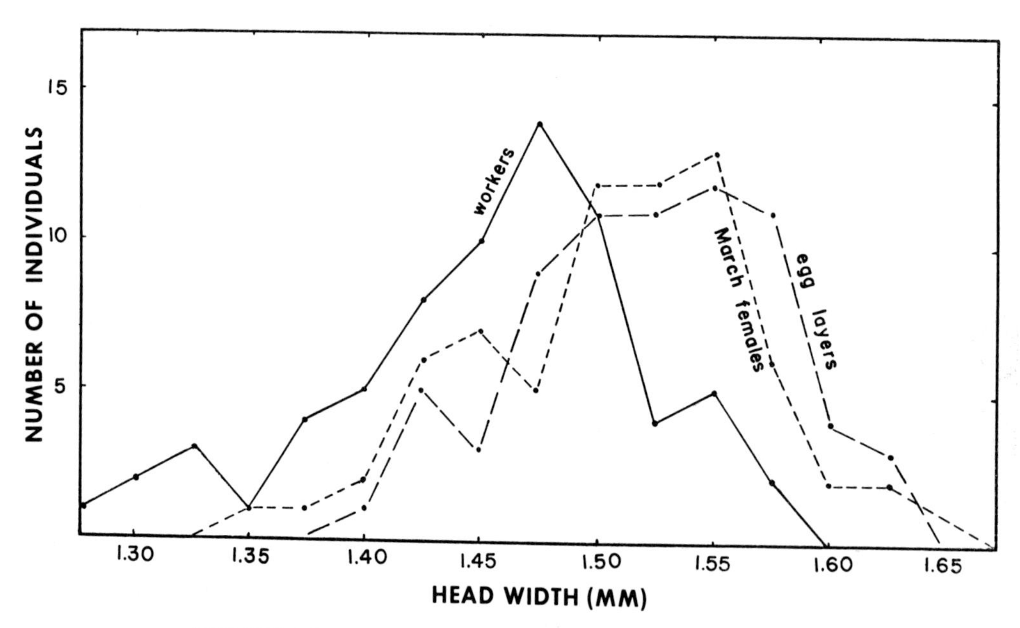

FIGURE 9.9. Frequency distribution showing sizes of workers and of egg layers or queens of *Exoneura variabilis,* an allodapine bee from Australia. "March females" were taken in the autumn (month of March) and are presumably gynes that would overwinter. (From Michener, 1965b.)

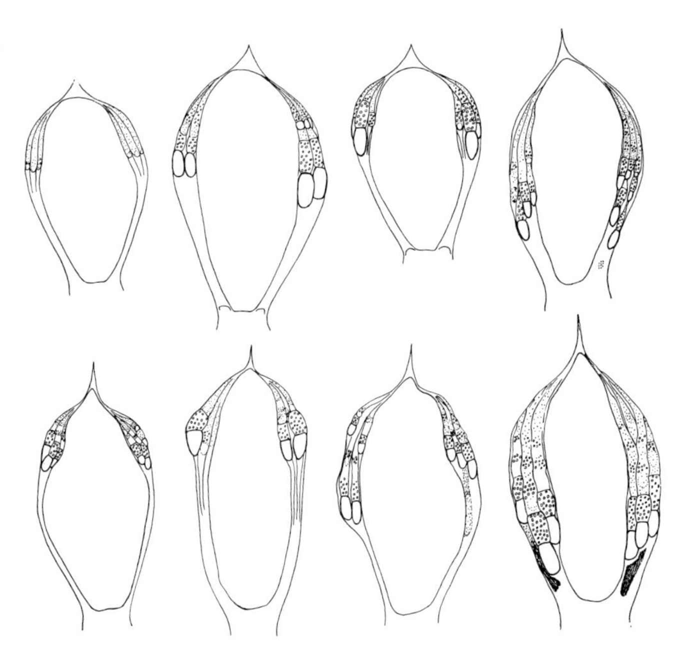

FIGURE 9.10. Ovaries of *Exoneura variabilis,* an allodapine bee from Australia. *Upper row: left,* ovaries of pale young callow; *center two,* of other callows. *Upper right and entire lower row:* ovaries of overwintering bees, presumably all gynes. (From Michener, 1965b.)

In some species which are essentially solitary, the number of workerlike females is very small, as shown in Table 6.3. In such cases, because of their scarcity, little is known of their activities, although it is evident that some are fertilized and others are not (upper part of Table 9.2). In species with more abundant workers, from 5 to 70 percent of them are fertilized, as shown in the lower part of Table 9.2.

That the workers play an active role in care of the young is suggested by direct observations of foraging and care of young by various females in nests, as well as by the much larger numbers of young to be found in nests containing workers than those lacking them. High percentages of pollen-collecting *Exoneura* and *Braunsapis* have been found to be unmated, with slender ovaries (Michener, 1962, 1963b). For example, in samples totaling 48 pollen collectors of *E. bicolor* and *variabilis,* 83 percent were unmated and 94 percent had slender ovaries. Pollen-carrying foragers of *E. variabilis* returning to their nests were commonly unmated and with slender ovaries, in contrast to other inhabitants of the same nests which were the egg layers (Michener, 1965b).

The activity of workers is further shown by a sample of 24 females of *Braunsapis foveata* and *leptozonia* taken collecting pollen (pollen loads on scopas) on flowers (Michener, 1968a). Of these only two were fertilized. Most had slender to very slender ovaries. Half had unworn wings, but the rest had various numbers of nicks up to 15, and the only one with slightly swollen ovaries had 20 nicks.

Of mature (not recognizably young or callow) adult females from nests of the same two species of *Braunsapis,*

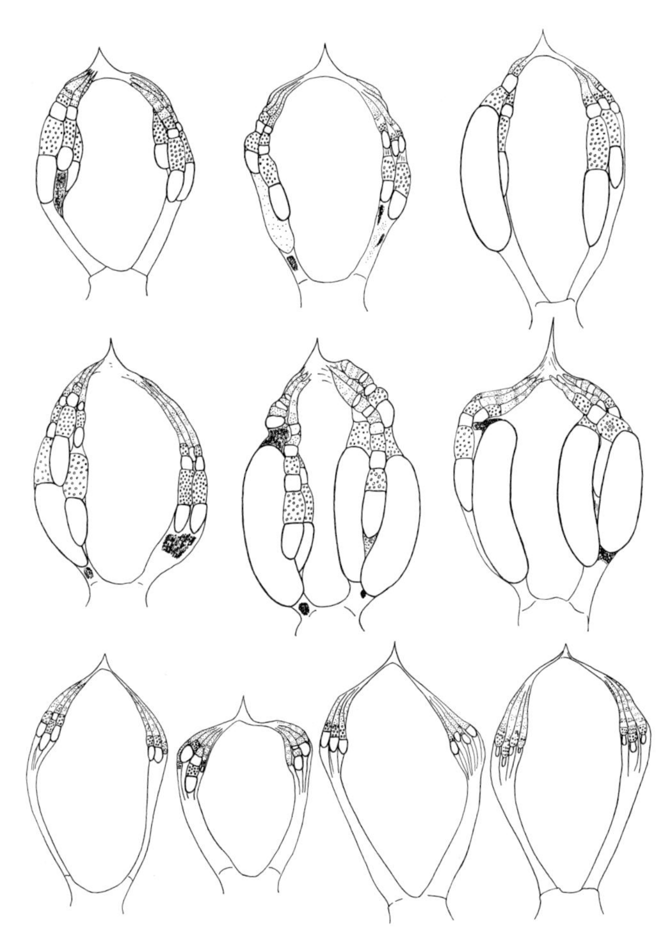

FIGURE 9.11. Ovaries of *Exoneura variabilis,* an Australian allodapine bee. *Two upper rows:* gynes or queens, the two in upper left perhaps workers or intermediates. *Lower row:* workers. (From Michener, 1965b.)

52 percent had been mated and 45 percent had ovaries enlarged as though active, recently active, or soon to be active egg layers. The low frequency of bees with enlarged ovaries or with sperm cells in their spermathecae in the sample of pollen collectors from flowers shows the relative inactivity of fertilized bees and of those with enlarged ovaries, and is surprising in view of the numbers of nests with only one adult female.

Since many allodapine workers are unworn (for instance, Tables 27.2 and 27.3), the question arises as to whether they constitute a distinct caste and die as workers or ultimately mate and disperse to establish new nests. The data strongly support the view that they are short-lived compared to queens and that at least most do not disperse and establish their own nests. (No data are available for forms in which supernumerary females—that is, probable workers—are scarce, such as those listed in the upper parts of Tables 27.1 and 9.2). In favor of this view is the smaller mean size of workers of various species; they belong to a different statistical population than the queens (Figure 9.9). Also, some of them do become worn and show moderate ovarian activity, unlike the young, usually mated gynes with slender ovaries that establish new nests. The abundance of immature stages compared to the small number of adults suggests that the workers are short-lived. The scarcity of new nests established during the active season when workers are most common indicates that they do not mostly disperse, develop into queens, and establish their own nests. Of course such dispersal is the rule in forms that do not usually have workers.

3. *Bumblebees*

In *Bombus* the castes differ strikingly in mean size and sometimes also in coloration. Color differences, when they exist, parallel size differences in that they are maximal between small workers and queens while large workers may be colored almost like queens. Even in *Bombus*, however, which has the maximal caste differences known in primitively eusocial bees, there are records of mated workers and of colonies headed by workers (caste recognized by size; see Chapter 28, B, and Plateaux-Quénu, 1961), while egg-laying workers are common.

In pocket making species the castes merge in measurements, as shown for *Bombus agrorum* in Figures 9.12 and 9.13. In gathering the data for these figures, based on material from nests taken from spring to autumn in England, Cumber did not attempt to distinguish workers from gynes except to note that the old queens that established the nests all fell into the second peak. Therefore one cannot show an overlap of separate curves, but merely a low point between the worker mode and the queen mode. Richards (1946), for a colony of the same

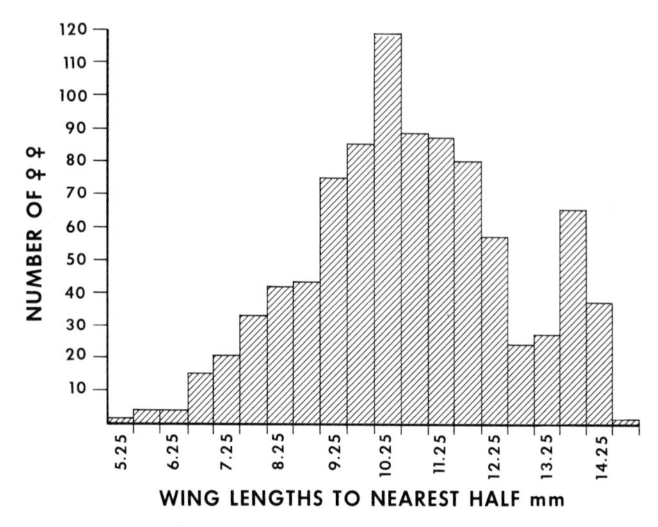

FIGURE 9.12. Frequency distribution of wing lengths of females of *Bombus agrorum*, a pocket-making species. (Original drawing by Ann Schlager; data from Cumber, 1949a.)

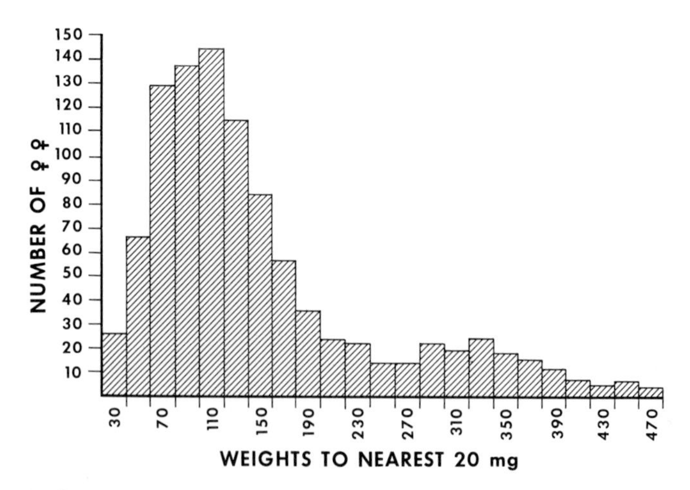

FIGURE 9.13. Frequency distribution of weights of females of *Bombus agrorum*, a pocket-making species. (Original drawing by Ann Schlager; data from Cumber, 1949a.)

species, showed both weight and wing lengths for individual bees (Figure 9.14). At least for the few individuals in this colony, wing lengths of the two castes overlapped broadly, but the gynes were more robust and weights did not overlap.

For pollen-storing *Bombus* such as *B. lucorum*, the castes differ distinctly in size without intermediates, as shown in Figures 9.15 and 9.16, data for which were gathered in the same way as for Figures 9.12 and 9.13. For Figures 9.12, 9.13, 9.15, and 9.16, the relative heights of worker and queen peaks depend principally on the amount of effort expended at various seasons in obtaining nest populations, not on bumblebee biology.

A more dramatic idea of the size differences among female *Bombus* in a single colony can be obtained from Figure 12.1.

For populations of individual colonies of seven Canadian species, Plowright and Jay (1968) give measurements which in general support the earlier findings. The distinction in caste differentiation between pollen storers and pocket makers is not complete, as has usually been suggested, however, and in addition there is one clear exception: *Bombus nevadensis* is a pollen storer which, at least under the artificial conditions of the Plowright and Jay study, produced queen-worker size intermediates as do pocket makers.

Behavioral differences between castes in *Bombus* are well known, not greatly different from those of certain

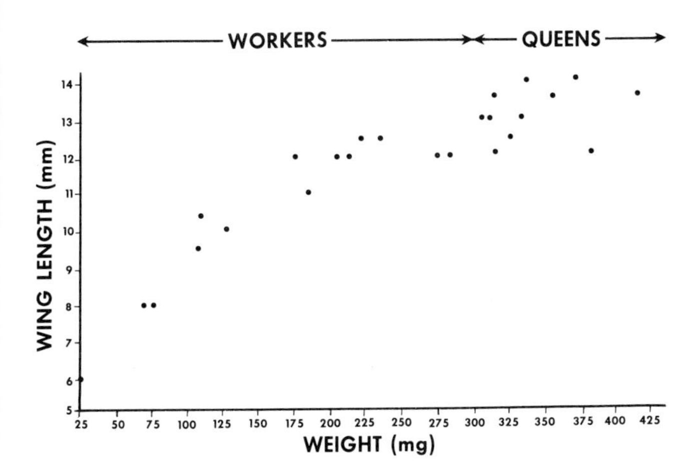

FIGURE 9.14. Wing lengths and weights of workers and queens of *Bombus agrorum.* (Original drawing by Ann Schlager; data from Richards, 1946.)

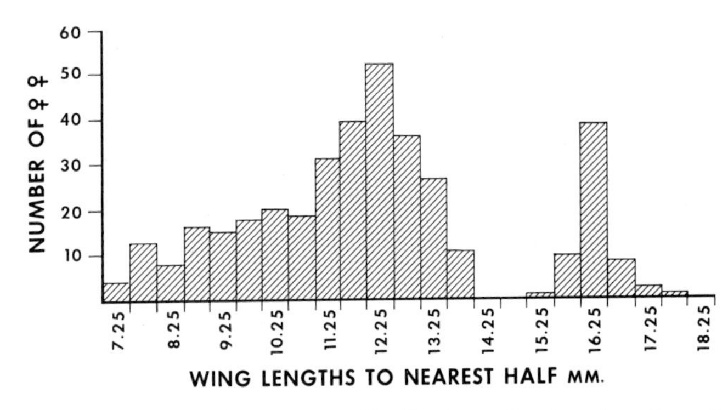

FIGURE 9.15. Frequency distribution of wing lengths of females of *Bombus lucorum,* a pollen-storing species. (Original drawing by Ann Schlager; data from Cumber, 1949a.)

halictines, and are explained elsewhere in this book (Chapters 10 and 28).

C. Caste differences in highly eusocial bees

The differences between the castes in the highly eusocial bees, as already indicated, are much greater than in the forms discussed above and intermediates are for practical purposes unknown as natural phenomena, although workerlike gynes and gynelike workers can be produced experimentally. The colonies almost always consist of a single queen and a large number of workers with extremely clear-cut division of labor between the two castes. Such complete caste differentiation would not be

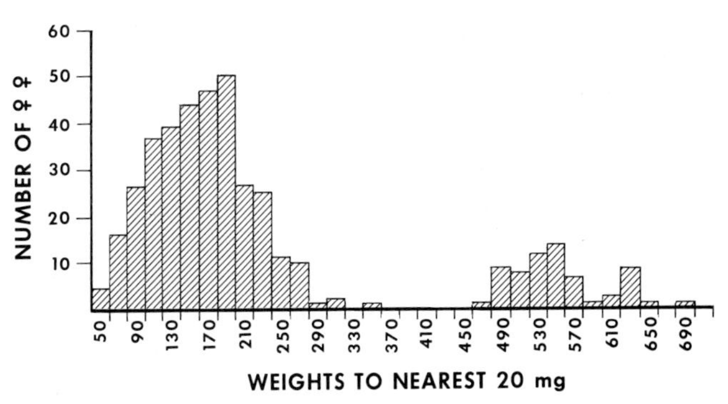

FIGURE 9.16. Frequency distribution of weights of females of *Bombus lucorum,* a pollen-storing species. (Original drawing by Ann Schlager; data from Cumber, 1949a.)

TABLE 9.3. Some structural differences between *Apis* workers and queens.

Feature	Worker		Queen
Relative area of antennal surface	2*		1
No. of chemoreceptive plates per antenna	2400*		1600
No. of facets of compound eyes[a]	4000–6300*		3900–4920
Antennal lobes of brain	Larger*		Smaller
Mushroom bodies (coordination centers) of brain	Larger*		Smaller
Hypopharyngeal glands	Large	*	Vestigial
Mandibular glands	Large*		Very large
Wax glands	Present	*	Absent
Nasanov's gland	Present	*	Absent
Ovaries	Small	*	Enormous
Number of ovarioles	2–12	4*	150–180
Spermatheca	Small or rudimentary	*	Large
Sting barbs	Strong		Minute*
Sting axis	Straight		Curved
Sting axis	Straight		Curved
Mandible with subapical angle	Absent		Present*
Mandibular form	Slender		Robust
Mandibular groove from near opening of mandibular gland	Present		Absent
Proboscis	Longer*		Shorter
Corbicula	Present*		Absent
Pollen press (auricle, pecten, rostellum, etc.)	Present*		Absent

SOURCE: Ribbands, 1953; Snodgrass, 1956.
Note: The probable ancestral condition is followed by an asterisk; an asterisk between the columns indicates an intermediate ancestral condition.
[a]Estimates differ, but all agree that the queen has fewer facets than the workers.

possible until the gyne, in the course of evolution, ceased to establish new colonies alone, since she must have workerlike structures and behavior, or more accurately, the structure and behavior of a solitary bee, to rear a brood of young by herself. The differences between the castes in *Apis* and the Meliponini are similar; in this respect, as in many others, the two groups of highly eusocial bees are about equally evolved socially.

Although the queen mates and lays enormous numbers of eggs, most of her modifications are structural and behavioral losses compared with workers and with females of solitary bees. She has no pollen-collecting apparatus or apparatus for producing wax or handling construction materials. As far as we know she does not have the behavior patterns that would lead her to flowers to collect food; she can feed on stored honey but during times of laying activity is dependent on products produced by the workers: bee milk, the larval food, in *Apis;* trophic eggs and larval food in Meliponini. She does not show aggressive responses against enemies, so far as known, except for those which are directed against other queens.

On the other hand, the workers of highly eusocial bees have specialized pollen-collecting apparatus, wax-producing glands, and apparatus on the legs for handling

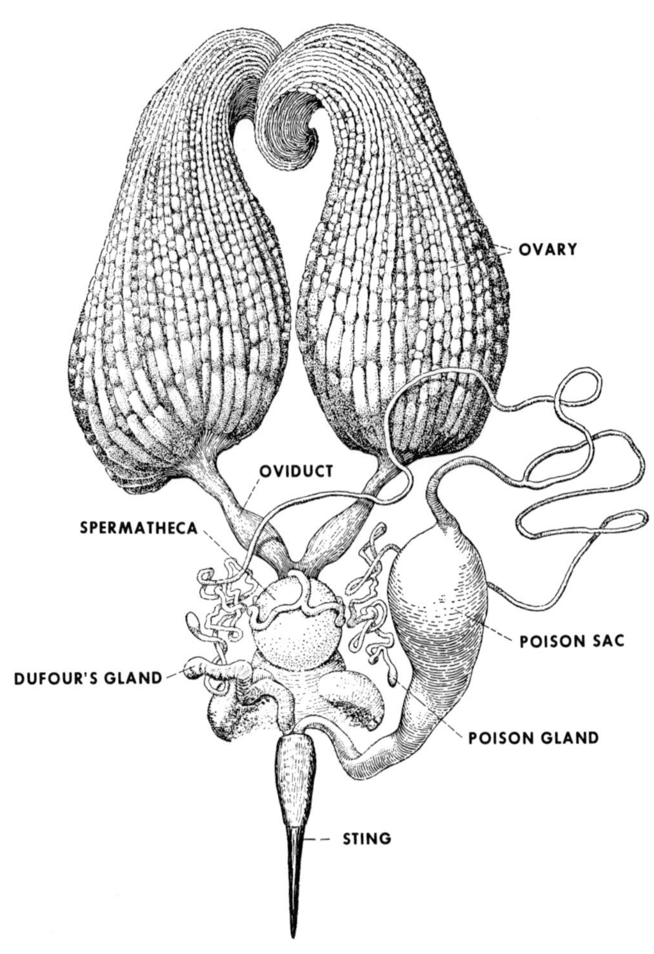

FIGURE 9.17. Ovaries and associated organs of a queen of *Apis mellifera.* (Modified from drawing by Snodgrass, 1956, © Cornell University; used by permission of Cornell University Press.)

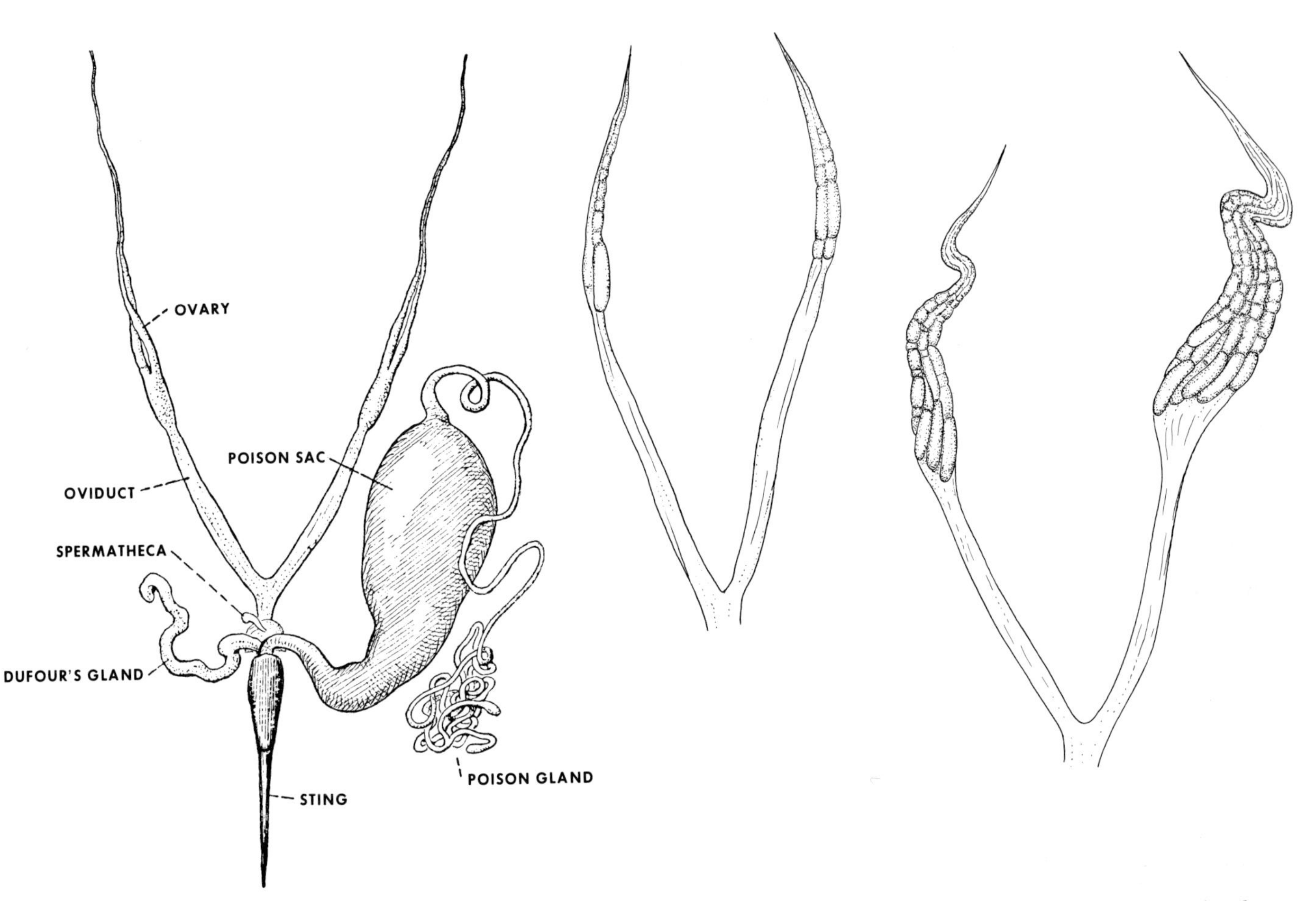

FIGURE 9.18. *Left:* Ovaries and associated organs of a worker of *Apis mellifera* showing no ovarian enlargement. (Modified from Snodgrass, 1956, © Cornell University; used by permission of Cornell University Press.) *Center and right:* Ovaries and oviducts of workers showing development of eggs. (Original drawings by C. D. Michener.)

the wax, mud, propolis, and cerumen. They also have behavior patterns that lead them to collect food from flowers and to construct nests from wax and resin or mud in places appropriate to the species, and they have aggressive reactions against natural enemies of many sorts. Workers, however, have lost the behavior patterns related to mating. Moreover, they do not have the proper structure or size to mate, at least in *Apis.* Their ovaries remain slender, or produce male eggs or trophic rather than reproductive eggs. In *Apis,* if eggs are laid, for example in the absence of the queen, they may not be properly placed in the cells or more than one may be placed in each cell instead of only one.

The structural differences between the castes have been studied in special detail in *Apis.* While some of the main features are outlined above, Table 9.3 lists further differences; Lukoschus (1956) details these and others, particularly those of the endocrine glands. The differences are of interest in showing the separate directions in which evolution has gone in the two castes, in spite of the lack of genetic caste differences (see Chapter 10, C2). The differences are in many ways comparable to the differences that exist between young and adult insects, where the same genome is able to produce very different forms. Ovaries and associated structures of the queen and worker of *Apis* are shown in Figures 9.17 and 9.18.

D. Concluding remarks

In semisocial and primitively eusocial bees a symbiotic relationship exists between the castes; nonetheless each caste is able to survive, rear offspring, and thus contribute to the next generation alone if necessary, without the aid of the other caste. The bees are not greatly different in structure from their solitary relatives. By and large, the more different are the seasons of their production, the more different are the castes. In some forms they do not differ externally, while in others there are striking size differences.

In the highly eusocial bees, on the contrary, the symbiosis between castes is obligatory and (with some exceptions, for example, the South African subspecies of the honeybee, *Apis mellifera capensis*) each caste must be associated with the other for successful reproduction. Particularly, queens are completely unable to survive or even to eat in the absence of the workers, although workers are able to live out their lives in the absence of queens. Although there may be certain seasons when most queens are produced, there is no period of exclusive queen production; workers and queens are reared simultaneously.

10 *Caste Determination*

We have already discussed sex determination (Chapter 8, A), that is, why it is that certain eggs develop into males and others into females. We here examine the question of why it is that some female eggs develop into gynes, others into workers.

In spite of the view (Chapter 5, H) that social evolution is often reversible, much of it must have involved increasing differentiation of castes. Hence one can say, as in the introduction of Chapter 9, that in a general way the more different are the castes, the more evolved is the society as a group of mutually dependent individuals. Alternatively, the more alike are the castes, the greater is the likelihood that individuals of one or both castes are viable without the other, and the social level is considered lower.

The importance of this for caste determination is that the more different are the castes, the earlier caste differentiation must set in, and hence the earlier determination must occur. Social evolution should, therefore, involve progressive advancement of the stage at which caste determination occurs. As explained below, the time of caste determination or at least caste bias ranges among social bees from after maturity, as in some halictids, to the time of the fertilization of the egg, in *Melipona.* When determination occurs in the adult state, all young females are technically gynes, queens becoming distinguishable from workers only later in adult life.

Since the process of becoming different, that is, the differentiation, is intimately related to the causation of the differences, differentiation and determination will both be considered below. In fact I do not know where to separate the two. If a certain amount of food (greater than that necessary to produce a worker) is required to produce a gyne, does determination occur when workers receive a stimulus to assemble that much food in a single cell, or when they do so, or when the larva eats all the food, or when it eats the food beyond the quantity needed to make a worker, or when that food is assimilated?

While the potentiality to develop into either caste must be provided for genetically, caste determination in most bees is not genetically controlled but is based on environmental conditions. At least in most cases this means it is based upon food. Since the food concerned is provided by other bees except perhaps in those cases where differentiation occurs in the adult stage, caste determination is usually socially controlled. (For reviews of caste differentiation see Brian, 1965a; Weaver, 1966.)

A. Caste determination in Halictinae

Males as well as females of *Lasioglossum imitatum* vary in size with the season, those produced in midsummer being smallest (Michener and Wille, 1961). The same is true for *L. zephyrum* (Batra, 1966b). D. R. Kamm has recently shown that in experimental nests, *L. zephyrum* produces larger cells and progeny at lower temperatures than at higher temperatures and under short photoperiod than long photoperiod conditions. Similar size variation has been verified for females only of *L. rhytidophorum* (Michener and Lange, 1958b). Such seasonal size variation could be a basic reason for the size difference between queens and workers found in many halictines, since by and large the two castes are produced at different seasons. The absence of a size difference between the physiologically and behaviorally very different castes of *L. marginatum* (produced at the same season but in different years for any given colony) suggests that for halictines, size difference is an unimportant aspect of caste difference, an insignificant by-product of the usual difference in season of production of the two castes. This view is supported by the observation that in relatively uniform laboratory conditions, female size in *L. imitatum* and its allies did not vary seasonally (Batra, 1968).

On the other hand, in semisocial groups of sisters or others of about the same age, the larger bees probably usually become queens. Knerer and Plateaux-Quénu (1966a) believe that this relationship exists in the semisocial spring colonies of gynes (queen and auxiliaries) of species that are later eusocial (but see note 1, Chapter 9, and material on *Augochlorella,* Chapter 9, A). It also exists in colonies of semisocial species of *Augochloropsis* and *Pseudaugochloropsis* (Michener and Kerfoot, 1967; Michener and Lange, 1959). This observation supports the view that size is a meaningful correlate of caste even in scarcely social forms. Laboratory experiments with queenless groups of *Lasioglossum zephyrum* support this view. The oldest bee will ordinarily develop ovarially and become a queen regardless of size; but if all are the same age, then the largest bee will usually do so (see Chapter 11, C).

Variation in size of food masses and cells according to the caste being produced is known in various Halictinae. For the primitively social *Lasioglossum* (*Dialictus*) *zephyrum,* Batra (1966b) says that when food is scarce in relation to egg production, egg layers will finish and oviposit upon food masses smaller than those used when food is abundant. The result is seasonal production of smaller bees presumably determined or biased as workers. While this idea of causation was based on impressions obtained in detailed field and laboratory studies and is probably correct, it is not documented by impressive statistics. It is similar to Legewie's view of *L. malachurum,* explained in the next paragraph. In a number of other halictines the cells that will produce the queens are known to contain larger food masses than those for workers (Figure 10.1) (Knerer and Atwood, 1966; Knerer and Plateaux-Quénu, 1966b; Wille and Orozco, 1970). Even social parasites like the bee *Sphecodes,* whose larva uses the stored food, differ in size to about the same degree as the castes of the host, depending upon whether a cell that was to produce a gyne or one that was to produce a worker was parasitized.

Such differences in the size of the food mass would be expected to be maximal in the species with greatest caste size difference; for halictines this is *Lasioglossum* (*Evylaeus*) *malachurum.* Legewie (1925a) noted that in central Europe the size of the cells as well as of the food mass varies with the brood in this species (Table 10.1). He notes that the largest bees (gynes) make and provision

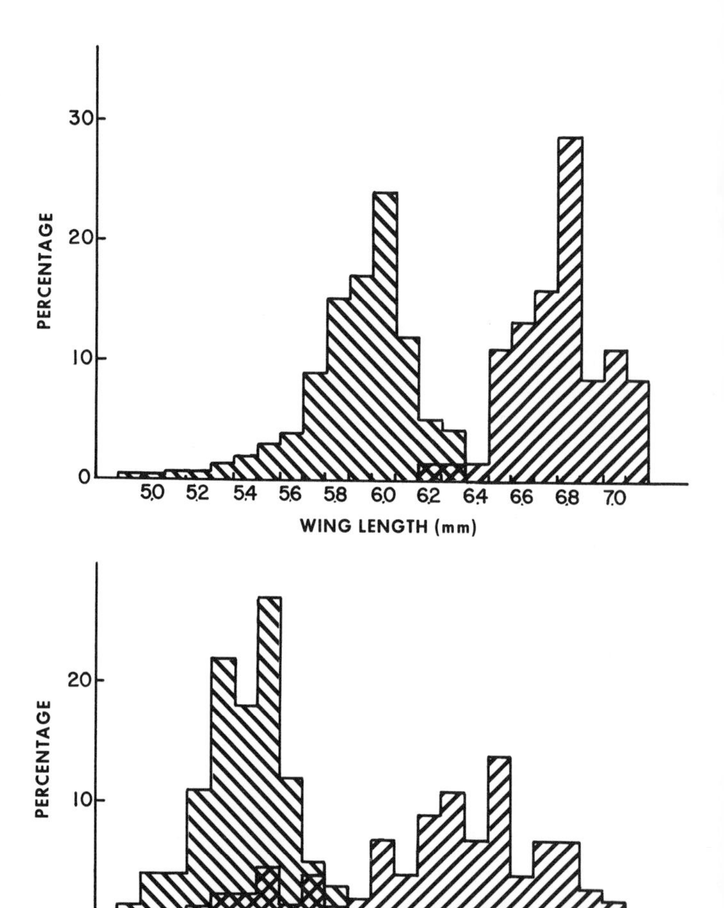

FIGURE 10.1. *Above,* histograms showing relative sizes of workers and gynes of *Lasioglossum* (*Evylaeus*) *cinctipes. Below,* histograms showing weights of pollen balls of the same species. The bimodality in pollen-ball weights presumably causes the bimodality in size of the female bees. Mean pollen-ball weights for the two groups were 33.9 and 61.7 mg. (The published data are obscure about the overlapping of the histograms, but the reproductions above seem satisfactory for practical purposes.) (From Knerer and Atwood, 1966, © American Association for the Advancement of Science.)

TABLE 10.1. Relation of cell and food mass size to number of bees and number of cells being provisioned in *Lasioglossum malachurum.*

	First brood	Second brood	Third brood
Usual number of bees making and provisioning cells	1	6–7	19–20
Cell lengths	10–11 mm	10–12 mm	11–12 mm
"Average" diameters of pollen balls[a]	3–4 mm (76)	3–4.5 mm (104)	5–6 mm (123)
Usual number of cells	6–7 (196)	19–20 (162)	60–65 (113)
Mean number of cells/bee making and provisioning them	6.5	3.0	3.2

SOURCE: Legewie, 1925a. Legewie does not give detailed statistics, and my interpretations are subject to his biases. Numbers of observations are given in parentheses.
[a] See also Figure 9.5, which shows relations among sizes of adult bees similar to those shown here for pollen masses.

small cells (to produce the first brood) and that relatively small bees (workers) make and provision large cells (for the third brood). Moreover, some of the food balls (for males?) are markedly smaller in the third brood than the "average diameters" noted in the table. The size of a food ball, according to Legewie, seems related to the number of females active in the nest at the time that the ball (and cell) is constructed. This is shown in a very general way by Table 10.1 (compare first brood with others) and more specifically by the observation that the cells and pollen balls manufactured first by each brood (except the first) are smaller than those manufactured later. He attributes this to the small number of workers present during the early activity of the second and third broods. Obviously, for each of these broods one or a few workers appear before the rest, and it is these which construct the first cells and provision them, when the total nest population of adults is small. Later, when the population of workers increases, the sizes of the cells and of their provisions increase. Lack of an appreciable difference in the number of cells being made and provisioned per bee for the second and third broods contrasts with striking differences in the size of provisions for these broods, and may be related to food availability. Pollen is less available in midsummer, when the small provisions for the second brood are being made, than in later summer, when provisioning for the third brood occurs.

One population of *Lasioglossum* (*Dialictus*) *umbripenne* also has discrete rather than overlapping caste sizes. In this case not only are gyne-producing cells larger (mean length 11 mm) than those producing workers and males (mean length 9.5 mm) but also the food mass in gyne cells is larger and of a rounded rectangular shape rather than flattened spherical as for workers, males, and all forms of all other known species in the genus. In another population of seemingly the same species the castes overlap in size, and gyne larvae have food masses of the usual shape (Wille and Orozco, 1970; Eickwort and Eickwort, 1971).

It appears that caste bias or determination in bees with discrete castes occurs during the larval stage and is based at least in part on the amount of larval food, in which case size is an important caste attribute, an idea which seems probable in view of the higher fecundity of large female insects generally as compared to conspecific small ones.

In *Lasioglossum* (*Evylaeus*) *marginatum,* on the contrary, caste is not determined until the adult stage. According to Plateaux-Quénu (1960b), females of a worker brood, if given the opportunity to mate, become queens. As explained in Chapter 9 (B1a), queens and workers are reported not to differ, even in average size, at the time of maturation. They are produced at the same season of the year, but the workers have no opportunity to mate since worker-producing nests are not open to the surface at the time when the males are active. Thirteen adult workers from a third-year nest were marked and transferred in the autumn by Plateaux-Quénu into a nest in its final year, containing males. Subsequent excavation recovered nine of these individuals, five of which were fertilized, showing that workers can be fertilized if given the opportunity. At least one such individual functioned as a queen; a year later she had a nest and four workers, a normal number for a queen of that age. Apparently fertilization or participation in the final phase of colony life actually converted a worker into a queen.

The mechanism of caste determination must differ among different halictines. This is scarcely surprising since sociality arose independently in the different groups. It seems likely that in those halictines having slight mean size differences between castes, the smallest individuals may be incapable of becoming queens, the largest possibly incapable of becoming workers, but the great majority may have both potentialities and may not be determined as to caste until after the adult stage is reached. Mating or lack of it cannot be the decisive factor among many halictines, as it seems to be for *Lasioglossum marginatum,* for one finds unmated females with large ovaries, for example, in *L. zephyrum* (Batra, 1966b), as well as mated workers in a number of species (Chapter 9, Blb). In *L. zephyrum* unmated queens have the same behavioral and physiological attributes as mated ones (Michener, Brothers, and Kamm, 1971a); their progeny being males, the colony does not last long.

Presumably in *Lasioglossum zephyrum* there is no efficient physiological switch mechanism causing individuals to become either queens or workers; intermediates are permitted and effectively contribute to the colony by both foraging and egg laying. In certain other species switch mechanisms are more effective, so that most (in the case of *L. imitatum*) or all (*L. malachurum*) of the females clearly belong to one caste or the other. And if it is a simple stimulus (mating vs lack of mating), the switch mechanism may be highly effective even if it does not operate until adult life, as in *L. marginatum.*

It has in the past seemed likely that, in primitively eusocial halictines, workers inhibit foraging by queens (since the gyne usually stops foraging when she acquires worker associates), and that the queen somehow inhibits ovarian development in the workers. These inhibitions probably really occur in various species and serve as mechanisms which either reinforce the caste differences or cause such differences to appear in the already adult bees. In *Lasioglossum* (*Dialictus*) *zephyrum,* however, overwintered queens deprived of their first brood of workers do not resume cell construction and foraging or are long delayed in doing so. Some other factor, presumably internal, seems to prevent resumption of such activities.

B. Caste determination in bumblebees

In view of the great size difference between castes (and among workers) of *Bombus,* determination must occur in the larval stage. The larvae are fed progressively in two different ways. In the pollen-storing species, each larva tends to become isolated in an area more or less separate from that of the others in its clump, and is fed individually and directly by the adult bees, which open the wax covering at the top and inject semiliquid food. In the pocket-making species, the adults push food into large pockets at the sides of the cells, and the food then forms the floor on which larvae lie and from which they feed. A switch either in the feeding behavior by the adults or in the behavior and physiology of the larvae causes nearly all of them either to stop growth at highly variable worker sizes or to continue to the queen size, although some caste intermediates (at least intermediate in size) are produced in pocket-making species (Chapter 9, B3). In pollen storers worker size variation is less, and there is a marked gap between the size of large workers and small queens, as shown in Figures 9.15 and 9.16.

In pocket-making species, larvae that are to become workers (and males) are left largely to their own devices in obtaining food from stores in the pockets. At an age of about five days, before their period of most rapid growth, larvae of pocket-making species spin weak silk partitions around themselves, so that their positions become fixed in the common distendible cell. Larvae in less favorable positions, pushed away from the food by fast-growing, centrally located larvae, obtain less food and become small adults. Figure 10.2 shows how this would happen, the marginal larvae being small and hardly having access to the food supply, and Figure 10.3 shows the results in three selected cases. An obvious switch occurs when queen production starts, since queen larvae (except perhaps in the subgenus *Subterraneobombus,* Hobbs, 1964) are fed directly by opening the top of the cell, just as are all larvae of pollen storers. This switch does not seem to account for caste determination, however, although it is associated with a change in the caste

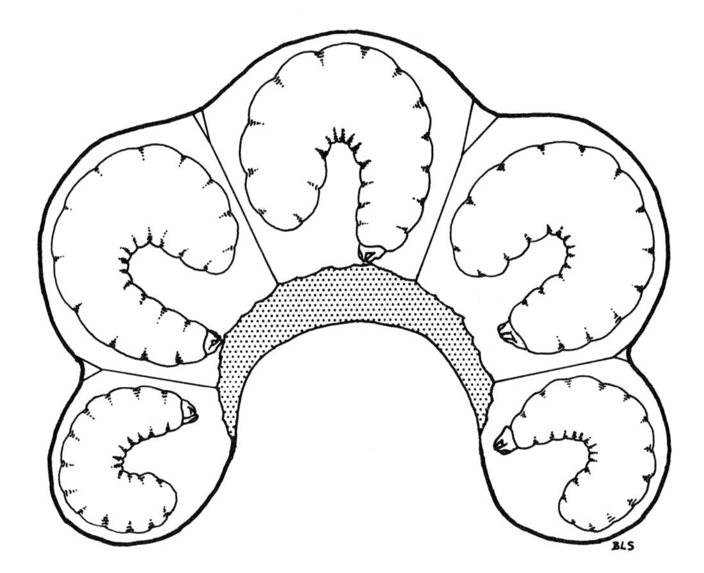

FIGURE 10.2. Vertical section through a cluster of larvae (a distendible cell) of a pocket-making bumblebee, *Bombus agrorum. Shaded area* represents pollen pushed into the pocket (see Figure 28.4) and available as food for the larvae. The lower larvae, which have little access to the pollen, are smallest. *Heavy line* represents the wax covering or wall of the cell. The larvae are separated by thin silk partitions, which they have spun. (From Cumber, 1949a.)

being produced. Even when pocket-making species are forced to feed all their larvae directly rather than through pockets (by giving indoor colonies only a liquid pollen-honey mixture), queens are not produced until late in colony development (Plowright and Jay, 1968). Moreover, the older larvae that will become workers or males in colonies having normal food supplies are occasionally fed by regurgitation through a temporary hole made in the top of the cell, as in the pollen storers (Chapter 7, D4).

The number of adults in the colony in relation to the number of larvae to be fed (adult–larva or worker–larva ratio) influences the amount of food received by each larva. A recent study of this point is that of Plowright and Jay (1968), who showed that in the presence of excess food for the colony, the size of the workers produced is dependent upon the adult–larva ratio, presumably upon the amount of food transferred to the larvae by the adults. Thus under their experimental conditions, the mean size of first-brood workers reared by a queen was inversely related to the number reared. And for second-brood workers, mean size was not only negatively correlated with the number of second-brood larvae but positively correlated with the number of first-brood workers present.

There is, by now, considerable other evidence of the importance of the worker–larva ratio, not only in deter-

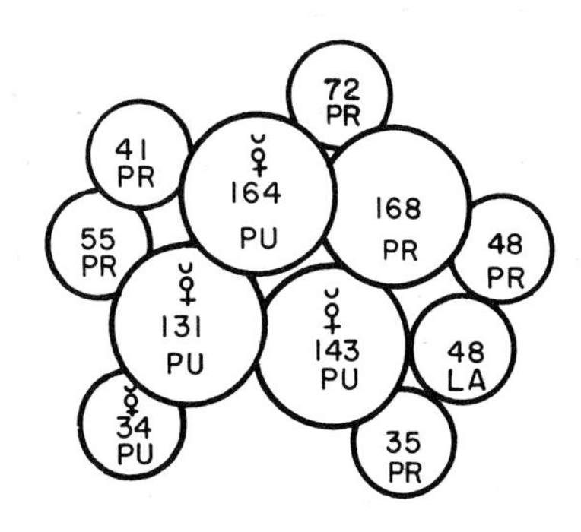

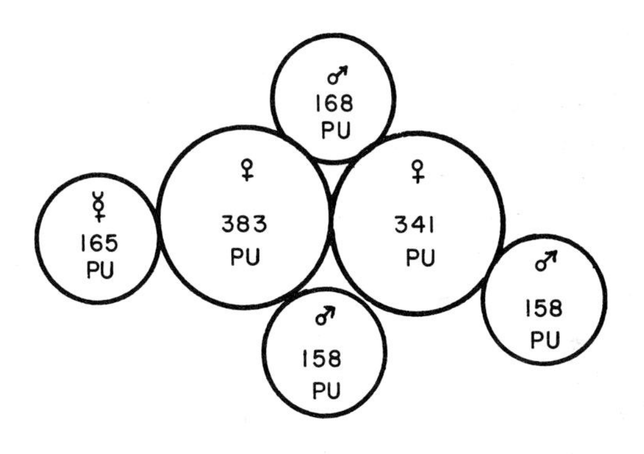

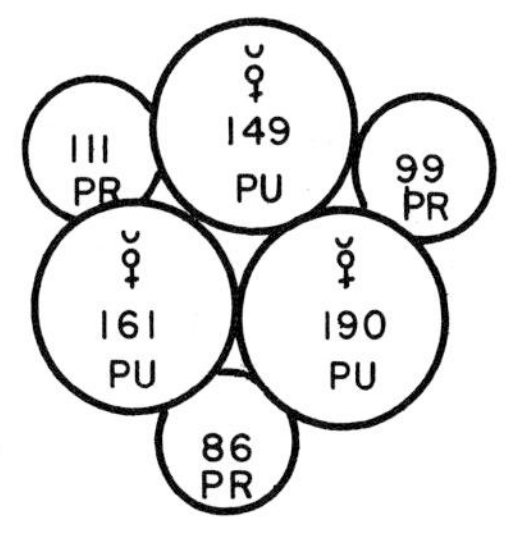

FIGURE 10.3. Diagrams of the brood in three clusters (cells) of pocket-making bumblebees, *Bombus ruderarius* (*top*) and *Bombus agrorum* (*middle and bottom*). Each cluster is derived from a single group of eggs laid almost simultaneously. *Circles* represent immature bees: *PU,* pupa; *PR,* prepupa; *LA,* larva. *Numbers* give the weights (mg) of each immature bee. (From Cumber, 1949a.)

mining size of workers but in caste determination. Richards (1946) was the first to gather data on the subject, and subsequent work (Cumber, 1949a; Free, 1955a; Röseler, 1967b, 1970) shows that there is a causal element in the relation between a high worker–larva ratio and gyne production. For example, Free put individual spring gynes of *Bombus pratorum* (a pollen storer) in boxes with abundant food and with varying numbers of workers. These queens soon laid their ordinary initial batches of eggs, as they would have in nature without workers. In the boxes with only one worker, or with none as in nature, all the progeny in this first brood became workers. But with two to five workers helping to care for the small number of progeny, part of the progeny in the experimental colonies became gynes instead. They were not only large in size but developed large fat bodies and showed slender ovaries even a month after emergence, all distinctive attributes of true gynes. (In nature some workers show partial ovarian development early in adult life, while gynes experience an ovarian diapause, their ovaries not enlarging until after winter.) Thus in all known respects such bees, produced at a time of high worker-larva ratio, were genuine gynes, even though developing in spring from the first laying of their mothers instead of at the end of summer when young gynes would be produced in the field.

Röseler (1970) got similar results with a pocket-making species, *Bombus hypnorum.* Although in such forms there is no gap between worker and queen sizes, he showed that when the adult–larva ratio was above 1 : 1.8, some progeny were in the size range for typical queens, and when the ratio approached 1 : 1, all the young were of that size.

The same author, working with a pollen storer, *Bombus terrestris,* got results interestingly different from Free's with *B. pratorum* and from his own results with *B. hypnorum.* The size of workers of *B. terrestris* reared by lone gynes is inversely related to the number of workers in the brood (Figure 10.4); when she has only a few larvae to feed, they become large. But in experiments in which queens were artificially provided with workers, even at the adult–larva ratio of 1 : 1, the mean size of the young bees did not exceed that shown for the largest bees recorded in Figure 10.4. The largest individuals produced were within the size range of workers, not that of gynes; the colonies seemed incapable of producing gynes.

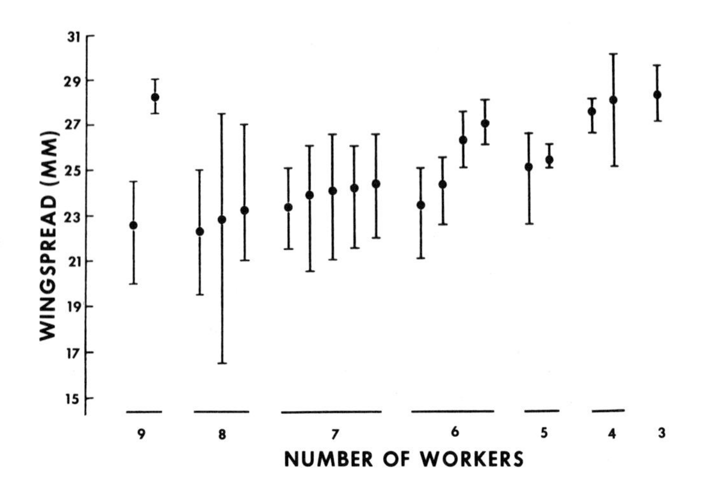

FIGURE 10.4. Sizes of workers produced by 19 lone gynes of *Bombus terrestris,* in relation to the numbers of workers in the broods. *Vertical lines* show the range of wing length; *dots* show means, for the workers. (From Röseler, 1970.)

Groups of workers without queens, however, could produce gynes of *B. terrestris* if the worker–larva ratio were 1 : 1.7 or higher. Apparently the queen in some way inhibits gyne production. But in two colonies made up with old queens from colonies that had already produced sexual forms, no such inhibition occurred. The adult-larva ratios were 1 : 1.3 and 1 : 1. Gynes were produced. Old queens evidently lack the inhibiting ability.

The young queen's inhibitory effect was shown to operate if the workers tending the larvae could go through a coarse screen (queen excluder) to an area containing the queen. Her inhibitory effect vanished if the screen was finer, preventing the workers from reaching the queen, even though contact through the screen was still possible. Workers exchanged every 24 hours from the queen's side of a fine screen to the larval side provided inhibition. Since bumblebees do not exchange food or lick one another appreciably, Röseler postulates a nonvolatile pheromone deposited in the nest by the queen, picked up or sensed by the workers, and having an effect on their feeding of the larvae. If his explanation is correct, this is the most elaborate social mechanism known among primitively eusocial bees.

Larvae up to 3.5 days of age from normal colonies, transferred to queenless groups of workers, showed the ability to become gynes. Older larvae all became workers in spite of high worker–larva ratios; their determination as workers was complete at that time.

This early determination means that the postulated inhibiting pheromone in *Bombus terrestris* probably has its effect before the *amount* of food could determine the caste. But the amount of food or at least the amount of attention received by larvae from adults also appears as a caste-determining agent, for even queenless colonies rear no gynes if the worker–larva ratio is low. The loss by old queens of the ability to inhibit, along with attainment of the needed worker–larva ratio, could explain the commencement of gyne production at the normal time in the life cycle of the colony.

How exceptional the *B. terrestris* caste determination may be among bumblebees is not evident. In the pocket makers and, to judge by Free's study of *B. pratorum,* in some pollen storers also, there appears to be no inhibition by young queens. In such forms (and this may be the majority of *Bombus* species), the requirements for gyne production are only a large food supply available to the colony and a certain minimum amount of attention given to the growing larvae. Such attention may result in nothing more than transferring large quantities of food from the source to the larvae.

The size bimodality of females even in pocket makers (Figures 9.12 and 9.13) and the discrete caste sizes in pollen storers indicate a sudden change in feeding or a larval switch mechanism. A relatively sudden change from the low, worker-producing worker–larva ratios typical of spring and early summer to high queen-producing ones of late summer might result from egg eating by workers, which is common in pollen storers at the time when reproductives are being produced. Hoffer (1882) gives an entertaining account of such egg eating and the attempts of queens to safeguard the eggs; a partial translation into English appears in Plath (1923). A decrease in the number of eggs laid per day by the queen would also increase the worker–larva ratio; such a decrease is shown by Cumber (1949a). A third factor was noted by Röseler (1967b) among bumblebees in controlled chambers, that is, destruction of larvae by workers. He theorizes that when bees become crowded in the nest, laying of unfertilized eggs by the queen is stimulated. The presence of the resultant male larvae, according to his theory, stimulates the throwing out of larvae, thus increasing the worker–larva ratio and producing queens from the females. These factors, together with whatever threshold or switch mechanism there may be, would promote the partial or complete size discontinuity that separates the castes in *Bombus.* That is, before queen production, workers rear fewer and fewer young per worker present until a point is reached where the larvae get sufficient food to become queens. As Plowright and Jay (1968) suggest, a larval switch mechanism is probably necessary to produce the size discontinuity noted in pollen storing species, but perhaps this is not so for pocket makers. Cessation of production of the pheromone postulated by Röseler at a time when the worker–larva ratio is already high enough to produce queens could have the same effect.

The fact that different larvae in the same cluster can become different castes shows that no fully general influence in the nest determines caste. The feeding advantages of central positions in larval clusters of pocket makers have already been noted and it was the central larvae in Free's experiments that became gynes. Temperature control is also better centrally. Adults, using either glandular food or honey and pollen, might in theory feed different larvae differently in the same cluster, but there is no evidence that this occurs.

C. Caste determination in the highly eusocial bees

In the highly eusocial bees effective mechanisms usually prevent development of intermediates between workers and gynes in nature (but see review by Plateaux-Quénu, 1961).

1. Trigona, Lestrimelitta, Meliponula, and Dactylurina

In the stingless honeybees other than *Melipona,* brood cells ordinarily of two very different sizes (Figure 29.12), both mass provisioned, are made by the workers. The large ones produce gynes, the small ones males and workers. There is no information as to the factors that cause the workers to build an occasional giant, gyne-producing cell. Such cells are in the combs or clusters of worker- and male-producing cells or more commonly at the margins of the combs. Darchen and Delage (1970) provided reasonably convincing evidence that the basis for caste determination is the amount, not the quality, of the larval food. They selected *Trigona* larvae that had almost finished the food in their own worker cells and fed them additional food from other worker cells. These larvae continued to feed and attained the size of normal gyne larvae. At that time these authors were unable to rear mature normal gynes from such larvae, but from intermediate-sized larvae they did rear dwarf gynes, recognized by the lack of corbiculae and other gyne features.

In a subsequent paper the same authors (Darchen and Delage-Darchen, 1971) verified that the mere addition of some larval food of the usual sort will convert a larva that would normally become a worker into a gyne. There is no qualitative, only a quantitative, difference between the nutrition of a worker and that of a gyne. The lack of specific qualitative differences in food is emphasized by the use in some experiments of food from cells of one subgenus of *Trigona* to feed larvae of another. Darchen (1973) even found that addition of a honey-pollen mixture to the provisions from a worker cell resulted in production of a queen. The source of food does not influence the caste produced. There is a switch mechanism activated by only a small amount of excess food that converts a worker into a gyne. The Darchens never saw an intercaste. A small food excess leads to a dwarf gyne, having the normal morphological features of a gyne even though it may be scarcely larger than a worker. The usual amount of food placed in a gyne cell leads, of course, to the normal large gyne. Since dwarf gynes of the species studied are not known in nature, another switch mechanism of an unknown type must cause the workers to construct and provision either the standard-sized worker cells or the large gyne cells, but not intermediates which would produce dwarf gynes.

Trigona is a large and diversified genus. The Darchens worked principally with two African subgenera, *Axestotrigona* and *Hypotrigona.* It remains to be seen whether all subgenera will show similar details of caste determination. Conceição Camargo (1972a, b), however, has shown that in the South American *Trigona* (*Scaptotrigona*) *postica,* approximately the same principles apply. She reared female larvae in artificial cells in which she placed various quantities of food taken from ordinary cells. There was high mortality, but of those that survived the ones reared on twice the normal amount of food became 50 percent slightly oversized workers, 20 percent intercastes, and 30 percent gynes. Similar percentages of intercastes but increasing percentages of gynes were found when larvae were given 2.25 and 2.50 times the normal amount of food for workers. Those receiving food from 2.8 to 5.0 times the normal all became gynes, the largest slightly larger than a normal gyne. One that received the contents of four worker cells was introduced into a queenless colony; she mated and functioned as a queen. The switch from worker to gyne production clearly occurred in the vicinity of two worker cell equivalents of food. Camargo also found that with the same quantity of food, resultant gynes are heavier than workers. Perhaps when sufficient food has been eaten to cause, for a given individual, the switch to gyne development, more efficient use of the food results, possibly mediated by hormonal switch that turns on the genes responsible for production of the characteristics of a gyne. As with the work on African subgenera of *Trigona,* the stimuli that cause workers to produce an occasional large gyne cell are unknown. That should be the next step in elucidating caste determination in *Trigona* and its allies.

A remarkable observation relative to sizes of gynes was made by Juliani (1967) in southern Brazil. In the same colony of *Trigona julianii* he found large gynes and dwarf gynes, the latter produced in cells mixed with those of workers. The biological significance of such dwarfs is unknown. This species may be intermediate in caste determination between *Melipona* (see below) and ordinary species of *Trigona.*

2. Apis

In the genus *Apis* gyne cells are also very different from and larger than those in which workers are produced. They are round in cross section and hang down, not being incorporated into comb (Figures 30.2 and 30.3). Those that are built for ordinary colony multiplication (swarming), or to produce a gyne to supersede an aging queen, hang from the edge of the comb; those that are for emergency production of a replacement when a queen has died are built out and down from worker cells and hence are on the surface of the comb.

In *Apis* it is known that qualitative as well as quantitative caste differences exist in the food of larvae (Jung-Hoffmann, 1966; Mitsui, Sagawa and Sano, 1964; Ribbands, 1953). Since the larvae are progressively fed instead of mass provisioned as in Meliponini, they must be watched to demonstrate the qualitative differences. The quality of the food given the larvae turns out to be of prime importance in caste determination.

The bulk of the food of larval *Apis* is here called bee milk, following Ribbands (1953). We have equivalent established usages, such as pigeon milk, for non-mammalian secreted or processed nutrients given by adults to young. The familiar term royal jelly, if used in this broad sense, is misleading, for most of it goes to workers, yet the term worker jelly has never been widely accepted. A reasonable series of terms, suggested by Smith (1959), follows:

Brood food—a general term applied to any food material supplied to growing larvae.

Bee milk—as proposed by Ribbands (1953), this term refers to the brood food secreted by the glands of nurse bees and presumably mixed with crop contents.

Royal jelly—this name applies only to bee milk which has been deposited in gyne cells.

Young worker food—the bee milk supplied to young worker larvae.

Mixed food of workers—the food placed in the cells of older worker larvae, characterized by an increased amount of pollen and thought to be a mixture of bee milk, nectar or honey, and pollen. It is sometimes called bee bread.

Jung-Hoffmann (1966) replaced walls and bases of worker and gyne cells with transparent material so that she could see what the larvae were being fed. Both gyne larvae and young worker larvae are fed bee milk, which consists of two visibly different constituents: (1) a clear component which is the product of the hypopharyngeal glands of nurse bees, presumably mixed with crop contents (honey or nectar plus digestive enzymes and water); and (2) a white component which appears to be the secretion of the mandibular glands of nurse bees, plus an admixture of the secretion of the hypopharyngeal glands.

Larvae in gyne cells are fed on a mixture of clear and white food in about equal amounts; this is royal jelly. During the first two days worker larvae are fed 20 to nearly 40 percent white food, the rest clear; this is young worker food. On the third day the amount of white decreases conspicuously, and thereafter they are fed on the clear component (about two thirds of the feedings) and a food that is yellow because it contains much pollen[1] and that presumably also contains and may consist largely of honey. All this is mixed worker food.

From these observations as well as from the studies of Smith (1959) and others, it is evident that differential feeding of gyne and worker larvae starts when they are very young, probably as soon as they hatch. Shuel and Dixon (1959, 1968) found differences in respiration of larvae fed in vitro on royal jelly and young worker food. Even during the first 24 hours after hatching, CO_2 production was greater with royal jelly as a food, and after 50 hours oxygen consumption was also greater.

One might assume that cues promoting differential feeding of the larvae by workers involve cell size and structure. Jung-Hoffmann, however, says that in a queenless colony larvae are already fed as gynes before construction of emergency gyne cells begins. Cell shape is therefore not decisive in determining the frequencies of the clear and white food components in early larval development, although presumably it has an important influence, for example, in swarm and supersedure gyne cells and probably in later developments in all cases.

The mechanisms by which royal jelly produces gynes and mixed worker food produces workers are not understood. Female *Apis* larvae have been reared in dishes of royal jelly, young worker food, and the like in various laboratories. Fresh royal jelly seems necessary to produce gynes in vitro, but gynes are not consistently produced—the number of workers and caste intermediates resulting from such experiments usually exceeds the number of normal gynes even if fresh royal jelly is provided repeatedly during larval growth. Weaver (1966) has reviewed this subject and believes that one or more highly labile constituents of royal jelly must be responsible for gyne production or alternatively for worker production.[2] The other side of the coin is that lack of some substance might be responsible for worker production or possibly for queen production. However, unless there are aspects to the feeding of young larvae not elucidated by Jung-Hoffmann, relative percentages of such constituents rather than their presence or absence must be responsible for the beginning of caste differentiation. The theory of Dixon and Shuel (1963, 1965) that a nutritional balance, perhaps acting through the endocrine system (Canetti,

1. Addition of abundant pollen is delayed until the fifth day (and until the fourth day for male larvae) according to Matsuka, Watabe, and Takeuchi (1973). The reasons for the discrepancy between their results and Jung-Hoffmann's are not clear.

2. Chemically, royal jelly is incompletely known, but no one of its known components seems to be responsible for caste determination. In addition to water, sugars, proteins, vitamins, all the common amino acids, and cholesterol, it contains large amounts of a fatty acid, *trans*-10-hydroxy-2-decenoic acid, related to the fatty acids of the queen-inhibiting pheromone but without pheromonal activity. It also contains singularly large amounts of the nucleic acids RNA and DNA.

Shuel, and Dixon, 1964), might effect the differentiation is still possibly correct.

Studies in which female eggs or larvae of various ages are transferred from worker cells to gyne cells or vice versa are another source of data on caste determination in *Apis*. One assumes that whatever larvae are in gyne cells are fed as queens, and in worker cells, as workers. Eggs and larvae through three days of growth seem totipotent; for example, transfer of larvae at three days of age or younger from worker cells to gyne cells results in essentially normal gynes (Becker, see Ribbands, 1953; Lukoschus, 1956). But larvae transferred at $3\frac{1}{2}$ days of age from worker to gyne cells produced variable adults, from gynelike although with some workerlike attributes to typical workers in the features studied (Weaver, 1957) (Figure 10.5). In Becker's study, for example, the mean number of ovarioles per ovary was 148 to 163 in gynes of different groups ranging from normal to those transferred to gyne cells at three days of larval age. But among bees resulting from transfers at $3\frac{1}{2}$ days, ovariole number varied from 7 to 156. The figures for workers range from 2 to 12. Comparable data for other characters were obtained by Weaver, who found that there is a strong tendency for such bees to be either rather workerlike or rather queenlike, and that individuals structurally midway between the castes are rare. The switch mechanism must be quite efficient. Few larvae four days old survive after transfer; they mostly die or are killed by workers. The few that have survived transfer became queenlike workers or almost typical workers.

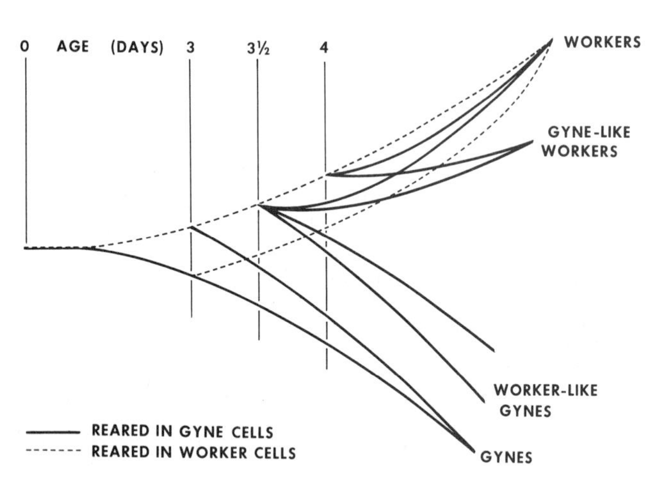

FIGURE 10.5. Diagram showing the results of experiments in transferring *Apis* larvae from worker cells to gyne cells (and in some cases back again) at various times from hatching (zero) to four days of age. Note the tendency to produce either workers and workerlike bees or gynes and gynelike bees; true intermediates midway between the two castes are very rare. (Original drawing by Barry Siler.)

More detailed study of "essentially normal" gynes like those produced by Becker and Lukoschus shows that the later the larvae are transferred from worker cells to gyne cups, the more reduced are their queenly attributes (Table 10.2). These observations support the view that differential treatment of larvae begins about as soon as they hatch.

Correlation of the caste characters is often influenced by transfers. This is illustrated, for example, by larvae transferred to gyne cells for one or two days starting at two days of age and then returned to worker cells. The resulting bees were basically workers, but in some characteristics they were more queenlike than any normally produced workers, while in other characteristics they exceeded the most workerlike extreme of normal worker variation. Thus the integration of various growth patterns is broken down by such transfers. The rarity of true caste intermediates may be due in part to the poor coordination of a potential intermediate's growth processes, some of which are directed toward producing queens, others toward workers. The result may be high mortality of the bees that are potentially midway between the castes. It is also true that hive bees often destroy such intermediate larvae.

There is no way to define a particular critical period when caste determination occurs. Differentiation starts very early and the differences increase progressively. As shown above, up to 3 days of age transfers result in nearly normal individuals of either caste. On the other hand, larvae hatched in worker cells, transferred for less than a day into gyne cells, and returned to worker cells before 3 days of age, became workers with somewhat gynelike ovaries or other attributes. One therefore cannot say that the critical period begins at three days of age. But it is evident that the third day is relatively critical and that by day 4 determination is nearly complete.

The belief that workers arise because of partial starvation is not sustained by these experiments, for worker larvae until over three days old, when differentiation is at least well started, are surrounded by food. Food quality, as indicated above, must be an essential part of caste

TABLE 10.2. Mean characteristics of queens of *Apis mellifera* produced by transferring eggs or larvae to gyne cell cups at various stages.

Stage of transfer (N)	Egg (27)	1-day larva (27)	2-day larva (27)	3-day larva (27)	4-day larva (6)
Weight (mg)	209[b]	189[b]	172[b]	147[b]	119
No. of ovarioles	317[a]	308[b]	292[b]	272[b]	224
Spermathecal diameter (mm)	1.310[b]	1.276[b]	1.212[b]	1.159[b]	1.033
Spermathecal volume (mm^3)	1.182[b]	1.093[b]	0.936[b]	0.821[b]	0.586

SOURCE: Modified from Woyke, 1971.
[ab] $P < 0.05$ and 0.01 respectively for difference between this mean and subsequent one to the right.

determination. Of course it is true that workers do receive, in total, less food than gynes, and starvation of gyne larvae is known to produce some worker attributes in the resulting bees.

The problem of why worker honeybees sometimes construct gyne cells, while otherwise not doing so, and therefore sometimes rear gynes, will be discussed in Chapter 11 (A).

3. Melipona

In contrast to *Apis* and to other genera of Meliponini, *Melipona* cells which produce workers (and males) and those which produce gynes are indistinguishable and intermixed (Figures 10.6 and 29.11). On emergence from their cells the gynes are about the same size as workers, although different structurally. The abdomens of queens swell later with ovarian development, so that older queens are larger than workers, as in most social bees. Gynes are produced in relatively high frequencies; most are killed by workers. As in other Meliponini, the cells are mass provisioned; there are no obvious differences correlated with caste in the provisioning of the cells. In view of the apparently identical feeding of gyne and worker larvae, as early as 1946 Kerr postulated that caste determination in *Melipona* is genetic instead of trophic as in other social insects. In certain seasons and colonies, however, expected ratios of queens to workers were not obtained, and it now seems clear that environmental (probably trophic) conditions also play a role in caste determination. Darchen (1973), in fact, maintains that trophic conditions can explain caste determination in *Melipona,* as in *Trigona,* but since in all of his experimental conditions both workers and gynes were produced, the likelihood of a genetic component still seems good.

Kerr, Stort, and Montenegro (1966) found that cells of *Melipona quadrifasciata* provided by the workers with small amounts of provisions produced workers, while those with larger amounts produced both castes. (Incidentally, amount of provisions put in a cell is *inversely* related to the number of workers that regurgitate into it; when food is plentiful as few as four or five can provide the needed material.) All larvae eat all the food in their cells. Therefore it was reasonable to conclude that pupal weight (taken at a certain stage of coloration) provides an index of the amount of food eaten by the larva. These authors found that no gynes occurred among pupae weighing less than 72 mg (Figure 10.7); above that weight

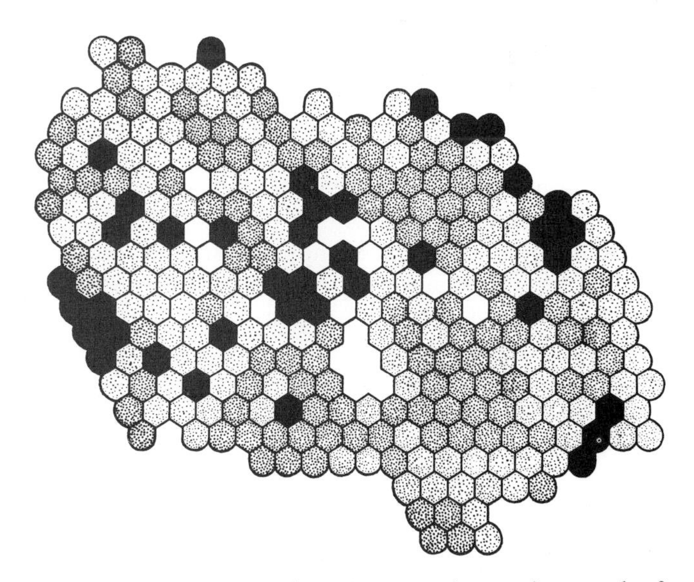

FIGURE 10.6. Distribution of sexes and castes in a comb of *Melipona pseudocentris. Black,* cells that produced gynes; *dark stippling,* males; *light stippling,* workers. (From Kerr et al., 1967.)

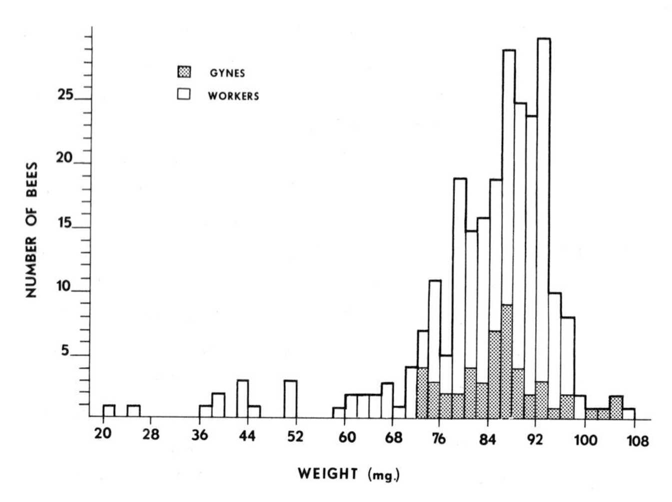

FIGURE 10.7. Distribution of weights of worker and gyne pupae of *Melipona quadrifasciata.* No gynes had pupal weights less than 72 mg. (From Kerr, Stort, and Montenegro, 1966.)

Kerr and Nielsen (1966) noted differences in the abdominal ganglia of gynes and workers, gynes of *Melipona marginata* having four ganglia and most workers having five due to separation of two apical ganglia that are fused in gynes. However, they further noted that a certain number of workers have the ganglionic arrangement of gynes (Figure 10.9). They believe that these workers are genetically gynes which, because of unfavorable trophic conditions, have developed morphologically as workers, except for the abdominal ganglia. When such individuals are counted as gynes, genetically expected ratios are obtained for *Melipona marginata* even under suboptimum trophic conditions. For *M. quadrifasciata* a large excess of workers with five ganglia in some colonies and the production of gynes with five ganglia in some colonies led Kerr and Nielsen (1966) to postulate, on the basis of data from only four colonies, another pair of alleles (N, n), which determines the number of ganglia. The theory of these authors is summarized by Kerr (1969).

(72-108 mg) 22.1 percent of 276 pupae were queens. Their data suggest a sharp change at about 72 mg, for even in the lower part of the gyne-producing weight range, from 72 to 84 mg, 24.7 percent of 73 pupae were gynes. Apparently colonies vary with external conditions, reaching the stage when all their pupae weigh over 72 mg during the warm season when there is plenty of food. At such times there is 3:1 segregation of workers to gynes. At any time when there is not adequate food many of the pupae weigh less than 72 mg, and a lower percentage of gynes is therefore produced. Figure 10.8 shows weights of two samples of worker pupae, one from a colony producing 10 percent gynes, the other from a colony producing 25 percent gynes. The greater weight of the latter group is obvious.

The 3:1 ratio can be explained for forms in which queens mate but once (as is the case in *Melipona,* see Kerr and Nielsen, 1966) and in which males are haploid, if one makes the following assumptions: To be a potential queen, a female must be heterozygous at two X loci (Chapter 8, A1). Thus she could have a genotype $x_a{}^1x_a{}^2x_b{}^1x_b{}^2$. Homozygosity at either or both loci produces an individual that becomes a worker. But doubly heterozygous individuals become queens only if trophic conditions are good (as shown by heavier weights of pupae); they become phenocopies of workers when such conditions are bad.

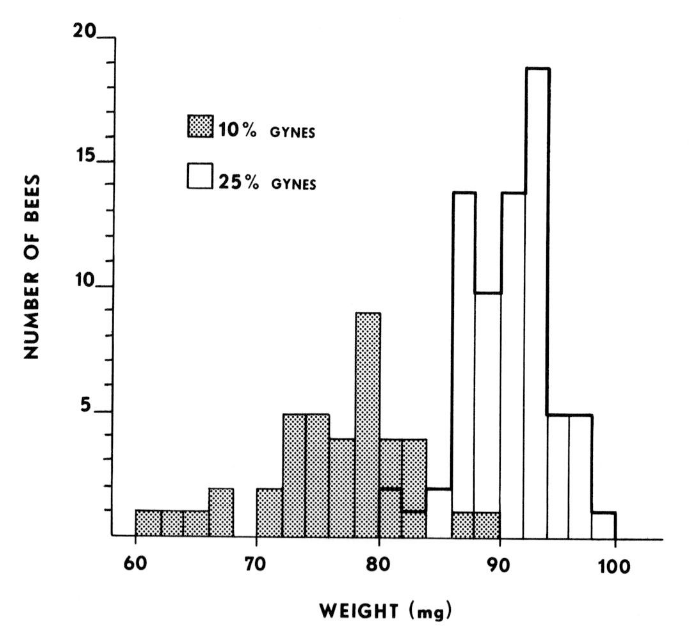

FIGURE 10.8. Distributions of worker pupal weights in two colonies of *Melipona quadrifasciata.* In the colony with lighter pupae, 10 percent of the female pupae become gynes. In the colony with heavier pupae, the equivalent figure was 25 percent. (From Kerr, Stort, and Montenegro, 1966.)

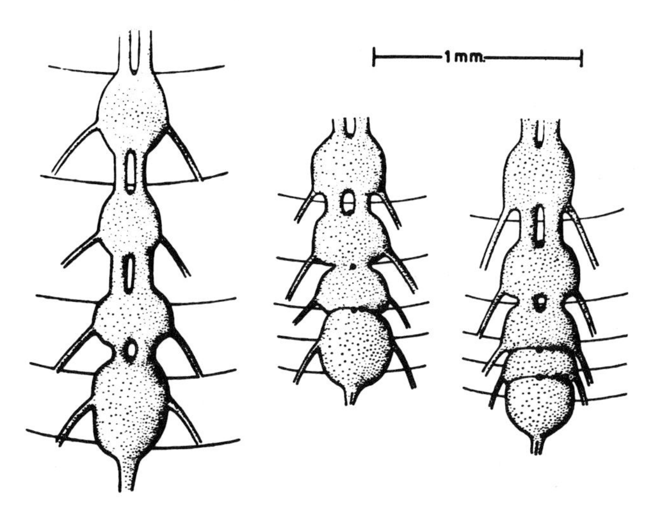

FIGURE 10.9. Posterior abdominal ganglia of females of *Melipona marginata*. *Left,* a gyne; *center,* a gyne-like worker, thought to be genetically a gyne but phenetically (externally and in behavior) a worker; *right,* a worker. (From Kerr and Nielson, 1966.)

D. Concluding remarks

Except in part for *Melipona,* caste differences are environmentally determined. As might be anticipated, forms with very similar castes have relatively inefficient mechanisms for determination, and many intermediates result. Such intermediates play effective roles in their colonies, having some workerlike and some queenlike activities.

The more different the castes, the less useful will a caste intermediate be—she is likely to be neither an effective forager and nurse bee (worker) nor an efficient egg layer (queen). Therefore switch mechanisms arise which assure that nearly every female develops either as a typical worker or as a typical queen. Such mechanisms may often be double or multiple, as though to assure that if one failed, another would probably function. Perhaps the real reason for multiple switch mechanisms, however, is that after one arose, another evolved which enhances caste differences or causes determination at an earlier age. The earliest mechanism to function ontogenetically is the one that ordinarily does the job, but experimental manipulation may reveal the existence of another, often evolutionarily older, but acting later ontogenetically.

I have noted before that in general the earlier the determination, the more different the castes. A partial exception is found in the Meliponini, in all genera of which caste differences are about the same and very noteworthy, about as great as in *Apis.* In *Trigona* and its allies determination occurs near the end of larval life when the larva either does or does not get enough food to grow past the worker and into the gyne developmental pathway. In *Melipona* the same is true, except that among those that receive enough food to become gynes, only some genetically favored individuals do so. Thus the time of caste determination in *Melipona* seems to be moving from late larval life to the time of fertilization of the egg.

11 *Control of Gyne Production and Worker Laying*

Gyne production and worker laying are under related control mechanisms in some social bees. Such a relationship might be expected, since if a worker starts to lay, it has in a sense become a queen. Breakup of the colony because of conflict among queens, too many young for the adults of a colony to support, or excess male production by unfertilized workers are among the possible results of uncontrolled gyne production or laying by workers.

In primitively eusocial bees the colony is ordinarily terminated by cessation of worker production and development of gynes and males. the number produced is probably maximized by selection, with constraints because gynes must be strong enough not only to survive and mate but also to produce colonies and become queens. In the final plase of colony life gynes are ordinarily produced in considerable numbers.

In highly eusocial bees, on the other hand, gyne and male production must be limited, or the gynes killed after they mature, to prevent repeated swarming and resultant small colonies. At the same time enough reproductives must be produced to replace aging ones and to start a reasonable number of new colonies. The colony should seemingly be able to produce gynes or laying workers as needed—for example, in an emergency if the queen dies, or at the proper season for normal reproduction. But one would expect evolution of mechanisms to limit or prevent their production at other times, when the energy of the colony should go into such things as worker production and food storage. Unfortunately the required controls are not fully known for any species, even *Apis mellifera.* The sections that follow are therefore rather fragmentary and concern primarily the highly eusocial bees, where the problems are both most acute and best studied.

A. Queen replacement and supersedure

During the life of a colony of social bees, there is always the possibility that the queen might die or be killed. In this case the "orphaned" colony can survive only if the queen can be replaced. At other times it happens that, while the old queen is still present, but presumably as a result of her senility, a new queen is produced. The old queen is then often killed. This is called supersedure. Most of this section deals directly with replacement, although parallels with supersedure are generally apparent. The two phenomena are much alike, for supersedure probably occurs when the old queen's production of social pheromones or other signals of her presence becomes so reduced that the workers act more or less as though she were absent.

For many years practical beekeepers as well as biologists wondered how it happens that replacement queens usually are produced when needed, yet in *Apis* are not produced at other times. The field has been reviewed for all social insects by Plateaux-Quénu (1961). Circumstantial evidence—for example, the finding of deteriorated colonies lacking queens and often producing only male offspring from eggs presumably laid by workers—shows that in some halictines and *Bombus,* if the queen dies, the colony may go to male production and die. The same occasionally occurs in *Trigona* and *Apis,* but is unreported in *Melipona,* probably because of the large numbers of young gynes more or less continuously produced through the genetic caste determination of this genus.

It is very likely that in those halictine bees where caste differentiation occurs among early season semisocial groups of adult gynes, if the queen dies or is lost, one of the remaining gynes, until then an auxiliary, is able to replace her. Such replacement is known to be possible in the socially similar groups of gynes established in the spring by certain species of *Polistes* wasps (Eberhard, 1969). In primitively eusocial colonies like those of *Lasioglossum zephyrum,* whose queens normally die and are replaced during the season of colony activity, queens must regularly be replaced as well as produced abundantly late in colonial life (see Chapter 6, D). In *L. zephyrum* evidence from artificial colonies suggests that the oldest of a group of young adult females has the best chance to become the replacement queen, but that there is also an interaction with size, large individuals having a better chance than small ones to become queens (Michener, Brothers, and Kamm, 1971a, b). The mechanisms whereby the original and replacement queens prevent other bees from becoming queens are unknown, although work now in progress indicates that the queen or the bee that is to become the queen moves about the nest more than the rest, often nudging the others and apparently stimulating them to work in the nest. Her influence is probably associated with this high level of activity.

In stingless bees queen replacement and supersedure commonly involve so much killing of young gynes that the whole process is dealt with in a separate section (B) below.

In *Apis* death of a queenless colony is ordinarily prevented by production of a replacement queen. If the queen of a colony dies or is removed, emergency gyne cells are built around existing eggs or young larvae in worker cells. A cell is enlarged at the expense of two adjacent cells and wax is added around its rim; ultimately it projects and hangs from the surface of the comb instead of hanging from the comb margin as do the ordinary gyne cells that precede either swarming (for colony multiplication) or supersedure.

What stimulates emergency gyne production or prevents gyne production in *Apis* from being continuous? It is in this context that most of the experiments on the subject have been performed. The results permit some insight into what may happen before swarm or supersedure queen cells are made.

The queen must continually communicate her presence and physiological activity to workers. That she does so is evident, for within 15 to 30 minutes of the time a queen is removed from her colony, the workers show agitation and some appear to search around the hive entrance. At any disturbance they expose the Nasanov gland and fan the air as though to attract home the lost queen.

It is important for the lack of the *Apis* queen to be perceived and reacted to promptly if the queenless colony is to survive. This is because, unless there happen to be gyne cells already present, the colony must construct emergency gyne cells around a few eggs or larvae less than three days old. If this is not done within a few days the opportunity will be gone, for the lost queen's female progeny will all be committed toward workerhood. Workers, in the absence of the queen, will lay eggs, but they will be unfertilized and (except in the South African race of *A. mellifera*) nearly always will produce only males; the colony will die with the expiration of the last workers.

The manner in which a queen makes her presence known to the workers has been the subject of a considerable series of experiments and numerous publications. The problem is how one bee, the queen, can keep up to 60,000 or more other bees aware of her presence, so that if she disappears they will take the needed action. The idea that the queen inhibits gyne cell construction as well as ovarian development of workers is old, but development of appropriate experimental methods has involved various difficulties, as summarized by Allen (1965a). The experimenter must consider such things as (1) the use of bees of an appropriate mixture of ages, (2) the normal seasonal variability in the tendency to build gyne cells and in the ovarian development of workers, (3) the fact that the ovarian development of workers is supposedly (but not according to Butler, 1959) related to the size of the group and declines if the group is too small, as does comb building and probably gyne cell construction, and (4) the failure of ovarian development in workers not given plenty of pollen. Recently a strong inhibitory effect of exposed brood on workers' ovarian development has been recognized. Apparently it is a more potent influence than the queen's pheromones

(Kropáčová and Haslbachová, 1971). It is likely that quantity of unsealed brood is an uncontrolled variable in some of the experiments on the inhibitory influence of queens and may explain some discrepancies in the results. No doubt other influences will be found, for there remain unexplained variations in results, perhaps partly due to genetic factors.

The effectiveness of a queen (or of alternative experimental materials or objects) in preventing signs of queenlessness among workers is usually assessed not in terms of the prompt agitation or searching mentioned above, which are difficult to quantify, but in terms of (1) production of emergency gyne cells and (2) enlargement of ovaries of the workers. Recognition of the queen by workers has also been assessed in terms of her ability to maintain a court of bees around herself.

Various authors independently (Butler, 1954b; Pain, 1955; Voogd, 1955) postulated a substance or substances on the queen's body that inhibits gyne cell construction and ovarian development of workers. Actually all three, in Britain, France, and Holland, published their first notices in 1954. As summarized by Allen (1965a) and Gary (1970), these and other authors showed that extracts of queens, whether presented in drinking water or on various objects (pith, dead workers, and the like), had an inhibitory effect similar to that of the queen herself. Even the daily addition of six workers that had recently licked a queen to a queenless group of 60 workers reduced ovarian development in that group (Butler, 1959). The bodies of queens from which fatty materials had been extracted did not inhibit unless reimpregnated with the extract. For the active principle Butler suggested the name "queen substance." Verheijen-Voogd (1959) and Butler and Simpson (1958) showed that the principal source was the queen's head, and more specifically, the mandibular glands and their reservoirs. An inhibitory substance was isolated, identified as *trans*-9-keto-2-decenoic acid, synthesized, and the inhibitory effect of synthetic material tested (Butler, Callow, and Johnston, 1961; Barbier and Lederer, 1960; Barbier, Lederer, and Nomura, 1960).

This material is relatively nonvolatile, although sufficiently so to be an attractant for drones in the field high in the air (Gary, 1962). In the nest, however, it is seemingly not attractive to drones or workers. The same material is available in similar quantities in heads of queens of *Apis mellifera, A. cerana,* and *A. dorsata* (Shearer, Boch, Morse, and Laigo, 1970).

Although produced in the mandibular glands, it appears to become distributed over the body of the queen by her grooming (Velthuis, 1970). According to Butler and his co-workers, it is then licked from her body by the court of workers that always surround her (Figure 11.1). Velthuis (1972), however, gives good reason to believe that it is mostly not licked from the queen but adheres to the bodies of workers that contact her. The queen is continually moving on the combs, and different bees from different parts of the nest form the court and either feed or lick the queen or both, as well as contacting her with their antennae, legs, etc. Allen (1960) found that 73 percent of the total number of workers in the circle around the queen remain there for less than a minute and that about three fourths of this 73 percent are there for less than 30 seconds. A few remain with her for long periods, up to 41 minutes. Bees of all ages and activities are involved in the court.

Assuming that the mean number of members of the queen's retinue is 11 and that each is with the queen, on the average, for a minute, and that no bee repeats, 660 bees would contact the queen per hour, 15,840 per day. Although in a small colony this means that nearly all bees could contact the queen daily, this would not be true for large colonies. Moreover, even in a small colony, it would not account for the very prompt response to loss of a queen. Therefore it has been postulated that worker bees that obtain queen substance from the queen's body pass it on to others via their food exchange, and that this process continues until every bee has a more or less uniform amount of the material.

Such dispersal of a relatively nonvolatile pheromone is consistent with the finding that more pheromone is needed to inhibit a large number of workers than a small number (Butler, 1954b), as would hardly be true for action of an odor. Moreover, this kind of dispersal seems reasonable in view of our knowledge of the extensive food transfer in a colony of *Apis*. But contrary to the theory of spread of an inhibiting substance by way of food are experiments of Verheijen-Voogd (1959) and Erp (1960) indicating that queen extract mixed with food containing over 5 to 10 percent sugar has no inhibiting effect, although the same material offered on bodies of dead workers has the usual effect. Verheijen-Voogd believed

FIGURE 11.1. Queen of the common honeybee *Apis mellifera* inspecting a cell, that is, her head and thorax in a cell so that only her wings and abdomen are exposed. The court or retinue of workers facing her is clearly recognizable. (Photograph by E. S. Ross.)

that the substance is perceived by taste and that sugary material masks its perception. Kaissling and Renner (1968), however, have identified olfactory receptor cells in the antennae that are specialized for perception of *trans*-9-keto-2-decenoic acid. Moreover, Velthuis (1972), in observations of the responses of queenless groups of workers to workers that had been with a queen, obtained behavioral support for antennal perception of *trans*-9-keto-decenoic acid. The material may also act in some direct physiological way as well as through the nervous system, for Butler and Fairey (1963) found that its injection into the worker haemocoel caused incomplete inhibition of ovarial development. This effect is likely to be via the "gonadotrophic" hormone from the corpora allata, for Lüscher and Walker (1963) found that queen extract significantly inhibits growth of these glands. Gast (1967) agreed with respect to early adult growth of the corpora allata and aded that the growth of nuclei in the neurosecretory cells of the *pars intercerebralis* is inhibited in workers by the presence of a queen. However, he noted a second growth period of the corpora allata that occurs only when the queen is present. That at least some workers must touch the queen (or under experimental conditions, the extract) for her *trans*-9-keto-2-decenoic acid to inhibit has been clearly demonstrated (Butler, 1959).

From the time of the early studies of queen substance, the presence of more than one inhibitory material has been recognized. Pain (1961), Pain and Barbier (1963), and others spoke of pheromone I and II (I was later recognized as *trans*-9-keto-2-decenoic acid) and believed both necessary to achieve inhibition of both gyne cells and worker ovaries. Butler and his co-workers also recognized the presence of additional inhibitory substances, but they obtained partial ovarial inhibition with *trans*-9-keto-2-decenoic acid alone. A second substance has been identified as *trans*-9-hydroxy-2-decenoic acid, which, outside the nest, is the principal cluster- or swarm-forming pheromone (Butler and Callow, 1968; Butler, 1969). It is probably the inhibitory scent recognized by Butler (1961) and pheromone II of Pain, or parts of these materials.

Other materials seem to be primarily important as attractive scents. Presumably they draw workers to the queen, causing them to lick her and thus acquire the inhibitory pheromones; by being deposited as footprint pheromones, they may make workers aware of the presence of the queen (Butler, 1959). The major such attractant is also produced in the mandibular glands (Gary, 1961), but a lesser one comes from Koschevnikov's glands on either side of the sting (Butler and Simpson, 1965). Velthuis (1972) differs from most earlier workers in maintaining that *trans*-9-keto-decenoic acid is not only perceived by antennal sensilla but is itself attractive.

Inhibitory materials are not entirely limited in origin to the mandibular glands. Thus Gary and Morse (1962a) found that queens whose mandibular glands had been removed could nonetheless partly inhibit gyne cell construction and could control small or sometimes ordinary (Zmarlicki and Morse, 1964) colonies. Velthuis (1970), unlike Morse and his co-workers, found that queens with the mandibular glands extirpated could inhibit ovarian development in workers and gyne cell construction by workers in colonies of various sizes and could elicit courts of workers just as do intact queens. He located the source of the inhibitory material in the abdomen, and postulated that it is secreted by epidermal glands of metasomal terga 2 to 6. He concludes that the queen does not depend on the products of her mandibular glands alone; there are multiple factors as in so many biological interactions. This view supports the concept of Verheijen-Voogd that these pheromones act, at least in part, not biochemically but through chemoreceptors. But the findings of Gast and of Lüscher and Walker suggest a biochemical influence also.

Inhibition from sources other than mandibular glands or by the attractant scents may be responsible for the finding by Butler and Fairey (1963) and others that the live queen inhibits even more effectively than combinations of the two mandibular gland pheromones, *trans*-9-keto-2-decenoic acid and *trans*-9-hydroxy-2-decenoic acid.

The communicative substances described above are among the most important materials shared by the members of an *Apis* colony. Somehow loss of the queen (cessation of her pheromone production) is perceived almost immediately; gyne cells often are started within three or four hours of removal of the queen, and worker ovarian enlargement is evident in a day or so. The immediate response could be a reaction to disappearance of one or more of the queen's scents, but the later responses seem related to the two enoic acids, the most effective of which operates principally by contact, not scent. The titer of these substances would have to drop rapidly in the face of an excellent distribution system if the drop were to be perceived and distinguished from mere failure of some bees to be near the queen or to obtain the pheromones first- or secondhand from her.

The queen produces about 0.1 μg of *trans*-9-keto-2-decenoic acid per worker per day or a total of perhaps 5000 μg daily, yet at any given time she contains only about 100 μg. She must be synthesizing the material constantly, and it must vanish at an equal rate. It seems that the inhibitory pheromones are inactivated in the colony. Johnston, Law, and Weaver (1965) have found that *trans*-9-keto-2-decenoic acid, at least, is metabolized quickly enough in the worker's gut to rapidly reduce the titer in the colony and thus account for the behavior of workers separated from the queen for a short time. They also suggest that workers may not destroy the carbon chain of the acid, but may return it in an inactive form with glandular food to the queen who then, at little expenditure of energy, could resynthesize the pheromone.

Supersedure often occurs in *Apis* colonies and is discussed in terms of the queen substance theory by Butler (1957, 1960). He postulates that an inadequate secretion of the inhibitory pheromones by a perhaps senile queen leads to gyne cell construction and supersedure, and reports that a queen being superseded in a large colony successfully inhibited gyne cell production in a small colony to which she was moved. As with gyne cells associated with swarming, it is the old queen that lays the eggs in the new cells.

The production of gyne cells before swarming, that is normal colony reproduction, is also explicable in terms of the inhibitory pheromones. Butler (1960) found that at such times the queen's production of the pheromone mixture (queen substance, assayed by worker response) drops to about one fourth of the level of production maintained by queens not involved in colony reproduction. Swarming may occur in a very crowded colony, however, even though the queen is producing the normal amount of the pheromones; perhaps their collection and distribution is ineffective under crowded conditions.

B. Killing of excess gynes

As is well known, there is typically one queen per colony. In some semisocial and primitively eusocial forms there may be several, and caste intermediates may make the number rather arbitrary. In *Melipona nigra* there are typically two and sometimes several functional queens per colony, a situation unusual for highly eusocial bees.

In semisocial and primitively eusocial bees, young gynes start new colonies. Mortality may be high, but unless a nest is invaded by a second gyne of the same (or different) species, as is common in *Bombus,* the mortality is not socially based. In many species (some halictines) excess gynes become auxiliaries, that is, workers (in semisocial groups), and ultimately die or leave (Chapter 6, B2). In highly eusocial bees, however, excess gynes are killed in the parental colonies; destruction of excess gynes is a social problem. Only a small percentage of the gynes produced actually become queens in highly eusocial bees, above all in *Melipona.*

In many stingless bees young gynes, although killed and carried out by workers sometime after maturation, are tolerated for a while. In colonies of *Trigona postica,* according to Kerr, Zucchi, Nakadaira, and Botolo (1962), one to seven virgin gynes are usually present among the workers; such gynes typically are well treated for about a month, then killed (Kerr, 1969). In some small species of *Trigona* excess young gynes are imprisoned by workers in wax and propolis chambers (Juliani, 1962, 1967). Such chambers have been recorded for several South American species in the subgenera *Plebeia* and *Tetragona* (Kerr, 1969; Terada, 1972). The workers form a circle around a virgin gyne and hold her in place or trap her in an incomplete pollen or honey pot. Then they construct a chamber of cerumen (wax and resin) around her or close her in the pot. Sometimes the gyne helps in building the walls which will enclose her. Such an enclosure has a small opening through which the gyne is fed for a time; in some cases the opening is closed between feedings. Sometimes a worker opens an enclosure and goes in, and presumably feeds the gyne. The gynes in their enclosures are usually attacked and killed by workers when there are mature gyne cells from which new gynes will soon emerge. Sometimes the time is short, but Juliani records a case in which a gyne of *Trigona julianii* was imprisoned from May to August.

Maintenance of young gynes appears to be a device that keeps them in reserve in case of death of the queen. It thus has much the same effect for *Trigona* as continuous production of excess gynes in *Melipona.* The equivalent for *Apis* is the mechanism of producing replacement gynes. The Meliponini use the relatively expensive method of continually producing gynes in case they are needed, while *Apis* produces new gynes only at certain seasons or at a time of an actual need.

As indicated above, it is usually workers that kill excess gynes in Meliponini. Kerr (1969), however, says that, at a time of queen supersedure in *Melipona quadrifasciata,* the newly accepted virgin queen helps the workers to deal with other virgin gynes that may emerge. Moreover Silva (1972) noted that in a colony of *Trigona* (*Plebeia*) *droryana* in which a young gyne emerged but was not imprisoned, she attacked the old physogastric queen in about three hours, biting her abdomen. Attacks by the gyne alone continued for about an hour, after which workers too attacked the queen, who was killed and partially dismembered within a few hours. Eighteen days after emergence, the young queen made her mating flight, and not until 42 days after that did cell construction resume and egg laying by the new queen begin.

Silva, Zucchi, and Kerr (1972) observed queen replacement in *Melipona quadrifasciata* and *quinquefasciata.* In four cases in which the laying queen disappeared or was removed, they found that several young gynes, which emerged one after another over a period of days, were killed before one was allowed to live. Others were killed before the successful one mated and began laying. The basis of selection of the successful gyne could not be determined. The mating flight occurred four days after acceptance of the young gyne by the colony. Thereafter her abdomen swelled in a few days to the size of that of a laying queen.

In *Apis,* as is well known, a young gyne stings others in the colony to death, often even before they are mature. Should two young gynes meet as they wander about, they grapple and attempt to sting until one kills the other. The workers pay little obvious attention, but carry out the corpse, as they would any other dead bee. Commonly a young gyne will tear open other gyne cells that she finds. Workers then destroy the cell and its contents, even if the gyne has not already killed the larva or pupa within.

Young gynes often produce shrill piping sounds with the flight musculature and skeleton, but without moving the wings (Simpson, 1964). When she does this, a gyne crouches so that the thorax is pressed against the comb, which is thus vibrated. Other young gynes perceive the vibration, presumably through the substrate, and respond. Piping seems to excite the gynes and probably helps rivals to find one another.

Workers also play a role in destruction of excess gynes. This was demonstrated by Darchen and Lensky (1963) in the case of colonies with multiple queens that had had their stings cut off so that no one queen was able to dispatch the others. But even in ordinary colonies, workers sometimes kill new gynes. Gary and Morse (1962b) showed that neither swarming nor supersedure necessarily follows production of a new gyne. The workers may kill her. These authors record one colony in which 21 gynes were reared and killed in one season. Killing of introduced gynes by workers has been studied by Yadava and Smith (1971), who propose that a disturbed gyne secretes a stress pheromone that stimulates attack on her by workers and thus her own death. The possibility of evolution of such a mechanism is discussed in Chapter 20 (D).

C. Castes among orphan groups of workers

In the absence of queens, there are forces leading to "caste" formation and division of labor within groups of workers. Such groups are often spoken of as orphaned colonies, although they can sometimes be made artificially simply by putting a group of worker pupae or of adult workers together. Plateaux-Quénu (1961) has reviewed the happenings in these groups not only for bees but for other social insects. Such groups are plainly semisocial, although they have been studied only in species that are typically eusocial. Their study should provide insight into the factors that prevent or reduce egg laying in normal colonies in which the queen is present.

1. Primitively eusocial bees

Interactions among individuals in orphaned colonies are readily demonstrated. In *Bombus* and in *Lasioglossum zephyrum* ovarial development is related to the number of workers living together (Table 11.1). Nothing is known of the causation of this relationship, but it may well be that in *Bombus* workers in groups are more often disturbed and activated by other individuals and hence eat more and are more likely to engage in other activities.

TABLE 11.1. Mean number of eggs per bee per day laid by orphan worker bees living individually or in groups of two to ten.

	Numbers of workers per group					
Species	1	2	3	4	5	10
Bombus pratorum	0.09[a]	0.56[a]	—	—	0.63	0.87
Bombus agrorum	0.13[b]	0.27	—	—	0.34	0.34
Lasioglossum zephyrum	0.87[a]	0.57[a]	0.36	0.42	—	—

SOURCES: Data on *Bombus* from Free, 1957, based on seven days observation; on *Lasioglossum* from Michener, Brothers, and Kamm, 1971a, based on variable but longer periods of observation.

[a] Significantly different from the next figure in the row.

[b] Not significantly different from the next figure in the row, but different from the third figure.

In *Lasioglossum* the reverse is to be seen—the more bees the fewer eggs laid per bee. This is the more common pattern (Michener, 1964a), and the difference from *Bombus* may be related to the fact that in *Lasioglossum* a cell has to be made and provisioned before each egg is laid, while in *Bombus* eggs are laid and the resulting larvae fed later.

Laboratory observations of groups of workers have been made for two species of halictines. In June and July, when most female pupae of *Lasioglossum zephyrum* in Kansas give rise to bees that become workers, artificial nests can be made containing pupae, but no adult bees. In observing the colonies that result from maturation of these pupae, groups of two to six unfertilized adults, we have found that one in each nest undergoes more ovarian enlargement than any others, stays in the nest most of the time, and does not collect pollen. The others, with varying but less ovarian enlargement, gather pollen, guard the nest entrance, and provision cells (Michener, Brothers, and Kamm, 1971a). In short, one has become a queen, the others workers, even in a species whose castes are normally feebly differentiated and even though the queen has not overwintered or mated. (Of hundreds of progeny of such queens, all were males.) There must be interactions among the females that somehow influence their ovarian development and their behavior.

The female that becomes the queen is not necessarily the largest of the females of the artificial colony. She is usually the oldest bee. Seemingly a bee that matures a day before the others tends to inhibit their ovarian development and becomes behaviorally and physiologically the queen. The inhibitory mechanism may be that her activities, running about and nudging other bees, reduce the frequency with which the other bees forage during the young adult stage, when ovarian enlargement may most easily occur. The same thing could reduce their probability of mating. If all the females mature at about the same time, the one to become the queen is usually the largest. If the colony is kept in a room with males, the one that becomes the queen is considerably more likely to mate than the others, and female progeny are then produced. Such a bee is typical of queens produced in summer. These data give some insight into the process of queen replacement which in *L. zephyrum* occurs during the life of the colony.

Experiments to be described by D. J. Brothers and me, in which queens were removed from colonies of *Lasioglossum zephyrum* and subsequent developments observed, are now being analyzed. The results suggest that upon removal of the queen, an adult worker that is specialized neither as a frequent guard nor as a frequent forager is most likely to become the replacement queen. Within a few hours of removal of the queen, such a behaviorally unspecialized bee may show characteristics of a queen, such as an increased frequency of nudging of other bees and of backing away from them.

Study of two orphaned colonies of *Lasioglossum nigripes* also showed one queenlike individual arising among the workers of each colony, although the other workers also all had considerably enlarged ovaries and might have laid some eggs. In both cases the queenlike bee was the largest, but only slightly so and only in head width, not wing length, these being the two measures of size that were made (Plateaux-Quénu and Knerer, 1968).

The same occurs in queenless colonies of *Bombus,* in that one or more workers become egg layers. Such individuals in *Bombus* are presumably workers which had experienced partial ovarian development and taken on some other traits of queens even in the presence of the legitimate queen. Such workers are known in most primitively eusocial species of bees.

In *Bombus* certain workers may lay even in the presence of the queen, although her dominance tends to prevent it (Chapter 12), and others have enlarged ovaries but do not lay. The death of a queen therefore commonly releases several workers for laying. The animosity between them may be extreme (Brian, 1952, 1954), but the one with the most enlarged ovaries usually dominates (Free, 1955c).

It is this animosity that appears to prevent establishment of permanent colonies in most species of *Bombus.* Cumber (1963), in comments on *Bombus terrestris,* explains the situation as follows: "It is apparent that ovarian development in the young nest is confined to the queen. As the nest grows in size and the vigour of the foundress diminishes, it is usual for a number of . . . house bees (as opposed to the foragers) to develop ovaries and to lay eggs. Under such conditions, antagonism, which is the very antithesis of the social harmony in the earlier stages of nest growth, develops between such laying individuals. When the weakened foundress dies, the antagonism gathers greater momentum, and precludes the possibility of the development of perennial nests." A large nest found in spring in New Zealand "illustrates well the disharmony which occurs when a thriving colony loses the control exercised by the foundress. Portions of the brood appeared to become the domain of individual bees, in this case both workers and queens [young gynes, C.D.M.], but mainly the latter. Periodically quarrelling developed between the domineering individuals of adjacent domains and mortal conflicts developed. Throughout the period of observation, and with increasing momentum in the earlier stages, it was common to find dead and dying workers and queens in the outer chamber. In some cases the pile of dead bees was stuck down with honey. Dissections often showed evidence of the victims having been stung, and invariably there was considerable ovarian development. Some of the queens . . . were fertilized."

Free (1955c) found that in orphan groups of workers, quarrelling started within a day and in each colony one individual (α) always emerged as dominant, often a second (β) individual dominant to all but α, and sometimes a γ individual could be recognized. The dominant (α) individual of each group guarded and presumably laid eggs, acting much like a queen in every way, even to the disappearance of evident attacks on other bees within a week or two after she started laying (Figure 12.2). Presumably the bees can recognize one another, perhaps by an odor or behavior somehow related to ovarian enlargement, as suggested by the observation that the α bee singles out the β bee for its most frequent attacks.

Plateaux-Quénu (1959) found a naturally orphaned colony of *Lasioglossum marginatum* which seemed to have developed an enormous laying capacity: 12 workers with large ovaries like those of old queens, 8 with moderate-sized ovaries like those of queens in two-year-old nests, 17 with one large egg each like queens in first-year nests, and about 40 with slender ovaries like ordinary workers. The nest contained 590 cells. The ovarian development of so many workers in an orphaned nest suggests that the presence of the queen normally inhibits such development.

2. *Highly eusocial bees*

Even among highly eusocial bees, in the absence of the queen division of labor and castelike differences arise among the workers. It is well known that in a queenless colony of *Trigona* or *Apis,* male-producing eggs are laid by the workers. Not all the workers are involved in laying, although considerable numbers may have enlarged ovaries.

Commonly 10 percent or more of the workers in an *Apis* colony have become layers a week after dequeening. There are records of laying workers producing 19 and 32 eggs per day—enormous rates compared to solitary bees, but low compared to queens. It is not clear whether or not these rates refer to ordinary laying workers or to the false queens described below. Perepelova noted that the average amount of time spent in laying an egg was 78 seconds, about eight times that required by a queen (Ribbands, 1953).

There is an inverse relationship in *Apis* between the amount of brood comb containing open cells (larvae being fed) and egg producing by laying workers; reduction in the amount of such comb results in larger numbers of laying workers (Pain, 1961). Presumably the nutritional requirements of the larvae reduce the nutrients available for egg production by the workers.

Aggressiveness and fighting are common among workers in queenless *Apis* colonies containing egg layers. Sakagami (1954) showed in one study (but not in another) that in both *A. mellifera* and *cerena* workers mauling or attacking others might be very variable in ovarian development, but those being mauled nearly all had enlarged ovaries. This is contrary to the situation in *Bombus,* where individuals with enlarged ovaries are the attackers, and may suggest a reaction related to destruction of excess queens. If laying workers have, to other workers, some of the features of queens, then attacking them as supernumeraries is explicable. Laying workers are not killed in these quarrels, however.

Sometimes in a queenless colony of *Apis* a laying worker over ten days old achieves a level of queenliness clearly recognized by the colony (Sakagami, 1958a), and in such cases the aggressiveness described above disappears. A colony can have only a single false queen—a worker that is treated like a queen, having a small court of usually three to eight workers around her, feeding her and licking her, as do the bees in the larger court of a true queen. A false queen (of *A. cerana*) laid eggs at the rate of 82 per day, an extraordinarily high rate for a laying worker. Not surprisingly, abdomens of false queens are a little swollen. In the presence of a false queen, ovarian development and oviposition by other workers decreases, a fact also demonstrated with extracts of laying workers by Velthuis, Verheijen, and Gottenbos (1965). (It is likely that these authors extracted laying workers at least some of which had attained the false queen stage and that Pain (1954), who detected no inhibitory pheromones in laying workers, dealt with ones that had not attained that stage.) The false queen, then, is from the social viewpoint a queen, although of course she is not mated and in most forms of *Apis* produces only male offspring. Probably false queens produce the pheromones of true queens and it is because of these substances that the colony resumes rather normal behavior.

Dreischer (1956) has enumerated a number of additional influences of queenlessness in *Apis,* that is, differences between the workers of a normal colony and those of a queenless one. In a queenless colony brood care (including cell cleaning, feeding of larvae, and so forth) is diminished; social cleaning of workers by other workers is increased; hypopharyngeal and wax glands remained enlarged later in worker life (implying increased longevity, Chapter 30, D), and there were some size differences in the corpora allata. Thus the queen has a marked influence on many aspects of the behavior and physiology of workers of her colony in highly eusocial forms.

D. Inhibition of reproductivity of workers

This topic is the converse of production of replacement queens or of queens in orphan groups. Production of such queens involves relaxation of inhibition against reproduction by females. This section, therefore, is far from a complete account, but brings together certain materials not adequately treated in the preceding parts of this chapter.

Even after a female bee has developed as a worker, she retains, probably in all species, the potentiality for ovarian development. If the queen dies, some workers in the colony lay and male bees are the usual result. Prevention or minimization of egg laying by workers, however, is an essential part of the mechanism to maintain the distinction between workers and queens and thus, in the presence of the queen, to maintain the division of labor between castes.

Some mechanisms by which egg laying and associated ovarian development of workers are inhibited by the queen have been studied in *Bombus* and *Apis.* In *Bombus pratorum* the aggressiveness of the queen and perhaps her odor are involved in maintenance of her dominant position and perhaps in inhibiting ovarian development of workers. She is more effective in a small group of workers than in a large one (Free and Butler, 1959). In *Bombus atratus* young gynes or queens but not older queens are effective in inhibiting ovarian development of workers, according to Saraceni (1972). She suggests that a pheromone may be responsible for the success of the young queens, an idea that is reminiscent of Rössler's postulated caste-controlling pheromone, which seems also to be produced only by young queens (Chapter 10, B).

Queens in *Bombus* influence not only ovarian development but the general level of activity of workers. The mechanism is not adequately known, but Meidell (1934) made some interesting observations clearly showing the existence of such a mechanism in a colony of *Bombus agrorum* from which he temporarily removed the queen. He found that an hour after her removal, activity in the nest had diminished below normal, several house bees having stopped work. Half an hour later the field bees had also stopped and all workers rested quietly on the comb, occasionally going to the honey pot to feed. After three hours limited foraging began again but the few field bees that left returned with nectar only, not carrying pollen. After four hours the queen was returned to the nest and began her normal activities as though nothing had happened. House bees gradually began their usual behavior and in half an hour their work seemed normal, but foraging for pollen did not resume for an hour after the return of the queen.

In *Apis,* pheromones secreted by the queen inhibit ovarian development of workers; the orgins and functions of these materials are discussed in section A above.

E. Concluding remarks

Among the important interactions in semisocial and eusocial colonies are those related to the reproductive activities of females. This matter may be looked at from the level of caste differences, without regard to their origins, as in Chapter 9. Or the concern can be with the evolutionary origin of caste differences, as in Chapter 20. A more physiological viewpoint is that of caste determination, having to do with the environmental or genetic factors that cause a given bee, during its maturation, to become a gyne, a worker, or an intermediate. This view, having to do with the origin of differences during the lives of individual bees, is dealt with in Chapter 10. The viewpoint of the present chapter is more sociological. It has to do with the conditions within the colony that lead to production of gynes, to ovarial development of workers or prevention thereof, or to killing of excess gynes should they develop.

In semisocial and primitively eusocial species, production of gynes must be largely controlled by the same sort of selection that controls productivity of nonsocial forms of life. Each gyne can, in theory, start a new colony, and this is likely to be without influence on the parental colony whose genes are being thus dispersed. The number of reproductively successful gynes will be maximized. In highly eusocial species, with their permanent colonies, however, the reproductive success of gynes will drop drastically if too many are produced or survive to leave the colony with swarms. Among other things, repeated fission will result in colonies too small for success. Therefore social control of production and survival of gynes in highly eusocial colonies is the principal topic dealt with above. In section C1, however, it is shown that the potentiality for social control exists also in primitively eusocial bees.

The development of castelike differences among workers in orphan colonies is widespread (see section C). Examination of this phenomenon may give some insight into the sort of relationships that existed between castes in ancestral forms. For example, in orphan *Apis* colonies one sees aggressive interactions among workers, with the outcome often related to ovarian size. Such behavior is not otherwise found in highly eusocial bees but is of regular occurrence in primitively eusocial forms. Perhaps the *Apis* are thus showing behavior characteristic of a primitively eusocial ancestor, characteristics that are largely lost in the highly eusocial bees of today.

12 *Division of Labor among Workers*

Within the worker caste there often is behavioral specialization. In bees this is never so elaborate as in ants, with their morphologically different subcastes, but it is sufficient to significantly influence integration within colonies. In the highly eusocial bees it is largely temporal, varying with age, needs of the colony and environmental conditions, but in certain primitively eusocial forms, especially *Bombus,* there is sometimes, in addition, permanent although commonly incomplete division of labor among workers.

The expression "division of labor" is used here in a broader sense than is usual, to include any behavioral pattern that results in some individuals in a colony performing different functions from others, even if only temporarily. For example, differential flower constancy among workers is a kind of division of labor, for some workers of a colony are exploiting one food source while others exploit another. Moreover, division of labor need not be absolute. Thus if two workers both perform functions *a* and *b,* but with strongly different frequencies, one can regard each as specializing in a different function, so that division of labor exists. For example, function *a* might be performed 75 percent by one worker, 25 percent by a second worker, while function *b* is performed 25 percent by the first worker and 75 percent by the second.

A. Permanent division of labor among workers

This is division of labor in which bees, once having adopted different activities, usually maintain those activities rather than passing through some regular or irregular sequence of different behavior patterns. Such division of labor is largely related to the varying queenliness of the workers, those that are more like queens often having somewhat different functions from those less like queens.

In halictids like *Lasioglossum imitatum* the egg-laying workers (Figure 9.7) of a mean size intermediate between queens and other workers have at least one activity, laying eggs, that other workers do not.

More extensive division of labor is reported for *Lasioglossum calceatum* by Bonelli (1965b). He says that one workerlike individual out of the first (June) brood of workers remains in the nest instead of foraging and, like the queen, is active in guarding the nest. He speaks of a subdivision of the worker caste into two subcastes, one that forages and is short-lived, the other that guards and is long-lived. Behaviorally the latter is merely a queenlike worker, and Bonelli himself later (1968) concluded that such guard bees mate, survive the winter, and establish colonies in the spring. Whether or not he is correct in this, the guards are rather clearly physiological and behavioral intermediates between queens and workers, not a distinctive subcaste. The parallel to bumblebee guards which have enlarged ovaries is obvious.

Further evidence of such division of labor has recently been obtained by D. J. Brothers and me in the course of work on small laboratory colonies of *Lasioglossum zephyrum.* The youngest of the bees were several days old, thus avoiding the temporal changes that distinguish young adults from older ones. For several behavioral components, the bee that most frequently guards appears to belong to a different statistical population than the bee that most frequently forages. Thus, although two types of workers are conveniently called guards and foragers (after typical activities), they have other characteristic behavior. For example, in comparison with a forager, a guard less often turns around when a queen nudges it, less often follows the queen, less often works on cells, and less often passes other bees. Among workers, the guard usually has the largest ovaries, the forager, the smallest. In spite of considerable day-to-day irregularity in behavior, observations made during periods up to 25 days—the greater part of the life of a worker—showed no evidence of changes from guard to forager or vice versa. The division of labor appears permanent, at least so long as the same bees remain in the nest. Other workers occupied intermediate positions between the principal guard and the principal worker in that they more often engaged in both guardlike and workerlike behavior, but the total activity of such bees was often low. They seemed behaviorally unspecialized and somewhat inactive. We believe that from such bees a guard or a worker is recruited when needed. In spite of the fact that, as in *L. calceatum* and *Bombus,* the principal guard usually has ovaries next in size to those of the queen, new queens also are usually recruited from unspecialized bees.

Worker size in *Bombus* is highly variable, small ones often being little over half the size (in linear measurements) of large ones (Figures 10.4 and 12.1, and Medler, 1965). Size variation among workers is especially apparent in the pocket-making species.

Knee and Medler (1965), Röseler (1967b), Plowright and Jay (1968), and Pouvreau (1971) verify that in colonies of some species, at least, mean size of the workers produced (indicated by a wing measurement) increases with the age of the colony, but the increase is sometimes erratic or temporarily reversed, and is not statistically significant in the first half of colony growth. In fact Röseler and Pouvreau show that worker size usually decreases for a few weeks after inception of the colony. As noted by Free and Butler (1959), it has not been easy to show the increase in size and there have been differences of opinion about it, because of different activities and different average longevities of workers of different sizes. For example, Cumber (1949a) reports an increasing percentage of small workers in colonies as the season advances. This is now believed to result from the high mortality of the larger workers as field bees subject to parasitism (for example, by conopids) and predation and is not inconsistent with a progressive seasonal increase in the average size of newly produced workers. Progressive increase in mean worker size as the colony grows is paralleled in *Lasioglossum malachurum,* in which the second brood of workers averages larger than the first (Figure 9.5).

As will be shown in section B below, there are differences in the ages at which *Bombus* workers of different sizes take on different activities. The result is that there are average differences in the activities of workers of different sizes, small ones spending more of their time as house bees than large ones. The greater amount of foraging by large workers is illustrated by the data shown in Table 12.1. A further difference between workers of different sizes is that large foragers carry more pollen loads; small ones are likely to return to the nest with nectar only. Thus in one colony *B. agrorum,* of returning foragers weighing 150 mg or more only about 26 percent (of 349 records) came back without pollen loads, while

FIGURE 12.1. Selected females of *Bombus fervidus* from a single nest. Lengths of the marginal cell of each individual was (*left to right*) 2.4, 2.8, 3.2, 3.5, 3.9, and 4.3 mm (see Knee and Medler, 1965). The bee at the right is a young gyne, the others are workers. (Photograph supplied by J. T. Medler, courtesy of University Extension, University of Wisconsin.)

TABLE 12.1. Relative amounts of foraging by large and small foragers in three *Bombus* colonies of different species.

Species (and no. of bees in colony)	Percentage of bees that foraged on more than half the bee-days of observation: Bees below mean weight	Bees above mean weight
B. sylvarum (49)	30	67
B. lucorum (181)	43	60
B. agrorum (31)	50	80

SOURCE: Modified from Free, 1955b.

of bees weighing less than 150 mg 72 percent (of 162 records) returned without pollen loads (Brian, 1952). Free (1955b) obtained somewhat similar results for several other species.

As might be anticipated, there are differences in the kinds of flowers visited at the same time and place by *Bombus* workers of the same species but of different sizes. Cumber (1949a) reports differences in proboscis length (therefore also in bee size) of workers of *B. agrorum* visiting different flowers. These differences could be understood in terms of flower structure. The statistically significant portions of his data are shown in Table 12.2. Such division of labor as to flowers exploited among size classes of workers is unlikely among other bees, for size variation among workers is far less that in *Bombus.*

Meidell (1934) notes that in a small colony of *Bombus agrorum* under intensive observation, no two workers used the same amounts of time for the four chief activities listed: "housework" (construction, cleaning, incubating), brood care, foraging, and guarding. Free (1955b) classified bees as "consistent foragers" if they went out on 75 percent or more of the days of observation and as "consistent house bees" if they were never seen to leave the nest. Over periods of several days about two thirds of the workers fell into one or the other of these categories, only one third falling in neither category, being intermediate.

It is evident from the preceding paragraphs that within the worker population of a *Bombus* colony there may be considerable division of labor based in large part on whether the workers are more or less queenlike in size and other attributes. Perhaps greater efficiency is attained by small bees moving about in the small spaces within the nest and by large bees in pollen carrying. But although commonly more or less permanent, the division of labor or at least the labor is not immutable. Free (1955b) removed foragers from colonies and found that house bees that probably had never been foraging began such activity and sometimes even the queen resumed it. Inconsistent foragers became more consistent. Conversely, removal of house bees caused other bees to increase their household activities.

In bumblebees, while there is little obvious interaction and no aggression among most workers, the existence of a few individuals that are aggressive toward other workers, toward robbers, or toward other enemies that try to enter the nest is well known. These are commonly house bees, former foragers with somewhat enlarged ovaries. They play the part of a functionally different group. They are workers that partake of certain attributes of queens without, however, necessarily being caste intermediates as judged by size.

Free (1955c) has studied such guard bees in colonies of *Bombus pratorum.* He found that a social hierarchy arises among some of the bees in a colony. Normally the queen is dominant; there may be two or three bees, the guards, arranged in a dominance order such that α can attack β and γ, β can attack γ, but not α, and so on, but most of the bees are in an amorphous subordinate group. Aggressiveness in a general way is related to the degree of ovarian development; young gynes tolerate one another, as do ordinary workers, but with ovarian enlargement aggressiveness develops. Any group of young gynes breaks up when they become aggressive, as the ovaries enlarge, and ovarially developed workers likewise

TABLE 12.2. Length of proboscis among workers of *Bombus agrorum* at two English localities.

Flowers	Mentum length (mm): Mean	Range (and N)
Locality 1		
Balleta nigra	2.36	2.15–2.59 (20)
Lamium album	2.14	1.77–2.43 (27)
Locality 2		
Symphytum officinale	2.69	2.46–2.95 (20)
Epilobium hirsutum	2.22	1.91–2.61 (32)

SOURCE: Cumber, 1949a.
Note: Specimens were taken at approximately the same time on two different flower species at each locality.

become aggressive and take high positions in the social hierarchy. Whether these aggressive workers have special significance to the colony other than in defense and sometimes in laying unfertilized eggs is not known.

Brian (1952, 1954) made similar observations on *Bombus agrorum.* She notes that the workers against which aggression is directed are most commonly others which also have ceased outside activities and probably are also developing in a queenlike direction by ovarial enlargement. Such bees may be forced to leave the nest; it is not clear whether their queenlike tendencies are in this way inhibited. When an attack was against a forager just returning from the field, that forager frequently discharged nectar which either fell to the floor or was drunk by the attacker. Such occasions were the only times when liquid was seen to pass from bee to bee in a nest. The attacks were not motivated by hunger, however, as there was honey present in the nests at all times (see also Chapter 11, C1).

Aggressiveness is prevented in any group if pollen is withheld; under such circumstances there is no ovarian development. Animosity is not, however, perfectly related to ovarian development; sometimes the β individual has larger ovaries than the α, which perhaps maintains its position because of experience in past encounters. The dominance order may change, possibly with changes in ovarial size. Animosity is maximal at the time of first egg laying and diminishes later, even if laying continues (Figure 12.2). The increase in tolerance after first laying also occurs in queens; multiple queen colonies can be made artificially after the early stages in colony development are past.

B. Ontogenetic division of labor among workers

In addition to the more or less permanent division of labor among workers described above, there is diversity of activities among different age groups. In the highly eusocial bees and even in *Bombus,* this is exceedingly important, involving not only behavior but physiology—such things as the secretions of various glands.

1. Primitively eusocial bees

In *Lasioglossum* (*Dialictus*) *imitatum, umbripenne,* and others, young adult females remain in the nest for a period of several days, in one case at least six days. During this time they move about in the tunnels and guard the entrance; the youngest, at least in *imitatum,* are timid and often descend into the burrow without blocking it; guards with a day or two of experience are much more effective (Michener and Wille, 1961; Wille and Orozco, 1970). Young adults of *Dialictus* species also do much of the excavation of burrows (Batra, 1964, 1968). Later they become foragers, and while they may guard the nest entrance, they do not usually do so.

In *Bombus* there is also temporal division of labor; young workers remain in the nest while older ones forage. Sakagami and Zucchi (1965) as well as Brian (1952) emphasize, however, that except for the youngest workers, any bee may engage in any activity. Thus, when in the nest, foragers attend to house activities. There is a report that the oldest workers of *B. muscorum* are active only as foragers, but this requires verification, especially since *muscorum* is a close relative of *agrorum,* the species that Brian studied. The age at which foraging begins depends on size, the larger workers becoming foragers at an earlier age. For example, Brian (1952) noted that in *B. agrorum* the small workers do not begin to forage until about 15 days old, while large ones begin at the age of five days. Thus, at any time the average size of foragers is greater than that of nest bees, and small workers are often among the latest in the season to forage (Miyamoto, 1963). Some of the smallest workers in some colonies of some species never leave the nest (Brian, 1952; Free and Butler, 1959). This, of course, involves permanent division of labor as compared to workers that do forage.

Brian (1952, 1954) reports that the age when foraging begins was not modified by removal of existing foragers from the colony or even by starvation. She therefore concludes that the change of behavior is a result of inherent ontogenetic changes. She even found that the amount of pollen collected is not influenced by the number of larvae in the nest. Kugler (1943) found that removal of syrup from the honey pot in the nest increased foraging activity but did not necessarily advance the age at which foraging begins.

On the contrary, Free (1955b) found that activities depend significantly upon the needs of the colony, as indicated in greater detail in section A above. The worker normally passes through a certain ontogenetic behavioral sequence, but its rate can be altered or direction reversed

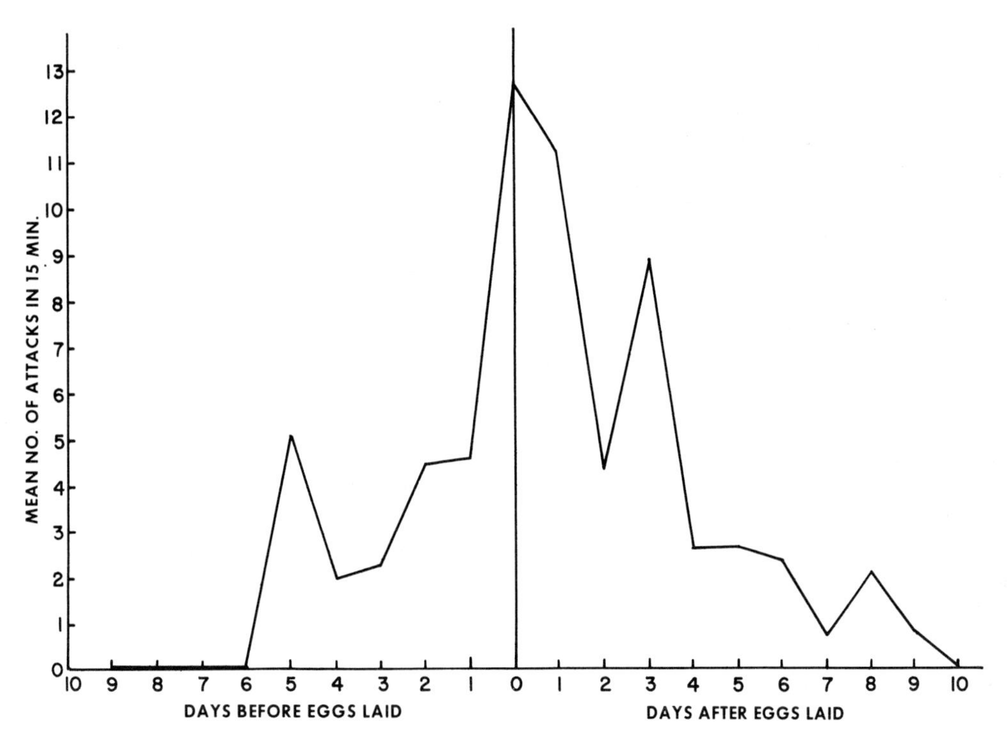

FIGURE 12.2. Graph showing the mean number of attacks on nestmates by the dominant worker per daily 15-minute observation period in queenless colonies made up of ten workers of *Bombus pratorum*. (From Free, 1955c, copied with the permission of the publishers of *Animal Behaviour,* Baillière Tindall.)

by the needs of the colony. For example, when the foragers were removed from a *Bombus* colony, younger bees that would not normally forage so soon began foraging. Likewise, removal of house bees resulted in foragers carrying out household activities.

Röseler (1967a and in MS), working in climate-controlled rooms, has verified and elaborated upon these results. In studying *Bombus hypnorum, terrestris,* and others, he showed that the hypopharyngeal glands and wax glands are already functional on the second day after emergence of workers, providing abundant pollen feeding is possible. The hypopharyngeal glands enlarge rapidly until about the fifth day, and typically remain about the same size, sometimes with slight further enlargement, until about the twentieth day, after which there is usually slow reduction in size. The wax glands start diminishing earlier, about the tenth day. The hypopharyngeal glands normally degenerate on about the tenth day, however, if younger workers have emerged and taken over the secretion of larval food and of wax. In the long-term absence of younger workers, these glands can remain functional until the fiftieth day. Thus under proper nest conditions a worker may be kept for life in one function, whereas an *Apis* worker perhaps always changes its functions (Sakagami, 1953b) in spite of the great flexibility also found in its ontogenetic behavioral sequence. If there are no larvae to receive the glandular food, or if younger workers of *Bombus* take over feeding, the pharyngeal glands recede in spite of abundant pollen feeding, and the corpora allata and ovaries enlarge. Thus the feeding of larvae and egg development by workers are mutually exclusive. This is illustrated by small orphan groups of *Bombus* workers, where eggs are laid only after any larvae being cared for have pupated.

2. *Apis*

In the highly eusocial bees there also is an ontogenetic sequence of activities which is related to a sequence in glandular development. In spite of relatively uniform-sized workers, the heaviest workers, as in *Bombus,* are those whose ontogenetic progress is most rapid (Kerr and Hebling, 1964).

An excellent review of the activities of workers of *Apis* has been given by Free (1965). It has been the usual view, based especially on the important works of Rösch (1925, 1930), that each worker *Apis* carries out the same programmed sequence of activities, and that flexibility is achieved by variations in the durations of the various tasks in response to conditions in the colony. Much subsequent writing and the studies of temporal division of labor in meliponine bees reported below have been carried out with this view in mind, although in fact Rösch recognized great flexibility, alternation in activities, and the like.

Lindauer, Perepelova, and Ribbands (references in Ribbands, 1953), however, emphasize the continued diversity of behavior in preforaging life, each bee changing from one hive duty to another at short intervals. Presumably the general sequence of activities noted by Rösch and others results from preferences that change with age. But all hive tasks can still be carried out, as in the solitary ancestors, which must have been able to undertake all nesting and also foraging tasks throughout much of their lives.

Table 12.3 gives some of the available data on ages at which *Apis* workers have been seen performing various activities in "normal" colonies, that is, colonies containing workers of all ages. It can be seen that striking differences exist among the studies of different authors. For example, although the lines attributed to Rösch in the table show predominant activities at certain ages, Lindauer's data show the same activities predominating at much greater ages. As another example, Rösch and Perepelova demonstrated that older larvae are fed on the average by much younger workers than those engaged in feeding young larvae. Lindauer's data do not support these conclusions.

TABLE 12.3. Ages (in days) at which various activities are carried out by workers of *Apis* in colonies containing individuals of all ages.

Activity	No. of records	Age range	Mean age	Reference
Feeding larvae > 4 days old	24	3–11	4.6	Rösch
	40	3–12	5.2	Perepelova
	161	2–28	11.5	Lindauer, 1952
Feeding larvae < 4 days old	25	6–13	8.6	Rösch
	47	6–16	9.2	Perepelova
	79	2–26	12.8	Lindauer, 1952
Receiving nectar from foragers	19	8–14	11.2	Rösch
Cleaning debris from hive (and	7	10–23	14.7	Rösch
removing remains of cell caps and cleaning cell bottoms)	330	2–20	13.9	Perepelova
Smoothing edges of cells and cleaning cell walls	321	2–20	8.7	Perepelova
Comb-building cluster	736	2–52	15.8	Rösch
First flight from hive	41	5–15	7.9	Rösch
First foraging trip	34	10–34	19.5	Ribbands
	47	10–32	20.1	Ribbands
	52	9–35	19.2	Ribbands
	390	20–41	30.2	Lindauer, 1952
		5–39	18.3	Sakagami, 1953a

Note: For references without dates, see Ribbands, 1953; Perepelova's data were grouped into three-day units; the figures assume that bees in all such units had the age of the middle day.

It is not suggested that these data are wrong or that the differences are not statistically significant. The point is that differences in the average activities of individuals of a given age exist among different colonies or in the same colony at different times. Presumably differences in internal and external conditions, such as the amount and kind of food stored or available in the field, the age distribution of the workers, and the previous activities of the workers (such as overwintered or not), lead to the observed variations in behavioral ontogeny. Investigators in different areas and at different times, therefore, inevitably obtain different results.

Studies of the hypopharyngeal glands, which produce a major component of bee milk, and of the wax glands show that size and activity of these structures, like behavior, are highly variable, according to the conditions in the nest. Young workers have well-developed hypopharyngeal glands, as is to be expected from the activity of such bees in feeding larvae. Later the glands become less enlarged, and it is at this time that they are most active in secreting invertase (Simpson, Riedel, and Wilding, 1968), which is involved in converting nectar into honey. In young workers the wax glands are usually thin, but Rösch found wax glands thickening to an average maximun of 53 μ in workers 16 to 18 days old, then rapidly diminishing to 19 μ at 22 days and 3 μ at 24 days of age (Figure 12.3). But there is much variation. Not all workers developed thickened wax glands, a fact which is not surprising, since when brood and storage space is adequate, old cells are being re-used and no new ones need to be built. One bee 24 days old had wax glands 60 μ thick. Presumably it lived in a colony where wax was needed and was sensitive to the need. Some individuals have hypopharyngeal glands and wax glands simultaneously enlarged and apparently active, and Lindauer found that even young nurse bees with wax glands only 20 μ thick can secrete a little wax. Under Lindauer's conditions wax secretion and manipulation were always interspersed among other activities, but his bees had no large spaces to fill with new comb, and it is likely that when such spaces exist, for example in a new nest site, some bees may become, for parts of their lives, full-time wax secretors and manipulators.

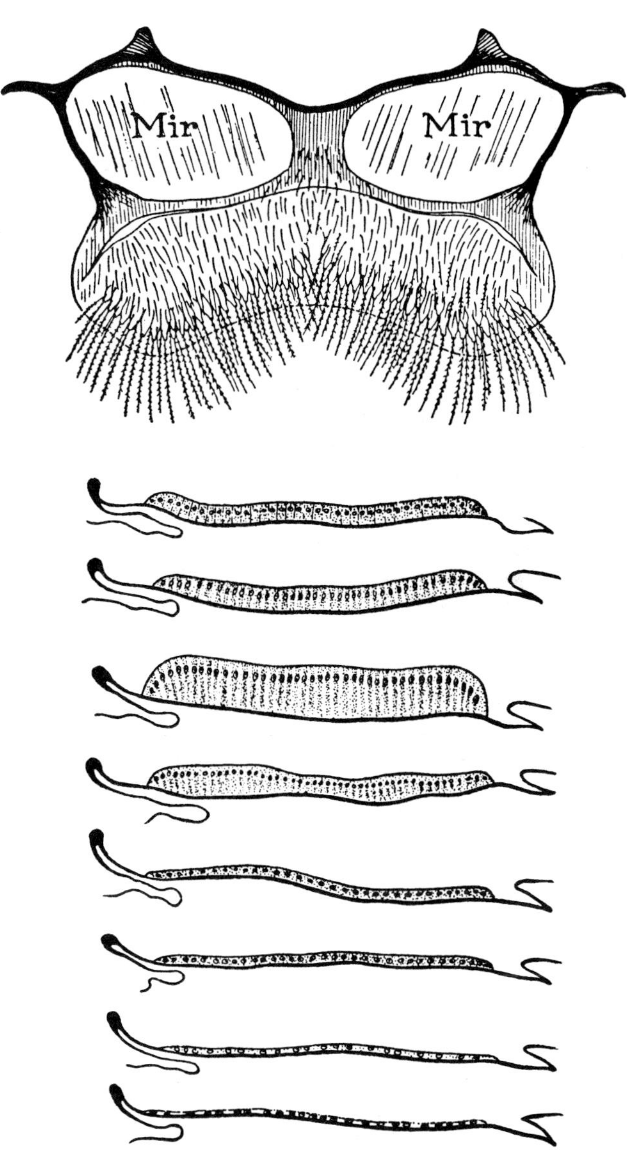

FIGURE 12.3. *Above:* Ventral view of a sternum of a worker *Apis mellifera,* showing the thin shining "mirrors" (*Mir*) that cover the wax glands. (From Snodgrass, 1956.) *Below:* Longitudinal sections of such sterna, showing stages of development and regression of the wax glands of *Apis.* (From Snodgrass, 1956, after G. A. Rösch.) (Both © Cornell University; used by permission of Cornell University Press.)

The mandibular glands also pass through important ontogenetic changes during the life of an adult worker. While she is a house bee, a worker's mandibular gland secretion is a whitish material, the chief lipid component of bee milk, *trans*-10-hydroxy-2-decenoic acid. Later, as the bee enters the guard and foraging phases of her life, the secretion becomes clear and less copious, and contains 2-heptanone, one of the alarm pheromones (Boch and Shearer, 1967).

The amount of the principal alarm pheromone, isopentyl acetate, produced in the sting chamber is maximal (up to 3 μg per bee) in bees of the age when some of them are guards, that is, at about the time that they change from house bees to foragers (Boch and Shearer, 1966). The production of the attractants by Nasanov's gland also varies with the age of the bee. They have no function in house bees and are essentially absent, while from the glands of foragers Boch and Shearer (1963) obtained up to about 1.5 μg of geraniol (one component of the Nasanov secretion) per bee.

Probably there are variations in the sensitivity of workers in a single colony to the stimuli that lead to the ontogenetic changes. Thus all authors (recently and strongly, Sekiguchi and Sakagami, 1966) emphasize that even among workers of *Apis* living in the same colony at the same time there is much variation in the ages of the bees when they change from one activity to another.

In spite of all the plasticity that exists in worker behavior in normal *Apis* colonies, there is a general temporal division of labor, as indicated in Table 12.3. Young adult workers are at first involved largely in cleaning the hive, removing waste products, and in nursing larvae. Later, as the hypopharyngeal glands are reduced in size, the abdominal wax glands of many workers become thickened and the bees are particularly active in wax secretion and in building. Somewhat later, as the wax glands diminish in activity, a few bees become guards and in any event most soon become foragers, foraging and guarding often occurring alternately. The distinction between house bees and field bees is the sharpest in the temporal division of labor of *Apis* workers, as it is also in that of other bees.

Sakagami (1953b) noted the various peculiarities in a colony made up of about 240 *Apis* workers that had emerged on the same day, plus a queen. The workers tended to pass through the usual sequence of activities in spite of the colony's need for all activities at all times, but they tended to specialize, that is, there was an incomplete division of labor among them.

As implied above and as in *Bombus*, the normal ontogenetic sequence in *Apis* can be greatly modified by the needs of the colony (Butler, 1954a; Free, 1965; Hoffmann, 1961; Ribbands, 1953). For example, if the nurse bees are largely removed, older workers resume nursing and experience a redevelopment of the hypopharyngeal glands. Conversely, removal of the older bees from a colony results, after some mortality, in foraging bees only 4 to 12 days old, with still well-developed hypopharyngeal glands, like those of nurse bees (Lindauer, 1961).

It is important to examine individual activity in greater detail with the hope of gaining more insight into the cues that enable worker bees to adjust their behavior and physiology to the needs of the colony. Lindauer (1952, 1961) recorded the activities of a single worker *Apis* during 177 hours of the first 24 days of her life (Figure 12.4). One can recognize the shift from cell cleaning to brood tending, comb building and capping, and ultimately to foraging in this figure, when one looks beyond the enormous amounts of time used in resting and patrolling.

Figure 12.4 does not give a full idea of the frequency with which worker *Apis* change their activities while they are house bees. Lindauer (1952, 1961) watched one bee for ten hours during the eighth day of her life, and the nearly continuous record is shown in Figure 12.5. The frequent changes from one type of activity to another are obvious. The matter is dealt with further in Chapter 7.

The activities of workers are somehow arranged to meet the needs of the society. Ribbands (1953) suggested that food supply somehow directly influences worker activity and division of labor while Lindauer (1952, 1961) emphasizes the long periods that workers spend moving about the nest (patrolling), apparently finding out what needs to be done, such episodes often ending when the bee does something seemingly useful (Figure 12.4). Cues received during patrolling are thought by Lindauer to lead to changes in the normal ontogenetic sequence. For foragers, which do little patrolling, responses of receivers provide comparable cues (Chapters 16, B and 17, D2). The viewpoints of Ribbands and Lindauer are quite possibly both right. Stimuli in the food, the nest, and the brood probably all influence the ontogeny of worker behavior.

3. Meliponini

Studies of behavioral ontogeny in meliponines show sequences similar to those of *Apis*. Plasticity, variation, and changes brought about by abnormal age distributions of workers have been little studied behaviorally for

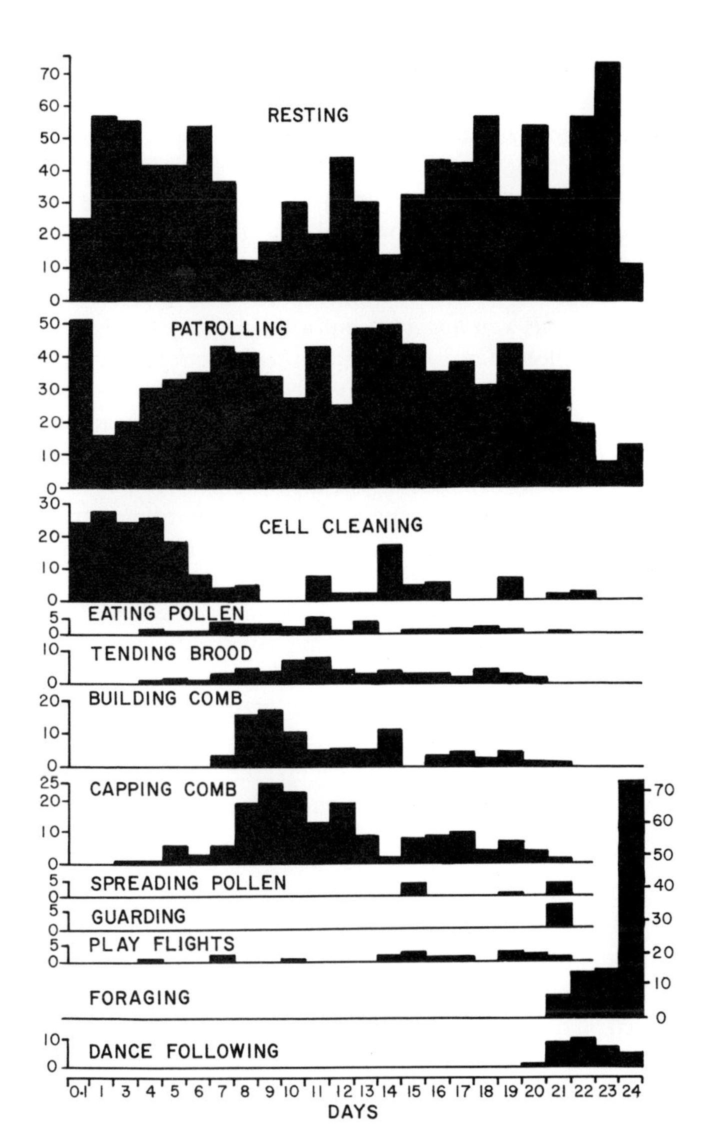

FIGURE 12.4. Time occupied by various activities during the first 24 days of life of a single marked worker honeybee living in a colony. The columns of figures at the left and right represent percentages of its time during which the bee engaged in each activity, out of the two to ten hours of observation daily. (From Ribbands, 1953, after Lindauer, 1952.)

meliponines, but it is a good guess that these bees are as plastic as *Apis*. In a colony of *Trigona* (*Scaptotrigona*) *postica* artificially made up of field bees, Dias and Simões (1972) found that hypopharyngeal glands, but not wax glands, were enlarged; in a normal colony, among bees of mixed ages, the field bees would have had small hypopharyngeal glands.

Melipona quadrifasciata differs from *Apis* in having wax production more scattered, and most intense before instead of after the period of providing food for larvae; moreover, flights do not start until workers are much older than in *Apis* (Kerr and Santos Neto, 1956).

Temporal changes in activities of *Trigona* (*Scaptotrigona*) *xanthotricha* were studied in Brazil by introducing about 60 cocoons into an observation nest inhabited by a colony of the closely related *T. postica* (Hebling, Kerr, and Kerr, 1964). The workers of *T. xanthotricha* emerged over a period of four days, and their subsequent behavioral ontogeny could be observed with time errors no greater than four days because they differed in color from their associates, workers of *T. postica*. In *T. xanthotricha* workers manipulated wax and resin throughout life (up to 94 days); patrolling was common, especially during the first half of their lives; they often associated with and sometimes fed the queen, forming a court around her, from about days 9 to 30; at the same time they provisioned cells. Secretion of wax was limited to days 31 to 36. However, working with wax was by no means so limited; closing of cells was seen on days 8 to 49 (especially 8 to 27), and construction of cells was a common

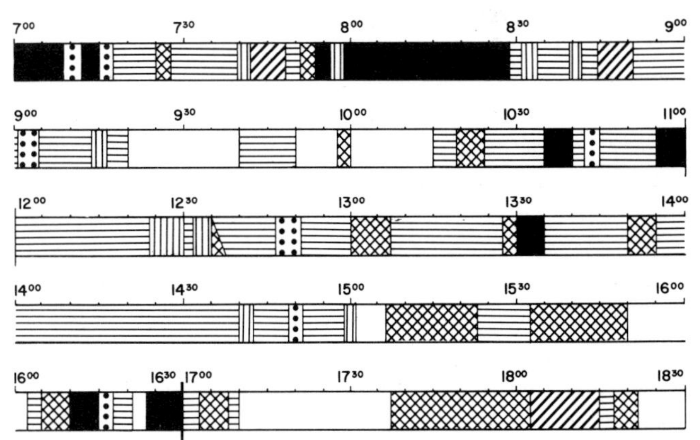

FIGURE 12.5. Record of the activities of one worker honeybee during ten hours on the eighth day of her adult life. *Black* = resting; *horizontal lines* = patrolling; *dots* = eating pollen; *slanting lines* = cell cleaning; *vertical lines* = tending brood; *white* = building comb; *crosshatched* = capping cells. (Drawing by Ann Schlager; data from Ribbands, 1953, after Lindauer, 1952.)

activity from days 12 to 25. Dehydration of nectar and fanning for ventilation started on days 29 and 35, respectively, and continued when conditions required such activity until the bees died. Guarding started on day 33, foraging on day 42; none followed scent trails until day 52, and none was seen to make such a trail until day 56. Thus foraging seemed to pass through three stages: (a) individual activity, (b) responding to communication, and finally (c) providing information to other bees.

For *Trigona postica,* workers laying the trophic eggs on which the queen feeds range from about 8 to 28 days of age (Akahira, Sakagami, and Zucchi, 1970). Dias and Simões (1972) noted, for this species, that the maximal development of the hypopharyngeal glands and abundance of wax production coincide with the period of ovarial enlargement of workers. For the same species, that part of the ontogenetic sequence involving foraging behavior has been studied by Cruz Landim and Ferreira (1968). Especially investigated was the development of the mandibular glands, whose secretion is used in marking foraging routes. Young field bees, at ages of about 30 to 43 days, do not and presumably cannot lay trail spots because the mandibular gland reservoirs are not full. The secretory cells, however, are large, and the reservoirs become full in middle-aged field bees, about 43 to 54 days old, and it is such bees that can communicate by odor-marking of spots along the route to a food source. During this period the glands are diminishing, and bees 60 days old had reduced glands and empty reservoirs and again could not communicate. The data are few; more work is needed, but it does seem that in *T. postica* and *xanthotricha* there is an age-related division of foraging activities.

Bassindale (1955) describes the perhaps somewhat different sequence of activities for workers of the African *Trigona braunsi* (called *gribodoi* by error), a minute species in the subgenus *Hypotrigona* quite unrelated to *T. xanthotricha.* Bassindale lacked data on the longevity of most of the worker stages, but he recognized five stages by the increasing extent of abdominal melanization. In the first stage (3 days in duration) the workers groom themselves and walk about (patrolling?). In the second stage, the bees do everything related to cell construction and brood production: for example, wax secretion; building, provisioning, and sealing cells; helping emerging adults; removing empty cocoons; and fanning. They also pack pollen into storage pots and do some general constructing, such as building the entrance tube. In the third stage the workers ripen nectar and build honey pots. They also, as in the preceding stage, help in general construction, remove empty cocoons, help emerging adults, and occasionally secrete wax. In the fourth stage they are largely involved in removing debris and fanning at the entrance, and in the fifth they forage, guard, and seal cracks with propolis. It is difficult to relate this account to that for *T. xanthotricha,* but a much longer period of wax secretion than for *T. xanthotricha* is probable.

By means of workers marked when very young, Darchen (1970) studied the behavioral ontogeny in another African *Trigona, T. nebulata komiensis,* a species of the subgenus *Meliplebeia.* Perhaps the outstanding fact is that Darchen reports the beginning of guarding, carrying trash to the outside of the hive, and foraging at only 13 days of age. The problem with these and many other observations is that we do not know the condition of the colony. If it had recently been moved, and thus had lost most of its older foragers, the early beginning of foraging is not surprising; but if it was an established colony including old workers, then the beginning of foraging by bees 13 days old shows a striking difference from the ontogeny of workers in the South American Meliponini that have been studied.

C. Division of labor among foraging workers

Even when living solitarily, some female halictines which show no evidence of parasitization nonetheless do not work and produce no offspring. There must be strong selection against this type of behavior, but it nonetheless occurs, apparently with surprising frequency. Possibly in all social bees, even primitive ones, there are workers that likewise do very little work or that at least do not forage. In *Lasioglossum zephyrum* in the laboratory, some workers are very active, others not, and some probably never leave the nests. Whether they do anything useful for the colony is unknown. In *Bombus,* as noted in the preceding sections, the smallest workers in a colony often never leave the nest. In *Apis,* also, there appears to be a considerable number of workers which, even after making one or more flights from the hive, do not forage or rarely do so (Sekiguchi and Sakagami, 1966). Seemingly their threshold for stimulation by dances and food odors is high. Their importance may be in defense, or as a backup force for foraging in the case of great field mortality or opportunity. In view of the relation noted by Kerr and Hebling (1964) between worker size and rate of progress through the normal behavioral ontogeny, it would be interesting to know if nonforaging *Apis* workers are of smaller than average size.

The above paragraph merely shows that among workers of foraging age, some forage very little or not at all; this may be looked at as an aspect of more or less permanent division of labor. But among workers that do forage there is often specialization of activities. We have already noted size-related specialization in flowers visited by *Bombus* workers (section A) and age-related specialization in foraging activities by *Trigona* workers (section B). The following paragraphs concern other kinds of division of labor among foragers. The differential foraging behavior here described is neither permanent nor ontogenetic in the sense that the bees pass through a sequence of stages. The differences among the groups of workers of a colony are often temporary and the sequence in behavior of a given worker is probably determined by foraging opportunities and the food sources which the bee happens to find.

Flower constancy, discussed in Chapter 14 (A), results in a sort of part-time division of labor among workers for almost all social bees. Flower constancy is the tendency for a given bee, on a given trip or for days at a time, to specialize in collecting from a certain kind of flower. Likewise the tendency for a given worker to specialize on either pollen or nectar collecting on a given trip or for days at a time causes a similar part-time division of labor, although in all species other individuals, or the same ones on other trips, collect both. Such specializations in activities doubtless result in improved efficiency (Chapter 14, A), the presumed advantage of division of labor of all kinds. Even in solitary bees there is often flower constancy, and there are times when both pollen and nectar are collected as well as times when the bee takes only nectar or only pollen.

In *Apis* and probably also in meliponines, further division of labor among foragers can be recognized. Studies have been made only for *Apis*, and are summarized as follows. Workers that forage from food sources that are useful only at certain times of day (such as a flower that secretes nectar only at a certain time of day, or a saucer provided with syrup daily at a certain hour) learn when to go to these sources. They tend to form a group of foragers that rest in the hive when their food source is not productive and that is little disturbed by the foraging, dancing, and other stimuli engendered by other food sources.

When honeybees are using different, simultaneously available, and differently scented food sources (different kinds of flowers or artificially scented syrups), bees visiting a particular source tend to group themselves in the hive. Frisch (1967a) set up two feeding stations 80 m from a hive in different directions, and fed odorless syrup. He marked the foragers going to each station with different colors, white and yellow. During breaks in the feeding he examined the hive and found the white and yellow marked bees intermingled. He then scented each food source differently, using lavender and orange blossom oil. Thereafter the bees of each group tended to be together; they formed separate foraging groups, doubtless recognizing one another by odor even when in the hive.

That specialists in exploiting a food source are grouped together during interruptions when that source is temporarily inaccessible (possibly during a rain) or not productive makes good sense. Then when a bee does return, having found that source again productive, it moves among others of its group and dances, stimulating its group companions to resume foraging. The members of the group are often stimulated, presumably by the odor, to approach and antennate such a returnee. Mere contact of a returned successful member of the group with another member in the hive resulted in a return to the food source by the latter in 40 percent of recorded meetings, while dancing was effective in about 90 percent of the cases. Because of the grouping in the hive, the forager returning after an interruption can quickly activate large numbers of bees to resume work on known food sources having the same odor. Even the artificial introduction of the odor of a feeding place into a hive will stimulate some of the bees of that group to inspect the feeding place (see Chapter 15, D2a).

Membership in such foraging groups is not permanent. As a particular flower wanes, individual bees change their foraging to other kinds of flowers. A bee may simultaneously belong to two or possibly more foraging groups. Frisch, in one of his experiments, noted two bees that were not very constant in their attachment to either of two flowers with which he was experimenting. When one of these bees returned to the hive from linden flowers, she activated members of the linden-foraging group.

When the same bee came back after visiting locust flowers, she activated members of that group. Neither the particular bee, nor the location of the food source, but only its odor is important in this matter. (All the experiments were done with food sources close enough to the hive that the dances were round. More distant sources, eliciting waggle dances, might have resulted in some distinction as to the location of the food.) For more extensive accounts of foraging, orientation, and related communication, see Chapters 14 and 15.

A different sort of division of labor among foragers, which is likewise known in *Apis,* probable in Meliponini, and doubtless lacking in other bees since they do not communicate about food sources, involves the distinction between scout bees and others. Ribbands (1953) has summarized the work of others, especially Opfinger (1949) and Oettingen-Spielberg (1949), on this subject. Both of these authors show that among the foragers there are a very few restless individuals that are always searching here and there, and that therefore discover new food sources. These are the scouts. They do not constitute any particular age group; both inexperienced and long-experienced individuals can be scouts.

The relative number of scouts was indicated in a series of experiments by Oettingen-Spielberg. She placed syrup on a scented, colored card, and during four days marked 280 bees that came to it. Nearby was a differently scented and colored card. On the fifth day she provided syrup on the second card instead of the first. The bees milled around the first card; after a time a bee found the food on the second card. Dancing and scent soon attracted bees to the second card in the same numbers that had formerly been at the first. After repetitions of this experiment, with the same results, she modified it by killing all bees that arrived at the second card. In two different experiments, she killed only two bees each. This means that only two out of 280 were acting as scouts at that time; preventing their return to the hive prevented the arrival of large numbers of others.

It should not be thought that all field bees are permanently divided between scouts and other foragers. When there is no food, as in early spring, when the weather is fine but no flowers are yet in bloom, many bees may be seen searching. Or again, when a rich nectar and pollen source is in full bloom, perhaps there are no scouts. Scouts are one extreme in a behavioral continuum ranging through ordinary foragers to the most conservative ones that continue working a crop even until it is very sparse and poor. Scouts are those whose behavior is most easily altered by changing opportunities. In a changing environment their importance to the colony is obvious.

D. Concluding remarks

Division of labor among workers has probably existed from the very inception of a worker caste. Young adult females of solitary bees have certain activities different from older ones. Foraging for food to be brought back to the nest, for example, comes later in the life of a solitary bee than some other activities. Workers show the same phenomenon—nest making and nest guarding normally precede a foraging stage. In a nest with workers of different ages, this results in division of labor among them.

Even in reproductivity, division of labor among workers appears to date from the origin of the caste. Thus in *Lasioglossum zephyrum,* which can reasonably be taken as an example of a bee whose worker caste is in the process of differentiation, variation in ovarian development and related behavior results in division of labor among the workers. Some are ovarially almost queens—indeed would be queens if bees with still larger ovaries were absent—while others have slender ovaries, and all intermediates exist. Thus reproductive division of labor among workers exists. *A priori* one might suppose that forms having division of labor among workers must be derived by further specialization from forms in which the only meaningful division of labor was between the queen and worker castes. But such forms probably never existed.

In view of this history, the universality of division of labor among workers of bee colonies is not surprising. The common elements in this division of labor among unrelated bees are also understandable.

Temporal and permanent division of labor both become more prominent with higher levels of social development. The former is the one which achieves impressive significance in the highly eusocial bees, where workers at different ages or in different physiological states do entirely different things. Permanent division of labor among workers usually involves statistical rather than absolute differences among the individuals, and is in general not very impressive. Probably it reaches its fullest development, among bees, in the bumblebees.

13 *Colony Multiplication*

In the semisocial and primitively eusocial bees, there is no evidence that large colony size ever leads to division of the colony. Rather, at a certain time of the year, dependent upon the life cycle of the species, workers are no longer produced and instead, males and gynes develop. Subsequently, usually after a period of hibernation or estivation, fertilized young gynes establish new colonies. In highly eusocial forms, however, if the reproductive rate exceeds the death rate of workers, a colony will grow larger and larger until the stage is reached when, probably due to social factors or to the limited space available in the nest, the colony divides.

New nests are established among semisocial and primitively eusocial bees in much the same manner as among their ancestors—solitary bees. In most cases, only one bee is involved in the initial stages, exactly as in solitary bees. New nests are established by lone gynes, sometimes joined later by others to form a colony. Sometimes a group of gynes that hibernated together may remain in the same nest to start a new colony. Nevertheless, in these bees, colony multiplication is usually a reflection of the number of the gynes produced.

Establishment is different in kind among the highly eusocial bees, where swarming is the device for colony multiplication, because lone individuals are unable to survive and gynes lack the ability to forage. Since a swarm is a group of workers in addition to a gyne, much more is involved than the production of gynes by the parental colony. As the number of young gynes may greatly exceed the number of swarms produced by a colony, the excess gynes are killed off by the workers or by one another (Chapter 11, B). The production of daughter colonies therefore reflects not only the potentiality of the mother colony to produce gynes but also its potentiality to produce the workers for the swarm and whatever food and building materials are carried away by them.

A. Swarming in stingless bees

In the Meliponini, workers from the mother colony locate suitable nesting sites, often hover about them for considerable periods of time, and carry propolis and cerumen to them. Ultimately the bees settle on one such site, seal the cracks if it is a hollow, and make a nest entrance normal for the species. The species that nest in exposed situations must make a whole nest. Whatever the species of bee and the site, workers from the mother colony carry to the new site not only cerumen and propolis but mud, animal excreta, or whatever building materials are characteristically used by the species. The new nest may attain reasonable completeness, except for brood combs. The food pots may be provisioned with honey and pollen. The process of building and provisioning may go on for weeks, the building materials being carried in the corbiculae, food in the crop. Alternatively, as in *Trigona julianii* and *varia,* there may be very little construction until arrival of the new queen (Juliani, 1967; Terada, 1972). One to several young gynes finally leave the mother colony and go to the new site with workers. This is the swarm proper. At this stage males (from a few to hundreds) may swarm about the entrances of both the old and the new nests. (The same thing may happen at an established nest, probably when an old queen has died and is replaced by a young, unmated one.) A young gyne makes a mating flight from the new nest and returns mated, carrying the genitalia of the male with which she mated in the apex of her abdomen. Unlike *Apis,* she mates only once. Thereafter the swarm of males diminishes, the supernumerary young gynes are killed by the workers if this has not already happened, and brood production begins. Sometimes all these things happen rather quickly, so that brood comb is being constructed before the brood chamber is fully enclosed by the usual layers (involucrum and batumen) of cerumen. There may still remain a period of as much as a month after the mating flight during which materials are carried from the mother nest to the new one; the first workers to mature at the new nest may be pupae before carrying ceases. The long-continued relation between mother and daughter colonies was demonstrated by Nogueira-Neto (1954), who artificially colored honey in mother nests and observed the appearance of color in the honey stores of daughter colonies.

The establishment of a new colony in the Meliponini is a rather gradual process, impossible at a great distance from the mother nest (see also Kerr, 1951; Kerr and Laidlaw, 1956). Note that it is a young gyne that leaves the old colony, where her mother remains as the queen; the mother is not only too large to fly, due to physogastry, but her wings are so tattered and shortened that she could not fly even if her weight were reduced to that of a young gyne.

B. Swarming in *Apis*

In *Apis* the division of the mother colony into two is abrupt and complete. Swarming in temperate climate *Apis* usually occurs in late spring at a time of plentiful food supplies. This makes biological sense, for the swarm must establish itself, build combs, store food, and rear young workers, all before the onset of dry (flowerless) or cold weather. It is at least possible that among the stimuli that set off production of new queens and preparation for swarming are the fullness of food storage cells and resultant lack of adequate space for expansion of the brood area of the nest. Simpson (1963), however, found lack of space for adult bees to be the critical factor in some circumstances. Certainly crowded colonies do swarm, apparently because of crowding. There may be other causes—many have been proposed, such as too many young workers and therefore too much brood food being secreted, diminution of the production of queen substance by the queen, etc. The latter is known to occur (Butler, 1960), but is probably a result of some prior stimulus to which the queen responds. The causation probably is no single stimulus but a combination of various ones, each probably having a different weight.

In any event, gyne cells are constructed hanging from the margins of some of the combs and a few young gynes are reared (see Chapter 10, C2). Even production of a new gyne is not always a necessary prelude for swarming; Simpson (1963) produced swarming from crowded colonies without gyne cells. Later the orphan bees left in the nest made emergency queen cells, that is, they converted certain worker cells into gyne cells to produce a new queen. At roughly the time the gyne cells are constructed, in the ordinary case when this precedes swarming, foraging diminishes. The bees act lazy. Those that do forage have trouble getting the house bees to accept provisions that they have collected—it takes minutes to find house bees that will accept nectar. This is often because there is not much space for more provisions; it is a regulatory mechanism or social interaction of considerable importance (see Chapters 16, B and 17, D2). When house bees are slow to accept provisions, they turn off the foraging.

Lindauer (1955, 1961, fully reviewed by Frisch, 1967a), the principal recent student of *Apis* swarming, found that at this time foragers, unable to continue foraging, start flying and walking about dark holes, cracks, empty boxes, and the like. They become scouts and are already house-hunting. The scouts, having changed their behavior in view of circumstances, play an important role in the swarm activities. They still dance, but the dances are not associated with food exchange, for the bees are not returning from food sources. The dances indicate positions of possible nest sites.

At the same time the workers stop feeding the queen. She lays fewer and fewer eggs, ultimately lays none, and loses weight. She is often disturbed by workers forming a ball around her or striking her. The actual swarming out from the nest is associated with buzzing runs (also known as breaking dances, or *Schwirrlaufen*) by workers. The queen does not lead, but goes out and takes wing along with the thousands of workers.

Esch (1967a) has shown that when a bee makes a buzzing run it produces a series of short buzzes with its wings at the usual flight frequency of 180 to 250 cycles per second while moving in a straight line. Such a bee apparently intentionally touches other bees, and while in contact buzzes continuously at 400 to 500 cycles per second. Bees contacted in this way themselves start on buzzing runs, which therefore quickly spread like a chain reaction through the entire colony and lead to the departure of many of the bees as a swarm.

Unless delayed by bad weather, roughly half the workers, along with the old queen, leave the nest in a swarm a day or two before emergence of the first young queen. This may amount to 30,000 bees, although many swarms are far smaller, especially afterswarms, which may leave during the weeks after the main swarm, and whose departure can hardly be related to crowding. The workers fill their crops with honey before leaving and the new colony is thus able to carry on for some time without foraging. Swarming workers are bees of all ages, although the percentage of old workers is reportedly less than in the parental hive population. The swarm, having issued into the open, assembles on some support, often a tree (Figure 13.1). Contrary to superficial appearances, the swarm is an organized mass. Inside, it consists of loose chains of bees hanging one to another, with enough space between them so that bees can move about. On the surface the older bees form a solid covering about three bees thick, with one opening in it for a flight entrance to the more loosely spaced interior.

From this location scout bees continue to investigate the potential nesting sites in the area, and perform dances on the outside surface of the swarm like those occurring in the nest earlier, and like those which at other times indicate food sources (Chapter 15, D2). Lindauer shows that workers finding the best nesting sites dance most vigorously and for the longest periods. In fact, a noteworthy feature of the dancing is its long duration. Unlike foragers, these bees do not have to stop dancing to go and get more food. By marking dancers, Lindauer showed that they do, however, from time to time return to the sites concerning which they are dancing. The dancers also mark potential nest sites with the scent from the Nasanov gland, which attracts other bees.

The dances recruit other bees to go to the various potential nest sites. The more vigorous and persistent dancers even recruit scouts dancing less vigorously, those that presumably found less favorable sites. Gradually more and more of the dancers come to indicate one favorable location. When all are in agreement, the swarm leaves for the new nest location. On some occasions Lindauer was able to recognize the time of agreement, read the dances of the scouts, and find the site before the bees flew to it; he was waiting for them on their arrival. Figure 13.2 shows the development of agreement in the case of one swarm, and Table 13.1, with the same data, shows the small number of dances during the first days, the subsequently active dancing, and the increasing percentage of dances indicating the place to which the bees ultimately flew.

Occasionally agreement is not attained; competing groups continue strongly. In one such case studied by Lindauer the bees, unable to "decide," built exposed combs on the branches of a tree at the swarming site. Such colonies do not survive cold winters. In another case the bees tried to leave, but started in two directions. But with only one queen, and with queen pheromones important in maintaining the swarm, the bees reassembled and ultimately left together as in the normal case.

By providing artificial cavities of various kinds for swarms in an area lacking natural nest sites, Lindauer was able to investigate what factors make a good site, that is, one that the scouts consider good. Factors included temperature, exposure to wind, odor, size, and the like. A most interesting one was distance. For equally good sites, the more distant one receives higher acclaim by the scouts. This makes good sense, as it should tend to disperse honeybee colonies; the swarm usually hangs close to the parental nest. The remarkable aspect of it is that the very same bees, using the same kind of dances, react in the reverse manner for food—the nearer a food source the more vigorous the dances.

The departure of the swarm is quite sudden; it seems stimulated by the buzzing runs of numerous bees, including scouts. Bees thought to be the scouts fly through the flying swarm and forward, then slowly back, and forward again, appearing to guide the swarm to its future home. It is the only place in *Apis* biology (except presumably in absconding, see below) where guidance of this sort is evident.

Coordinated activity like *Apis* swarming clearly requires communication. The following are some parts of the communicative behavior that are known: (1) reduction in the inhibitory pheromone production by the queen, releasing workers to construct gyne cells; (2) reduced willingness of house bees to accept food from foragers, presumably stimulating nest searching by the former foragers; (3) buzzing runs, which say in effect "let's go," when the swarm leaves the parental hive and

FIGURE 13.1. A swarm of *Apis mellifera* assembled in a tree. (Photograph by E. S. Ross.)

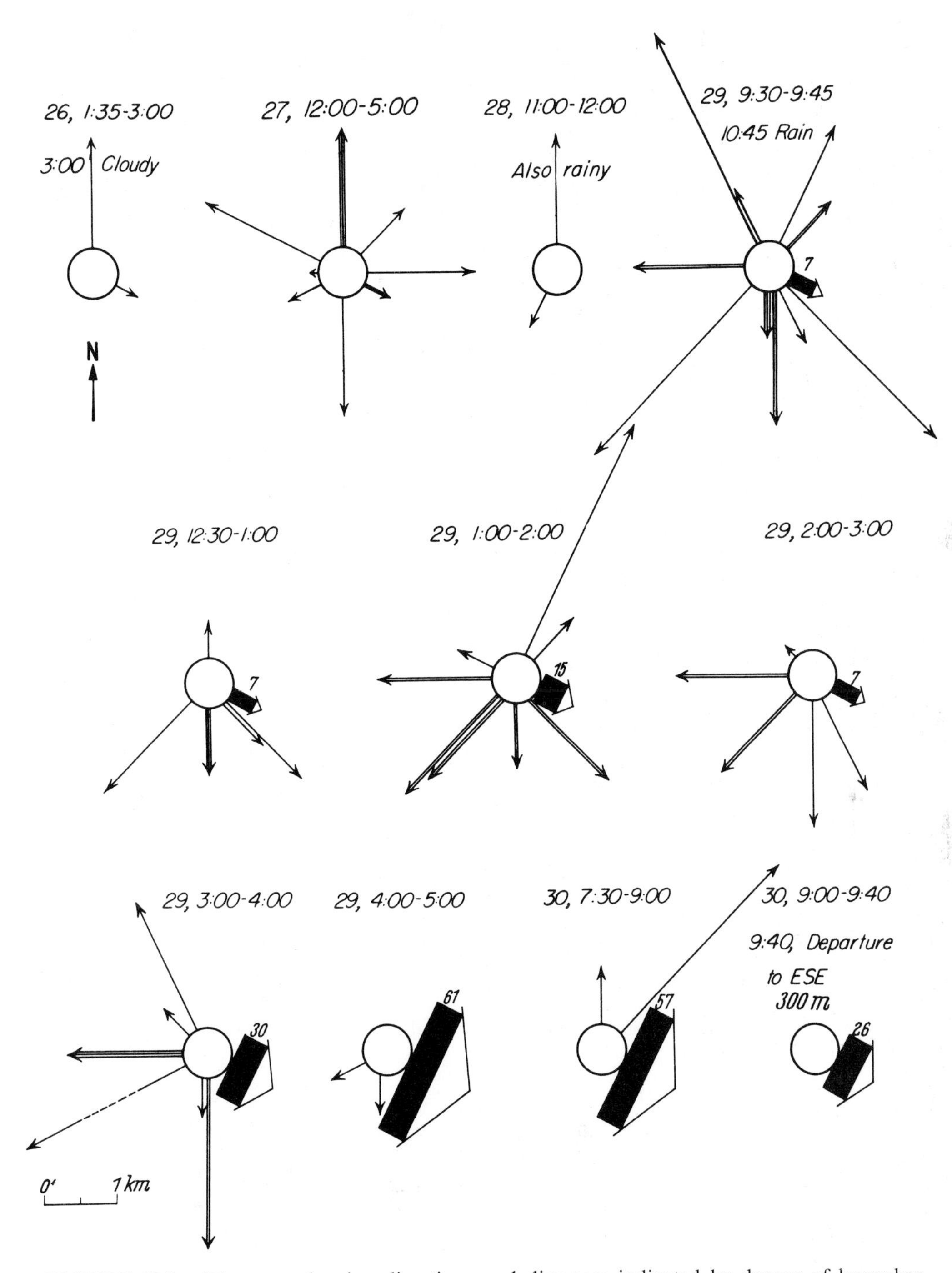

FIGURE 13.2. Diagrams showing directions and distances indicated by dances of honeybee scouts on the surface of a swarm, June 26–30, and achievement of "agreement" on one nest site. The arrows indicate the directions of the sites; lengths of arrows are proportional to distances indicated by the dances; and widths of arrows are indicative of the number of scout bees dancing, also indicated by numbers beside the broader arrows. (From Frisch, 1967a, after Lindauer, 1955.)

TABLE 13.1. Dances on the swarm indicated in Figure 13.1.

Date	Observation period	Total dances/hr	Percent of dances for place ultimately selected
VI-26	1.4 hrs (cloudy)	1.3	50.0
VI-27	5 hrs	2.2	18.2
VI-28	1 hr (rainy)	2.0	0
VI-29	0.25 hr	124	33.3
VI-29	0.5 hr	28	50.0
VI-29	1 hr	28	53.6
VI-29	1 hr	14	50.0
VI-29	1 hr	39	76.9
VI-29	1 hr	63	96.8
VI-30	1.5 hr	39.4	96.6
VI-30	0.7 hr	39	100

again when it leaves the place where it has hung awaiting a decision about the new nesting place; (4) the dances concerning nesting sites. There is much not known about swarming communication, especially that occurring in the preparatory phase in the parental hive. What sets off reduction in the level of queen substance? What triggers termination of feeding of the queen by workers?

Pheromones play a role in gathering and holding the swarm together around the queen. If the queen is removed or kept behind, the swarm returns to its nest or to the place where the queen is. Thus the queen plays an essential role in causing cohesion of the swarm. She does this through two pheromones, both produced in her mandibular glands. One is *trans*-9-keto-2-decenoic acid (often called queen substance, Chapter 11, A), which in a nest inhibits gyne cell construction by workers. It is also a sex attractant and an aphrodisiac. The other is reported by Butler and Simpson to be *trans*-9-hydroxy-2-decenoic acid, a minor sex attractant which reinforces the first in influencing queen cell construction and worker ovarian enlargement. In the context of the swarm, the odor of the first attracts flying workers but does not cause them to settle, while the odor of the second causes bees to settle and form a quiet cluster. Butler and Simpson (1967, reviewed by Butler, 1969) found that the odor of the two materials, mixed, was as effective in settling a swarm as a live mated queen. Barbier and Pain (1960) noted that the same materials attracted and caused clustering of caged worker bees; evidently they also function in contexts other than swarming. Morse and Boch (1971), on the other hand, report that the second substance listed above is ineffective in swarm stabilization but that an unknown substance extracted from queens' heads has this effect. Further work is needed to explain this discrepancy.

The queen-derived pheromones are not the whole pheromonal story of swarm behavior. If they were, the queen would have to select the place of settling of a swarm, which she is probably ill-adapted to do. Morse and Boch (1971) have shown that it is the Nasanov gland pheromone liberated by workers that attracts other workers and the queen to the settling places. They got excellent attraction and swarm formation with a mixture of the queen pheromones with synthetic Nasanov pheromone (geranic and nerolic acids, geraniol, and citral in proportions 266:133:50:100). Mautz, Boch, and Morse (1972) have further shown that swarm settling depends not only on the interplay of the queen's pheromones and the Nasanov pheromone of workers, but also upon buzzing runs. If a queen is separated from her swarm, scouts that find her either expose the Nasanov gland, perform buzzing runs, or both. Some bees move back and forth from the queen's location to the cluster of workers. Nearly all the scenting is at the queen's location, but buzzing runs are about equally common at the two sites. They lead many bees to fly and to be attracted by the pheromone complex at the queen's location. There is no obvious explanation for the performance of runs at her site.

The paradox that swarming results from crowding, yet swarms often issue from an *Apis* colony after an earlier swarm has relieved the crowded conditions, deserves further comment. Departure of the first swarm with the old queen from a colony in spring or early summer is commonly not associated with piping. (Piping is a sound produced by the flight mechanism of a gyne, when the wings are held still over the back. It consists, usually, of a long—±1 sec duration—sound followed by a series of bursts of about one-quarter second in duration, and is performed with the thorax pressed against the substrate, thus communicating the vibrations to the comb.) Piping is often heard, however, in connection with afterswarms. Simpson and Cherry (1969) appear to have verified a relation, reported on various occasions in earlier centuries, between piping and swarming. Piping usually occurs when two or more young gynes are present in the same colony. Worker bees respond by being quiet when they perceive piping or similar sounds. It is at least possible that prolonged piping stimulates the buzzing runs by workers that lead directly to swarming. Colonies that lack gyne piping produce no afterswarms; some colonies with piping do produce afterswarms. Simpson and Cherry had one colony with only one young gyne (no old queen) in which a recording of piping seemed to cause departure of a swarm. Swarms are sometimes reported with more than one queen. These are afterswarms; flight sometimes occurs before excess queens have been killed off.

C. The origin of swarming behavior

The origin of swarming behavior as it occurs in stingless bees can be readily imagined. In primitively eusocial bees such as *Bombus,* young gynes establish new colonies alone or sometimes enter existing young colonies, usually of their own species, and take over, or attempt to do so. In stingless bees the situation differs in that the existing young colony was established by workers from the parent colony and is called a swarm; it is not the victim of an intruding young gyne. In *Apis,* however, a sharp difference exists and no intermediate conditions are known, so that understanding of the origin of *Apis* swarming is not so easy. It is the old queen that leaves, and she is accompanied by workers which have not prepared a nest in advance.

It seems probable that swarming as seen in *Apis* arose, as suggested by Ribbands (1953), along with the ability of the whole colony to migrate in response to unsuitable conditions. Such migration, or absconding as it is called in honeybee literature, is extremely common in tropical *Apis* (*dorsata, florea, cerana mellifera adansonii*) and relatively uncommon only in the presumably derived, cool temperate zone races of *A. mellifera,* where large and nonmovable stores of food are vital to survival because of winter conditions. Absconding, however, does not lead to colony division; for this reason various authors contend that it has nothing to do with swarming (Kerr and Laidlaw, 1956). But the similarities are so impressive that a relation, at least the evolution of common behavior patterns for both swarming and absconding, seems evident. For example, for both, the bees fill their crops with honey before leaving the nest. Smoke causes the same response, even in the northern races of *A. mellifera* which do not usually abscond. Presumably this response, much used by beekeepers in handling bees, arose because of the improved probability of survival of bees carrying food in the event that they are forced from their nest by fire (Newton, 1968).

It is noteworthy that for swarming (or absconding) to work as it is practiced by either Meliponini or Apini, communication about geographical locations must be possible. This method of colony multiplication occurs only in those bees which have such an ability, also used with respect to food sources. In those Meliponini (such as smaller forms of *Trigona*) in which communication as to food sources is by odor, with neither odor trails nor cues indicating distance and direction, one would expect virgin queens going to new nests to be accompanied by workers familiar with the new site. It would be interesting to learn if queens go alone in those forms having a better system of communication.

An interesting aspect of swarming in all highly eusocial bees is the problem of changing the orientation of foragers to a home site. They must "forget" the old site. As is well known, if a beehive is moved a short distance to a new location within the area already familiar to the foragers, many or most of them after a trip will return to the old site instead of the new one. This can result in decimation of the foraging population of a colony that has been moved. But when bees swarm they establish a new site, usually well within the familiar area, yet foragers return to the new site. Taranov (1955) has considered this problem for *Apis.* He notes that return to the old site is much more likely for bees with empty crops. Swarming bees fill their crops with honey before leaving the old site, hence more easily return to the new one.

14 *Foraging and Orientation*

The food of bees is nectar and pollen. The highly eusocial bees, as well as bumblebees, convert the former into honey for storage through inversion of the sugar and partial dehydration. Many Apidae use much food of glandular origin, and various bees eat eggs, but even in such cases the ultimate raw food resources are exclusively nectar and pollen.

Bees of both sexes (except reproductives of highly eusocial bees) obtain food by flying to suitable flowers, where they take nectar. Females also take pollen. Female bees that are making nests or working in them take excess quantities of these foods, beyond the immediate needs of the individual. These excesses are used to provision cells or to feed larvae, or are stored for future use. Female bees must have appropriate harvesting and carrying equipment for both nectar and pollen, as well as an ability to find the flowers and then return to an often small and hidden nest entrance. The structures used in harvesting and carrying foods are described and illustrated in Chapter 1 (Blb, c). The present chapter concerns related behavioral matters. Additional material on foraging is found in Chapter 12.

This chapter has been prepared to give a substantial amount of information about how individual bees get about in the environment outside their nests and find food. Such bees can be at any social level. Success in their foraging flights, of course, is intimately related to their nesting success, whether they are solitary or social bees, but this chapter is only indirectly related to social behavior. It serves, however, as the basis for Chapter 15, which concerns social aspects of foraging.

A. Foraging flights

One of the noteworthy attributes of social bees is their ability to bring considerable amounts of food to a nest from vast numbers of minute sources—flowers. Gary (1967) has estimated up to 163,000 trips from a single colony of *Apis* in one day. Free (1955d) found that after 24 hours of starvation, workers of *Bombus* would drink amounts of syrup averaging about 50 percent of their body weights, but sometimes attaining 90 percent, and furthermore that they could fly with such loads in their crops. Larger workers could more quickly fill their crops than small ones. Pollen loads of *Bombus* average about 20 percent of the body weight, but reach 60 percent.

Almost the same figures have been obtained for nectar with *Apis,* the average load being about 50 percent, the maximum 92.5 percent, of the body weight of a worker. For pollen loads, weight varies greatly for pollen of different plants, probably with density, from 10.5 percent to over 36 percent of the worker's weight, these figures being means for heather and hard maple pollen respectively. Probably the size of the pollen load, rather than its weight, stops a bee from collecting and stimulates it to return home.

The size of the nectar loads of *Apis* and probably other bees is influenced by various factors. Núñez (1966, 1970), in a study using sugar syrup, found that the size of the load increased with sugar concentration and with air temperature. Moreover, when the flow of syrup to the feeder was slow, the bees took smaller loads, took longer to collect them, spent more time in the hive between trips, and abandoned the effort at a flow rate varying inversely with the sugar concentration. Similar results have been obtained by Pflumm (1969a, b). Most studies of these variables have dealt with sugar concentration because it is easily controlled in experiments, but Núñez and Pflumm have emphasized the volumetric aspects also.

The number of flowers which a bee visits to get a load of nectar or of pollen varies greatly according to the kind of flower, time of day, soil moisture (which influences nectar concentration and volume), etc. An *Apis* worker may fill its crop from a single flower of some species of *Eucalyptus*. These flowers are visited but probably are not often pollinated by small creatures such as *Apis;* birds and bats are more important, since they go from flower to flower. On the other hand, most flowers produce little nectar. Ribbands followed one nectar-gathering *Apis* on a 106-minute trip that included 1446 *Limnanthes* flowers. In dry summers when nectar secretion is restricted, many trips probably involve thousands of flowers. Foraging flight durations of several hours have been noted at such times.

The number of flowers visited by an *Apis* worker to get a load of pollen is usually less than for a load of nectar; rarely, one flower provides a full load, but several dozen or a few hundred are usual.

The durations of trips for almost any species of bee varies from a few minutes to hours. For *Lasioglossum imitatum* durations are shown in Figure 14.1. In *Apis* there have been many studies of duration of trips, often showing mean durations for nectar gatherers of about 30 minutes, much less for pollen-gathering trips (Ribbands, 1953). For *Bombus,* while the ordinary trip is of about the same length, some workers may stay away the whole day and even overnight. Presumably they are not working the whole time.

The number of trips per day varies greatly among individuals. In *Lasioglossum imitatum* a female makes from one to seven trips for pollen (Michener and Wille, 1961) and sometimes one or two others for nectar feeding. *Augochlorella* makes even fewer trips per day, probably ordinarily only one or two (see Table 23.1). For *Apis,* there are studies with marked bees (Ribbands, 1953) showing mean numbers of trips per day per worker collecting nectar to be in the range of 7 to 13, although to syrup dishes a single bee may make 150 trips. Presumably the number is ordinarily small because of the time and energy needed to reach and collect the minute quantities of nectar in most flowers. Pollen-collecting bees may make more trips per day than nectar gatherers if the pollen is available all day. Ribbands observed one marked worker take 47 loads of pollen from California poppy flowers in one day.

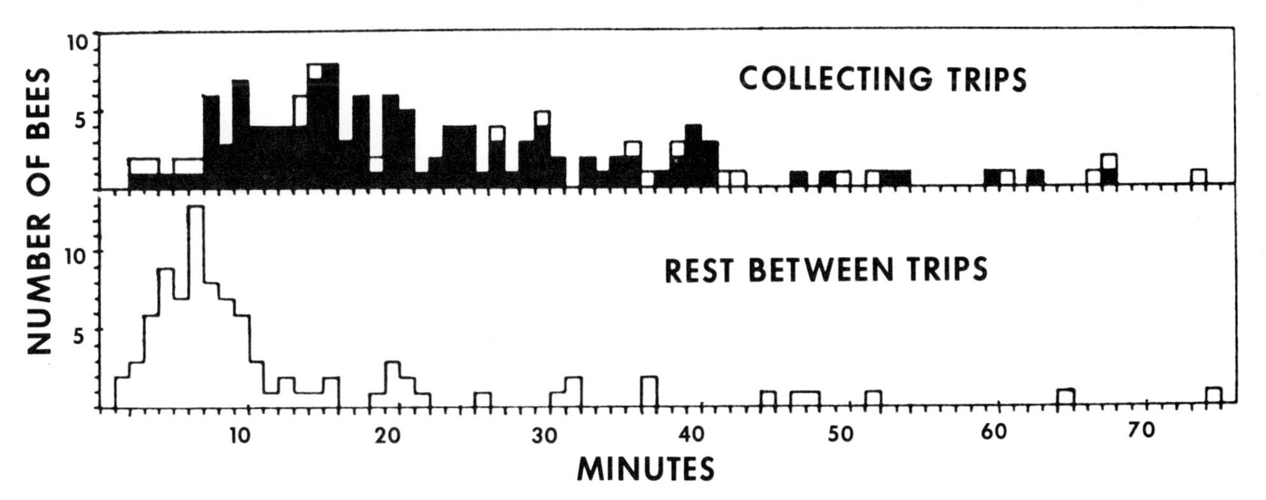

FIGURE 14.1. Durations of trips away from the nest, and of intervals in the nest between trips, for the minute North American halictid *Lasioglossum imitatum. Black areas* indicate bees returning with pollen loads. (From Michener and Wille, 1961.)

As indicated in Chapter 12 (C), a foraging trip may be for pollen, for nectar, or for both. In halictines such as *Lasioglossum imitatum* there is a tendency for afternoon trips to be for nectar only, and since these bees lack nectar storage facilities, this food is probably used by the individual bee concerned. In *Bombus* and the highly eusocial bees, nectar-gathering trips are not concentrated in the afternoon, and the nectar is stored in the nest for use of all nest inhabitants. Bees carrying both nectar and pollen often have a full load of one and a partial load of the other. In both *Bombus* (Free, 1955d, 1970a) and *Apis*, a foraging bee tends to have a temporary constancy to either pollen or nectar. For example, if it is collecting pollen on a given trip, several consecutive trips will usually also be for pollen. But on other days the same worker may collect nectar, and still later, pollen again.

Workers of *Apis* seem to prefer nectar gathering; changes from pollen gathering to nectar gathering are much commoner than the reverse. There is no definite ontogenetic sequence in types of foraging such as occurs for types of house activities. Some bees gather only pollen for their entire foraging lives, others only nectar. Most foragers, however, undertake both, simultaneously or in sequence. The mixture takes many forms, even among bees in the same colony at the same time. Some collect both pollen and nectar on each trip. Some individuals observed by Ribbands (1953) collected pollen each morning from poppy flowers that produced no nectar, but later each day (even in the continued presence of abundant poppy pollen) visited other flowers where they collected nectar only, or both nectar and pollen. Still other individuals, as indicated previously, collect nothing but pollen for days, then change to nectar; less commonly the reverse change occurs. The individual differences are related to conditions in the hives (see Chapter 7, D5), to different experiences of the individuals, and, among the various studies, to differing outside environmental conditions. Bees would be an interesting subject for further study of the diversity of behavior attainable in similar environments by genetically similar individuals that are capable of learning but not of intelligent responses.

Probably all polylectic bees—bees that gather pollen from a wide variety of flowers—show a tendency for constancy to a given kind of flower on a particular foraging trip. Among social bees as different as *Lasioglossum imitatum, Bombus,* and *Apis,* when workers in the same colony are going to patches of mixed flowers, observation of a given worker shows that on a given trip and often also on consecutive trips she goes only to a particular kind of flower. Such flower constancy is incomplete; mixed pollen loads show that it often breaks down. Different studies of *Bombus* showed 44 to 65 percent of the pollen loads to represent a single species (Free and Butler, 1959), and in many of the mixed loads over 90 percent of the pollen was of one species; lower percentages were obtained in a more recent study of *B. lucorum* (Free, 1970a). Bumblebees show less constancy than does *Apis*. In nectar collecting, constancy is less marked than in pollen collecting.

The known mechanisms that promote flower constancy are (a) learning of shape, color, and scent of the flowers; (b) loyalty to a particular vicinity in foraging; and (c) learning of a time of day for foraging. Flower constancy probably promotes efficiency in foraging. Even in a field of mixed and equally attractive and productive flowers, a bee can probably work more efficiently if its sense organs and behavior are temporarily "set" for a particular kind of flower from which it obtains food in a particular way. Another bee may equally efficiently work another kind of flower. Charles Darwin in 1876 put forward much the same idea, suggesting that constancy for a particular kind of flower enables bees to work faster because they learn "the best position on the flower and how far and in what direction to insert their proboscis."

Numerous subsequent investigators (such as Weaver, 1956) have noted individually different behavior of workers on flowers of the same kind at the same time and place and concluded that bees learn, as individuals, how to manage flowers from which they are foraging. Interesting support of this view is provided by Reinhardt (1952), who studied *Apis* workers foraging on alfalfa flowers. These flowers are traps which, when tripped, catch the proboscis of a bee. It was quite clear that some bees learned to avoid the trap by using either wilted or already tripped flowers. Many avoided tripping by learning to probe the flower from the side. Changes in behavior and increasing rates of nectar gathering were noted for individual bees, presumably as a result of experience and learning.

Improvement of foraging efficiency must also result from the tendency of individual bees to be constant to localized foraging areas, even though the same flower species may be widely dispersed. Such behavior is known in *Apis* (Free, 1966), also in the unrelated and nonsocial

Megachile, apparently in *Bombus* (Free and Butler, 1959), and probably in many or most bees. Thus for several consecutive trips, or even until the flower's season passes, a particular bee returns to the same vicinity to forage. An *Apis* worker may return, trip after trip, to the same few square meters of an alfalfa field, particularly if there are distinctive landmarks (Levin, 1966). This can only be because, once habituated to a given location that provides a satisfactory quality and quantity of food, it is more efficient for a bee to return repeatedly to that place than to search for other sources. This may be true even if the place being exploited is farther from the nest than certain other food sources, because a bee flying to a known site moves fast—11.3 to 19.3 km per hour (homeward) for *Apis,* corrected arithmetically for winds (Ribbands, 1953). Searching, on the other hand, is a slow and uncertain process.

Apis workers tend to forage within a few hundred meters of the nest if adequate food sources are available within that distance. While interactions among bees of different colonies away from their nests are not usually evident, they do occur and influence foraging. Thus bees from different hives tend not to forage in the same parts of an alfalfa field, although there is by no means clear division of the field into territories for each hive (Levin and Glowska-Konopacka, 1963).

Many kinds of flowers are open, or secrete their nectar, or (more often) liberate their pollen only at certain times of day. Such an arrangement is doubtless advantageous to the plant because it increases the chance that pollen will reach stigmas of the same species, for all flowers of a given species are likely to be open at the same time. Moreover, closure at other times protects nectar and pollen from excess drying or from rain. Bees that get to the flowers at the proper time will have a competitive advantage over those that arrive too early or too late. It is therefore not surprising that bees have a time sense that enables them to return to a food supply at the same time of day when they utilized it previously. Probably all bees can do this; at least many kinds do arrive at flowers daily at appropriate times. But only for *Apis* have experiments been done to show that the bees respond to an internal timing mechanism in addition to external stimuli such as light intensity or flower odors.

The first detailed experiments on time sense in *Apis* were those of Beling (1929); various subsequent observations by other authors are reviewed by Ribbands (1953), Renner (in Chauvin, 1968) and Beier and Lindauer (1970). If bees are allowed to visit a syrup source for only a brief time, such as 1 or 3 hours, the same individuals will return to that place in maximum numbers 24 hours later, even if no food is put out for them. Repetition of the training improves their timing, so that most of them arrive during or just before the time to which they are trained. The same individual bees can be trained to at least three different times during the day (Figure 14.2), and they can be trained to go to different places at different times: for example, one place at 10:00 A.M. and another at 4:00 P.M. If the concentration of sugar syrup at

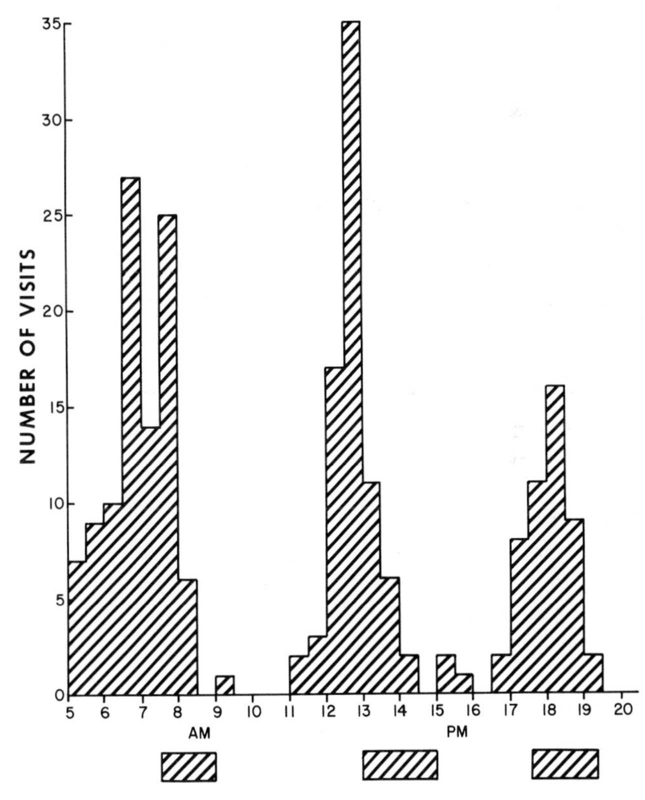

FIGURE 14.2. Training of 19 individually marked *Apis* workers to visit the same place for food at three different times of day. Bees were trained for six days by providing syrup only at the times indicated by the rectangles below the histograms. Visits were recorded on the seventh day, when no food was provided. The same bees came during each of the three peaks of visiting, showing that a given bee can be trained simultaneously to a food source for at least three different times of day. (Modified from Ribbands, 1953, after Ingeborg Beling.)

a feeding place is varied on a regular schedule, they learn to come in greatest numbers at the time when it is most concentrated. The advantage of this ability in foraging is obvious; patches of different kinds of flowers in different places open at different times of day or secrete nectar or liberate pollen at different times.

Apparently the bees cannot be trained to longer or much shorter periods of return than 24 hours. If trained on a 48-hour cycle, they still return every 24 hours. That the cycle is metabolically mediated is indicated by pharmacological experiments with substances that accelerate metabolism, iodothyroglobulin and salicylic acid, which cause bees to arrive at food sources too soon, and with euquinine, carbon dioxide, and other depressants, which cause them to arrive too late. That the bees' timing is not due to some cyclic environmental stimulus not appreciated by us is further shown by displacement studies, in which bees trained in constant illumination to a 24-hour cycle in Paris were taken to New York and vice versa. In the new, constantly illuminated environment they retained their 24-hour cycle, unmodified by the five-hour time differential between the two places (Renner, 1957). Other studies show that under field conditions they gradually adjust their rhythm to the new location. The bee's actual cycle in continuous light is 23.4 hours, according to Beier, Medugorac, and Lindauer (1968); it must be reset daily by external stimuli, so that it does not drift ahead of clock time. These stimuli are not provided by dawn and dusk, as one might suppose, but appear to be an effect of the sun's movement in relation to landmarks. Beier and Lindauer (1970) speak of it as a "sundial" effect.

The influence of the environment on foraging intensity is conspicuous. Ribbands (1953) reviews the effects of various environmental factors on *Apis* flight. There are obvious interactions among factors such as temperature, light intensity, wind, rain, dew, time of day, colony size, stores of food in the colony, distance to food sources, ease of obtaining food, kind of food, concentration of sugar in the nectar, abundance of nectar, and so on. Some studies of each of these factors, taken independently, have been made with more or less the results that would be anticipated *a priori* (see Frisch, 1967a). This area of investigation obviously needs the gathering of much data in which many possibly important environmental factors are included, followed by a multivariate analysis, instead of the univariate studies such as have been made. Such an investigation would probably have practical as well as scientific implications.

Jaycox (1970) has investigated some of the influences of the workers' nest environment on foraging by *Apis*. The queen and her pheromones stimulate nectar collection. Because nectar-collecting bees often also gather pollen, there is also an increase in the pollen collected, as a by-product of the availability of nectar-bearing flowers and the collecting of nectar by the worker bees. In addition, as noted in Chapter 7 (D5), the presence of brood stimulates pollen collecting.

General field observations clearly indicate that the highly eusocial bees and *Bombus* can forage under less favorable meteorological conditions than other bees found in the same areas—lower temperatures, less light, more wind, light rain, etc. This is especially true of *Bombus*, perhaps because of its large size and insulating fur; these bees sometimes are active even when there are snowflakes in the air, or at twilight, or in moonlight. Heinrich (1972c) has shown that *Bombus vagans*, at environmental temperatures below 24° C, warms up by contraction of flight muscles, without wing movement, to 32 to 33° C. During foraging the flight muscles are regulated at this temperature over a wide range of environmental temperatures below 24° C. At higher environmental temperatures, thoracic temperatures rise—to 40° C at an ambient temperature of 30° C with activity in the sun. *Apis dorsata* also forages on moonlit nights (Diwan and Salvi, 1965). Perhaps it is significant that it is the largest *Apis* species. Other social bees, such as halictines and allodapines, are less tolerant of unfavorable conditions, and various solitary bees, especially some of the small ones, are active only in the most favorable weather. (There are a few crepuscular and nocturnal solitary bees, but these are mostly foraging specialists, not generalists like the social bees.)

Foraging range probably increases in a general way with size. There are hints from homing experiments that the range of minute halictines like *Lasioglossum imitatum* (4 to nearly 5 mm in body length) does not exceed 150 m (Michener, unpublished), but in Kansas nests have been observed being provisioned as much as 100 m from the nearest flowers. Wille and Orozco (1970) estimated the flight range of *L. umbripenne* in Costa Rica as 100 m. Bees that were marked and released about a kilometer from the nest usually did not return, suggesting that they were not familiar with the landmarks at such a distance. Bumblebees fly considerable distances if necessary. Padre J. S. Moure and I saw *Bombus morio* flying regularly across Guaratuba Bay in southern Brazil, an over-water distance of 2.5 km, which may have been only a fraction of the total flight. Janzen (1971) concluded from homing after artificial displacement that some of the larger euglossine bees have flight ranges in excess of 20 km.

Many studies of the flight range of *Apis* have been made (Ribbands, 1953). Maximum flights of 13.5 km have been recorded (Frisch, 1967a), but a colony cannot maintain itself with food from such distances, for most of the nutrition obtained has to be utilized in making the flights. Moreover, in spite of the speed of their flight, significant losses in efficiency result from the required flight time. Ribbands, studying weight gains of colonies 0, 0.6, and 1.21 km from nectar sources, showed reduction in nectar intake at the last distance of 14 percent due to flight time alone. His calculated reductions of nectar intake for trips of varying durations are shown in Table 14.1. Percentage reductions due to distance are much greater for trips of short duration, for example pollen-collecting trips, than for long ones, and much greater in bad weather than in good weather.

Data on flight ranges of Meliponini are given in connection with studies of communication (Chapter 15, B).

B. Pollen gathering and transport

The use of pollen as a protein source that must be carried to nests, which sometimes are miles from the flowers where it is collected, has resulted in development of a whole series of structures for handling pollen. Few counterparts to the pollen-handling structures of bees are found in other insects; I know of no other groups that collect and carry the dry, dustlike material. Nectar is brought back to the nests of bees in the crops of foragers. Pollen as food for the individual bee is often also found in the crop and, at least in some Meliponini, may be carried in this way for use of the colony. But pollen is ordinarily transported on the pollen-carrying hairs or scopa of females. It is carried either in the dry, somewhat powdery form (Halictidae or allodapines) or in a firm mass stuck together with a little nectar (as in Apidae).

In the former case, as in many solitary bees also, pollen is carried interspersed among the coarse, often plumed hairs of the scopa. In the Megachilidae the scopa is entirely on the underside of the abdomen. In the halictines (Figure 1.7) the scopa is on the hind coxae, trochanters, femora, and to a lesser extent on the tibiae, basitarsi, and in some species on the sides of the propodeum and undersurface of the abdomen. In allodapines scopal hairs for pollen transport are limited to the outer- and undersurfaces of the hind tibiae, and to a lesser extent the basitarsi. In the Apidae the scopa, which is on the outer side of the hind tibia, is modified to form, along with the tibial surface, a corbicula or pollen basket. The outer tibial surface is smooth, polished, and gently concave. Around this area is a row of coarse, in-curved hairs, usually a very few also arising from the concave surface (Figure 1.8). Pollen, moistened with nectar, is packed into this basket and held in place by the hairs.

Bees do not ordinarily gather pollen from flowers with the scopa. Loose pollen may be trapped by the hairs on the body of the bee as it moves about on flowers and subsequently combed off, usually by the front basitarsi. More frequently, these basitarsi, which are usually equipped with a comb or brush of hooked hairs, are used to remove the pollen from the anthers. A multitude of intricate symbiotic relationships exists between bees and flowers; coevolution has produced mechanisms that provide both for pollination of the plants and nutrition of the pollinators. The details of the techniques used by the bees to get pollen vary greatly with the flower and with

TABLE 14.1. Calculated effect of foraging distance on weight of nectar brought into an *Apis* hive.

Time for complete trip, including unloading (min)	5	15	30	60	120	240
Percent reduction from distance of 0.60 km	30	18	10	5	3	1
Percent reduction from distance of 1.21 km	56	30	18	10	5	3

SOURCE: Ribbands, 1953.

the bee; since they are of no special concern in relation to social behavior they are not dealt with here.

In *Lasioglossum imitatum* and other halictids the bees brush anthers with the fore tarsi, transfer it to the middle legs, and thence to the scopa on the hind legs and perhaps to the hairs of the abdominal sterna, which also carry considerable pollen (Michener and Wille, 1961).

In the Apidae the pollen is brushed toward the mouthparts and some of it moistened with a little regurgitated nectar or honey before being passed back to the corbiculae. The details of transferring the pollen to the corbiculae have been studied in *Apis*. Accounts of the process have been published by various authors, but an excellent source is Hodges (1952). After becoming well dusted with pollen the bee usually hovers for a few moments and while on the wing, or less commonly at rest, brushes her body rapidly with the hairs of the basitarsi; the head, appendages and front part of the thorax are brushed by the front legs, the other parts of the thorax by the middle legs, the abdomen by the nine combs of hairs on the inner surface of each hind basitarsus. Pollen from some kinds of plants is dropped, but if the pollen is to be saved, that from the front and middle legs is passed back to the same nine combs on the hind basitarsi. When enough pollen is accumulated there, the inner surfaces of these basitarsi are rubbed rapidly together in a pumping, lengthwise motion; this is the most easily seen part of this activity of a hovering bee. At the downward stroke of one leg, the tibial comb or rake across the end of the tibia (Figure 1.8) removes pollen from the opposite basitarsus into the space between the auricle and the comb. Movement of the basitarsus with respect to the tibia then pushes the sticky pollen mass upward onto the smooth surface of the corbicula. Pollen from the inner surface of one basitarsus is thus transferred to the outer surface of the opposite leg.

C. Building materials and transport

Most bees do not bring nest-building materials from outside the nest. Megachilids, however, carry them with the legs or with the help of mandibles. The Apidae also commonly use building materials from outside sources; they carry them in the corbiculae like pollen. Indeed, *Bombus* uses pollen mixed with wax as the building material.

The Euglossini, Meliponini, and Apini use resin, alone or mixed with wax or mud, for building. They carry both resin and mud in corbiculae. In *Apis*, dance communication may indicate the location of resin sources, just as it serves to facilitate foraging for other resources. The euglossine genus *Euplusia* sticks bits of bark, used in cell construction, to resin in the corbiculae for transport.

The extremely smooth, shiny corbicular surfaces are probably an adaptation or preadaptation for resin carrying. Bees frequently carry materials so sticky that one wonders how they could ever clean themselves; fresh pine pitch and rather fresh, tacky, oil paint do not easily come off the human skin, but both are carried to the nests at times by bees and come off the corbiculae apparently easily and cleanly. Figure 14.3 shows the way in which *Trigona postica* cuts out pieces of resin (or in this case cerumen, a resin-wax mixture) with the mandibles and transfers the material with the middle tarsus to the corbicula for carrying home. Figure 14.4 shows competition among individuals for the material, even though both bees are standing on an abundance of the cerumen.

Bassindale (1955) gives an interesting account of the essentially identical method by which a minute African *Trigona, T. braunsi,* loads and carries propolis. The foraging worker bites off a piece of the resin with her mandibles. It is then transferred to the inside of the middle leg with the aid of the first leg. With the middle leg, it is then carried back and patted into place on the corbicula or pollen basket of the hind tibia. The next piece is passed back on the opposite side. Subsequent pieces go back, usually on alternate sides, until six or eight pieces are stuck in a lump on the outer side of each hind tibia. The bee is then ready to fly with its load of sticky material.

Unloading is accomplished without the help of other bees in spite of the stickiness of the material. The bee presses the sticky load against the wall of the nest and then helps to separate the material from the corbicula by pushing the basitarsis of the second leg along the smooth surface of the corbicula from the proximal to the distal end. The process is then repeated on the other side; unloading is thus accomplished in only a few seconds.

D. Orientation

The need to fly from a nest, find food or nesting materials, and return to the nest is common to virtually all bees and nonparasitic wasps. The evidence based on wasps, bumblebees, and *Halictus* and casual observations

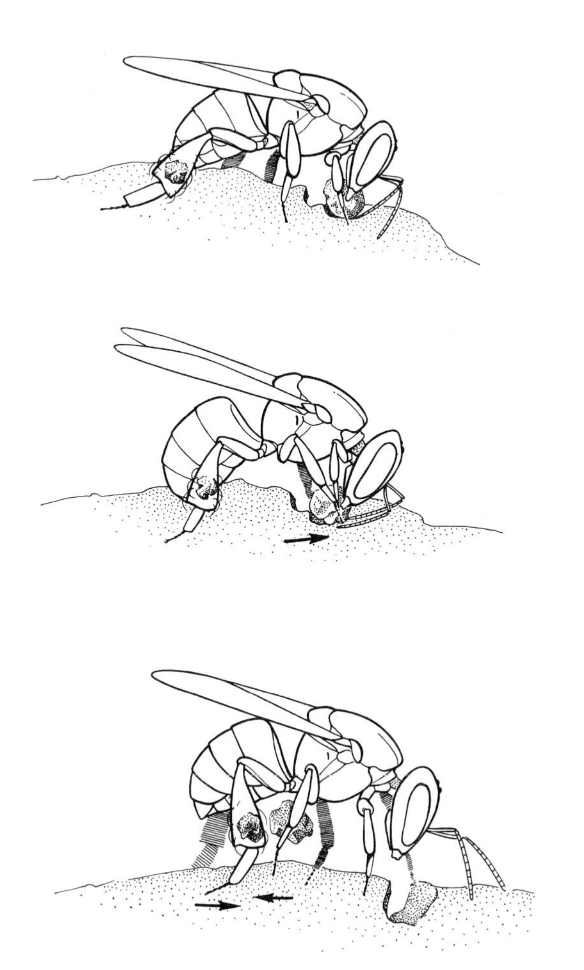

FIGURE 14.3. *Trigona postica* collecting cerumen (wax and resin). *Above,* particles are cut out by the mandibles; *middle,* manipulated by the forelegs and middle legs; *below,* transferred to the corbicula on the hind tibia by means of the middle leg. *Arrows* indicate movements of mid and hind legs. (From Sakagami and Camargo, 1964.)

of many other forms including allodapines, suggests no important differences between the orientation mechanisms of solitary and of social bees and wasps (Jacobs-Jessen, 1959). Orientation of *Apis* has been studied in detail (Frisch, 1967a) and summarized in various works on the honeybee. With appropriate adjustments for the size and flight range of the insects, the following account is probably applicable to almost any bee. It is based, however, entirely on worker honeybees, particularly apt subjects for such studies because they are relatively unexcitable (not disappearing for hours at every disturbance) and because each foraging worker will make numerous trips per day for syrup, so that her behavior under experimentally modified conditions can be observed. The summary below is brief not only because of the excellent, detailed accounts already available but because the orientation ability is in no known respect unique to social forms.

Flight orientation involves primarily visual stimuli both from the ground and the sky. Bees discriminate with regard to form, color, and light intensity. Stimuli from the sky include the sun's position and the direction and magnitude of polarization of light. Besides visual stimuli, time of day is important. Odors are also utilized, more often in short-distance than in long-range orientation. Airborne sound is probably not involved. Maze experiments indicate that learning of the route back to the nest is independent of learning the route out from the nest.

As has been suggested to me by Dr. Rudolf Jander, foraging in bees involves two types of orientation, (a) object orientation, and (b) topographic orientation. Object orientation is illustrated by responses of a bee to flowers. The bee flies toward an appropriate flower no matter where it may be recognized. Sight, odor, or both mediate the response. Topographic orientation is orientation toward a place. Thus flight toward a known food area or to the nest involves orientation with respect to a particular place. The remainder of this chapter is largely concerned with topographic orientation.

The topographic orientation of bees is itself divisible into two subtypes. The first is based on knowledge of localities—the constellation of landmarks as related to the nest site, food and water sources, and the like. The second is vector orientation—the use of information about distances and directions.

1. Short- and moderate-distance orientation

This section largely deals with knowledge of localities, although the close approach to a food source such as a flower may often be object orientation. Young bees become familiar with the environment outside their nests by making orientation or exercise flights—short flights from which they return without foraging. As they age and become foragers, they make longer flights. Bees that have made no orientation flights are often lost if liberated away from the hive; they seem to return only if, by chance, they fly close enough to react to its odor. More experienced bees are progressively more effective in returning to the hive; a few seconds hovering in front of the hive followed by a flight of from three to seven minutes is

FIGURE 14.4. Two workers of *Trigona postica* on cerumen (wax and resin) of an abandoned nest of another stingless bee. The upper bee has bitten off and transferred to its corbiculae large lumps of the cerumen. The lower bee is starting to rob one of these lumps from the upper bee, which has raised its hind leg in ineffectual avoidance. (From Sakagami and Camargo, 1964; original drawing provided by J. M. F. de Camargo.)

sufficient to enable most bees to return promptly if taken 50 m away. The first such flights may extend a few hundred meters and subsequent ones a few thousand.

To determine what sorts of information bees learn on their orientation and other flights, many observations and experiments have been made. They may be summarized by saying that bees take note of all of the sorts of things that we would be likely to notice—position with respect to landmarks, height, color, and so forth. The low-acuity insect eye, however, will respond to less detail and fine structure in the environment than could a high-acuity vertebrate eye. If a hive is moved several kilometers the bees usually orient to the new position before leaving, and can then return directly to it. If a hive is moved a meter or two, however, bees often hover over or clump at the old site before finding the new one.

Like people, bees do not necessarily use all the cues that they could use in their topographic orientation. For example, in a row of like hives, most bees find their way to and from their own, although more bees do drift from hive to hive in the middle of the row than at the ends. This shows that position or small landmarks on the boxes or ground surfaces serve to guide most bees correctly. However, if movable colored fronts are put on the boxes or transferrable designs of certain sorts are used, interchanging of the color (Figure 14.5) or design results in bees returning to the wrong box. Such conspicuous features must override the subtler features which serve in the row of unadorned hives.

There are limitations in man's use of landmarks, due to sensory and nervous limitations. A man (or a bee) can get lost in an environmentally uniform area—a uniform forest will serve even though no two of the tree trunks are in detail alike. But the bee's sensory and nervous limitations are not like man's, as seen from the paragraphs on color vision and on form perception in Chapter 1 (B3a).

Ribbands (1953), in reviewing previous work, refers to the discussion in the literature early in the century as to whether *Apis* workers are attracted first, that is, from a long distance, by color and then by odor or vice versa. It now seems perfectly clear that either may happen; bees are quite flexible in this aspect of their behavior. For bees that are not scouting but know roughly where they are going, landmarks (including flower color and form) typically operate at greater distances than do odors. But the many flowers which are inconspicuous and greenish, yet fragrant, are likely to draw scouting bees first by scent and only on close approach by appearance. Odor plays an important role in stimulating bees to settle, as on a flower or perhaps a nest entrance.

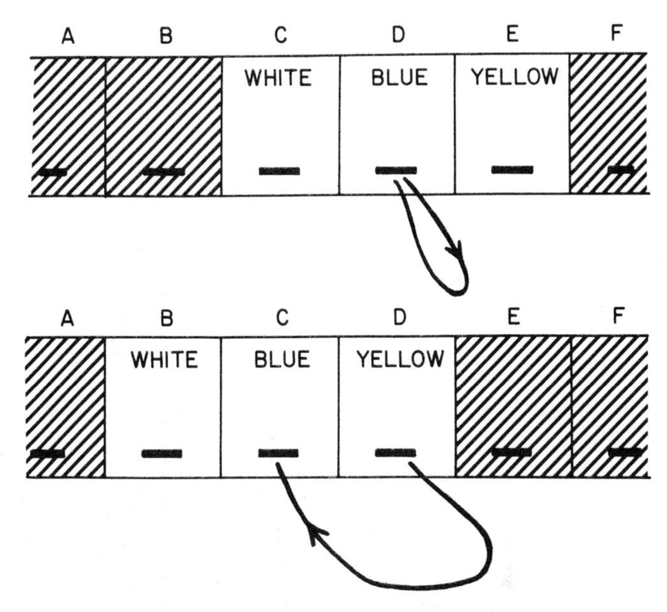

FIGURE 14.5. Two arrangements of colored hive fronts used on a row of contiguous hives. *Above,* the original arrangement (*C* was an empty hive). *Below,* hives and bees not moved, but colored fronts shifted; many returning bees from *D* entered *C*, having become accustomed to the blue color. (From Ribbands, 1953, after Frisch.)

With respect to food sources and presumably other places visited away from the hive, there is excellent evidence that *Apis* workers make observations useful for subsequent orientation as they approach the site. Opfinger (1931, 1949) conducted experiments showing that the return of a bee to a food source depended on the color, position, and odor at previous arrival times. These features, in her experiments, could be changed while the bee was feeding, but the changed features seemed not to be memorized. On the departure flights, however, many bees did note more distant landmarks. Pertinent details of Opfinger's experiments are given in English by Ribbands (1953).

Olfaction plays a significant role in *Apis* orientation. The sensory organs are the pore plates on the antennae. Frisch considered the sense of smell to be similar to that of humans in the area of spicy, flowery, or fruity smells;

bees could be trained to distinguish numerous such odor substances at concentrations more or less similar to those perceived by man. He found that for a few substances, the bees' own products, secretions of mandibular glands, and Nasanov's gland and wax, bees were considerably more sensitive than man. Ribbands (1955), Vareschi (1971), and others, however, using more sensitive techniques, show that for various substances honeybees have scent thresholds of a lower order than had previously been recognized (Chapter 1, B3b). For the most part, materials that smell similar to us are the ones likely to be confused by bees, suggesting that surprising as it may be, olfactory principles may be similar for vertebrates and arthropods. Training has been difficult with respect to foul odors, although bees perceive them, as shown by avoidance and ventilation responses and sometimes successful training. Curiously, however, honeybees prefer water sources that are to us malodorous, for example, water polluted with cow dung or sewage, and Butler (1954a) describes experiments showing that it is in fact the odor that influences the choice.

The question of bee memory of landmarks and color and odor cues in their complicated environment is of interest. Recent studies of the subject are by Koltermann (1969) on scent training and by Menzel (1967, 1968, 1969) on memory of colors. Finally, Menzel and Erber (1972) reported on the influence of quantity of reward on learning. This sort of memory influences the success that a bee has, after leaving a food source, in returning to it. Rate of learning is influenced by the kind of stimulus. Scent is learned more easily than color and color more easily than form. A single experience with a scent, if rewarded by sugar solution, produced 90 percent constancy to that odor in subsequent choice experiments. If rewarded at a second scent, the bees still chose the first when given a choice, until they had had several experiences with the second. Thus after one feeding at scent I and 10 feedings at scent II, 14 percent of the bees still selected scent I in choice experiments. That the other 86 percent still remembered scent I was shown by offering a choice between I and a new scent, III. In this case 98.3 percent chose scent I over III.

Memory of scent decreased with time, but was most efficient at 24-hour intervals—an interesting relation to the time sense. The decrease in efficiency with time was strongly influenced by the input of other information. Bees confined to the hive after training remembered best, while those that flew and acquired other information through days from the training time to the testing time performed less well in the tests.

The study of memory of colors was done with spectral colors and with rewards of sugar solution. As little as three seconds of colored light was sufficient for training. One reward caused a strong preference for the associated color for a short period, but three rewards established a preference that was not reduced over a two-week period. Menzel (1967) and Lindauer (1969) noted that speed of learning varies with the color; violet is learned more quickly than other colors tested, and bluish green more slowly. Speed of forgetting also varies. Violet offered only once was not forgotten for six days; yellow was forgotten after two days. The longer the first color used in training was offered, the longer it took a bee to learn a second or replacement color.

The preceding paragraphs deal with learning ability, but something that might be called motivation also influences learning rate. Over a wide range (15 percent to 60 percent) it seems probable that sugar concentration in syrup does not influence the rate of learning of the cues that will permit successful return to the food. The quantity of syrup available, however, that is, the yield of a source per unit time, influences learning rate. A minute amount of syrup that can be sucked up in one second leads to short-term memory, so that a bee can return immediately to the site. But continued uptake of syrup is needed for long-term memory of the location. It turns out, therefore, that a supply of syrup provided slowly over 30 seconds, requiring continued sucking, leads to better memory of the site than a large supply at which the bee quickly satisfies itself and stops sucking (Menzel and Erber, 1972).

2. *Long-distance orientation*

Those cues for orientation so far dealt with are all of importance at moderate or short distances; they are the counterpart of landmarks used by a ship along the coast or a small plane approaching an airport. The visual acuity of bees is not good compared with that of man; details of form often do not bring ready responses. Olfaction is usually valuable for flight orientation only at short distances. As with an airliner flying between distant cities, navigation of bees across miles of relatively uniform country is scarcely practical without distance and direction information. The same is presumably true for a tiny *Lasioglossum* flying across 150 meters of prairie. The following paragraphs, therefore, deal with navigation for long-distance flights, during which only distant landmarks and celestial navigation, together with measures of distance travelled or work expended, can be expected to serve.

Many insects, including honeybees, have the ability to go in a straight line by keeping the sun in a particular direction as they move. This is called sun compass movement. Having a time sense, bees can use the sun for orientation throughout the day even though it seems to move across the sky. Thus, to briefly describe an experiment by Frisch: Honeybees were trained to a certain scented syrup source 200 m west of the hive (Figure 14.6), and in the evening 29 of them were marked. The hive was then closed and during the night taken to a broad, level area unfamiliar to the bees, and similarly scented syrup sources were placed 200 m from the hive in all four directions. Of the marked bees, 20 appeared at the western feeding site, 5 at the southern, and 1 each at the eastern and northern sites. Further experiments clearly showed that at least on sunny days most bees could, at any hour, select a particular compass direction. As to the few that went to the southern dish, Frisch notes that the evening before, in the original location, the sun was 40° to the left of the direction from hive to syrup. The following morning in the new location about the same relation existed when the bees arrived at the southern dish. He suggests that perhaps these were inexperienced foragers that had not yet learned the 24-hour solar cycle and that took the angular direction with respect to the sun remembered from the evening before, instead of the correct compass direction as determined by relating time of day and angular difference between the food source's and the sun's directions. Figure 14.6 shows the relationships of hive, dishes, and sun, based on Frisch's account.

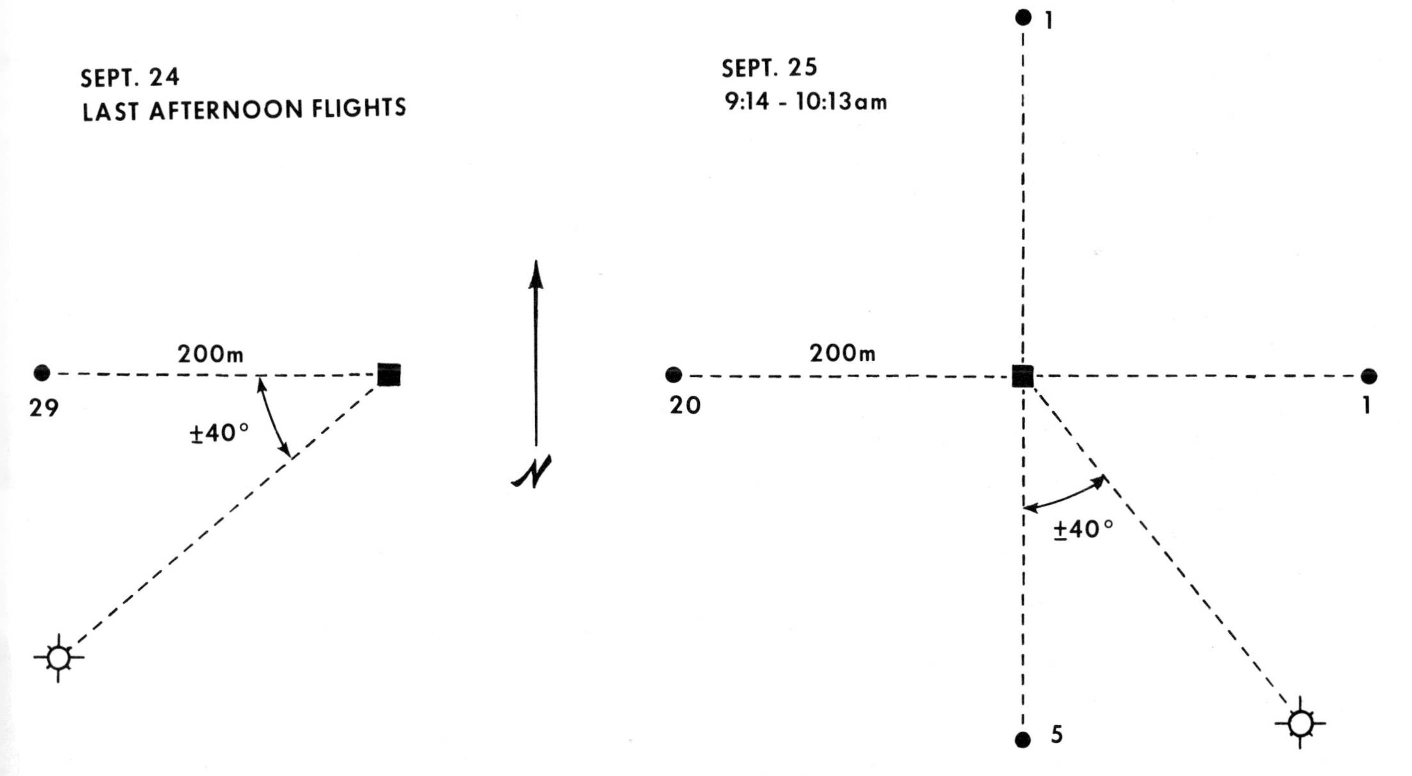

FIGURE 14.6. Spatial relationships of a hive of *Apis* (*square*), food dishes (*dots*), and azimuth of the sun in an experiment by Frisch (see text). The sun compass orientation learned at one location (September 24) has been displaced and corrected for time of day so that at a new location on September 25 most bees (20) still go west to the food. A few (5) failed to make the correction and went to a food dish in the south, having the same angular relation to the sun as the dish at the original location the afternoon before. Still smaller numbers (1 each) arrived at similar dishes to the north and east. (Original drawing by Barry Siler.)

The ability of experienced bees using sun compass orientation to adjust their flight direction according to the time of day is further illustrated by Meder (1958), who held bees for an hour at a food source in a dark box. His exhaustive study shows that such bees leave the food source in the same direction as undelayed bees (Figure 14.7). The delayed ones must have corrected the direction of their flight with respect to the sun's position by 15°.

In experiments in which striking landmarks, visible for long distances, were compared in effectiveness with sun compass orientation, it was found that landmarks usually were followed if continuous and running between hive and food source. Thus the edge of a woods was followed after moving a colony to a new site, if the route to the syrup at the old site was along such an edge, even though the compass direction of the edges at the two places was different. But in experiments with more irregular or finely divided habitats, the bees followed the compass direction.

That the sun's course across the sky is learned by each individual bee has been shown by Lindauer (see Frisch, 1967a; Lindauer, 1957b) in an ingenious series of experiments. Bees in the Southern Hemisphere that have learned that the sun moves counterclockwise across the sky (as one faces the sun), relearn that it moves clockwise when taken to the Northern Hemisphere. In the Tropical Zone such relearning must occur twice annually, without movement of the hives, when the sun passes the zenith. Bees inherit remarkable abilities to learn the sun's movement. After seeing only the afternoon part of the sun's path, they seemingly know the whole path and orient properly when freed in the morning.

Sun compass orientation associated with the time sense, so that the bees can find a compass direction at any time of day, must be of utmost importance. Flights or periods of time away from the nest are often long, especially in *Bombus*, which may be out for a whole day; the bees must take the right direction home in spite of the sun's advancement. Moreover, the sun compass is involved in communication which, thanks to the abilities mentioned, can continue as long as is appropriate without the communicator going into the field and exposing herself to the field dangers.

Orientation when the sky is fully cloud-covered is possible if the clouds are not too thick. Ultraviolet light penetrates light clouds, and bees can recognize the sun's position when we cannot (except with ultraviolet-sensitive equipment).

Orientation as though by sun compass is also fre-

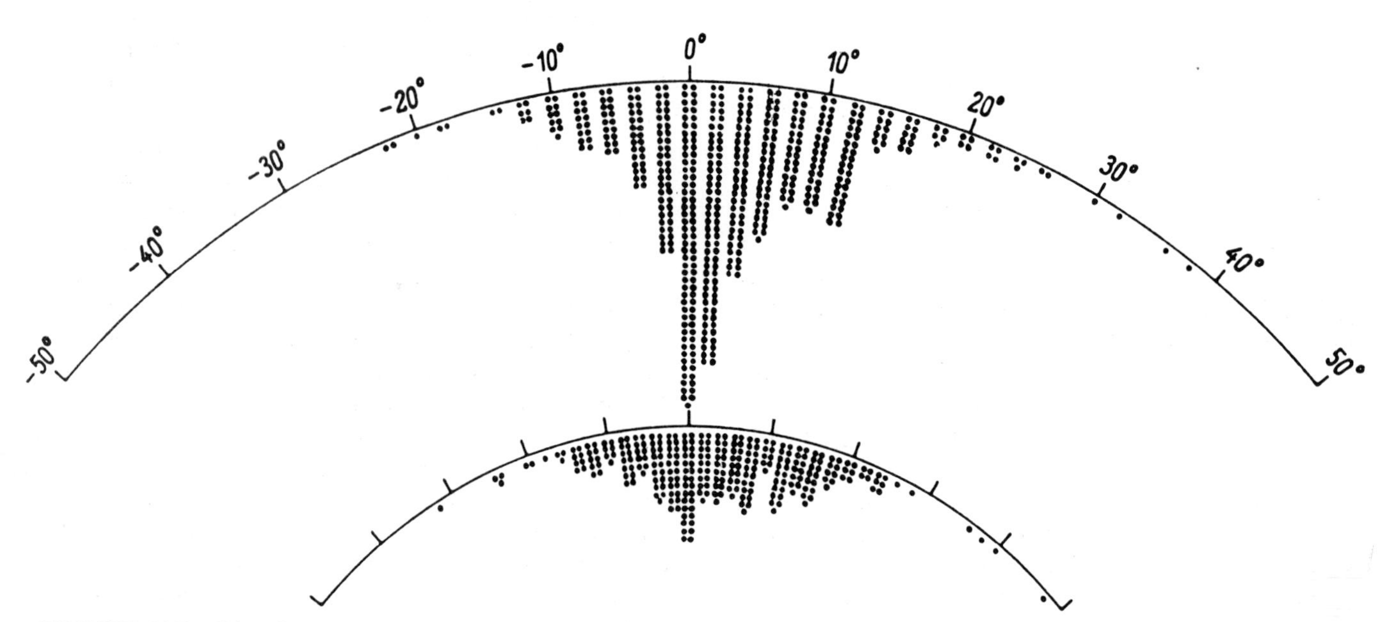

FIGURE 14.7. Directions of departure of workers of *Apis* from a feeding place. The direction to the hive is represented by 0°. Each dot represents the direction taken by one bee. The upper group represents bees departing without interference. The lower group represents bees kept at the feeding place for an hour in the dark. There is no meaningful difference between the two groups in directions taken. (From Frisch, 1967a, after Meder, 1958.)

quently possible when any blue sky is visible, even if the sun is hidden by thick clouds or by a mountain. Our understanding of this matter is a monument to the persistent and ingenious work of Frisch and his students, brought together by Frisch (1967a). Bees, like most arthropods, can perceive the direction of the plane of vibration of polarized light, something which humans ordinarily cannot do without instruments. The light from the blue sky is polarized in a distinctive pattern which moves across the sky with the sun; this polarization is destroyed by cloud cover, so that orientation by this means is possible only with blue sky visible. Bees presumably learn the apparent movements of the sun and of the associated polarization pattern together, so that sight of the latter allows them to respond as though to the former. Thus in the afternoon shadow of a mountain, bees at a new site were trained to go to a feeding station. The next day at another new site they were tested in the sunlight. The bees that had learned the direction to the food when the mountain hid the sun flew in the same direction when orienting by the sun the next day. Most of the numerous other experiments on perception of the polarization of light concern dances and communication, not orientation of bees in the field, and are discussed in Chapter 15 (D2d).

The ability to select the proper direction for flight toward a goal such as a known food source or a nest may not be all that is needed to arrive there. An odor perceived on such a flight could cause a bee to stop at the goal, but knowledge of how far to fly before starting local searching would be extremely important, particularly if the odor of the goal was not very distinctive or widespread, and if there were small errors in the direction taken. Bees do have a "distance sense," as demonstrated by Wolf (1927). As shown in Figure 14.8, he trained bees to go to a food source 165 m from a colony in a flat open area "without optical orientation marks." Bees captured at the food source were fed and released there and at three other sites, as shown in the figure. The flight paths shown were not verified, but corresponded to times required to return to the hive, as well as to subsequent data verifying a distance sense used in connection with communication and based on the amount of energy required for the flight rather than a genuine measure of the distance covered.

A bee, of course, makes use of all of its various orienting abilities in finding and shortening its way. There are observations of bees which, in going to and from a food

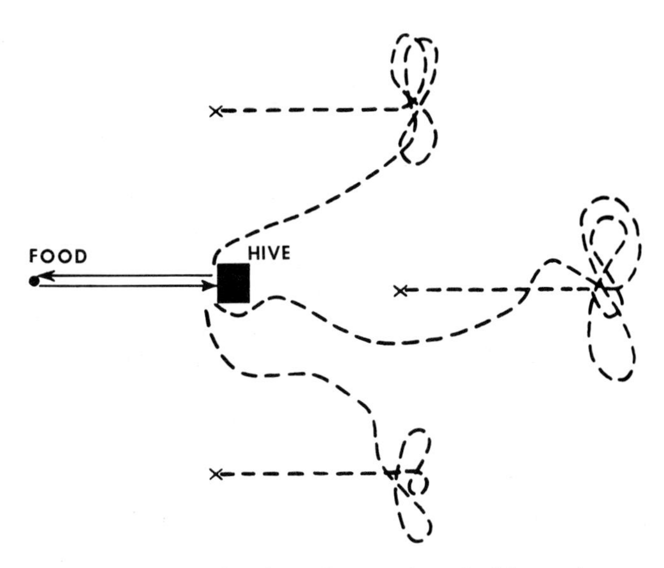

FIGURE 14.8. After honeybee workers had learned to go back and forth from the hive to the food dish in an open area without landmarks, some were captured at the feeding place and liberated at sites marked *x*. *Broken lines* represent the assumed routes, first to the place where the hive would have been had the bees not been moved, and then, after wandering, to the hive. (From Ribbands, 1953, after Wolf, 1927.)

source, initially used a circuitious route following known landmarks, but after several trips cut off the angles and soon were flying the straight route. "Bee line" is not an idle expression; bees really do commonly fly the shortest route to and from a food source.

I suppose that most of what is written above concerning orientation and learning applies to bees in general, whether social or solitary. One must not imagine, however, that all bees are identical in these respects. Presumably the principles are similar; all bees must learn and orient themselves with respect to locations and odors of nests and food sources. But probably no two species are identical in the details. Indeed, Lauer and Lindauer (1971) have shown that two races of *Apis mellifera* differ in the details of their orientation and learning capacities. For example, with striking optical cues near the food source, Carniolan bees (*A. mellifera carnica*) find and learn the location of the food faster than Italian bees (*A. m. ligustica*). In an optically uniform environment, however, the learning curves for the two races are alike. Carniolan bees can use less prominent visual cues than Italian bees. These authors also cite various other differences in orientation and learning between the two races, all presumably genetically controlled.

15 *Communication Concerning Food Sources*

Orientation as described in the preceding chapter has to do with activities of individual bees—how one bee finds its way back to the nest once a food source is found, and how it then returns to the food source for more. Probably any female bee of whatever species (and many or all males as well) can do this sort of thing. Among some social bees, however, a further ability important for foraging exists, namely communication of the locations of food sources to other bees. There is no way in which bees can describe a location to other bees, unless transmission of the odor of the site on their own bodies is so categorized. Therefore locality knowledge, except odor, is not communicated or is communicated only by a bee leading recruits. Vector orientation (Chapter 14, D) is communicated, however, in some highly eusocial bees, and object orientation, as toward a kind of flower with a particular odor, may also be communicated.

The biological significance of communication with respect to food supplies in highly eusocial bees is considerable. First, it takes no imagination to recognize the advantage to the colony in the use of distant food resources if the finder can guide or direct inexperienced colony members to the source, or can notify experienced bees that a resource with which they are familiar is again productive. Secondly, the colony rather than the individual becomes the foraging unit. This promotes distinctiveness of the colonies, because even nearby colonies commonly take in different foods or the same foods in different proportions. If each bee made its own search for flowers, the mixture of foods brought in by neighboring large colonies would be much the same. Communication, however, leads to distinctive colony odors and mutual recognition, which permits defense against robbing (see Chapter 18, C).

Since the colony is the foraging unit, and different units even in the same area forage differently, the colonies partly escape competition with one another. Their coexistence should be facilitated by the differences in foods taken.

Nearby colonies of a bee that does not communicate about its food supplies, where every individual searches independently, will be in direct competition. The nest sites will therefore need to be dispersed, as they are in *Bombus.* Nest aggregations are wholly unknown for *Bombus,* even though nearby conspecific *Bombus* nests sometimes show some differences in food sources. Highly eusocial bees, however, can take advantage of numerous nearby nest sites and form aggregations of colonies, because the foragers will interfere less with one another. Aggregations of nests of *Apis dorsata* (up to 156 nests in one tree, Chapter 5, D) are well known. Some *Trigona,* such as *T. cupira,* often build several nests less than a meter from one another, or only a few meters apart.

A. Nonspecific communication

There is little or no communication about food sources in the parasocial or primitively eusocial bees. Michener and Wille (1961) showed that for *Lasioglossum imitatum* there was no relation between return of foragers to the nest and departures from it, except that when bees were working rather rapidly and therefore staying for relatively short periods in the nest, the return of a forager was often soon followed by a departure, presumably that of the same bee. There is thus no evidence of social facilitation with regard to foraging trips in this species.

In *Bombus* the return of a forager increases the likelihood that other individuals will soon leave the nest; thus social facilitation occurs. When young gynes forage to augment the worker force, passage of a gyne through the nest entrance causes a disproportionate increase in passage of other bees during the next five seconds (Blackith, 1957). How the increase is mediated is not known. Schneider (1972) reports different sounds from arriving than from departing bumblebees. This fact, taken with the importance of sounds in communication of highly eusocial bees, suggests that sounds may lead to the social facilitation of *Bombus.* Even if not, sounds made by primitively eusocial bees like *Bombus* may have been precursors of the communicative sounds of highly eusocial bees. But in *Bombus* there is no dance, no communication of food location, and only recently has a response to the odor of the food source been noted (since there is, with rare exceptions, no transmission of food among adults).

As summarized by Brian (1954), various authors have observed a single bumblebee returning again and again to a rich food source like a dish of syrup without ever recruiting other individuals to the same source. But Free (1969, 1970a) states that the odor of the food stores of a *Bombus* colony can influence the choice of flower species visited by its foragers. Nearby colonies of the same species may exploit the local flora in different ways, not explicable by random foraging of individuals. This is the only record of a social interaction influencing the choice of food sources in primitively eusocial bees.

In *Apis,* also, the odor of the food stored in the combs may influence the bees' choice of food sources. Free (1969) showed this by exchanging the stored food (but not the bees) of colonies that had been collecting from two differently scented syrup sources. The bees were then offered the choice of syrup scented with the two materials; a high percentage went to dishes scented like their new food stores. Scents introduced into the hive but not into the food had no such influence. There are no data on the practical influence of this sort of simple interaction in highly eusocial bees; presumably it is usually overridden by the individual-to-individual chemical and mechanical communication described below.

B. Relation between flight range and communication

In highly eusocial bees the colony size is usually relatively large, and the problem of maintaining the biomass of the colony on food brought in minute quantities from dispersed sources must be considerable. Few data are available concerning foraging distance or flight range, and there is much variation according to local circumstances. Most of the data are based on experiments in which a source of syrup to which bees have been trained is moved progressively farther from the nest. Bees accustomed to it continue to visit it; a distance is reached where communication ceases, as shown by failure of bees to recruit additional visitors (and in *Apis,* failure to dance), and finally a distance is reached at which even regular visitors fail to return to the syrup. Such data are shown in Table 15.1. Understandably, the presence of competitive food sources seems to be important in determining these distances, as is perhaps shown by the relatively low figures for subspecies of *A. mellifera* obtained in a Brazilian savanna with many flowers. The sizes of the bees range somewhat erratically from smallest at the top of the table to largest below.

TABLE 15.1. Maximum flight and communication ranges for some bees.

Species	Communication ceases[a]	Flights cease[a]
Trigona iridipennis (Lindauer, 1957a)	100	120
T. mosquito	300	540
T. spinipes	630	840
T. amalthea	800	980
Apis cerana indica (Lindauer, 1957a)	700	750
A. mellifera mellifera	2,430	2,500
A. mellifera adansonii	2,430	2,800
A. mellifera (Knaffl, 1953)	11,000	12,000

SOURCE: Kerr, 1959, except as indicated after species names.
[a] Distances are in meters. Data are based on experiments in which syrup sources to which bees have been trained were moved progressively further from the nest. Cessation of communication was shown by failure of bees to recruit additional visitors, and, in *Apis,* failure to dance. Longer ranges are those at which even trained individuals failed to return. Size of bees increases erratically from smallest at top to largest below.

There is in general a positive relationship between biomass of colonies, probable foraging distance (reflected in bee size), and effectiveness of communication. Thus, among the highly eusocial bees, and particularly among those living in large colonies consisting of moderate-sized (rather than small) bees, good communication concerning food sources exists. The greater the flight range, the greater the need for such communication, as the foraging area involved is a squared function of the flight range (πr^2). Hence communication is better developed in the larger Apinae than in the minute ones. Poor communication in minute species, considered to be primitive by Kerr and Esch (1965), quite possibly represents instead a loss associated with small size and the adequacy of olfaction for finding food in forms with limited flight ranges.

C. Communication in stingless honeybees

In the following account methods of communication are arranged in order of increasing complexity.

In *Trigona* (*Frieseomelitta*)[1] *silvestrii*, a South American species, returning foragers only have the capacity to alert their associates to go to syrup in an artificial source if perfume is added to the syrup (Kerr and Esch, 1965). The implication of this observation is that the odor of the returning foragers is utilized by the new foragers in finding the food source. No information is provided about distance and direction. Returning foragers do produce a weak sound on entering the hive and pass nectar to receiving bees, but nothing has been established as to the communicative significance of this sound, if any. The sounds are made by wing vibrations; Kerr and Esch postulate that the sounds owe their origin, evolutionarily speaking, to bees vibrating their wings between trips and thus maintaining the thoracic muscles at flight temperature. In nearly all other highly eusocial bees such sounds have communicative significance. *T. silvestrii* is a small species, perhaps specialized in that respect; whether its poor system of communication is also a specialization (loss) or a primitive feature is not known. Kerr (1969) says that communication in two other species of *Frieseomelitta* is equally rudimentary.

In a series of other small to minute species of *Trigona*—*T.* (*Plebeia*) *droryana, T.* (*Trigonisca*) *muelleri, T.* (*Hypotrigona*) *araujoi, T.* (*Tetragona*) *jaty,* and probably *T.* (*Tetragona*) *iridipennis*—from South America, Africa, and Asia, and also in *Meliponula bocandei,* a moderate-sized African species, a somewhat more elaborate communication occurs. Returning foragers alert potential foragers not only by a strong sound but by zigzag running, often striking other bees (Kerr and Esch, 1965). The zigzag running is suggestive of the bumbling dance or jostling run of *Apis* (Schmid, 1964), the function of which remains unknown. In the experimental nests of Lindauer and Kerr (1958, 1960), containing relatively few bees, the tendency of the returned forager to seek out and jostle other bees was obvious. Return of such foragers resulted in a general increase of activity of bees in the colony, with the exception of the incompletely pigmented, callow individuals. From time to time the run is interrupted by a sharp semicircular turn, at which point a small amount of nectar is given to another bee. Figure 15.1 shows the path of one zigzag run. For the species listed above, the returning bees convey no information as to distance and direction; alerted bees are merely stimulated to leave the nest. They arrive in equal numbers at test food sources in any direction from the nest if equally distant from it, providing the odor and taste are like that of the original source (Figure 15.2).

A modification of this system of communication occurs in *Trigona* (*Axestotrigona*) *ferruginea* and *T.* (*Nannotrigona*) *testaceicornis.* In these bees, soon after a successful forager has returned to her colony and produced her characteristic sound, nearby workers repeat it; the sound spreads and in less than a minute the whole colony buzzes and all the responsive bees leave to forage.

Trigona testaceicornis is also of historical interest as the first meliponine whose communication about food sources was proved. Various persons had noted the rather sudden appearance of large numbers of stingless bees on sweets that had long been exposed without attracting bees, and had therefore assumed communication. However, in 1953, W. E. Kerr put a dish of sugar syrup a few meters from a series of nests of *testaceicornis.* No bees found it in over a day. But after one worker was taken to it and allowed to return to its hive, two more appeared within 30 minutes, and within a couple of hours 11 were

1. In the past and in Chapter 29 I have included *Frieseomelitta* with the subgenus *Tetragona.* However, it has distinctive features of nest structure and communication, and distinctive minor morphological features. For present purposes I find it convenient to use the name in spite of morphological proximity to *Tetragona.*

FIGURE 15.1. Sketch of a nest of a stingless bee showing the combs (*shaded*), involucrum (*light lines*), and the path of a returned forager in a zigzag run (*heavy line*). The tubular nest entrance is indicated above. (From Lindauer and Kerr, 1958.)

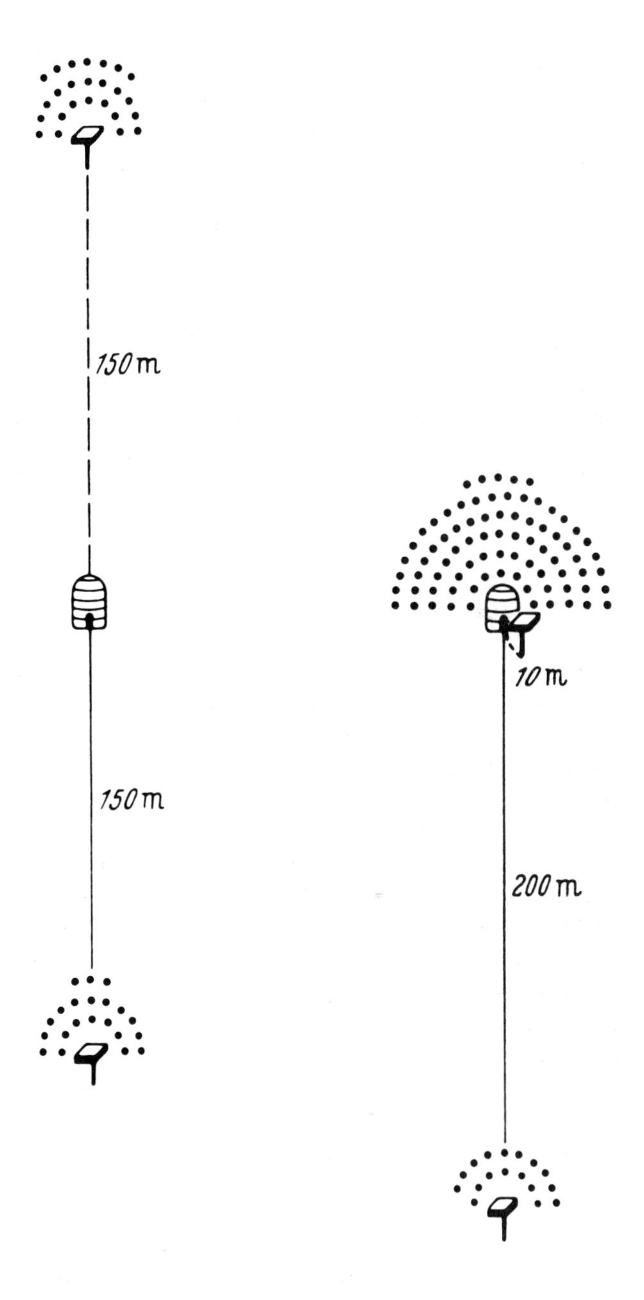

FIGURE 15.2. Diagram showing, for *Trigona* (*Plebeia*) *droryana,* that bees returning to the nest (*center*) from a food source (*bottom*) alert other foragers, which arrive in about equal numbers (shown by dots) at equidistant food sources (*left*) and in greater numbers at a nearby food source (*right*). Returning foragers stimulate others to activity but transmit no information as to particular food sources other than, probably, odor. (From Lindauer and Kerr, 1958.)

visiting the dish. All were marked and found to be inhabitants of a single hive; none from the other hives had found the food (Kerr, Ferreira, and Mattos, 1963). This does not mean that *T. testaceicornis* indicates distance and direction to food sources; so far as known, odor and proximity of the dish to the nest must have explained the find of the food by the ten alerted recruits. Of course there may be other and still unknown communicative abilities in this species.

In the bees discussed above, the communication involves a stimulus for foraging, with olfactory guidance to any food source having the odor of the original source; in the species discussed below communication can contain information directing recruits to a specific food source.

Figure 15.3 shows the counterpart, for such a bee, of the experiment diagrammed for a species of *Plebeia* in Figure 15.2. One aspect of such improved communication involves marking the source or its vicinity with an odor. This was noted even in the African *Meliponula bocandei* mentioned above. Kerr trained workers to a syrup source 50 m from the nest. Many workers, on leaving the food, alit after a flight of only 50 cm or a meter toward the nest and deposited mandibular gland secretion, leaving in the vicinity of the food a bee odor that attracts workers. In this species recruited foragers are provided with no indication of distance or direction to the food source (Kerr and Esch, 1965). By way of comparison, *Apis* workers may also mark the vicinity of a good food source, but it is done with a secretion of Nasanov's gland.

The more extensive use of odor was noted by Kerr (1969) for *Trigona* (*Partamona*) *cupira,* and he believes that it is widely important in the genus. He placed two food sources 36 m from a nest and trained a bee to one of them. In one hour 53 other bees (all killed so that they played no role in further communication) had come to that place, only 14 to the other. He says, "The newcomers apparently followed the scout bee, since they arrived together. I found out that the scout bee was emitting a very strong odor." He postulates that the bee, excited by finding food, "opens its mandibular gland orifices in such a way that the whole animal becomes enveloped in odor and releases maximal quantities of odorous material into the air." This idea of aerial odor trails is understandable only for almost windless weather; they may function, however, in the calm of tropical forests, especially if the followers fly close behind the scout.

In the larger species of American *Trigona* (various species of the subgenera *Trigona* proper, *Oxytrigona, Tetragona,*[2] *Cephalotrigona, Scaptotrigona*) and probably in the robber bee *Lestrimelitta limao,* after a few flights to and from a food source, a forager makes a return flight on which she not only marks the vicinity of the food with odor but also stops at short intervals on the way to the nest and leaves a series of odor marks on the ground or on vegetation (Kerr and Esch, 1965; Lindauer, 1961, 1967; Lindauer and Kerr, 1958, 1960; Blum et al., 1970). Figure 15.4 illustrates the positions of such a series of marks made by a forager of *T.* (*Scaptotrigona*) *postica.* The secretion from the mandibular glands is used to form these scent marks, which seem to be an elaboration of the food site marking as done by *Meliponula.* Such a trail is formed only if the food source is over a certain distance from the nest—9 m for *T.* (*Scaptotrigona*) *filosofiai,* 26 to 30 m for *T.* (*Trigona*) *amalthaea*—and the trail is said to be polarized (heavier scent near the food source) so that

2. The bees here involved are in the group of the subgenus *Tetragona* that is sometimes called *Geotrigona.*

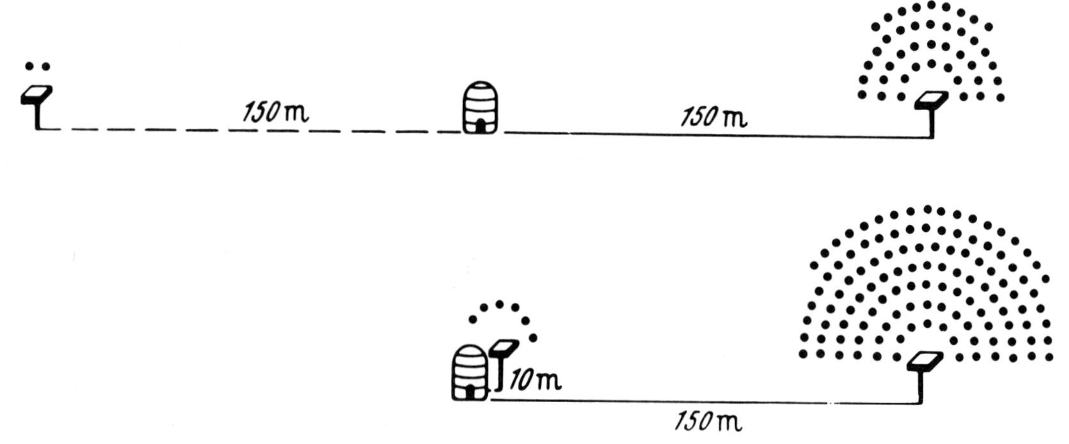

FIGURE 15.3. Diagram showing, for *Trigona* (*Scaptotrigona*) sp., that bees returning to the nest (*center*) from a food source (*right*) alert other foragers, which arrive preferentially (as shown by dots) at the original food source even if, as in the lower diagram, a nearer source is provided. Returning foragers thus transmit information with respect to position of a particular food source. (From Lindauer and Kerr, 1958.)

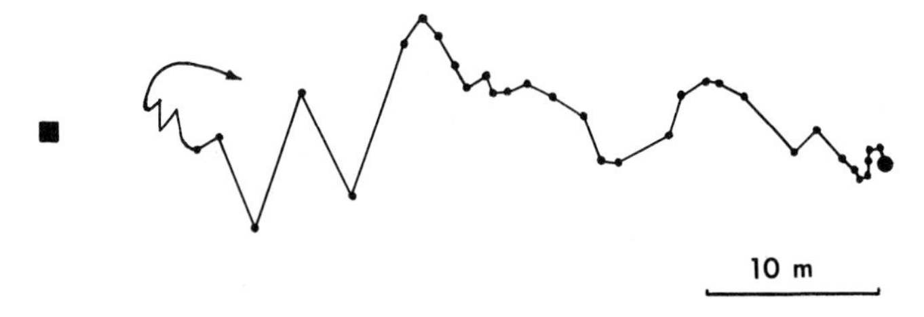

FIGURE 15.4. Flight path taken by a forager of *Trigona* (*Scaptotrigona*) *postica* returning from a food source (*right*) toward its nest (*left*). Each dot represents a place where the bee stopped and left an odor spot. Near the nest the bee turned and started back toward the food, presumably followed by other bees. (From Lindauer and Kerr, 1960.)

bees can tell which way to follow it (Kerr, Ferreira, and Mattos, 1963). The average distance between marks varies with the species: for example, 2 m or less for *T.* (*Scaptotrigona*) *postica* and *bipunctata,* 7 m for *T.* (*Trigona*) *spinipes.* Perhaps to avoid confusion near the nest, trails are not marked in its immediate vicinity; the closest odor spots vary from 2.7 m from the nest in *T.* (*Tetragona*) *mombuca* to 35 m in *T. spinipes.* Having marked a trail in this manner, the forager may enter the nest and make zigzag runs and characteristic sounds, but sometimes she only approaches the nest as in Figure 15.4; then, with a group of other workers, she follows the marked trail to the food source. For example, Lindauer and Kerr (1958, 1960) record that a worker of *T. postica,* after 12 ordinary trips to and from a dish of syrup 50 m from the hive, marked the route, and on reaching a point of few meters from the hive, was followed by other bees, turned back toward the food source, and arrived there two minutes later accompanied by nine newcomers. Such a forager may then make additional ordinary trips, followed by another scent-marking flight later in the day. The amount of material that can be secreted in a short time by mandibular glands is large; Nedel (1960) reports for workers of *T. postica* that after giving off large amounts of the secretion the supply can be replenished in about 20 minutes.

Cruz Landim and Ferreira (1968) postulate that for *T. postica* the bees that commonly wait at the nest entrance are the ones usually recruited by a route-marking forager, and that they are old foragers which have themselves lost the ability to mark routes (see Chapter 12, B3). Sometimes the newcomers follow the guide bee from odor mark to odor mark and to the feeding place so closely that when the guide slows down to land, other bees strike her in the air. Sometimes the guide bee leads the others back and forth in a group for several trips. The evidence seems strong that in addition to the odor marks the guide bee is necessary to lead the newcomers to the food; neither alone suffices. The trips of the guide bee before she makes the odor marks show her ability to fly back and forth using visual cues. To what extent she uses odor cues when leading the newcomers is not clear.

Scent marks retained their effectiveness for 8 to 19 minutes, depending on weather, numbers of bees that had marked the sites, species, and perhaps other factors; with bees going back and forth frequently, and scent marks being replaced, a sort of aerial scent trail should exist, unless the wind is strong. If two series of scent marks from nearby nests cross, bees of one colony may follow a trail made by a bee from the other colony. It is not evident whether or not guide bees must pass close together to give the opportunity for some of their followers to become attached to the wrong guide, but if the necessity of the guide is correct, as indicated above, then such transference of followers must happen. Kerr, Ferreira, and Mattos (1963) were even able to show, in such experiments with crossed trails, that *Trigona* (*Scaptotrigona*) *xanthotricha* follows trails of *T.* (*S.*) *postica,* but not *vice versa,* and that the more distantly related *T.* (*Scaptotrigona*) *postica* and *T.* (*Trigona*) *spinipes* did not follow one another's trails.

The scent pheromones in at least four species of *Trigona* are known chemically, and appear to be the alarm pheromones or parts of the alarm pheromone mixtures; at low concentration, as in the field away from the nest, the response is entirely different from the alarm response produced by high concentrations at the nest (see Chapter 18, A3). The compositions of mandibular gland secretions of two species of the subgenus *Scaptotrigona, T. postica* and *T. tubiba,* are shown in Table 15.2. Both 2-heptanone and 2-nonanone are attractants and at high concentrations release attack by both *Trigona* species. Benzaldehyde is attractive to *T. tubiba* but does not release attacks by either species. This substance, which dominates the secretion of *T. tubiba,* is the essential volatile trail-making material. For *T. postica* benzaldehyde is also attractive, but the C_{11} to C_{13} ketones increase its effectiveness for trail marking. These closely related species may be able to distinguish one another's trails in view of the chemical differences.

TABLE 15.2. Chemical compositions of the mandibular gland secretions of workers of *Trigona postica* and *T. tubiba.*

Compound	Species	
	T. postica	*T. tubiba*
2-Heptanone	+	+
2-Nonanone	+	+
2-Hendecanone	+	−
2-Tridecanone	+	−
2-Pentadecanone	+	−
2-Heptadecanone	+	−
2-Nonadecanone	+	−
2-Heneicosanone	+	−
2-Tricosanone	+	−
2-Pentacosanone	+	−
Benzaldehyde	+	+
n-Hendecane	+	+
n-Tridecane	+	+

SOURCE: Blum, 1970.
Note: + indicates presence of compound listed at left; − indicates its absence.

According to Blum (1970) the mandibular gland secretion of a third closely related *Trigona, T. xanthotricha,* contains, in addition to aliphatic hydrocarbons, 2-heptanone and benzaldehyde, but like *T. tubiba* lacks the extensive methyl ketone series of *T. postica.* The fact that *T. xanthotricha* can follow trails of *T. postica* can be explained by the presence in the secretion of the latter of key substances in the secretion of the former. The methyl ketones seemingly provide no information for *T. xanthotricha,* but *T. postica* is dependent on them for part of the message and does not follow trails of *T. xanthotricha.*

In *Trigona* (*Tetragona*) *subterranea* the secretion of mandibular glands of foraging workers is largely neral and geranial, the isomers of citral (Blum et al., 1970). Its trails are therefore chemically very different from those of other stingless honeybees so far studied, except for the robber bee *Lestrimelitta limao.*

Communication of the type described above functions well for food sources high above the surface of the ground, for example in tree tops, since the trail of odor spots and the guide bee can go up through vegetation as well as along the surface of the ground. The communicating system of *Apis* seems to lack an ability to direct bees up into tree tops or other high objects (Lindauer and Kerr, 1958, 1960; Frisch, 1967a). *Apis* workers do lay odor trails where they walk, reinforcing routes into a hive entrance, for example; the material also serves as an attractant that encourages bees to alight, but it has no significance in long-distance communication (Butler, 1967a; Butler, Fletcher, and Watler, 1969).

The above paragraphs should not be taken to imply that the larger *Trigona* do not also employ zigzag running and buzzing to alert bees in the hive; in fact these types of nonspecific communications were studied early in some such species (Lindauer and Kerr, 1958, 1960). The larger *Trigona,* however, have the additional ability to use trails of odor spots.

The robber bee *Lestrimelitta limao* is said by Stejskal (1962) to mark the route from its nest to nests being robbed with a liquid from the mandibles that smells like citrus trees; the material has been identified as citral (Blum, 1966). It comes from the mandibular glands and is presumably used like the scent marks of the large *Trigona* species; it is also used in subduing the hosts (see Chapter 19, C).

In the genus *Melipona,* which so far as known makes no use of odor spots, the sounds produced in the nest by returned foragers are more distinct, and the duration of the bursts of sound varies directly with distance to the food source, as in *Apis* (Figure 15.5). The sound may be used, therefore, for communication more elaborate than mere alerting or stimulation to activity (Esch, Esch, and Kerr, 1965; Kerr and Esch, 1965). Esch and Kerr discuss

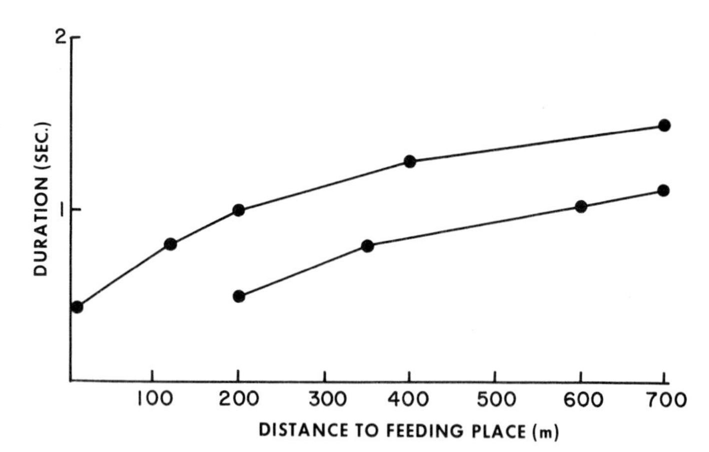

FIGURE 15.5. The duration of the communicative sounds made by returning foragers is related to distance from the nest to the food source. *Above, Melipona quadrifasciata; below, Apis mellifera.* (From Kerr and Esch, 1965.)

the similarity of the presumably communicative sounds of *Melipona* and *Apis,* and their probable evolution from flight sounds. Playing back a tape of sounds made by returned foragers causes bees that know the food source, but not necessarily others, to go to the food.

Returned foragers guide hive mates toward food sources, but only after much repetition do the followers actually search out the food (Esch, 1967b). Kerr and Esch note that a forager, for example of *Melipona seminigra,* leads recruits toward a food source for only 10 or 20 m, and suggest that progressive abbreviation of such a direction-giving flight could lead by ritualization to the straight run of *Apis* dances, which in its primitive form in *A. florea* is also a movement toward the food source.

Thus in *Melipona,* morphologically the most *Apis*-like of meliponine bees, we find a system of communication in which distance and direction are indicated without deposition of an odor trail and without foragers necessarily leading recruits all the way to food sources. Duration of the burst of sound indicates distance, and direction of flight indicates direction to the food. Whether recruited bees actually respond to these stimuli that are obvious to human observers is not certain.

The efficiency of leading for a short distance from the nest entrance is thought to be improved by the broad, flowerlike nest entrances of *Melipona seminigra* (Figure 29.6), on which six to ten bees can rest and be recruited simultaneously by a forager; in other species of the genus only a single bee fits into the nest entrance at a time (Kerr and Esch, 1965; Kerr, Sakagami, Zucchi, Portugal-Araújo, and Camargo, 1967).

The efficiency of meliponine communication has been compared with that of the well-known system of *Apis* by Lindauer and Kerr (1958, 1960). More new recruits per hour were attracted to food sources by some of the large species of *Trigona* than by honeybees. However, the relative sizes of the *Apis* and *Trigona* colonies used in the experiments are unknown. The results only indicate considerable effectiveness for both systems.

D. Communication in Apis

In *Apis,* communications concerning the distance and direction to food sources is in part by means of the famous dances first elucidated by Karl von Frisch, whose great work on the subject (1967a) explains the dance communication in elaborate detail. Lindauer (1967) summarizes some of the recent findings. The mechanism serves not only to alert or recruit potential foragers and get these new foragers to food supplies that have been discovered by other members of the colony, but also serves to get them there in approximately the right numbers to exploit those supplies (Núñez, 1971). Honeybee dances appear to communicate information for vector orientation and sometimes locality knowledge (by odor). Although the effectiveness of the dances in transmitting vector information among bees has recently been questioned, as is explained below, the existence of some sort of communication is not in doubt. It is perfectly plain that the number of bees going to a food source after a bee has found it and returned to the hive varies with the concentration and accessibility of the syrup and inversely with the density of bees already exploiting it. Thus the number of bees that goes to a food source is not vastly in excess of the number that can be accommodated there. No one questions that a food source may go unobserved by bees for days and then, when a bee finds it or is artificially introduced to it and is allowed to return to its hive, more bees soon appear there. The only question is whether the physical features of the dances provide vector information. The alternative is that odors, providing for object orientation and locality knowledge, are responsible for such behavior.

1. Olfactory communication in Apis[3]

The importance of floral scents in *Apis* communication has been known for years, as indicated in the reviews of Ribbands (1953) and Frisch (1967a) (see also Chapter 12, C). Returning foragers closely approach, contact, and even feed bees that are ready to go foraging. The latter thereby learn the odor and taste of the returning foragers, probably are stimulated to leave promptly, and once in the field are frequently attracted by the odor and taste learned from the returning foragers.

Examples of such communication include the following, in addition to those mentioned in Chapter 12 (C). In one of Frisch's earlier studies he took an *Apis* hive to the Munich botanical gardens, where some 700 kinds of plants were then in flower. Everlastings (*Helichrysum*) were one of these kinds, and bees did not visit them.

3. Such communication is also likely to function among stingless bees, but not among primitively eusocial forms.

Frisch trained 10 marked bees to a syrup dish surrounded by cut everlastings. In the following hour 13 unmarked bees went to the everlasting bed in the garden, presumably in response to odor carried on the body surfaces of the bees that had visited the dish surrounded by cut everlastings.

It seems that odors carried on the body surface as well as taste of nectar are involved. Frisch allowed bees to suck phlox-scented syrup from a container in a cyclamen flower close to a hive. Most of the recruits went to phlox flowers, but some to cyclamens. Comparable experiments with the feeding site located 600 m from the hive led to recruits going exclusively to the flowers having the odor of the syrup. Apparently the odor on the bees' body surfaces was lost during the longer flights home. This finding, however, is not applicable to all odors used in other similar experiments.

Such studies invite practical applications. Frisch (1967a) discusses various cases where yields of seed crops have been raised, sometimes greatly, by giving appropriately scented syrup to bees at the time their hives were first placed near a crop to be pollinated. The method is widely used in Russia, far less so elsewhere (Glushkov, 1958). Changing *Apis* foraging from one good food source to another is often difficult, so that starting them on the desired one when hives are first placed in a new locality is important.

Probably all species of *Apis* can increase the attractiveness of food sources to recruits by supplementing the natural floral odors with their own odors. At very large and concentrated food sources, rarely at flowers but commonly at dishes of rich syrup, workers of *A. mellifera* that have already made one or more trips to the food often expose Nasanov's gland and liberate its secretion (geraniol, nerolic and geranic acids, and citral, the last (which is a mixture of the isomers neral and geranial) present in small amounts but an important part of the attractant) (Butler and Calam, 1969). While hovering over the food or usually on alighting beside it, they raise the abdomen and expose the gland openings (Figure 15.6). Such scent marking is especially important for odorless sources of food or for water. *A. florea* and *cerana* are much more likely than *A. mellifera* to scent-mark flowers.

Frisch (1967a) redescribes his well-known experiments in which the rear abdominal terga of *Apis mellifera* at a rich food source were sealed together with shellac. Bees at this source attracted no more recruits than those at a nearby poor syrup source where bees did not expose their Nasanov's glands. Of course in the absence of the shellac, the bees returning to the good source marked it with the volatile gland secretion, attracting recruits there in large numbers, about 10 times as many as to the poor source or to the good source when not chemically marked by the bees. Thus bees that have found a good food source communicate its location chemically to others, not only with the odor of the source but with their own secreted attractant. The latter material is also used to say "come here" under other circumstances, as in swarming or when a swarm enters a new nest, or when a queen bee has been lost from a colony (Renner, 1960).

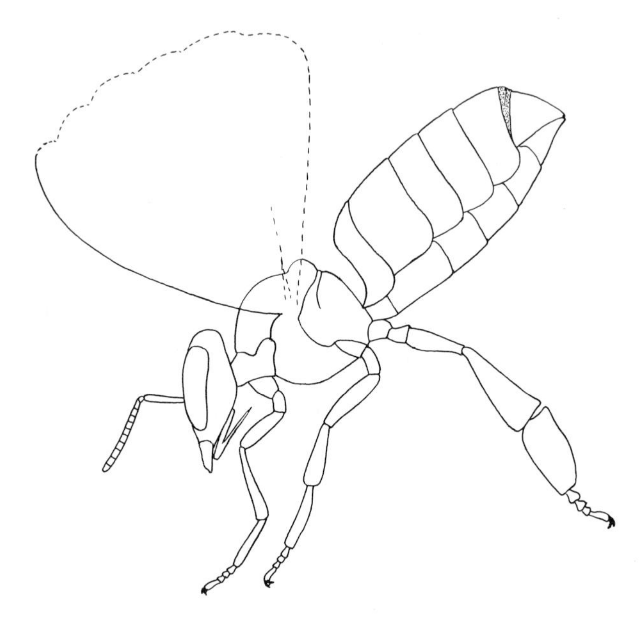

FIGURE 15.6. A honeybee worker with the Nasanov gland (*stippled*) exposed, fanning and thus dispersing the scent. Compare Figure 18.8. (Original drawing by C. D. Michener.)

The marking of food sources is not by mandibular gland secretions as in *Meliponula* and the trail-marking species of *Trigona*. There is the remarkable parallel, however, that odor marking of a food source both in these meliponines and in *Apis* does not occur until a bee has visited it several times (Free, 1968). Perhaps such delay reduces the likelihood of sending bees to transitory sources. Communicative dancing also does not usually occur until after a bee has made several trips to a food source.

According to Free and Williams (1970), the original use of the Nasanov gland secretion was probably to attract bees to water, especially scentless water. Most foods have natural odors and are also commonly dispersed; recruiting should be relatively easy. Water, however, is often localized and not strongly scented; recruiting to it must present special difficulties. In any case, after several trips to water a bee commonly exposes its Nasanov's gland.

Wells and Wenner (1971) note that Nasanov gland odor is released by bees that have just been in the vicinity of disoriented bees, whether recruits to food or water, or to a nest site or a swarm-settling point. These bees that release the attractant are not disoriented, but other bees are. These authors object to calling the Nasanov gland product an attractant because it affects only the disoriented bees and does not attract others. But most pheromones operate only in certain contexts, and there seems to be little justification for this objection. Free and Williams (1972) note that there is a significant positive correlation, although not a high one, between communicative dancing in the hive and exposure of the Nasanov glands on food sources. Thus a bee that exposes the gland is likely also to be one that dances. This could merely mean a higher responsiveness to a variety of stimuli on the part of some individuals than others.

The sight of other worker bees also tends to increase the attractiveness of a food source, even if it is an artificial bee or a mirror image that could involve no odor (Kalmus, 1954). At the nest entrance, also, the sight of other workers leads bees to alight (Butler, Fletcher, and Watler, 1970).

2. Dance communication in Apis mellifera

Mere random searching by *Apis* workers should get some of them close enough to food sources that the chemical as well as visual cues would prove effective. However, both social facilitation to stimulate foragers to activity and specific communication providing information on distance and direction to the area of the food source, together with the odors there, should increase the effectiveness of foraging.

The emphasis in this section is on transmission of information about vector orientation, that is, distance and direction. Such communication, as noted earlier, is provided by the dances of worker honeybees, observed long ago but first understood by Frisch and his students and described in detail by Frisch (1967a). The dances are not performed in flight; the bees walk or run over the surfaces of the combs of cells, that is, they dance on vertical surfaces in the dark of the nest, surfaces also supporting thousands of other bees. Admitting weak light for the observer's needs does not seem to influence behavior although bright light and light direct from any polarized source (including the blue sky) should be excluded. The dances themselves are small and mostly more or less round in shape.

a. The round dance. We will first consider the type of dance communication that indicates a food source only a short distance away and that contains no distance or direction information. This is the round dance. A returned successful forager enters the hive and promptly offers her crop contents (unless she has collected only pollen) to other bees waiting about on the combs (Figure 15.7). Many of these are young workers who accept nectar from returning foragers and either pass it on or process it into honey and place it in cells where it will be stored. These workers are not yet ready to forage. Others, how-

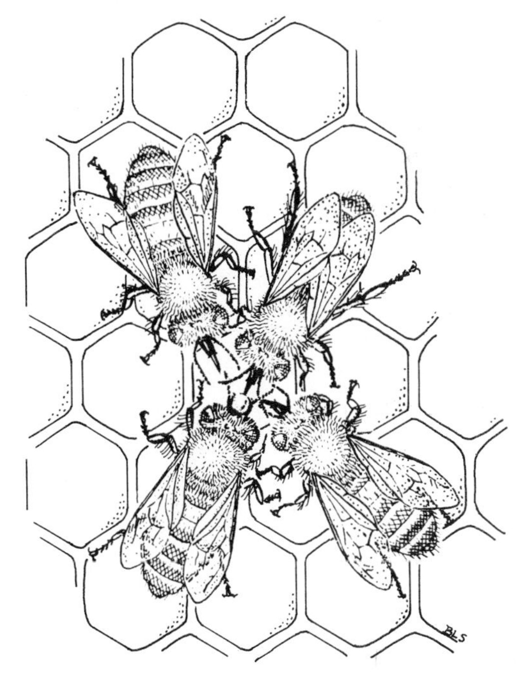

FIGURE 15.7. Liquid transfer in *Apis mellifera.* The returned forager (*lower left*) is giving nectar to three other workers. (From Frisch, 1967a.)

ever, are older workers, some of which are in the proper state to forage. Such bees not only accept bits of syrup from the returned forager but follow her on the combs, often being attracted to her by odor (Chapter 12, C). She now dances her dance.

In the round dance the bee runs in a small circle, reversing and going in the opposite direction after every turn or two, or sometimes after less than a full turn (Figure 15.8). The dance may stop after one or two reversals or may go on through twenty or more, after which it stops abruptly, often to be resumed once or twice more by the same bee at the same place or elsewhere in the nest. Young bees seem repelled by the dance and move away, clearing a little space for it, while some bees of foraging age are attracted, touch the dancer, and may be recruited to the food source about which she is dancing. After a short period of cleaning herself, removing pollen if she carried any, and feeding, the forager then leaves the nest again.

Those bees that were stimulated by the dancer touch her with their antennae, often trailing after her as she dances or cutting across the circle to catch up with her. These followers often pay special attention to her pollen loads if she has not got rid of them before dancing. Recruited foragers leave the dancer, clean themselves, take honey as food needed for the prospective trip, and

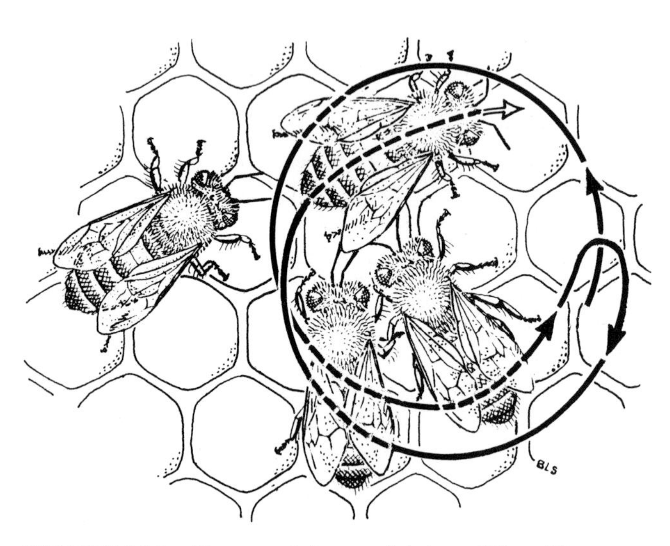

FIGURE 15.8. The round dance of *Apis mellifera*. The upper worker is dancing in the pattern indicated and is followed and antennated by other workers. (From Frisch, 1967a.)

then, usually within a minute, leave the hive. Such activities can be observed only with marked bees in an observation hive, otherwise any one bee is lost among myriads of others.

Knowing of the odor of the nearby source from antennating the dancer herself and from tasting some of her nectar, the recruited bees, even if unfamiliar with this particular source, are able to find it. If the bees recruited are already familiar with the food source, recruitment works particularly rapidly, and may occur even without a dance, simply as a result of stimulation by odor and, commonly, touch. These are bees of the same foraging group, as explained in Chapter 12 (C). They only have to be stimulated to forage and, in effect, told that food is now available at the particular site, which has an odor which they recognize. Foragers from pure odorless sugar-water dishes stimulate only bees with which they make contact in pushing their way among bees on the combs. Bees from fragrant flowers or scented dishes often influence more distant bees. Thus individuals from as far away as 2 or 3 cm may push their way through to the returned forager and examine her with their antennae. Presumably these are bees that have collected from the same source before and react to its familiar odor.

Dancing occurs only when food is abundant, attractive (for syrup this means high sugar content), and not already overcrowded with bees; the more desirable it is, the more likely are dances, and the more vigorous and long-lasting will they be. The more vigorous and long-lasting the dances, the more bees they recruit to the food source. Table 15.3 illustrates the relationship, as shown in one experiment, between sugar concentration in syrup, existence and duration (and vigor) of dances, and number of newcomers at the feeding place. In this case there were only four marked foragers. Recruits arriving at the syrup were removed. Therefore the exponential increase in recruiting that occurs if recruits also return to the nest and communicate did not occur.

Vigor or liveliness of the dance is not easy to measure and is not indicated in Table 15.3, but according to Frisch it increases with sugar concentration and is easily recognized by an observer; at its extreme it leads to side-to-side abdominal vibrations similar to those that occur in the straight run of the tail-wagging dance described below (Wittekindt, 1960). Presumably vigor adds to the communicative effectiveness of the dance. Interestingly, there is a social control of vigor. For the same food source, bees

TABLE 15.3. Experiment on *Apis* showing relationship between sugar concentration in syrup at the feeding place, dancing of returning bees, and number of newcomers arriving at the feeding place.

Sugar concentration (molal)	Four marked foragers: No. of returns	Four marked foragers: Percent of returns with dancing	Mean duration of dancing (sec/bee)	No. of new recruits
$^3/_{16}$	29	0	0	0
$^1/_4$	39	7.7	0.25	0
$^3/_8$	49	53	10.5	3
$^1/_2$	52	73	17.6	10
1	55	92.7	14.9	15
2	48	100	23.8	18

SOURCE: Frisch, 1967a.
Note: Each observation period was 30 minutes in duration.

will dance more vigorously if few are returning from it than if there are many. Presumably vigor is related to the promptness ("eagerness") with which house bees react to the loads of food being brought in. When many bees are returning, there is competition among them for the attention of house bees, hence vigor diminishes (Steche, 1957).

The concentration of sugar that leads to dances is not constant but varies with internal conditions in the nest, availability of other sources, and taste. Lindauer (1949) studied variation in the concentration threshold over a three-month period, with results shown in Figure 15.9. In the early summer season of major nectar production, a high concentration (up to 2 M) was required to cause dancing. In midsummer when nectar in the field is scarce, a sugar concentration as low as $^3/_{16}$ or even $^1/_8$ M causes dances. Flowery odors in the syrup lowered the concentrations at which dancing occurs. Pollen substitutes would stimulate dancing only if the colony lacked pollen stores, showing the influence of internal hive conditions on dancing. Further evidence of the importance of nest conditions comes from water gathering, which at times of high temperature in the nest and water shortage leads to dancing (see Chapter 17, D2); low sugar concentration rather than high stimulates it in this case.

These studies have been made with experimental conditions that provide an abundance of the syrup or water. In nature, syrup is usually supplied in small quantities, and Núñez (1970), working with sugar solutions of uniform concentration, has shown the importance of flow rate in eliciting dances. Bees returning from adequate sources (high flow rates, 8.4 μl/min) danced mostly before and during food-giving contacts. Bees returning from meager sources (low flow rates, 1–2 μl/min) danced mostly after giving food to receiving workers, as though the additional stimulation of so doing were necessary to elicit dances. Presumably even more meager food sources would elicit no dances.

Communication of food sources (at least the dances) does not usually begin until a bee has made several trips, commonly as many as ten, back and forth. Unlike the large *Trigona* species, *Apis* workers do not lead recruits to food sources, but send them there. In the case of the round dance, inexperienced recruits fly about in all directions near the nest. If several scented sources like the one to which dancers are trained are put up in different directions from the nest, new recruits responding to round dances appear in equal numbers on all of them. The bees do not know *where* to look (except nearby[4]) but do seek a particular odor.

b. The tail-wagging dance. At increasing distances from the nest something better than the round dance, with its social facilitation and added odor guidance, has evolved. This is the tail-wagging dance. This dance permits transmission of chemical information (for example,

4. For the meaning of "nearby," see section D3 below on variation in dances among species and races of *Apis*.

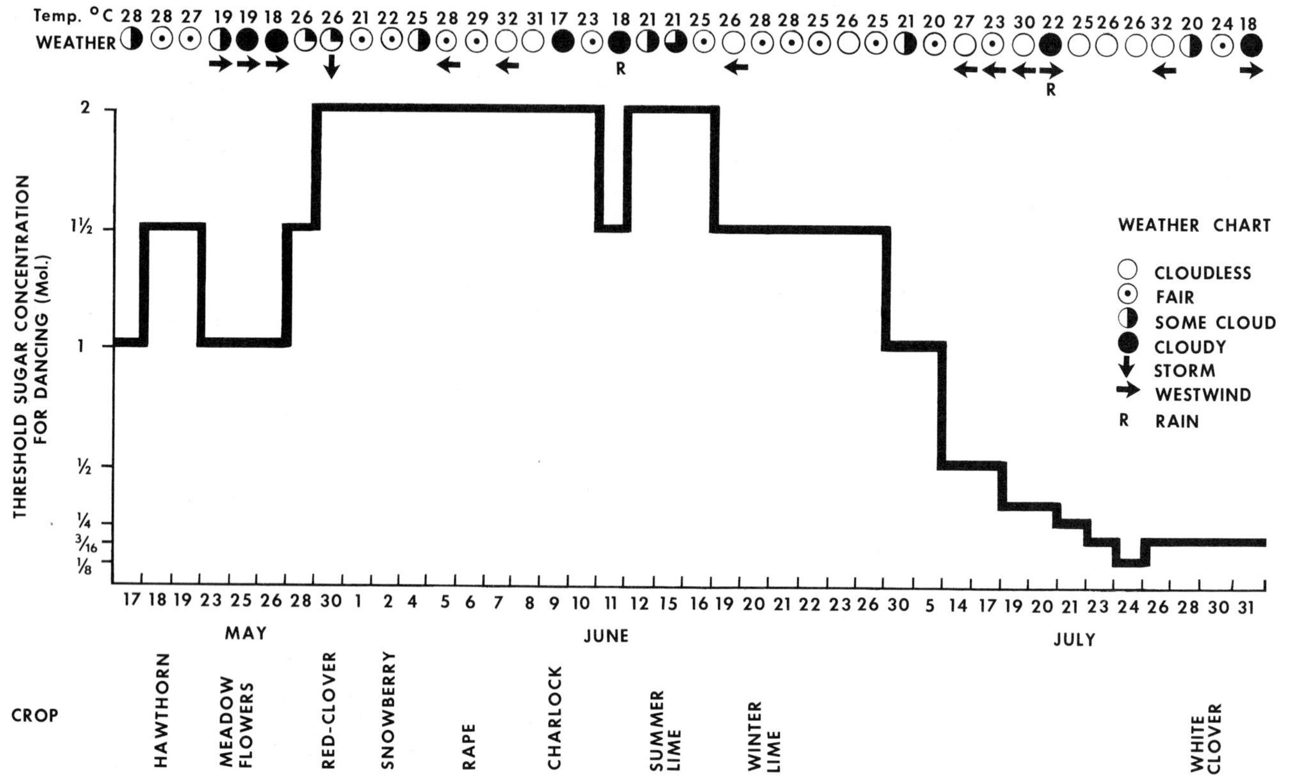

FIGURE 15.9. Seasonal variation in the minimal concentration of sugar syrup necessary to stimulate dancing by foraging of *Apis*. (From Ribbands, 1953, after Lindauer, 1949.)

floral odors) exactly as does the round dance, facilitated at extraordinarily rich food sources by the secretion of Nasanov's gland, and has the same controls based on food richness, abundance, and exploitation, so that about the right number of foragers is recruited for each food source. In addition, it contains information on distance and direction to the food source. Thus bees often are sent to locations kilometers away. This is a performance of which one sees a hint in *Melipona,* where recruits are led part way and sent off to a distant food supply, but is accomplished in *Apis* inside the nest and by a dance little larger than a round dance.

In the Carniolan race of *Apis mellifera,* Frisch's chief experimental subject, round dances give way to intermediate dances at a distance of 15–25 m from the nest, and typical tail-wagging dances are not used until distances of 85–100 m are being indicated. The intermediates are 8-shaped in the race *carnica* (top row, Figure 15.10). These distances vary among other species and races of *Apis,* in which the intermediates are sickle-shaped (lower row, Figure 15.10) (section D3 below).

Typical tail-wagging dances consist of a short straight run, with the bee turning to one side and returning by a semicircle to the starting point, followed by repetition with the bee ordinarily making the second semicircle on the opposite side of the straight run (Figure 15.11). Thus it is a roughly circular dance consisting of two halves. The straight run is emphasized by vigorous shaking of the abdomen from side to side (about 15 wags per second) and usually by a buzzing sound made by the flight musculature and skeleton but without noticeable wing beating.

At low sugar concentrations the brief and rather feeble dances are silent; 15,000 such dances observed by Esch uniformly failed to recruit bees to the food source (Esch, 1963). This could have been due as much to brevity and

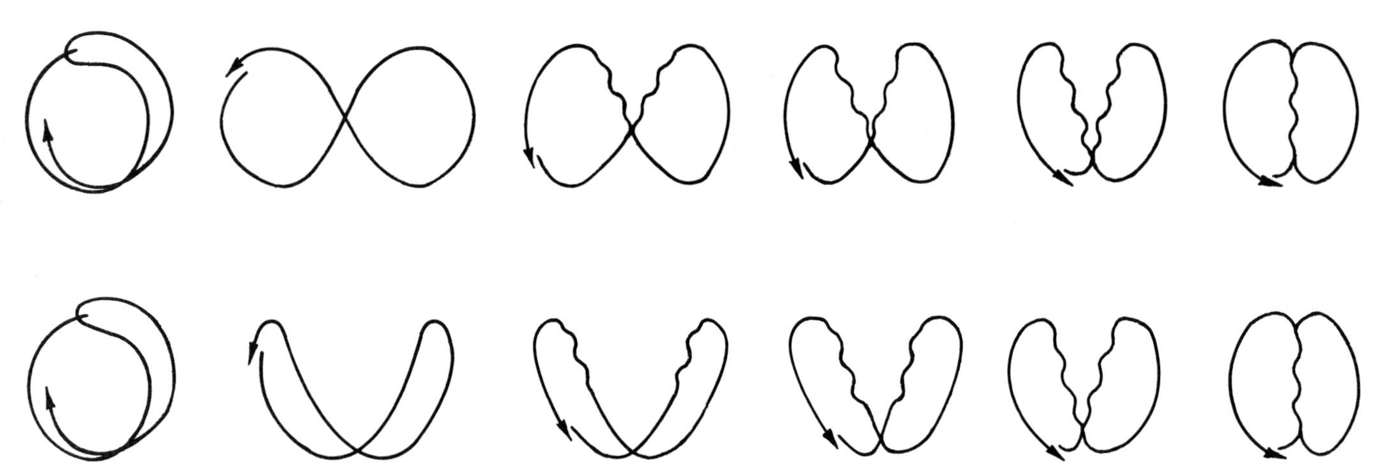

FIGURE 15.10. Transitions from the pattern of the round dance of *Apis mellifera* (*left*) to the tail-wagging dance (*right*). *Upper row* shows the transition via 8-shaped patterns, as is found in *A. m. carnica; lower row* shows transitions via sickle dances, as in most races of *A. mellifera.* (From Frisch, 1967a.)

lack of vigor as to lack of sound. The sound appears to be an important part of the dance communication of *Apis,* however, in spite of the fact that bees seemingly do not hear airborne vibrations. They respond readily to vibrations in the substrate or in objects touched, presumably through sense organs in the legs and antennae.

Like the round dance, the tail-wagging dance may be repeated at one place, but commonly a bee repeats it at different places. Also as in the round dances, bees being recruited follow and antennate the dancer who, before and between dances, gives them and others nectar or syrup from her crop. According to Esch (1964a), movement of the dancer and nearby bees can be halted by a brief sound apparently produced by a follower "asking" for a food sample. This may occur with round dances also; at certain seasons it was not noted and more needs to be learned of this seemingly two-way communication.

c. The indication of distance. Certain often interdependent attributes of the dance vary with the distance from food source to nest (Table 15.4). The figures shown in that table are approximate or accurate means (or ranges) based on variable but usually large amounts of data. Statistics showing variation, such as standard errors for various attributes, are provided by Frisch and Jander (1957) and by Frisch (1967a). There is, of course, variation around each mean. Some attributes, such as length of the straight run, vary in successive dances of the same individual, so that an alerted recruit must follow several

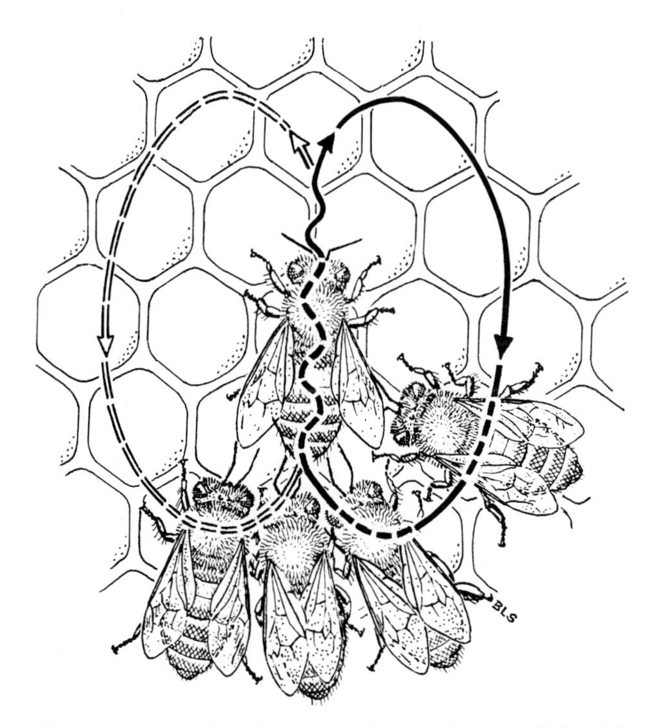

FIGURE 15.11. The tail-wagging dance of *Apis mellifera.* The upper worker is dancing in the pattern indicated; she is followed and antennated by other workers. (From Frisch, 1967a.)

TABLE 15.4. Some of the factors in *Apis* dances that vary with distance from food source to nest.

	Distance from hive to food (m)							
	<25	25–90	100	500	1000	2000	3500	4500
Dance form	Round	Intermediate	Tail-wagging					
Length of straight run in comb cell diameters	—	<2	2	2–3	3	3–4	4	4–5
Length of straight run (mm)	—	<7	7.3	9.6	11.2[a]	—	—	—
Duration of waggling and buzzing (sec)	0	<0.4	0.40	0.95	1.34	2.08	3.12	3.98
No. of waggling movements/dance	0	<5	5	10	18	27	40	50
No. circuits/15 sec of dance	7.5[b]	8.5[b]	10	6	4.5	3.3	2.5	2.3
Percent of tail wagging without semicircles	—	—	—	—	3–4	14	—	33
Average distance (cm) of dances from hive entrance	—	4	—	8	10	—	12.5	20

[a] The increase in mean lengths of the straight run with increasing distance to the food source as shown here is deceptive, as shown by the data for intermediate distances. The mean length of the run for 200 m is 7.9 mm; 300, 10.3; 400, 11.0; 500, 9.6; 600, 10.4; 700, 10.3; 800, 11.2; 1000, 11.2; 1100, 10.7. There is no obvious trend for distances between 300 and 1100 m. Between 111 and 195 dances were measured to obtain each mean (Frisch, 1967a).
[b] These figures may not mean much, as there is considerable scatter, at least partly because fractional circuits are often made, as well as circuits that are more than a complete circle. Tail-wagging dances are more rigidly structured.

dances if it is to obtain useful information about distance to the food. Other features are less variable and presumably more useful as distance indicators. Probably the bees integrate information from all or several of the features of the dance that vary with distance.

The earlier work of Frisch and his associates put emphasis on the number of circuits per 15 seconds. (A circuit is from the beginning of a straight run, through that run and back to the beginning point—in other words, half of the typical tail-wagging dance.) The dances are larger the greater the distance to the food source; hence the number of circuits per 15 seconds is smaller. Figure 15.12 shows the widely known relation of mean number of circuits per 15 seconds to distance to the feeding place. It seems that this should serve as an excellent indicator of distance for shorter flights, like most *Apis* foraging flights, but would fail at greater distances because as the slope of the curve decreases, the number of circuits per unit time probably conveys less information about distance. When the same data are presented as circuit duration, however, a sloping line is obtained (Figure 15.13), suggesting that this feature might convey information at any distance. Since the duration of the curved return part of the circuit increases but little with distance to food, the duration of the straight run varies markedly with the distance (Figure 15.13). Moreover, this duration is closely correlated with distance. It is therefore the feature believed by Frisch (1967a) to be the most probable distance cue for the recruits.[5] This view is supported by the attention given by recruits to the straight run; they often lose contact with the dancer on the return part of the circuit, go to the place where the next straight run will begin, and touch the dancer again with their antennae while following the run. The duration of the straight run also coincides approximately with the duration of the buzzing sound emitted by the flight mechanism of the dancer; it is quite likely that the outstretched antennae of bees following the dancer are as much to perceive the vibration as the odor of the bee (Esch, 1961; Esch, Esch, and Kerr, 1965). Increasingly the belief is developing that the duration of the vibration is likely to be the most important distance-indicating cue in the dance because, as indicated above, silent dances do not recruit foragers. Eskov (1969) places emphasis both on duration of the sounds and the

5. Of features of dances measured by Frisch and his associates, duration of the straight run was best correlated with distance to food, but number of waggle movements per run was only somewhat less well correlated. Frisch argues that since few organisms are able to count, duration is a more probable cue.

number of impulses, which, however, is directly related to the duration.

As Frisch (1967a) remarks, bees are living organisms, not precision machines. All features of the dance of a given individual vary somewhat. Moreover, the workers of a given colony are not identical in their indication of distance, as shown in Figure 15.14. The tempo of the dances of a forager diminishes with age, so that she seems to indicate greater distances than when she was younger. The latter point is shown in Figure 15.15. Such variation may cause inaccuracies in the responses to dances. Variations due to temperature (the dance tempo increases slightly with the temperature) may be compensated for in the physiology and hence the responses of recruits.

Bees have no tape measures. Distances indicated by dances are not actually measured as distances, as shown by studies of bees flying in winds, bees flying up or down grades, bees artificially weighted or hindered by bits of material stuck to them to increase air resistance, and the like. Instead, the dance indicates the amount of energy required for the outward flight, or some factor correlated with this amount, somewhat modified by the requirement for the homeward flight.

In spite of potential sources of errors, most new recruits do go about the right distance, after which they no doubt search for a place having the proper odor, learned from antennating and taking nectar from the dancer. Accuracy of dances in transmitting distance information to new recruits is shown by Frisch's "stepwise experiments." In such an experiment marked bees are trained to a scented syrup source. Then, between the hive and the food source, and beyond in a straight line, a series of cards with the same scent are put out, with an observer at each to count the bees attracted. At the food source every newcomer is killed. At the cards there is no food. Therefore only the dancing of marked bees going to the food continues,

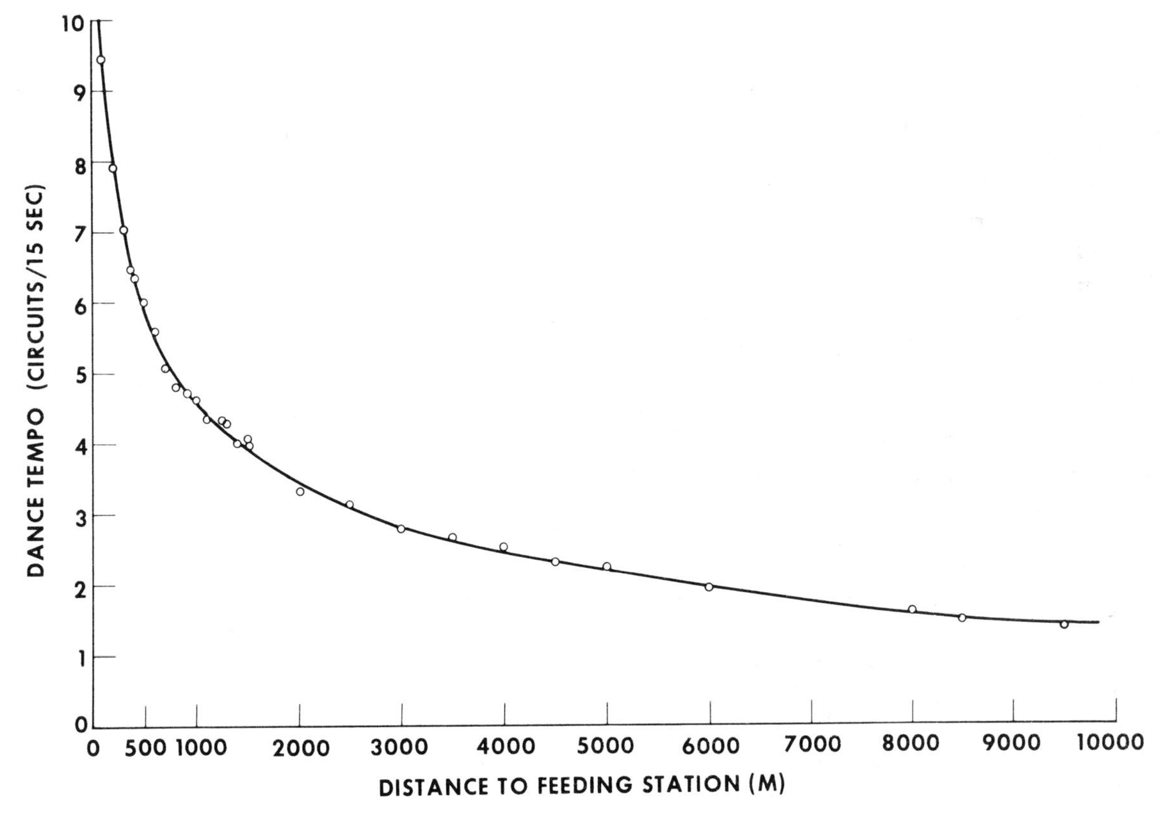

FIGURE 15.12. The relation, in the tail-wagging dance of *Apis,* between the average number of circuits per 15 seconds and the distance to the food source. (From Frisch, 1967a.)

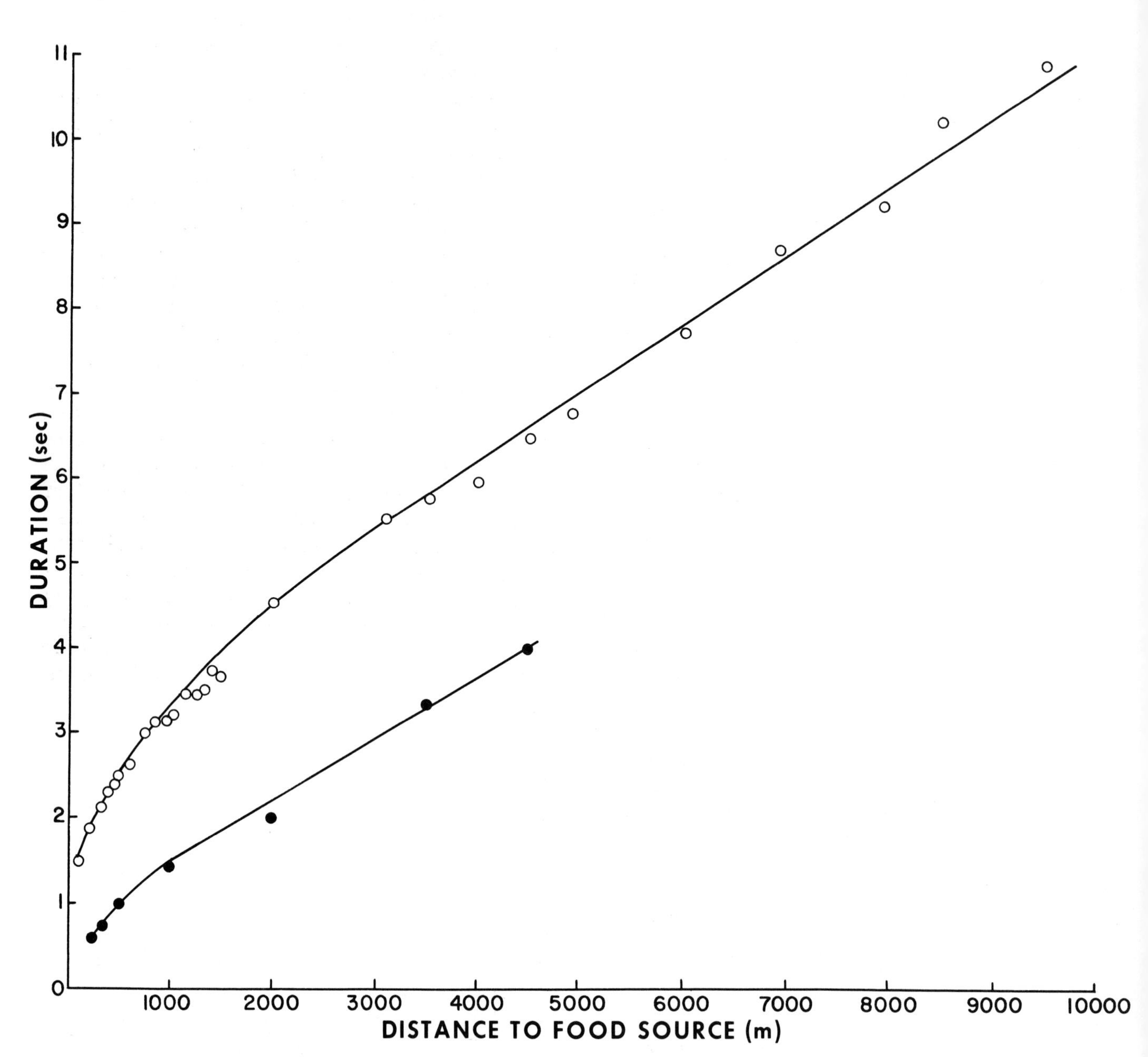

FIGURE 15.13. Relation between the dance tempo of workers of *Apis mellifera* and the distance to the feeding station. *Upper line* shows the duration of the dance circuit; *lower line,* the duration of the straight run. (Modified from Frisch, 1967a.)

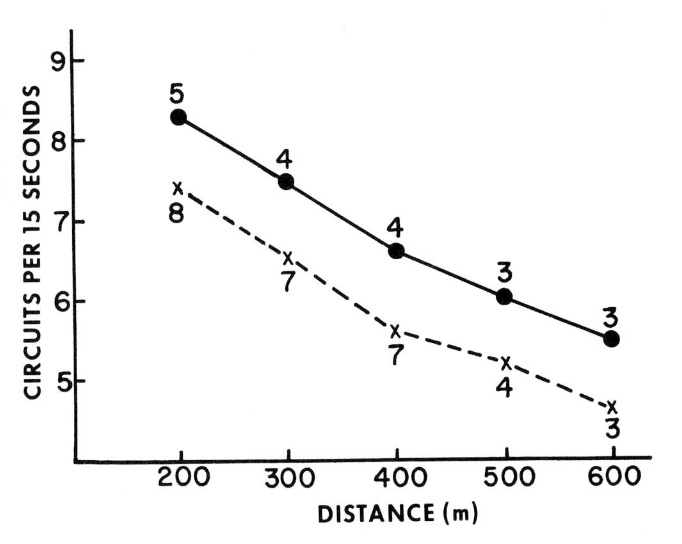

FIGURE 15.14. Individual differences in dance tempo among hivemates of *Apis mellifera* having equal flight experience. *Upper curve* represents average dance tempos based on 31 to 57 observations for each distance, 3 to 5 individual rapid-dancing workers. *Lower curve* represents average dance tempos based on 44 to 62 observations for each distance, 3 to 8 individual slow-dancing workers. (From Frisch, 1967a, after E. Schweiger.)

and stimulates recruits to go out in the proper direction (see below) for some distance where they start searching for the proper odor. The numbers of bees arriving at the various cards gives an indication of the accuracy of the distance communication. No card is placed at the food source, for conditions would be different there due to the coming and going of the marked bees and the possible presence of the secretion of Nasanov's glands. Cards are placed as close as 50 m to the food source. Figure 15.16 shows results of some such experiments. The scatter in distances indicated by durations of the dance runs is greater than the scatter in distances flown by newcomers, suggesting that a recruit pools the information from several dances, or responds to other features of the dance, or both.

d. The indication of direction. Mere knowledge of distance to a food source would not be very useful except close to the nest. But add a direction component to the information provided by the dances and it becomes possible to send a bee to a particular vicinity. The following paragraphs concern the method by which the tail-wagging dance is used to indicate direction. This kind of dance communication is an outstanding accomplishment of nature, and its elucidation by Karl von Frisch and his co-workers has been one of the truly outstanding events in the field of animal behavior.

Apis workers sometimes dance on horizontal surfaces, for example, just outside the hive entrance. This is probably the primitive orientation of the dance (see discussion of *A. florea,* section D3 below). Moreover, horizontal dancing can be observed experimentally merely by turning the vertical combs on their sides. When dancing on horizontal surfaces, if the sun or blue sky is visible, the straight waggle run is in the direction of the food source. The bee maintains the same angle relative to the sun as on the flight to the source. The relationship is indicated by Figures 15.17 and 15.18. Of 95 such dances whose directions were determined by Frisch (1967a), 51 deviated by 1° or less from the true direction to the food source 200 m distant. The average deviation was 1.07°, the maximum, 5°. Followers are recruited just as on vertical surfaces, and although they touch the dancer usually from an oblique position, after following her for a few straight runs, they are able to fly off in about the right direction. If heavy clouds obscure the sun and blue sky, or if the dancing area is covered in some artificial way by the experimenter, dances become disorganized as to direction. Curiously, they also become somewhat disorganized as to length of the straight run, so that they indicate

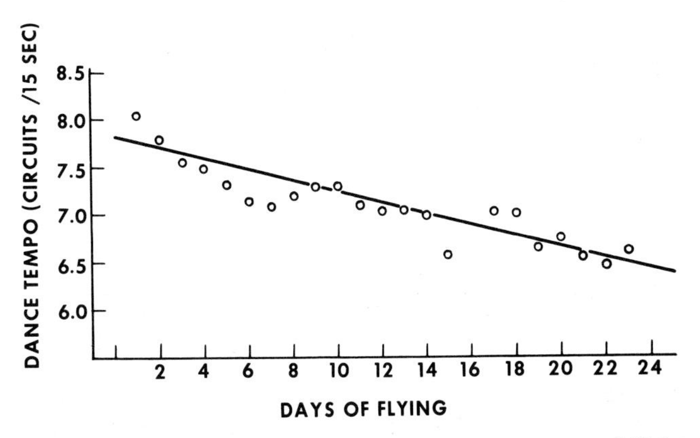

FIGURE 15.15. The relation between dance tempo and flight experience of workers of *Apis mellifera* visiting the same food source. The older the bee the more slowly she dances. (From Frisch, 1967a, after E. Schweiger.)

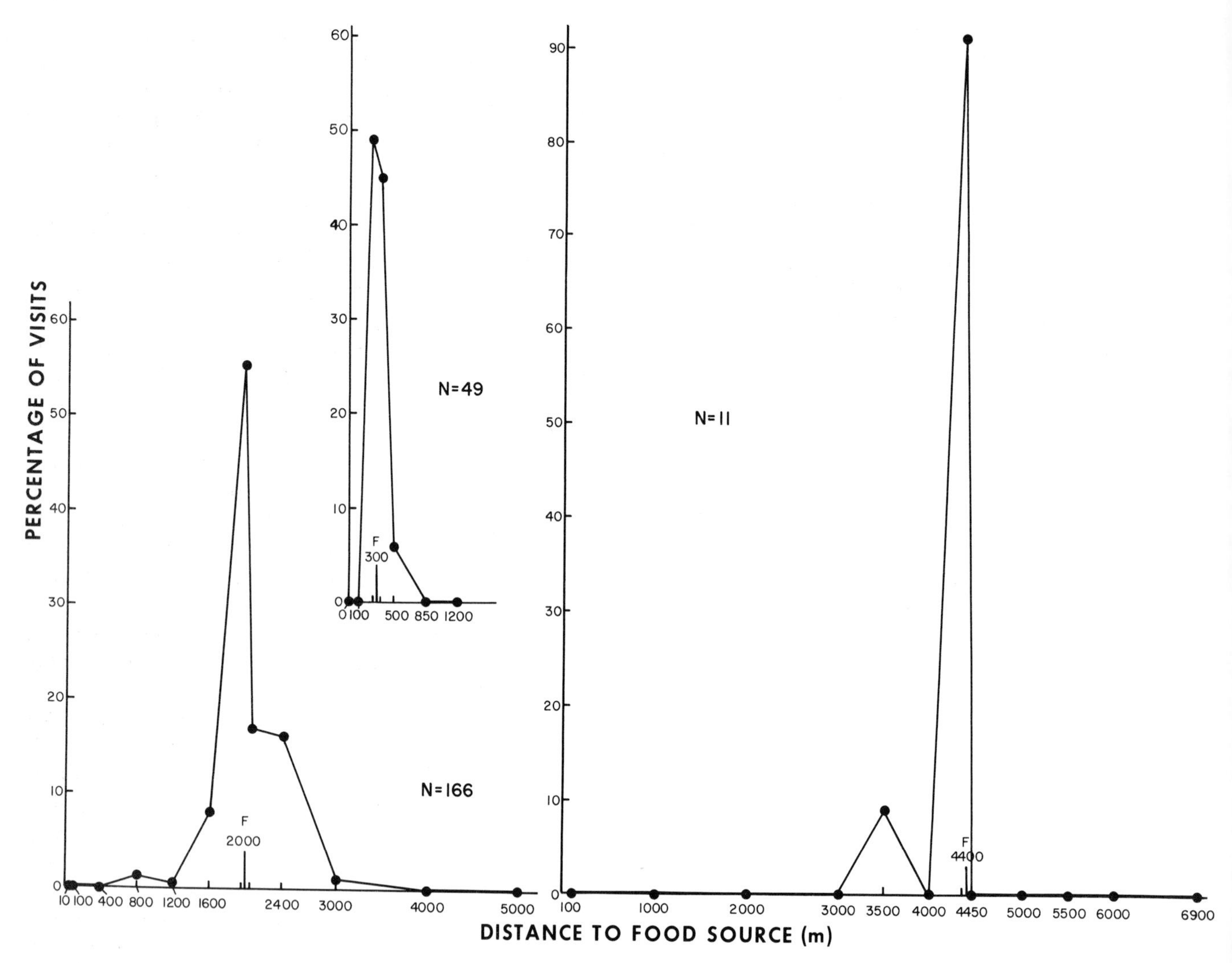

FIGURE 15.16. Results of three "stepwise" experiments with *Apis* workers. Return of workers to hives from food sources (distances marked *F*) 300, 2000, and 4400 m away results in visits by other bees to simulated food sources at approximately the same distances. The numbers on the abscissae represent locations, in meters from the hive, of simulated food sources placed in a row. (Data from Frisch, 1967a.)

neither distance nor direction properly. Recruited bees, however, follow them and then fly out, and fly farther than in response to round dances, but may go in any direction. Disorganized dances are promptly reorganized by an artificial sun, even a flashlight.

It is the azimuth of the light source, not its elevation above the horizon, to which bees respond, and in all subsequent paragraphs direction toward the sun refers to its azimuth, not its elevation.

As explained previously, dances typically occur on vertical surfaces in the dark of the nest. Here the dancer changes the angle with respect to the sun (Figure 15.18) to an angle with respect to gravity. She cannot see the sun or the blue sky, nor can she even see at all in the dark; the major orienting stimulus available on the vertical surface is gravity. Gravity is perceived by groups of minute hairs at the neck and at the narrow base of the abdomen, as well as by sensilla in the legs, as was discovered by Lindauer and Nedel (1959). The hairs are deflected by the head and abdomen, which, lacking legs to support them, tend to hang down. Cutting the nerves that innervate the groups of hairs, gluing head and abdomen immovably to the thorax, or artificially weighting these parts of the body produce bees whose dances are disoriented; for further details, see the review by Markl (1971). These groups of hairs are the same sensilla that are also necessary for gravity orientation for comb building (Chapter 7, A). Presumably they enable the bee to know which way is down, whatever she is doing.

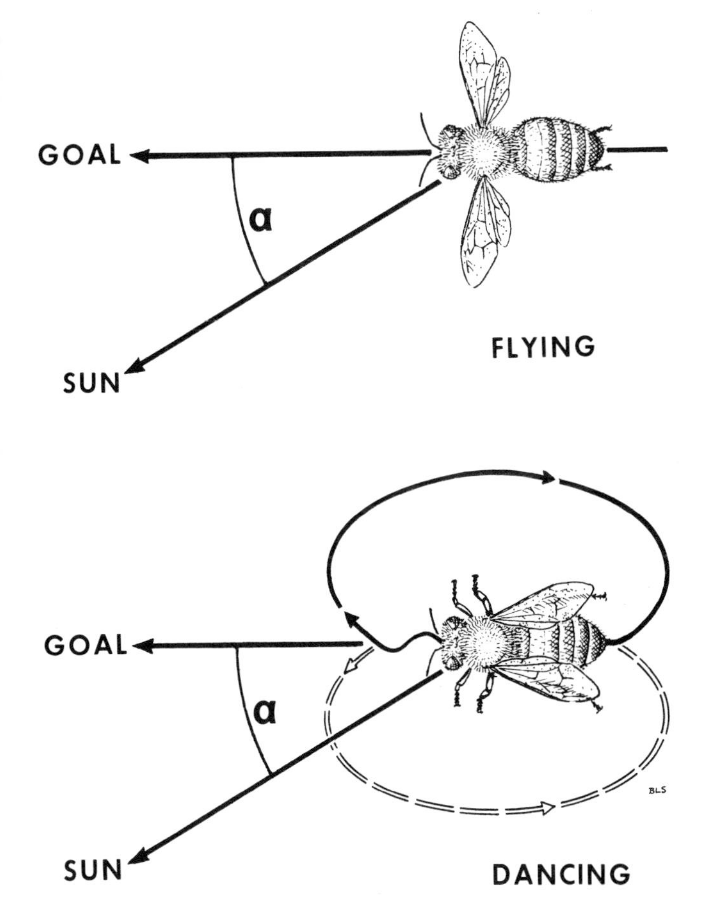

FIGURE 15.17 (*above*). A bee flying toward a food source keeps the sun at a certain angle (α) relative to the direction of flight. (From Frisch, 1967a.)
FIGURE 15.18 (*below*). During the straight run of a tail-wagging dance on a horizontal surface, an *Apis* worker moves in such a way that she sees the sun at the same angle (α) as during her previous flight to the food source. (From Frisch, 1967a.)

The normal dancer varies her tail-wagging run according to the direction to the food source. Thus a straight-up run means "fly toward the sun;" down means "fly away from" or "180° from the sun;" and 30° to the left of straight up means "fly 30° to the left of the direction toward the sun." By this code any direction in the 360° around the nest can be indicated (Figure 15.19).

The recruits obtain information by touch and odor on a dark, vertically oriented surface, then leave the nest and transfer the information back to the sunlit horizontal outdoors, the reverse of the change made on entering the hive by the dancer. The recruits fly in approximately the direction with respect to the sun indicated by the dancer. Of course bees dancing with respect to a given food source do so in progressively different directions as the sun moves across the sky. A straight-up run at sunrise, if the sun were straight east, indicates a food source in that direction. At noon (in the Northern Hemisphere) the same place would be indicated by a horizontal run to the left, at sunset by a straight-down run (see also Figure 15.19).

The flight direction elicited by sickle dances (lower row, Figure 15.10) is the bisector of the angle formed by the two waggle runs.

As indicated earlier, there is minor variation in dance components, such that successive dances are not alike. If recruits followed single dances and left, they would often fly in somewhat different directions. Esch and Bastian (1970) showed that recruits follow at least six dances, and Mautz (1971) found them following from 6.9 to 12.8 dances, before they successfully fly to a food source. Foragers that make fewer dances on returning probably stimulate some bees to activity but direct none accurately to the food.

In natural situations foragers returning from different food sources presumably ordinarily have different odors, so that recruits can easily follow sequential dances by a single bee or at least dances by bees returning from the same place. But under conditions where bees from two different artificial food sources carry identical odors, bees that follow dancers may follow first a dancer from one, then a dancer from the other. Such bees integrate information from the two dancers and fly in an intermediate direction (Lindauer, 1971).

The sort of accuracy of direction seen in flights of recruits in response to tail-wagging dances has been studied by fan-shaped experiments. These are the equivalent for direction studies of stepwise experiments for distance

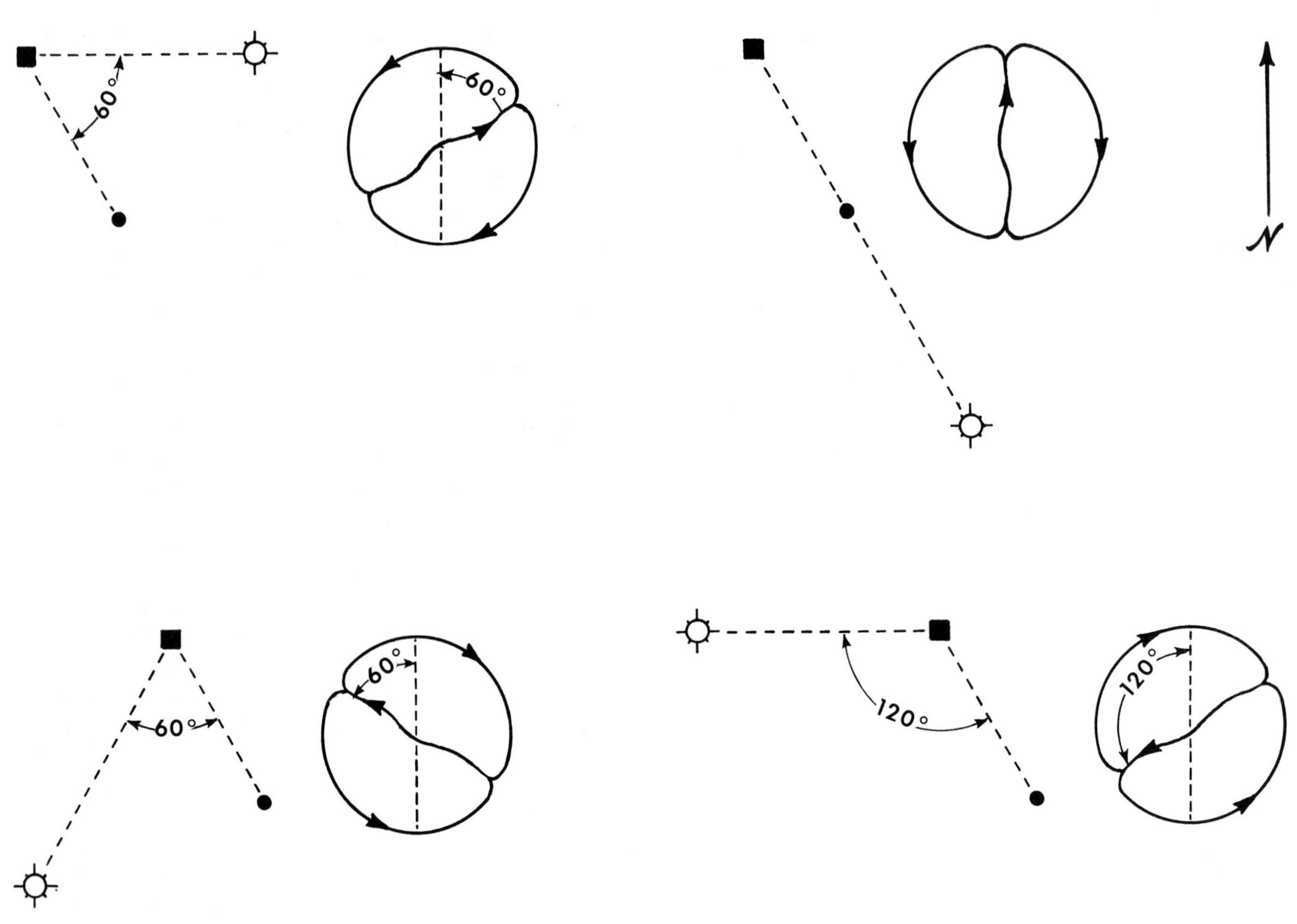

FIGURE 15.19. Diagrams showing the varying orientation of waggle dances of *Apis* workers on a vertical surface at different relative positions of hive (*square*), food sources (*dot*) and sun. The spacial diagrams can be looked at as viewed from above, with the sun rising in the east (*upper left*), moving across the south (*upper right, lower left*) and setting in the west (*lower right*). In this context, the dances shown illustrate changing dance orientation during one day at a single hive with one food source. (Original drawings by Barry Siler.)

studies. The methods are the same, except that the odor cards are equidistant from the hive and form an arc, as shown in Figure 15.20. Most bees fly toward the food source. There is a tendency to a windward deviation from it, however, presumably due to (1) drift of bees to the windward in correcting for wind in their flight and (2) downwind drift of the odor from the upwind cards.

Having an internal time sense and having learned the sun's movements (Chapter 14, A), the bees are able to change the direction of the straight run of the dance with the advance of the sun in the sky without actually seeing the sun or blue sky. Thus if a bee dances for a long time in the dark hive, it continually corrects the direction of its waggle runs so that recruits are sent in the right direction.

On oblique surfaces bees can transpose the visual angle to the gravitational one if the angle of the surface is over 15° to the horizontal, and show partial transposition down to 5°. This ability is of little practical significance in communication about foraging, but is presumably of importance in communication about nest sites at swarming time (Chapter 13, B) because the dances occur on the often sloping surfaces of swarms of bees.

In the tropics there are two times each year when the sun at noon is at the zenith—it has no azimuth. Lindauer (1961) watched *A. mellifera* at this time and found that disoriented dances started at 11:50 and continued to 12:10. The bees lost much of their ability to communicate during this 20-minute period. Thus bees were able to detect a deviation of only 3° from the zenith. Bees were able to continue dancing if the noonday sun passed the meridian over 2.5° south or north of the zenith. These data correspond well to one another and to the angular divergence of the ommatidia in the upper parts of the compound eyes, 2–3°, and the relationship may be biologically meaningful. In a more detailed study, however,

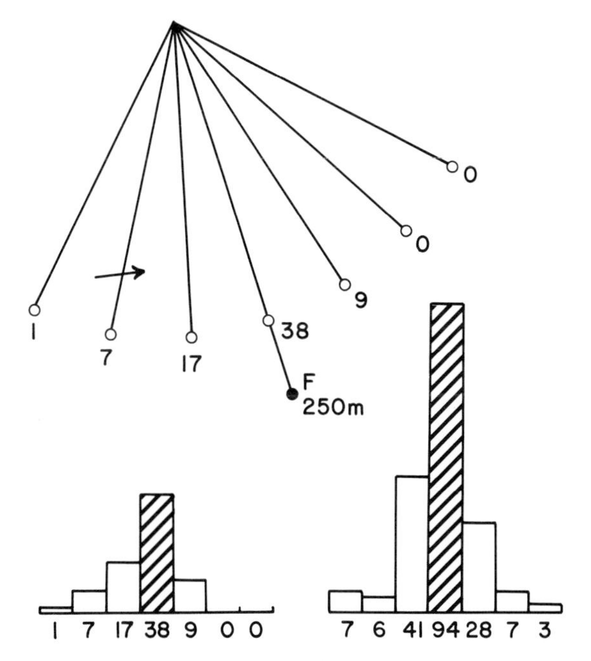

FIGURE 15.20. Results of a "fan-shaped" experiment with *Apis* workers. Return of workers to the hive (*above*) was from a food source (*F*) 250 m away and resulted in visits by other bees to simulated food sources (*circles*) 200 m away and arranged in an arc. As shown by the numbers beside the circles and by the left-hand histogram, most bees during a half-hour period flew in the direction of the original food source. The right-hand histogram shows the same thing for a second half hour of observation. *Arrow* represents wind direction. (Modified from Frisch, 1967a.)

New and New (1962) showed that while still seemingly able to communicate when the sun is near the zenith, bees' dance angles often differ widely from those that one would predict. They note that there are two mechanisms determining the angle of the straight run of the waggle donce: (1) observed sun azimuth, and (2) memory of changes in sun azimuths or dance directions. The second operates when the first cannot, as when in the hive for a long time or at night or when the sun is nearly straight up, and can indicate direction even when the sun is too nearly straight up for accurate perception of an azimuth.

Apparent errors in the direction of the runs on vertical surfaces occur under experimental conditions when the bees can see the sun or blue sky, for they then respond both to gravity and to the light stimuli. Bees following the dance are not misdirected, for they are influenced in the same way by both gravity and light. Other apparent errors, which Frisch (1967a) calls the residual misdirection, occur even when light is not an influence, because dancers seem to prefer vertical (up and down) or horizontal (to left or right) runs to intermediate oblique angles (Figure 15.21). Thus for certain directions of the waggling run, the mean dance directions recorded err in favor of vertical or horizontal runs by as much as five or six degrees. Although there may be individual errors, such mean deviations do not occur when bees dance on horizontal surfaces and are oriented by the sun. These deviations seemingly result from the conversion of visual to gravitational angles, not only in *Apis* but in other insects which make such conversions, and do not result in misdirection of recruits which have to make a comparable conversion on leaving the hive. Lindauer and Martin (1968) found that the residual misdirection disappears if the earth's magnetic field around the bees is compensated, and increases if the magnetic field is artificially increased. Thus it appears that the earth's magnetism somehow influences the bees' orientation to gravity.

Bees flying in a crosswind compensate for drift by turning their bodies obliquely into the wind. They thus maintain a correct course. But since the head and eyes must turn with the body, they see a different solar angle than when making the same journey in calm air. Indeed the angle may change with wind velocity during a single trip. The bees correctly indicate direction in spite of such perturbations. Frisch (1967a) likens this to the ability of a bee to indicate by her dances the true direction "through" an obstruction, such as a building, that she

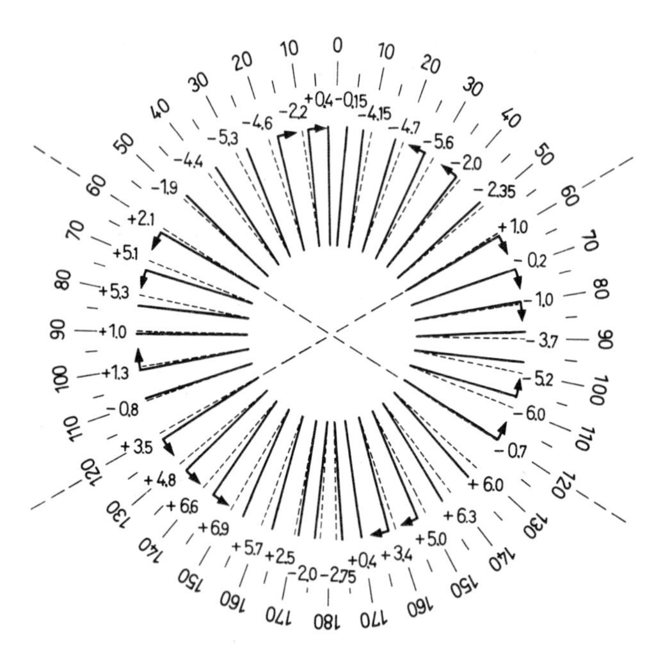

FIGURE 15.21. Average values of dance directions (based on 6759 dances) of *Apis* workers on vertical combs. *Dotted lines* show expected dance directions, considering the locations of the food sources. *Solid lines* show the observed directions, averaged for each 10°. *Arrows* show the deviations of these means from expected directions; *figures* show the amount of deviation in degrees for each solid line. This deviation is what Frisch calls the residual misdirection. The number of dances measured for each 10° ranged from 75 to 308. (From Frisch, 1967a.)

actually has flown around. Recruits follow the direction indicated, over the obstruction, only on subsequent trips learning the easier route around it. Waggling runs indicate the direction for the trip to the food source; only under unusual experimental conditions does the return trip influence the run.

It is not surprising that behavior as complicated as dancing is modified by sublethal doses of poisons. Schricker and Stephen (1970) and Stephen and Schricker (1970) found that doses of less than .03 μg of parathion per bee caused loss of ability to orient the dances properly on vertical surfaces and hence loss of ability to communicate direction. On horizontal surfaces poison-induced deviations were negligible. It thus appears that the poison influences the centers that control conversion of orientation on a horizontal plane to orientation by gravity on a vertical surface in the dark. This is an illustration of the subtle effects of minute amounts of an insecticide on apparently healthy organisms.

e. Individual improvement in dance communication. Although naïve bees of the proper age group dance, and respond to dances, the perfection of the response requires experience. Butler (1954a, after Lindauer) says it "must not be thought that a household bee, that is about to become a forager, leaves her hive on a foraging expedition immediately she has followed the movements of a dancer. Actually it takes her some hours, or even days, before she is able to follow a dancer's actions sufficiently well to be able to apprehend the information which the various dance movements quickly convey to experienced foragers. At first the recruit follows dance after dance without being able to follow the whole movement correctly, as she continually loses touch with the dancer, in the darkness of the hive, when the dancer makes one of the sudden and important turns which are an integral part of the dance. But eventually after a number of attempts which may be spread over many hours or even days, the recruit learns to follow the dancers satisfactorily and thus to obtain the information she requires. When she has done so, she ventures forth to forage and in almost ninety percent of observed cases returns with food of the same kind and from the same plant as that indicated by the dancer." After several trips the "recruit may then dance in her turn and so help to recruit and instruct further foragers." Such observations support the idea of function for the dances, although it must be admitted that they were made by persons convinced of the function. The important role of learning in the responses of individual bees to various dance components has also been emphasized by Lopatina and Chesnokova (1969).

Frisch (1967a), in writing of distance communication, notes that as a forager flies back and forth to the same place over a period of days, the variability in its dances diminishes, presumably with experience. Likewise, in indicating direction, there are individual variations in accuracy of directions of straight waggling runs, and even individual biases to the right or left. Accuracy increases with the age and experience of the dancer.

f. Doubts on and verification of dance communication. In spite of the extensive and convincing detail in which the dance communication of honeybees has been made known, Wenner, Johnson, and associates (see Wenner,

Wells, and Rohlf, 1967; Johnson, 1967; Wenner, 1967; Wenner and Johnson, 1967; Wenner, Wells, and Johnson, 1969; Johnson and Wenner, 1970) have recently thrown doubt on vector communication by dances. They incorrectly indicate that Frisch did not distinguish responses of new recruits from those of bees that had already been to the food source. (Actually he made this distinction as early as 1923 and emphasized it in his large volume, 1967a.) Wenner and his colleagues further indicate that the dances do not communicate distance and direction, that the bees find their way to the food sources by means of other cues such as odor, and that the stepwise and fan-shaped experiments of Frisch apparently demonstrating directional and distance communication lacked essential controls, for example, because the training dish with food might accumulate attractive odors having a longer distance influence that Frisch believed, and hence affecting the nearer scent cards. An experimental design that provided such controls did not demonstrate either directional or distance communication from foragers to inexperienced potential foragers. While inadequacy of controls in earlier experiments may be a justified criticism, their adequacy is strongly defended by Frisch (1967b, 1968) and Lindauer (1971). The differences between flights of recruits responding to round and tail-wagging dances and the differences in flights responding to disoriented waggle dances (such as occur on horizontal surfaces when sun and blue sky are not visible) and organized ones could not be accounted for in the absence of directional and distance information from the dances. There is such a mass of interlocking theory and observation in the work of Frisch and his co-workers that it scarcely seems credible that a major block of it could be wholly wrong. Frisch attributes the results of Wenner and his associates to the use of concentrated syrup (1.5–2.0 M sugar) in setting up feeding stations, with the result that at the time of the experiments bees are re-aroused and fly all about the vicinity. Frisch uses dilute syrup (0.5 M sugar) that scarcely elicits dancing in setting up the station and until time for the experiments. Lindauer (1971) further criticizes Wenner's experimental design.

No one doubts that the dances occur. They can be observed in almost any bee hive. No one doubts that they vary as the sun passes across the sky and also that they vary with the distance and direction to the food source and with the concentration of the syrup when this is the food. A human observer can determine from the dances the position of the food source (Lindauer, 1961; Frisch, 1967a). It seems most unlikely that the dances are nonfunctional performances on the part of the bees or that they serve only for transmitting odors or have some as yet unknown function.

Gould, Henerey, and MacLeod (1970) have carefully reexamined the experimental methods of both the Frisch and Wenner groups and found both wanting in some respects. It suffices here to point out that in field experiments of this sort with social creatures having diversified sense organs, the number of possibly important environmental variables and combinations thereof is vast. Gould and his co-workers conducted several experiments using the most careful controls yet applied to the problem, but only two stations on opposite sides of the hive in any one experiment, at distances of 120 m from the hive. Both were feeding stations. Bees danced with respect to only one station because of the higher concentration of sugar in the syrup there; otherwise the stations were identical. The number of bees trained to each station was the same and recruits were killed so that they could not add to the recruiting force. If recruited bees were merely stimulated to seek the scent on dancers, they should have appeared in equal numbers at each station. But 96 percent went to the station where the syrup was more concentrated, apparently as directed by the dancing bees. The possibility of some other, unimagined kind of communication directing the bees to the food source indicated by the dancers remains, but seems most unlikely.

Another series of recent experiments independently supports the communicative function of dances. Gonçalves (1969) took advantage of the fact that *Apis* workers that walk for a few meters through a narrow tube from the hive to a food source dance in a manner that indicates the true direction but a distance of 50 to 100 m. Thus they indicate a food source at a place that they have never visited. A circle around the hive of eight identical syrup dishes was so managed as to show at each dish the number of arriving bees recruited by dancers from the tube. While some recruits went to each dish, a significantly larger number went to the dish in the same direction as the tube (but far beyond it). Since this dish could have had no differential odor, and like the other dishes was regularly changed to prevent it from acquiring one, the idea of dance communication was supported.

Lindauer (1971) also was able to verify that bees can be sent by the dance alone to food sources which they have never visited, as noted in section d above.

The work of Lopatina and Chesnokova (1969) with a dummy dancing bee also seems to exclude olfactory stimulation. They report that 60 to 70 percent of the bees that contacted the dummy flew in the direction indicated by it. Operated at a given frequency, but at various angles to the vertical, the dummy stimulated bees to search for sources at the same distance but in different directions from the hive. Operated at different frequencies, the dummy caused bees to search at different distances from the hive. It would seem that, by modifying the movements and vibrations of the dummy, the way is open to analyze the relative importance and function of the various elements of the dance, to discover what the follower perceives in the movements of the dancer and how, and to ascertain the relative importance of information from dances and from odor in determining behavior of foragers.

The demonstration by Mautz (1971) that the success of alerted bees in finding feeding stations varies with the number of dances that they followed before leaving the hive also supports the view that the dance is an important part of the communication.

g. Other types of dance communication. Numerous other kinds of movements ("dances") and sounds are known in *Apis* colonies. Most are not adequately understood. Frisch (1967a) describes most of these. Wittekind (1966) studied a hive that was closed for long periods. After such a period, when scouts first found a food source, they did not take time to fully fill their crops or corbiculae but returned immediately to the colony with small loads. They went to the most crowded parts of the comb, pushed among the other bees, and following a winding course, offered food to them. Unoriented tail-wagging runs were frequent. All this seemed simply to alert bees and start them foraging. The scouts later danced in a normal way. Except for the tail-wagging runs, this was probably similar in appearance and function to the jostling run of Schmid (1964) and to the running and bumping of other bees by returned foragers seen in Meliponini.

Another communicative movement is the buzzing run, used in starting swarming, subsequent flights of swarms, and absconding flights (Chapter 13, B).

h. Summation of dance communication in Apis mellifera. In summation, one can do no better than quote (with figure references omitted or changed to correspond to present volume) two paragraphs from Frisch (1967a):

> It is worthwhile to devote a little thought to the chain of accomplishments that take place between the flight of the harvesting bee and the directed flight of the newcomer. In her dance the harvester transposes the solar angle that is proper to the goal into the gravitational angle—even when with a sidewind she has to set herself obliquely against the lateral drift and hence sees the sun at a different angle. The dance followers understand the angle correctly. In doing so their eyes are no help to them. In the dark hive the eyes are functionless, and the senses of smell and touch take over the transmission of information. The dance followers grasp the proper angle with astounding precision, although most of them stand oblique to the dancer—especially when she has a large following. Beyond this they take note of the floral scent that clings to the dancer and of her announcement of the distance by means of the dance tempo; in doing the latter they observe several dances and average the result. If then they fly out they go in search of this scent at the corresponding distance, transposing the gravitational angle back into the visual angle and scarcely letting themselves be borne aside from the direction striven for even by a violent crosswind. As for anyone to whom the feeling of reverence for nature's creations is foreign—it might well dawn upon him here.
>
> Though we may be pleased that we understand the dances so well, at the same time we must be impressed with how fussily and clumsily we make use of the information given by the bees. We must possess a stopwatch in order to determine the tempo of dancing. We need a distance curve pieced together from many single observations in order to estimate the distance with some degree of reliability. We need a protractor to determine the direction of dancing, the azimuth curve for reading off the position of the sun for a given place and instant, and the circular diagram of figure 15.21, derived from several thousand individual observations, in order to take account of the 'residual misdirection.' Finally we must have a compass, a surveyor's tape, and assistants, in order to apply in the open country the result we have calculated. How elegantly this task is mastered by the bees themselves, who need follow only a few dance circuits in the dark hive, then according to their innate behavior set forth in free flight, guided by the sun, and steer toward the goal. A truly splendid accomplishment!

One of the remarkable features of the dance communication is that it can be used not only to indicate the locations of food sources, but also to direct bees toward other things. Thus it can serve to direct bees to water (Chapter 17, D2), to supplies of resin (propolis), and to nesting sites (Chapter 13, B). Such multiple use of a single sort of communication is comparable to the multiple use of many pheromones, which have different meanings in different situations and concentrations. A given communicative element, once evolved, can develop a repertoire of functions and thus considerably enhance the interactions and the integration within a colony.

It must be emphasized that much of the foraging traffic of *Apis* workers consists of bees going to and from food sources already known to them. These may be individuals simply going back and forth carrying provisions. Or they may be bees that have been to a source previously and stopped visiting it, perhaps because it was no longer productive, but are stimulated to revisit it by odor or contact or dances of a successful forager from the same place (or another place with the same odor). In either case the foragers know the route by landmarks and sun compass, together with odor cues, and can fly accurately to the site, probably in a more or less straight line. The dance communication concerns bees that have not been to a given place before. It directs them only to the right vicinity, after which they probably must hunt for the food source. Thereafter they can go more accurately and quickly to the right spot because of having seen the local landmarks.

A noteworthy fact is that bees seemingly know in advance what they are going to do. Thus bees enter a hive in different directions depending on the direction to the location of the forthcoming dance and whether they will dance near the entrance (as for a nearby food source) or far in the interior (as for a distant one, see Table 15.4). More impressive is the relation between the length of a trip and the amount of honey taken as fuel into the crop of a forager setting out on the journey. According to Istomina-Tsvetkova (see Frisch, 1967a), amounts are as follows: for 5 m, 0.78 mg; for 500 m, 1.61 mg; for 1000 m, 2.20 mg; and for 1500 m, 4.13 mg. It is as though the bee planned its trip in advance and took on a fuel supply accordingly.

3. Comparative studies and evolution of dance communication

Direction-indicating flights of *Melipona* and dances or runs, with associated vibratory signals (which in *Melipona* are related to distance but which in other stingless honeybees merely stimulate recruits to activity or to follow chemical cues) may be the only extant precursors of *Apis* dances (see section C above). The differences between such activities and *Apis* dances are so great that one cannot be certain of the homologies, although they are highly suggestive.

Within the genus *Apis,* however, dances are known in all species; the various parts of the dances are easily homologized. Dances of the species of southern Asia are known principally through the work of Lindauer (1957a, 1961), and are considered here in the order of increasing complexity of the dance communication.

Apis florea, the smallest species, makes a single comb hanging from a small, more or less horizontal branch. The base of the comb, which surrounds the branch, is broadened by elongation of the honey storage cells and forms a broad, approximately horizontal surface (Figure 30.8) which is solidly covered with bees. It is on top of this layer of bees that the dances are performed. Therefore they are in a horizontal plane, exposed to the sky or sun, and the straight runs of the tail-wagging dances point toward the food source.

The giant honeybee, *Apis dorsata,* also nests in the open but under a large branch or overhang of rock. Its tail-wagging dances occur on the blanket of bees that covers the comb, on that side of the comb that is exposed to light. Unlike *A. florea,* the dances are on a vertical surface, but like *florea* they require vision of the sun or blue sky. *A. cerana* (*indica*) is like *A. mellifera* in nesting in dark holes and in its communication, except for the details explained below.

Jander and Jander (1970) have shown that *Apis florea* is the only *Apis* which agrees with various solitary and primitively eusocial bees and Meliponini in reducing the angle of directional orientation to gravity with increased slope of the substrate. *A. dorsata, cerana,* and *mellifera* maintain the angle constant. The evolutionary change involved must have been a prerequisite for, or must have accompanied, the evolution of dances on vertical surfaces with the transfer from light to gravity orientation (and back again for recruits).

There are differences in the number of circuits in the tail-wagging dance per 15 seconds for the species listed above, but the curves do not extend much beyond the zone occupied by curves for the various races of *Apis mellifera* (Figure 15.22). The lower the curve in that figure, the more accurate should dance rate be in discriminating short distances, but the less useful in distinguishing long ones. It is likely that the normal flight

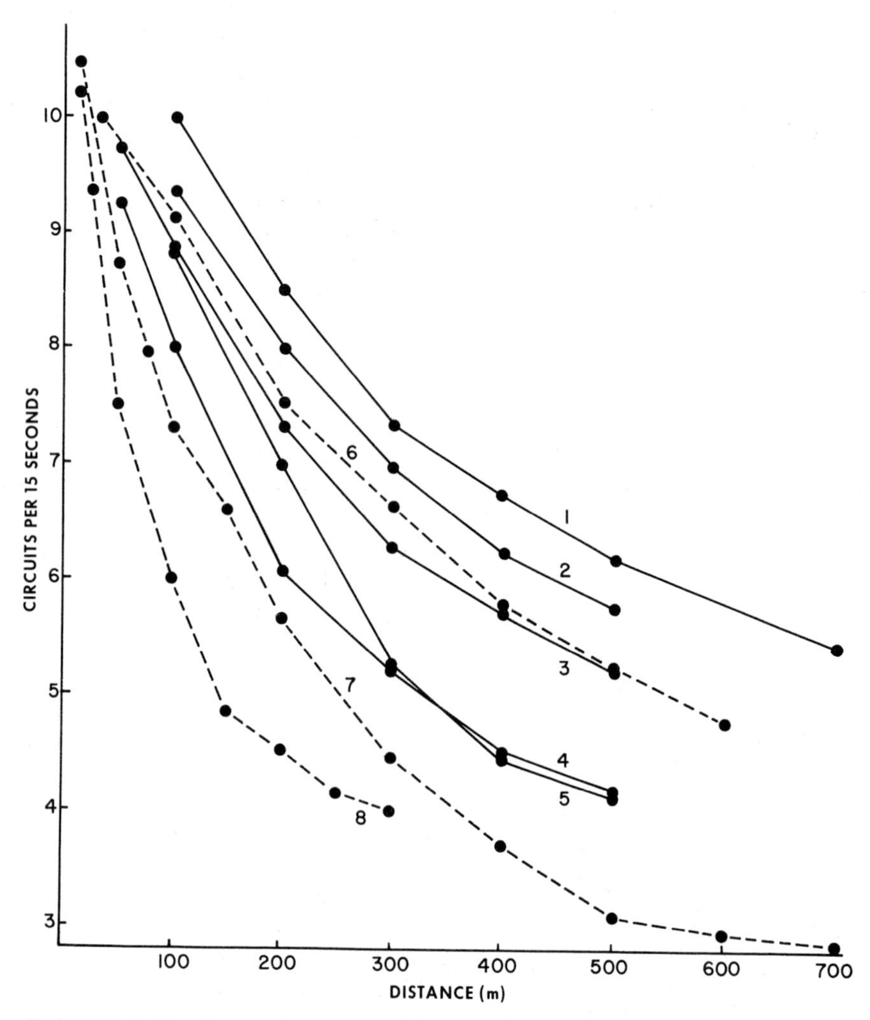

FIGURE 15.22. Curves showing specific and subspecific differences in tail-wagging dance rates for the genus *Apis.* The abscissa represents distances from nest to food source. *A. mellifera* is represented by solid lines, its subspecies as follows: (1) *carnica;* (2) *mellifera* and *intermissa;* (3) *ligustica* and *caucasica;* (4) *lamarcki;* (5) *adansonii.* The other species, represented by broken lines, are (6) *A. dorsata;* (7) *A. cerana indica;* (8) *A. florea.* (Data from Lindauer, 1957a, and Frisch, 1967a, somewhat simplified.)

ranges of the forms with the lower curves in Figure 15.22 are relatively short.

The various species and races of *Apis* differ in the distance from the hive to the food source at which occurs the change from round dances through intermediates (usually sickle dances) to tail-wagging dances. As can be seen from Figure 15.23, the Carniolan form of *A. mellifera* is at one extreme in this respect, *A. cerana indica* at the other. Interestingly *A. c. cerana* from Japan has dance distance communication more like that of *A. m. ligustica* than *A. c. indica* (Sakagami, 1960b).

Eskov (1968, 1969) and Levchenko and his colleagues (1969) have reported on studies of racial differences in tail-wagging dances in *A. mellifera.* They found differences in the durations of the dance cycle and of the vibration during the wagging run. They also found that certain races (Carniolan and Carpathian) dance for scarcer forage and lower sugar concentrations than others.

In mixed colonies (for example, of the Carniolan, *carnica,* and the Italian, *ligustica*), the bees of different races of *Apis mellifera* misunderstand one another and Frisch (1967a) gives meager evidence that among hybrid workers color and behavior segregate together. Thus yellow hybrids that look like Italian bees have the dance behavior and responses to dances of Italian bees while their gray sisters (same mother, perhaps different fathers) that look like Carniolan bees have dances and responses to dances like Carniolan bees. Weak as this evidence is from the standpoint of genetics, the dances and responses to them are evidently genetically mediated. An individual *Apis* is able to improve its language communication with use, but in all main attributes the language and its understanding must grow automatically with the aging bee so that it can function when foraging begins.

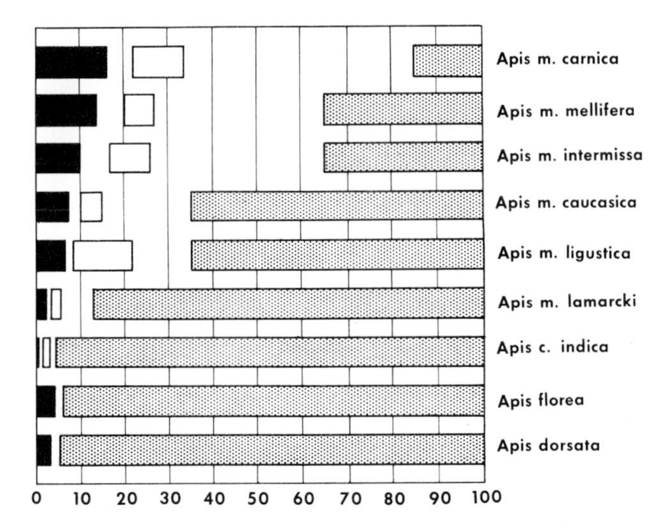

FIGURE 15.23. Distances from nests to food sources in meters and corresponding dance types for various subspecies and species of *Apis. Black* represents round dances; *open rectangles,* intermediate type dances (mostly sickles); *shaded areas* represent tail-wagging dances. *Open spaces* represent unclassified gradations in the continuum from round to tail-wagging dances. (Original drawing by Ann Schlager, based in part on Frisch, 1967a.)

I have already suggested something of the possible origin of dance communication of *Apis* via *Melipona,* but this says almost nothing about the basic origin of the nervous and sensory patterns, and may suggest that these arose *de novo* in the apine bees. This appears not to be the case. Rather, the genus *Apis* has seemingly taken advantage of behavior and sensory attributes that are very widespread through the insects. Some attributes, like the light compass reaction and responses to the light polarization pattern in the blue sky, are widespread in arthropods and in general use for orientation among wasps and bees. Frisch (1967a) has reviewed the more basic possible roots of the dance language. A summary follows.

Dethier showed that blowflies perform circus movements from which dances might evolve. If food is taken away from blowflies, they walk irregularly, turning to the left and right, as though searching for the food. If given a bit of food and then put elsewhere on a flat surface without food, the flies make the same movements there. This makes no biological sense—the lost food was never there. But it suggests *Apis* dances in that it is a delayed response to food. The bees, of course, are not searching for food where they dance, but the more or less circular movements of the dance could have evolved from such a basic behavior pattern. Better food results in longer and more vigorous searching movements by the flies, again suggestive of *Apis.* The movements of the fly are subject to modification from external stimuli, either light or gravity. And like an *Apis* forager on returning to its hive, the fly, when it enters a group of other flies, often regurgitates some of the liquid food (onto the substrate) which the other flies eat. No one suggests that bees evolved from flies, or that flies transmit information by their searching behavior, but only that, if such behavior is part of the common insect heritage, it could be the root from which dances arose.

Tail-wagging may also have its roots in common insect behavior. In various moths the abdomen is shaken from side to side after flight and there is a relation between the duration of the prior flight and the duration of the shaking.

The conversion of an optical angle to a gravitational angle is an ability with little biological significance so far as known except in *Apis mellifera, cerana,* and *dorsata,* but is very widespread among insects. It is not always done in exactly the same way as in *Apis.* For example, the dung beetle *Geotrupes* identifies the direction toward the light as down, not up as does *Apis.* But there is something in the nervous centers of insects that relates responses to one major environmental stimulus (light) to those for another (gravity). The genus *Apis* took advantage of this fact to evolve communication that would work on vertical surfaces and in dark cavities, presumably a prerequisite to the invasion of temperate regions from the tropics, since exposed colonies of *Apis* do not survive temperate winters.

16 *The Handling and Transfer of Materials within Nests*

Although a small part of this chapter concerns feces and building materials, it is almost all about the movement of food, and by implication nonvolatile pheromones, among adult members of bee colonies and from the adults to the larvae. Once food is brought back to the nest it can be used (1) for the growth of larvae, (2) for the maintenance of adults, (3) for production by workers of secreted nutrient materials (bee milk) for larvae and other adults, (4) for production of nonfertile trophic eggs by workers, or (5) for production of reproductive eggs by queens and workers. Alternatively, the food can be stored and subsequently used for any of the purposes listed.

A. Foods

Foods given to larvae of the solitary to primitively eusocial bees are nectar or honey, the principal carbohydrate source, and pollen, the principal protein source. These materials may be supplemented by secreted materials originating in the adult female bees, but there is no evidence for such food supplements except in *Bombus*. Even in *Bombus,* studies of pharyngeal and other glands of queens and workers before, during, and after the stages when larvae are being fed showed no variation, according to Free and Butler (1959). Moreover, such bees ate no more pollen when feeding larvae than at other times. This was taken as evidence that they were not secreting large quantities of a proteinaceous food like bee milk, although they might, of course, add enzymes or other materials in minute quantities to the larval food. Contrary to this, Röseler (1967a) reports larval food being produced in hypopharyngeal glands of young adult workers of *Bombus*. Under certain conditions, according to need in the colony, the glands remain enlarged for long periods or throughout life of the bee; this fact may explain the earlier results.

In the highly eusocial bees an important part of the larval food consists of secretions of young adult workers. In Meliponini it is difficult to tell how much of the viscous material with which cells are provisioned is honey and how much is secretions. The existence of large hypopharyngeal glands and their enlargement in the proper age group of workers, as well as the aspect of the material in the cells, all suggest that the brood cell content is mostly secretion, mixed with pollen except in the uppermost layer in each cell. In *Apis* secreted material (bee milk) is the only food of those larvae that will become queens and of other larvae during the first days after hatching (Chapter 10, C2). Worker and male larvae are thereafter fed on a mixture consisting partly of pollen and honey, but Simpson (1955) says that less than a tenth of the nitrogen requirements of growing worker larvae is from pollen. Most of it is from the secretion of the hypopharyngeal glands of nurse bees. The main lipid ingredient of the bee milk, however, is *trans*-10-hydroxy-2-decenoic acid from the mandibular glands (Boch and Shearer, 1967).

Eggs or young larvae are also foods of bees. Egg eating is widespread in social Hymenoptera. It occurs in allodapine bees, in primitively eusocial halictines, and quite possibly even in solitary bees. Eating of eggs and young larvae has social significance in *Apis* (for example, destruction of diploid males). In Meliponini, trophic eggs laid by workers are especially important, being one of the main foods of queens in many species.

B. Food handling and interadult contacts within the nest

The amount of contact among the adults in a colony varies enormously from the parasocial and primitively eusocial groups to *Apis*. In the halictines there are no contacts among the adults that appear to have anything to do with transfer of materials. When two bees meet, they commonly tend to retreat from one another. Current uncompleted studies in Kansas suggest that contacts among adults of *Lasioglossum zephyrum* may play a role in caste determination, but there is no evidence of transfer of materials among the bees. The best description of contacts among adult halictines in a colony is that of Batra (1968), based upon observations of *L. versatum* and *imitatum* in observation nests in Kansas. She says, "Various bees, sometimes including the queen, were often seen jostling one another while waiting in the burrow near the cell being provisioned. Fighting was seen only once when one bee bit another that was part way into a cell and blocking the burrow with her abdomen. The bitten bee departed, the biter entered the cell and oviposited. Another unmarked bee soon entered, ate the first egg and oviposited again in this cell. Foragers bringing pollen into the cell forced aside other bees in the burrow and sometimes pushed other bees out of the cells in order to deposit pollen loads." From such observations she concluded that "bees that were actively working or waiting to oviposit thus seemed relatively aggressive. Idle bees usually retreated when confronted by them. Trophallaxis and social grooming were never seen." Certainly there is no visible exchange of food, licking of one another, or any such contact among adult individuals. Those such as queens that do not leave the nest for food sometimes feed in cells that are being provisioned, but foragers never feed others directly.

When a foraging parasocial or primitively eusocial bee (except *Bombus* and the allodapines) returns to the nest with a load of provisions, the forager places the food in a completed brood cell, as do solitary bees. Except to obtain food for maintenance of the adult bees themselves, foraging does not occur unless a completed brood cell is available. Some of the food placed in the brood cell may be used as food by adult bees, especially those that do not forage, but most of it becomes the provisions for that cell and will ultimately be eaten by the larva which will grow in it. The report by Fahringer (see Friese, 1923) that *Halictus scabiosae* fills its cells full of pollen for storage is presumably an error, possibly based on misinterpretation of fungus-filled cells.

Allodapine bees, likewise, have no specialized sites for food storage, but they often do put pollen on the inside of the nest burrow. (They make no cells.) Usually only small patches of pollen are formed. For example, in species of *Braunsapis* studied in Cameroon (Michener, 1968a), many of the nests, opened near midday after an hour of sun following days of rainy weather, contained a mass of freshly gathered dry pollen perhaps 2 mm in diameter and 0.5 mm thick plastered on the burrow wall. The pollen was usually placed below the prepupae and pupae, immediately above the larvae, suggesting its use in feeding larvae. In stems that had been charred by grass fires at the level of the pollen deposit, larvae contained bits of carbon along with pollen, sometimes enough that the gut content was dark instead of yellow, supporting the view that the pollen stores are used to feed larvae. Adults must scrape up bits of charcoal along with the pollen before carrying it farther down the burrow to larvae. Pollen stores are also eaten by adults, however, for the pollen stores disappear not only from *Braunsapis* nests containing larvae but from those that do not.

In the species of *Allodapula* in western South Africa, pollen storage reaches considerably higher levels, and masses 10 mm or more in length are often formed. Such pollen is not used up from day to day, but in many nests remains for long periods, and becomes caked. At least in *A. acutigera* pollen is only stored in nests containing larvae, contrary to the situation in *Braunsapis* and related genera. In various species and genera in the arid western Cape Province, pollen is more often stored than in the humid climate of eastern South Africa (Natal), where flowers are more continuously available. (Michener, 1971).

Foraging allodapines, then, on returning to the nest, frequently or always deposit pollen on the nest wall instead of on or beside the larvae. Thereafter, as Skaife (1953) says in a study of *Allodape mucronata,* the adult female stands over the mass of dry pollen, regurgitates a liquid (nectar?) onto it, and mixes the two with her mandibles to make a stiff paste, which she then carries to the larvae. Occasionally, in *Allodape* and *Braunsapis,* larvae of any size may receive a drop of the nectar with little or no included pollen. As will be explained in more detail later, feeding of larvae in most genera of allodapines is sparing, and at any one time most larvae have nothing to eat.

In *Bombus* the returning foragers put the food materials into special storage pots or pockets in the nest, as described in Chapter 28 (A), instead of placing it in brood cells. If carrying pollen, a returned forager stands on the edge of the pollen receptacle, facing away from it, and with her middle legs pushes the pollen masses out of her corbiculae. She may then turn and pack the material down herself, or leave this to a house bee. In the case of a pollen pot, the food, if not eaten by adults, must be transferred to larval cells and fed with honey to the larvae. In the case of a pollen pocket, both larvae and adults have direct access to the pollen mass.

A returned forager with nectar stands on a honey pot with her head in it and with rapid abdominal contractions forces the nectar from her crop into it. As with field activity, foragers show temporary constancy, often a few days in duration, to particular pollen receptacles or especially honey pots. Perhaps, as in the case of foraging, such constancy enhances efficiency—a bee goes to a known unloading point instead of searching.

The presence of storage places for honey and pollen has an important influence on the flow of materials among members of the colony. Lecomte (1963) showed, for a *Bombus* colony of about 50 workers, that six hours after one forager was fed radioactive syrup, about 51 percent of the bees were radioactive. The importance of the storage pot as the point of transfer was shown by another colony of about the same size, abnormal in that it lacked a honey storage pot. After a forager was fed radioactive syrup, there was no transfer whatever to other members of the colony. The food pots normally provide the opportunity for extensive interchange of substances among individuals of a colony in *Bombus.* Except for the food itself, however, no social use of this transfer point is known in *Bombus.*

In the highly eusocial bees, nectar is not ordinarily put into the storage places or into brood cells or fed to larvae by the returning foragers. Instead, the foragers transfer the nectar to other individuals, most of them house bees. These receivers are workers which have not yet begun foraging activities.

The interaction between foragers and receivers is extremely important in the social organization of a colony. The following is based on *Apis,* but probably is more or less applicable to any of the highly eusocial bees. When a forager with a crop full of nectar arrives in the nest, she promptly approaches a receiver head-on, opens her mandibles, and regurgitates a drop of nectar onto the slightly projected base of her proboscis (upper surface of prementum and stipites), the rest of which remains folded back under the head (Figure 15.7). The receiver then extends her proboscis and, with the tip of it, takes nectar from the drop. Meanwhile the antennae of both bees are in continual stroking motion, an activity that probably keeps the two bees properly oriented for the transfer.

If the nectar is desirable relative to most of that being brought in at the time, the receiver shows excitement by rapid antennal movement and by quickly taking the nectar. Sometimes a single hive bee takes all of it in 30 to 50 seconds; if others have been attracted and reach out to take their shares, transfer may be much quicker. On the other hand, if the quality of the nectar is poor, antennal movement is slow and receivers often refuse the nectar after the first taste. The forager may have to try several different receivers to get rid of her crop load of nectar, little by little. In the extreme case she is unable to do so, and may crawl to a cell and regurgitate it.

The speed with which a forager can unload her nectar appears to be a cue for her subsequent action. If the delivery time is short (and of course if the source is qualitatively and quantitatively suitable), she dances, sending other foragers to the same place, and soon returns herself. If the delivery time is long (over 60 seconds), there is little dancing, so few or no more bees are recruited to the food source, and if the time is too long, the forager herself does not return but goes elsewhere or remains in the hive. This social interaction is an essential part of the control of the quality of the liquids brought into the hive. Much of the time the nectar with the highest sugar content is preferred, but when the water is needed, that with the lowest sugar content and even water can be favored by the action of receivers (Chapter 17, D2).

Receiving bees offer nectar to others, which in turn feed

others, and so on. Thus incoming nectar from all sources is mixed with the crop contents of various bees. When nectar is coming in good quantities, the hive bees accumulate more and more in their crops, all of them taking more or less equal quantities.

At the same time, the nectar is being converted to honey by treatment with invertase from the hypopharyngeal glands of the bees that handle it. The invertase causes hydrolysis of the sucrose. Simultaneously the nectar is ripened, probably by bees whose crops are nearly full. Ripening, usually done in the outer parts of the nest where bees are less crowded, consists of repeated exposure of the nectar by spreading it as a film between the parts of the proboscis; such exposure results in evaporation of water, increased concentration of the sugar, and along with the inversion of the sugar, converts the nectar into honey, which is regurgitated into storage cells. Often bees take it out of cells again for further ripening before the cells are capped with wax.

Ripening of nectar is not limited to highly eusocial bees. Halictine foragers, solitary as well as social, often rest, for example on a leaf, and extend or retract the proboscis, ripening nectar. Presumably the result is reduction in the amount of water placed in the pollen balls in the rearing cells. It is not known whether the sugar is inverted, but inversion seems probable.

In highly eusocial bees it may take two or more times as many bees to ripen honey as to collect the nectar. Nectar contains from 20 to 90 percent water. *Apis* honey usually contains about 18 percent water, varying from 12 to 26 percent. Most meliponine honey has similar percentages, 12 to 34 percent water having been recorded for various samples of various species, except that the group of *Trigona spinipes* stores very watery, acid, badtasting honey with 44 to 50 percent water (Kerr, 1959).

Pollen brought to the nest by foragers of the highly eusocial bees is taken to the storage cells or storage pots by the foragers themselves. Sometimes they go directly; on other occasions, as described by Bassindale (1955) for a minute African *Trigona,* some of them stop to feed or be fed by other workers before reaching the pollen storage pots. In *Apis* they may stop to regurgitate their nectar and even to dance before reaching the storage cells.

Pollen pots and storage cells have the potentiality of functioning for indirect transfer of substances other than food among members of the colony, for in the Apidae the pollen is mixed with small amounts of nectar before being carried back to the nest; such nectar might contain pheromonal or other secretions of the bees. There is no evidence of such an interaction, however.

The control of the flow of pollen into the nest of *Apis* appears to be similar to that for nectar, even though pollen is not accepted by hive bees. When there is a shortage of pollen in the nest, house bees show a great interest in the pollen loads in the corbiculae of a pollen-laden forager, touching them with the antennae and feet or even nibbling them. Pollen is an essential protein source for young workers that are secreting bee milk; some is also fed directly to larvae mixed with the milk. If the demand is great, the pollen is promptly eaten from the storage cells by the young workers. But if the supply is good, house bees press it into the cells for storage.

Food exchange among workers of highly eusocial bees is at the heart of the social system of these insects, and it is the social interaction which such exchange makes possible that sets these bees apart from all others. Colony defense, control of queen production, communication concerning food and water sources, and cooling of the hive are among the functions dependent upon rapid and extensive food exchange. The food transfer is even more extensive than indicated by the paragraphs above. There is constant interchange of food materials, together with whatever pheromones or other secretions are mixed with them.

The food exchange is demonstrated by studies using radioactive syrup. For example, six worker bees were allowed to carry radioactive syrup into a colony of 24,600 *Apis* for 3.25 hours. Within 3.5 hours about 20 percent and within, 27 hours over 50 percent of the workers were radioactive (Nixon and Ribbands, 1952; Delvert-Salleron, 1963). Ribbands (1956) says that one crop full of radioactive syrup will be shared among nearly all the bees of a large colony, and that the sharing is at random, the sugar being passed along irrespective of the recipients' age or activity. This is contrary to Nixon and Ribbands' actual results, however, for they clearly show that foragers become radioactive in largest numbers and first (62 percent in 3.5 hours; 76 percent in 27 hours), and house bees more slowly, those in the brood chamber, mostly nurse bees which eat largely pollen, slowest of all. The rapid spread of radioactivity among foragers is surprising, since syrup is offered by foragers to house bees, and more surprising is the observation that the large feeding larvae, which are fed by nurse bees, were all radioactive after

27.5 hours. Evidently there is more to be learned about food flow in *Apis* colonies.

Free (1955c) has provided a theory for the origin of the food transfer among adult bees, so characteristic of highly eusocial bees, from the dominance-subordinance responses found in small colonies in which a social hierarchy exists. Dominant bumblebees (queens or laying workers, nest guards) frequently attack others. Rarely a forager will thereupon regurgitate a drop of nectar, which the attacker sucks up either from the floor or from the mouthparts of the subordinate bee. Direct food transfer is otherwise not found in *Bombus*. In honeybees submissive intruders also regurgitate drops of fluid when examined by guards. In dequeened *Apis* colonies aggressiveness arises among workers when they start laying, just as in *Bombus*, and submissiveness is shown in the same manner as by intruders. A honeybee soliciting food, with mandibles open and movements toward the head of another bee, suggests the movements of an aggressor, and the response of the other bee, frequently regurgitation, suggests that of a submissive bee. It therefore seems quite possible that the food transfer system so conspicuous among highly eusocial bees evolved from dominance-subordinate responses. It is possible also that removal of food from subordinates and its transfer to aggressors helps to ensure the reproductive differences between workers and queens.

Outside of the Apidae, the only bees that are known to transfer liquid food directly from bee to bee are in the Xylocopinae—*Braunsapis, Exoneura,* and probably *Xylocopa* (Michener, 1972a). The manner in which it is done is more or less as in Apidae. It is possible that the Xylocopinae are related to the ancestors of the Apidae—these groups are reasonably close in Figure 2.2, and the details of the connections shown there may be wrong. The function of the feeding in Xylocopinae may be to increase the probability of survival of young adults. Such a function would extend and not necessarily negate the dominance-subordinance theory of food transfer.

C. Treatment of the queen

In semisocial and primitively eusocial bees the queen eats from brood cells that are being provisioned or from food stores in the nests, or she may even fly out and feed from flowers. She receives no special attention or feeding from workers.

In most of the highly eusocial bees, however, the queen is frequently or constantly the subject of special attention from the workers. In many Meliponini there is usually a "court" of workers surrounding the queen (Figures 16.1 and 16.3) more or less as in *Apis,* although commonly the bees are less numerous. The members of the court antennate and sometimes lick the queen, but only rarely is she fed directly by the worker bees. Her main sources of food are larval food from freshly provisioned cells and trophic eggs laid by workers (section D2 below). In *Apis* the court is almost constantly recognizable (Figure 11.1), consisting of about a dozen bees (Chapter 11, A). They touch and lick her continuously. When laying she is fed bee milk by nurse bees every 20 to 30 minutes. Feeding is direct, requiring 1.7 to 3.3 minutes. However, before swarming or in the autumn when she stops laying, feeding of the queen by workers also stops, and she then takes honey from cells.

FIGURE 16.1. Workers of *Trigona* (*Scaptotrigona*) *xanthotricha* making a court around a queen of the closely related *T.* (*S.*) *postica* in an artificially mixed colony. (From Hebling, Kerr, and Kerr, 1964.)

D. Egg eating

Many bees are egg predators on their own species or feed on young larvae. Sometimes such predation appears to be chiefly a matter of social control—limiting colony growth or preventing the growth of certain classes of individuals. At other times it also has an important nutritional function.

1. Social control by egg eating

Since most bees lay one egg per cell, the number of cells constructed limits egg laying and hence colony growth. So also does egg eating. Egg eating also may limit the types of young that develop, for example, by favoring eggs produced by the queen instead of those laid by workers. In bees such as solitary Halictinae, the occupant of a nest usually completes the provisions, perfects the shape and smoothness of the pollen mass, and immediately thereafter lays her egg on top of the pollen mass; then she closes the cell. In social Halictinae, the same is true; the bee that shapes and smooths the pollen ball oviposits upon it. However, Batra (1964, 1968) has shown that in *Lasioglossum* (subgenus *Dialictus*) one to five other females often wait in the burrow near the cell during oviposition. As soon as the egg layer has plugged the cell and left, one of these females often enters the cell and eats the egg, then, after resmoothing the pollen ball, lays another egg on it. There are no visible differences between the first egg laid and the second. A reasonable interpretation is that of Batra, that is, that at a time when more bees are ready to lay eggs than can be provided with freshly provisioned cells, there is competition for the latter, and some eggs are destroyed. No hostility among adults was associated with this behavior. In each of the two cases of egg eating by *Lasioglossum* (*Dialictus*) *zephyrum* in which castes of the bees were known, a queen destroyed and replaced an egg laid by a worker (Brothers and Michener, unpublished). It will be interesting to know whether this is the usual case in halictine oophagy.

Occasional eating of eggs is also known in *Allodape* (Skaife, 1953).

In *Bombus*, for example, *B. lapidarius*, egg eating is rather common, usually by workers that remove the eggs from cells in which they are being laid by the queen. This usually occurs late in the season, when the colony is large, laying workers are present, and males and young gynes are being produced. The queen attempts to defend the eggs and to close the cell before workers take them. She may remain near the cell for several hours, after which time the new cell and its contents seem to lose attractiveness to the workers (Hoffer, 1882; Plath, 1923; Free and Butler, 1959). Attempts to defend the eggs are often partially and sometimes largely unsuccessful. Sladen (1912) recorded an instance in which a queen of *B. lapidarius* made and oviposited in three cells, one after the other, each of which was destroyed and all the eggs eaten by the workers. Free, Weinberg, and Whiten (1969) have shown that in this species it is commonly cells built by workers that are involved in these contests. The queen opens them, eats the worker-laid eggs, rebuilds them, and lays her own eggs. Then the sort of behavior recorded by Sladen occurs. Egg eating in *Bombus* is not associated with lack of other food. It may be one of the factors leading to the high worker/larva ratio needed to produce gynes in *Bombus* (see Chapter 10, B).

Eating of workers' eggs by queens has also been seen in various species of *Bombus*. Thus Sakagami and Zucchi (1965) recorded queens of *B. atratus* aggressively eating eggs laid by workers. Such activity has also been recorded in nests of other primitively eusocial insects (such as *Polistes*), and must serve to decrease the reproductiveness of workers and to limit the production of males, since workers' eggs would frequently be unfertilized and male-producing. Of course worker-laid eggs often do survive and produce males, in spite of the queen's activities.

In *Apis*, as indicated in Chapter 8 (A2), young larvae of diploid males are eaten by workers. This bit of behavior strengthens the usual sex-determining mechanism. Eggs laid by the queen in the autumn, when workers have stopped feeding her, are also eaten by workers. These are examples of the sorts of social control achieved in the honeybees by eating eggs or young larvae. Worker honeybees also will eat eggs not properly located, for example, any outside the area where bees are dense enough to maintain proper temperatures.

2. Trophic egg eating in Meliponini

Egg eating and other activities related to oviposition are particularly interesting in the Meliponini, as shown in a series of papers by Sakagami and his associates on 18 species of Meliponini (Beig and Sakagami, 1964; Buschinelli and Stort, 1965; Sakagami, Beig, and Akahira, 1964; Sakagami, Beig, and Kyan, 1964; Sakagami, Montenegro, and Kerr, 1965; Sakagami and Oniki, 1963;

Sakagami and Zucchi, 1963, 1966, 1967, 1968; Camillo, 1972). The 1966 paper and Kerr (1969) provide summaries.

The following is a description of cell provisioning and egg laying by *Melipona quadrifasciata,* as an example for subsequent comparison. Figure 16.2 illustrates the processes. When a cell is ready, that is to say, when its upper margin is 1–1.5 mm higher than the surface of the comb, a concentration of workers in the vicinity of the cell can be noted. Each worker, one after another, inserts its head into the cell. Meanwhile and perhaps in response to the workers' activity, the queen, who has been elsewhere in the colony, approaches and takes up a fixed position at the side of the cell. The phenomenon of insertion of the head is executed by the workers one after another, more and more frequently and with increasing excitement. The queen touches with her antennae and anterior feet the workers that are inserting their heads into the cell. After a few minutes, perhaps as a response to the queen's fixation, a worker inserts her head into the cell and regurgitates larval food (honey, pollen, and salivary secretion of the worker); this is followed by a second, a third, and so on. At this point the queen, who was stationed beside the cell and touching the workers that were inserting their heads into it, moves 0.5-1 cm away from the cell, almost as though she "feared" the food material. This retreat occurs only in the genus *Melipona,* not in other meliponines. Four or five discharges of food by different workers into the cell are sufficient to provision it. Commonly, toward the end of the provisioning phase, one or more workers turns and lays an egg in the cell, standing it on its end in the middle of the surface of the provisions, just like the egg later to be laid by the queen. However, the worker egg is only about two thirds the size of that of the queen. After provisioning is complete or sometimes before it is complete, the queen approaches the cell, inspects it, eats any worker eggs that it contains, and also eats some of the larval food from the cell. After feeding, the queen turns and deposits her own egg in the center of the food surface in the cell; then a worker closes the cell, using the material of the cell rim to draw across the opening and form the closure. A further idea of the activities can be obtained from a study of Figure 16.3, which, however, is based on *Trigona postica* and shows the court of workers around the queen.

While in *Apis* ovarial enlargement of workers is inhibited in the presence of the queen, in *Melipona* some workers regularly have enlarged ovaries in the presence of the queen, and they are the ones that lay the trophic eggs upon which the queen feeds (as well as many male-producing eggs[1]). Ovaries were well developed in nurse-bee stages of workers of most meliponine species studied by Sakagami, Beig, Zucchi, and Akahira (1963). In some species as many as 100 percent of the workers in the proper stage were found to have enlarged ovaries, although in both newly emerged and older workers ovaries were slender (Figure 16.4).

There are many variations in the details of the cell-provisioning and egg-laying procedure among the many stingless bees. Almost all forms show quantitative differences in the amounts of time required for various activities. In addition, there are more interesting differences, as indicated below.

In the African genus *Meliponula* the queen does not become fixed before laying. As in *Melipona,* only one cell is provisioned and laid in at a time.

In *Melipona* there is almost always a single new cell about which the bees gather, to which the queen becomes fixed, and in which she ultimately lays an egg. After it is completed she may go on to another cell or wait until another is ready. Alternatively, in some of the South American subgenera of *Trigona,* as well as in the African subgenus *Hypotrigona* (Kerr, 1969), several cells are simultaneously constructed and worker excitement and insertion of heads into the cells takes place for several cells at about the same time. In some cases provisioning of the several cells occurs successively, one after another. This has been observed in the subgenera *Trigonisca, Frieseomelitta, Duckeola,* and *Friesella.*[2] In other cases such cell provisioning occurs initially but soon becomes synchronized, several cells being provisioned at the same time. Such behavior is found in the subgenus *Scaptotrigona.* Finally, cell provisioning may be wholly synchronized, several cells being encompassed by the group of excited workers and provisioned simultaneously. Such behavior is found in the subgenera *Nannotrigona* and *Plebeia.* In still other subgenera there are intermediate varieties of behavior: for example, cells are sometimes

1. For material on male production from eggs laid by meliponine workers, see Beig (1972) and Chapter 8 (A3).

2. In the classification used in this book, *Frieseomelitta* and *Duckeola* are groups within the subgenus *Tetragona* and *Friesella* is a distinctive *Plebeia.*

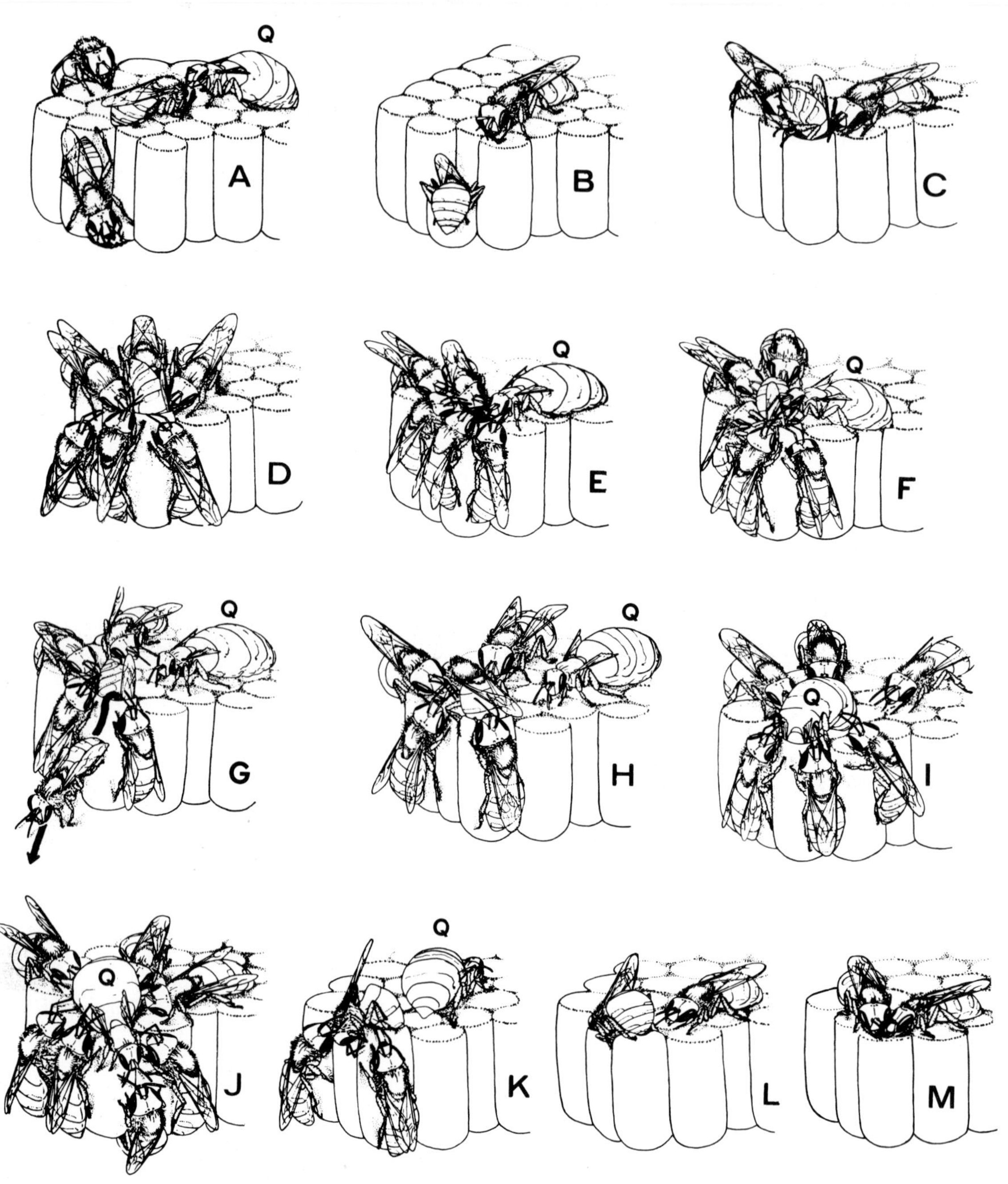

FIGURE 16.2. Behavior of queen (*Q*) and workers of *Melipona quadrifasciata* from the beginning of the construction of a new cell at the edge of a comb (*A*) until the cell is completed and closed (*M*). (*B*) Gradual growth of the cell; one worker has its head in the still incomplete cell. (*C*) The cell has reached the height of the complete cells; one worker has its head and thorax inside. (*D*) The cell now extends above the level of the comb surface; workers crowd around and appear excited as one of their number inserts her body into the cell. (*E*) The queen arrives at the cell. (*F*) The queen taps a worker that is inserting herself in the cell. (*G*) An inserting worker is discharging food into the cell, one that has just discharged is rushing from the cell (*as shown by an arrow*) and the queen is fixed nearby. (*H*) A worker is laying an egg in the cell; the queen remains fixed. (*I*) The queen is eating a worker's egg. (*J*) The queen is laying her egg. (*K*), (*L*) and (*M*) Closing of the cell, the queen leaving in (*K*). (From Sakagami, Montenegro, and Kerr, 1965.)

FIGURE 16.3. Oviposition process in *Trigona* (*Scaptotrigona*) *postica.* The queen is about to eat a trophic egg (large white object which she is antennating) from inside the rim of the cell. Workers form a court around the queen. *A* (on honey pot), *G* and *F* are exposing the base of the glossa, as in food transfer, and *F* is beating her wings; *B* is laying an egg in a provisioned cell to which the queen will soon go; *C* and *D* are working on nearly complete cells; *H* is provisioning a cell (and shows wax scales being secreted between the abdominal terga); and *E* and *J* are indifferent to the queen and to the other activities shown. (From Sakagami and Zucchi, 1963.)

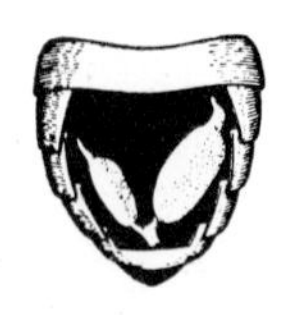

FIGURE 16.4. *Left:* Degenerated slender ovaries of a worker of *Trigona* (*Scaptotrigona*) *postica; right,* enlarged ovaries of another worker of the same species, presumably ready to lay an egg. (From Sakagami, Beig, Zucchi, and Akahira, 1963.)

provisioned in groups while at other times a cell is provisioned individually, as in *Melipona.*

Although, as described above, the queen in *Melipona* retreats quickly from the cell being provisioned, the workers that are provisioning it, after regurgitating into it, leave it relatively slowly. However, in the majority of species of *Trigona,* after putting her crop contents and salivary material into the cell, the provisioning worker leaves rapidly, sometimes falling from the comb, and in the subgenus *Partamona,* even flying for a short distance, as though to get quickly away from the cell or the queen (Figure 16.5).

Probably the most interesting aspect of the cell-provisioning and egg-laying process in the stingless bees is the laying of trophic eggs. In some groups worker ovaries are not enlarged and no such eggs are found. This occurs primarily in various minute species of *Trigona,* and in all Old World *Trigona* that have been studied (Yoshikawa, Ohgushi, and Sakagami, 1969). In most stingless bees of the Western Hemisphere and in the African *Meliponula,* however, trophic eggs are laid. In *Meliponula,* as in *Melipona,* they are shaped like eggs laid by queens, and are positioned similarly to those of queens, but are smaller. In most American species of *Trigona* the trophic eggs are as large as or larger than eggs of queens, less slender and sometimes nearly spherical. According to Akahira, Sakagami, and Zucchi (1970), in *T. postica,* a species with a very large trophic eggs, the nucleus degenerates before the egg is laid. Trophic eggs of *Trigona* are placed on the rim of the cell, as though to be more readily accessible to the queen (Figure 16.5). It is in such species, like those of the subgenus *Scaptotrigona,* that direct feeding of the queen by workers is rarest and that the trophic eggs are apparently the principal food for the queen, although no doubt she also takes some of the food placed in cells for the larvae.

Silva, Zucchi, and Kerr (1972) have studied the ontogeny of the queen's fixation and laying behavior in *Melipona.* In the first two days after the nuptial flight

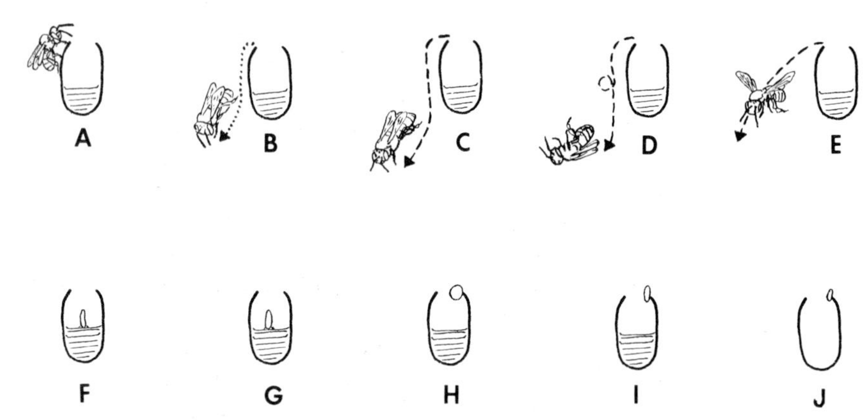

FIGURE 16.5. *Top row:* Sectional diagrams of cells, showing the ways in which provisioners of various kinds of stingless bees leave after regurgitating larval food into the cells. (*A*) *Hypotrigona* scarcely withdraw; (*B*) *Melipona, Meliponula* leave relatively slowly (*dotted line*); (*C*) most groups leave rapidly (*broken line*); (*D*) *Geotrigona,* in the subgenus *Tetragona,* leave very rapidly, falling in the process; (*E*) *Partamona* leave very rapidly, flying. (From Sakagami and Zucchi, 1966.)

Lower row: Sectional diagrams of cells, showing eggs laid by stingless bees. (*F*) Egg laid by queen; (*G*) trophic egg laid by worker of *Melipona* or *Meliponula;* (*H*) large trophic egg laid by a worker of the subgenus *Scaptotrigona;* (*I*) trophic egg about the size of queen egg—most *Trigona* subgenera; (*J*) like (*I*) but laid before instead of after provisioning—*Geotrigona* in the subgenus *Tetragona.* (From Sakagami and Zucchi, 1966.)

there is no fixation or laying. The queen's first fixation is disorganized. She does not withdraw after workers begin provisioning a cell, or she may wander away from the vicinity. The first fixation and the first laying take longer than usual.

Although cell provisioning and egg laying have been studied comparatively only by Sakagami and his co-workers, as listed above, some earlier workers had surprising amounts of information on the subject. Fixation, cell provisioning, and associated activities in *Melipona* have been known for nearly a hundred years (Girard, 1874), but the associated trophic egg laying is a discovery of the past decade. The following account of a minute African species of the subgenus *Hypotrigona* is given because this is apparently a species in which trophic eggs are not laid; it is one of the minute *Trigona* species referred to above.

Bassindale (1955) says of *Trigona* (*Hypotrigona*) *braunsi* that, after a number of cells have been finished, the queen walks onto the brood and, on finding a finished cell, rests at one side of it and holds her antennae high. In this species the queen is never surrounded by a "court," but nevertheless workers usually find her when she takes up this fixed position; four to six, one after another, dive into the cell and regurgitate brood food into it. The queen looks into the cell once or twice during the process, and when it is fully provisioned she turns around and rapidly lays an egg, which stands up vertically from the surface of the semiliquid provisions. As the queen moves away from the cell after oviposition, a worker starts closing it, using the wax in the margin of the open cell. The whole process, from the arrival of the queen to cell closure, may take as little as two minutes. Sometimes, however, it is delayed by repeated inspections by workers prior to regurgitation of the first of the brood food. After leaving the first cell, the queen wanders on until she finds a second, where the process is repeated. This is continued until the new cells have been provisioned, at which time she wanders away from the brood, to which she will return later when more new cells have been constructed. These observations were made in considerable detail, and it seems unlikely that either the backing into the cells required for laying of trophic eggs by the workers or the feeding upon such eggs by the queen would have been missed by the observer. He nonetheless notes that feeding of the queen by a worker was observed only once and, in fact, the workers seemed to keep away from her. It therefore seems likely that in this species the queen's food is almost exclusively larval food taken from cells before oviposition. More recent observations of another African species, *T.* (*Meliplebeia*) *nebulata komiensis,* also show the lack of trophic laying (Darchen, 1969).

Existence of relations between food for laying workers, food for the queen, and colony growth in Meliponini are obvious. An experimental study to determine the cues involved in this intricate communication would be most interesting. It seems reasonably clear that the newly completed cell (and sometimes perhaps workers waiting there) is the cue that causes the queen to first become fixed, but it is not clear in what way her presence then leads to provisioning of the cell, the laying of trophic eggs, and these to laying by the queen.

E. Feces

Defecation by adults probably occurs in nests in halictine bees; at least deposits that are probably adult feces sometimes occur in the burrows. Allodapine bees (at least *Braunsapis*) defecate outside of laboratory nests (Michener, 1972b). Bumblebees frequently defecate in the nests, often in a particular area. Meliponine bees have special areas in the nests where feces are deposited, except that queens defecate elsewhere and their feces are eaten by workers, perhaps a device for pheromone dispersal (Kerr, 1969). In *Apis* defecation by healthy adult workers never occurs in the nests. In winter the rectum becomes greatly distended and the bees defecate on short cleansing flights on early spring days.

In allodapine bees larval feces are carried out of nests by adults. They are removed from cells in some species of *Lasioglossum* (*Evylaeus*) that leave the cells open (Knerer, 1969b). Legewie (1925a) appears to have worked with a population of *L. malachurum* in which the cells are not closed (Chapter 25, B); he found feces in the cells if the cell cluster was removed from the nest and kept without adult bees. However, in cell clusters excavated from the ground at a time when they contained pupae, the cells were usually free of feces. In species of the same genus and in populations of the same species that close the cells, feces remain in the cells until after emergence of the adults. Then they may be removed along with other materials with refurbishing of the cells or construction in the nest.

In the Apidae larval feces are left inside the cell with the cocoon when the young adult emerges. They are normally removed with the cocoon if it is torn down.

F. Building materials

Movement of building materials within nests must occur at least to a limited extent in all nests. In allodapine nests bits of pith are moved about and made into the entrance collar. Soil is moved out of cells or from elsewhere deep in a halictine nest and dumped in a tumulus outside, used to fill old cells or other spaces, or used to line the burrow. Soil is also brought into newly excavated halictine cells and used to form an earthen lining. In other species of halictines which construct their cells instead of excavating them in the substrate, the homologous earthen "lining" is built in a chamber in the soil, using earth brought from elsewhere in the nest, frequently from the walls of the cavity in which the cells are being built.

In some solitary bees, material for building cells is brought into the nests. This is true especially in the Megachilidae (see Chapter 3). It is not true of allodapine or halictid bees.

The use of secreted material for cell linings occurs in many bees. In most solitary burrowing bees and in the halictines, such material is waxlike in appearance and applied as a thin layer on the inner surface of the smooth cell lining. It is never reworked or reused. The origin and the characteristics of this lining are explained in Chapters 7 (A) and 17 (A).

In the bumblebees as well as in the highly eusocial bees, cells are built at least in part of material secreted from epidermal abdominal wax glands. Little is known of the Euglossini; if they indeed do use wax, it is mixed with resins or mud to make the cells. In *Bombus* the wax is mixed with pollen. In the Meliponini it is mixed with resins and called cerumen. Only in *Apis* are the cells made of more or less pure wax. In all these forms the bees are capable of secreting wax in one part of the nest, manipulating it, and moving it for use to another, unlike the halictines, which place their waxlike secretion directly on the cell walls. In *Bombus* and the Meliponini the wax mixtures may be removed from old cells, mostly after cocoons have been spun by the larvae (Figure 16.6), and re-used in new situations, so that there is a movement of construction materials about the nest. Resins or other materials brought into the nest for construction purposes are also moved about as needed.

FIGURE 16.6. A part of a comb of *Trigona carbonaria.* The larvae had matured and spun cocoons, and had pupated when this photograph was taken. The workers have removed and re-used the cerumen (wax plus resin) material from the ends of the cocoons, but between the cocoons the hexagonal pattern of walls of cells remains. (Photograph by C. D. Michener.)

The methods of moving building materials vary considerably. Soil in nests of halictine bees can simply be pushed by the head or by the apex of the abdomen, and it is in this way that it is usually pushed out to form the tumulus. Details of its movement within the nest of social forms have been indicated, chiefly for the subgenus *Dialictus* of *Lasioglossum,* by Batra, and are described in Chapter 24 (A), or again, for forming the entrance constriction, in Chapter 18 (A1). Moving of building materials within the nest by *Bombus* and highly eusocial bees is largely by the mandibles, which especially in *Apis* but also in *Bombus* and the meliponines are formed for manipulation (or mandibulation) of pliable material like waxes and resins (Figure 1.5).

FIGURE 16.7. Diagrams showing the flow of major materials in nests of halictine and allodapine bees. Q = queen, W = worker. (*A*) A solitary halictid. (*B*) A primitively eusocial halictine, such as many *Lasioglossum* (*Dialictus*) *imitatum.* (*C*) A subsocial allodapine. (*D*) A primitively eusocial allodapine. For further explanation see text. (Original drawings by Barry Siler.)

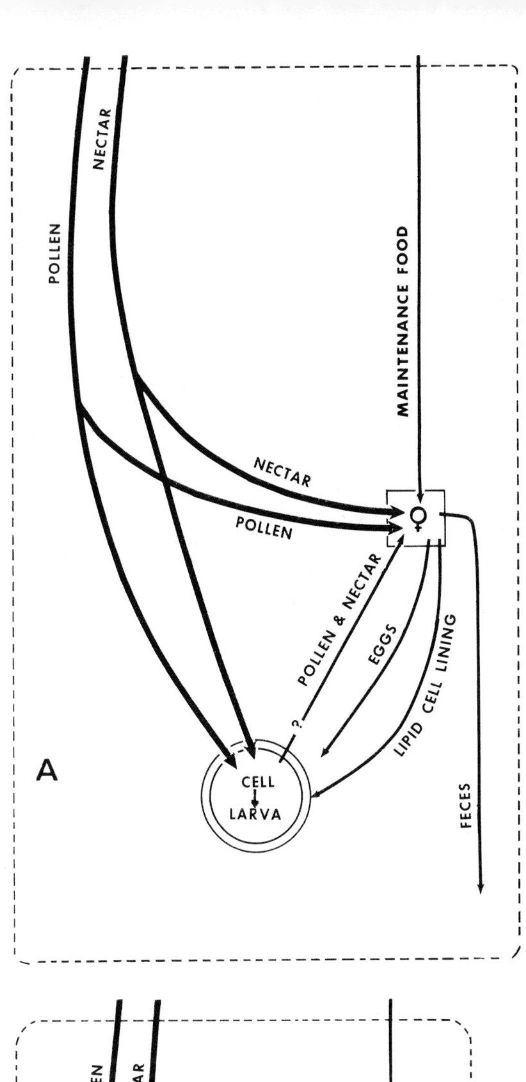
POLLEN
NECTAR
MAINTENANCE FOOD
NECTAR
POLLEN
♀
POLLEN & NECTAR
?
EGGS
LIPID CELL LINING
CELL
LARVA
FECES
A

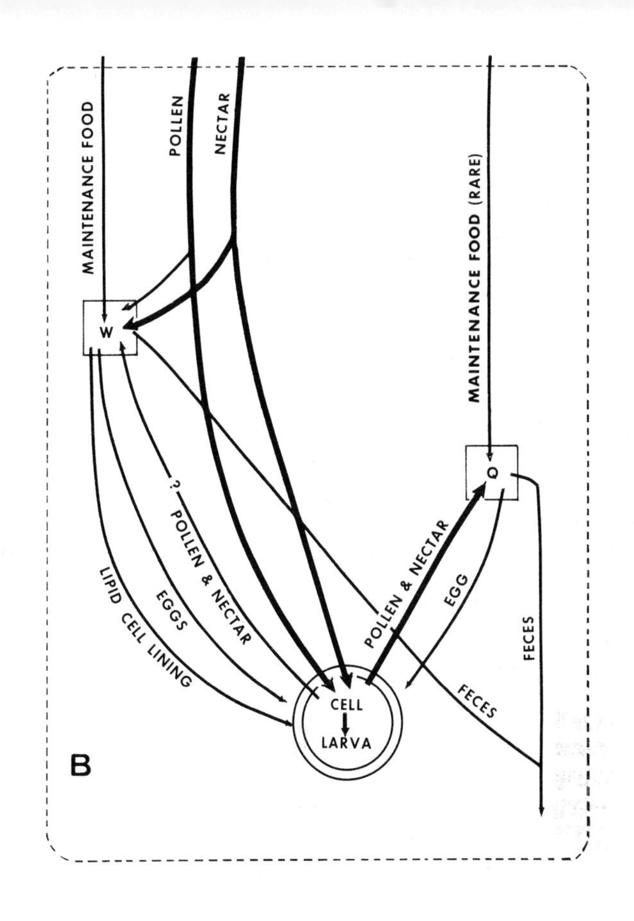
MAINTENANCE FOOD
POLLEN
NECTAR
MAINTENANCE FOOD (RARE)
W
Q
?
POLLEN & NECTAR
EGGS
LIPID CELL LINING
POLLEN & NECTAR
EGG
FECES
FECES
CELL
LARVA
B

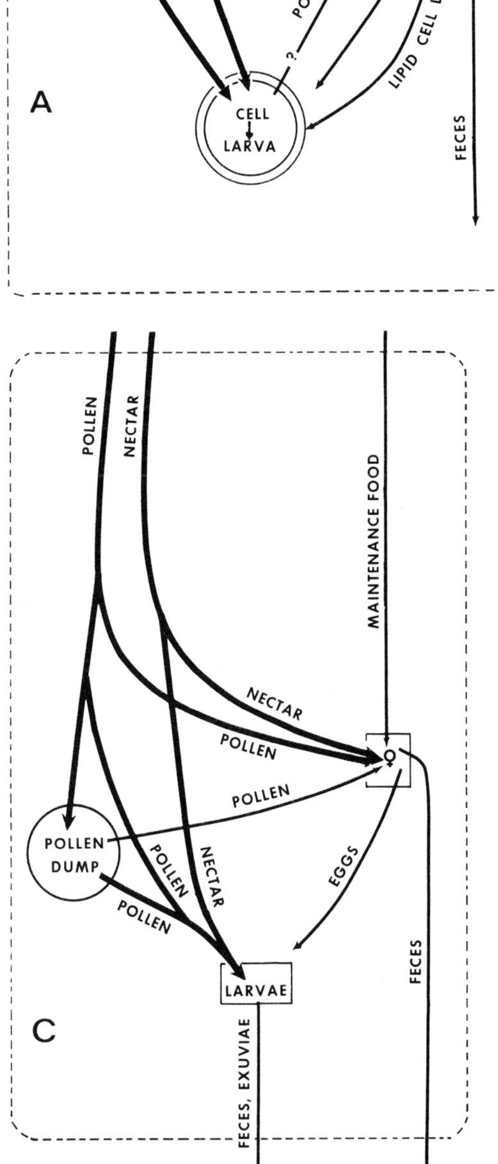
POLLEN
NECTAR
MAINTENANCE FOOD
NECTAR
POLLEN
♀
POLLEN
POLLEN DUMP
POLLEN
NECTAR
POLLEN
EGGS
LARVAE
FECES
FECES, EXUVIAE
C

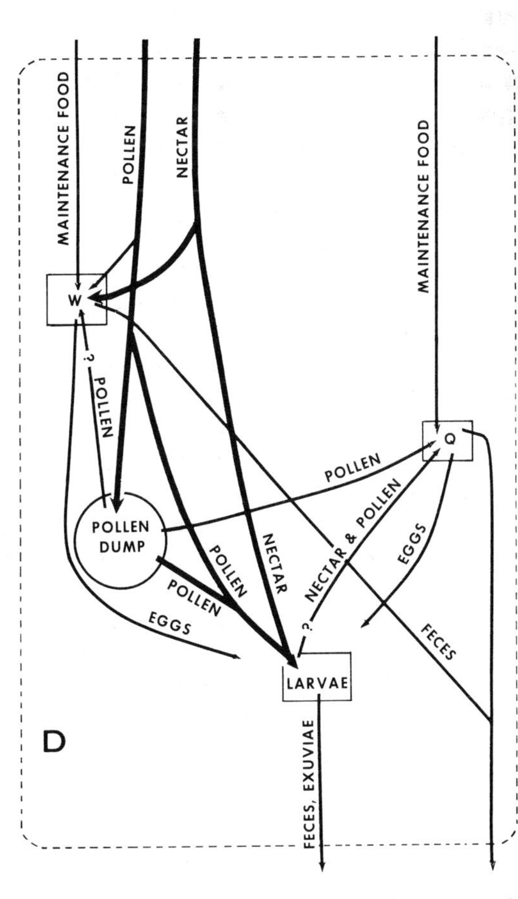
MAINTENANCE FOOD
POLLEN
NECTAR
MAINTENANCE FOOD
W
?
POLLEN
Q
POLLEN
NECTAR & POLLEN
EGGS
POLLEN DUMP
POLLEN
POLLEN
NECTAR
?
EGGS
FECES
LARVAE
FECES, EXUVIAE
D

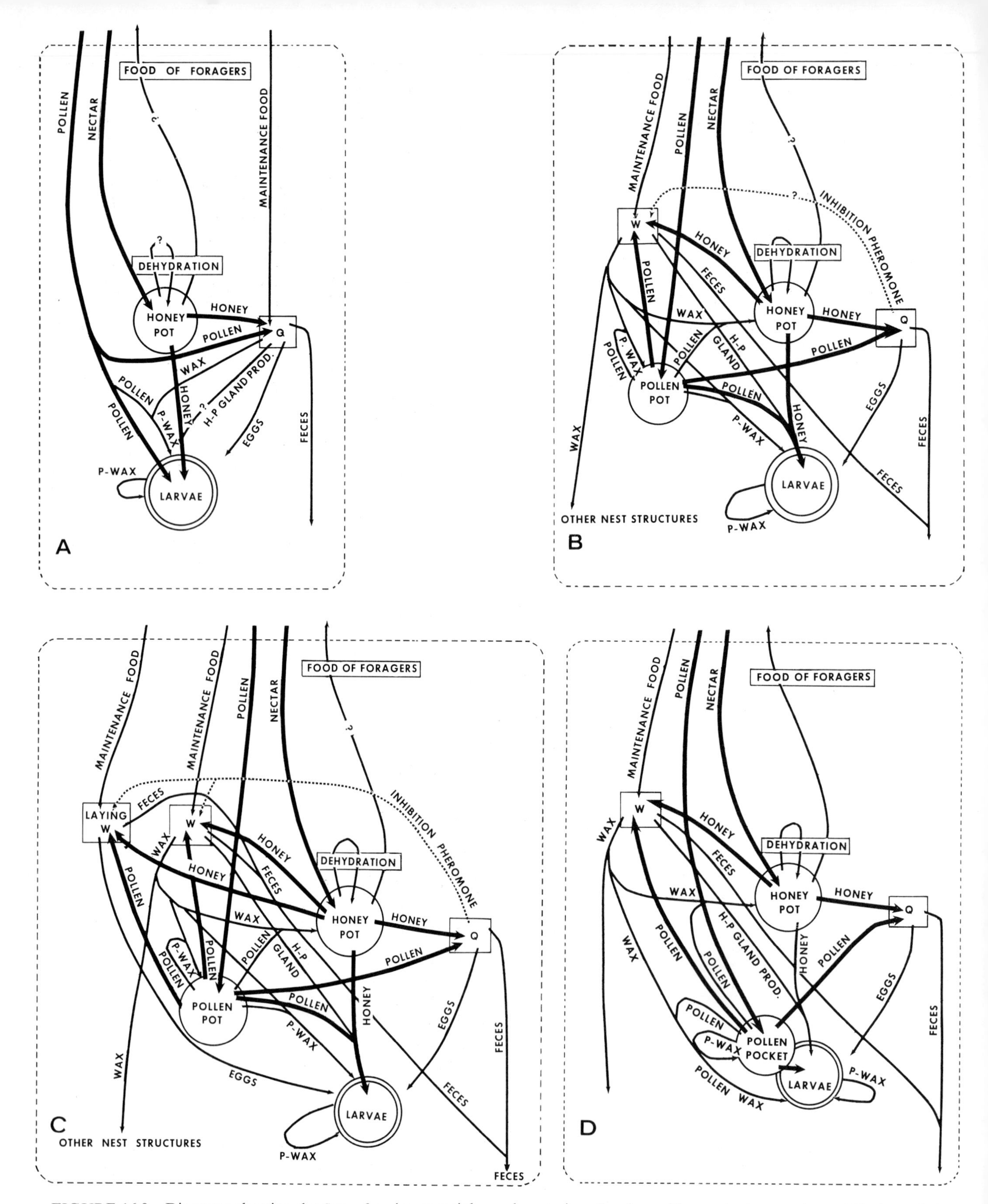

FIGURE 16.8. Diagrams showing the flow of major materials in colonies of bumblebees (*Bombus*). Q = queen, W = worker. (*A*) A young *Bombus* colony, still in the subsocial stage. (*B*) A pollen-storing *Bombus* without laying workers. (*C*) A pollen-storing *Bombus* with egg-laying workers. (*D*) A pocket-making *Bombus* without laying workers. *P-wax* = pollen-wax mixture, *H-P gland prod.* = hypopharyngeal gland products. For further explanation see text. (Original drawings by Barry Siler.)

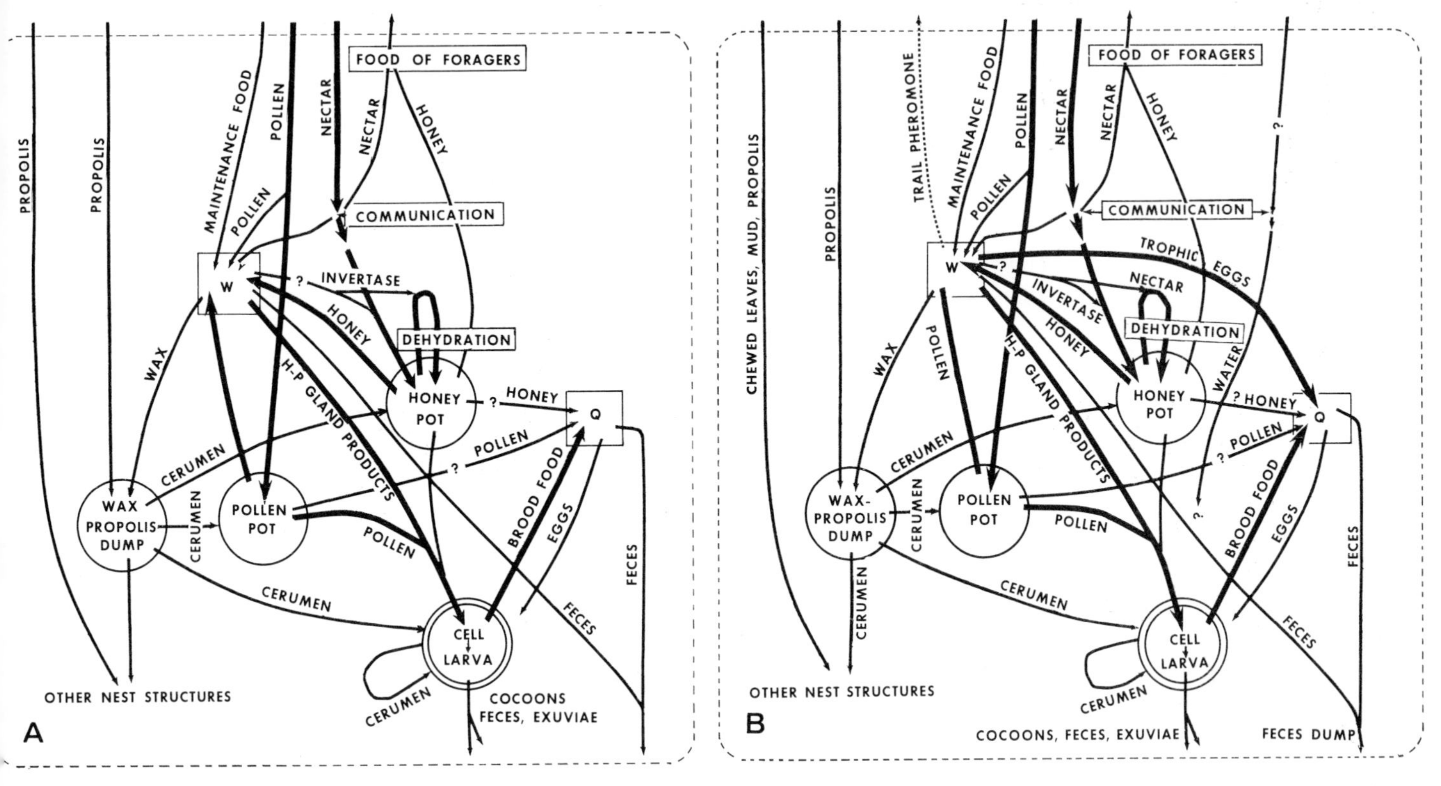

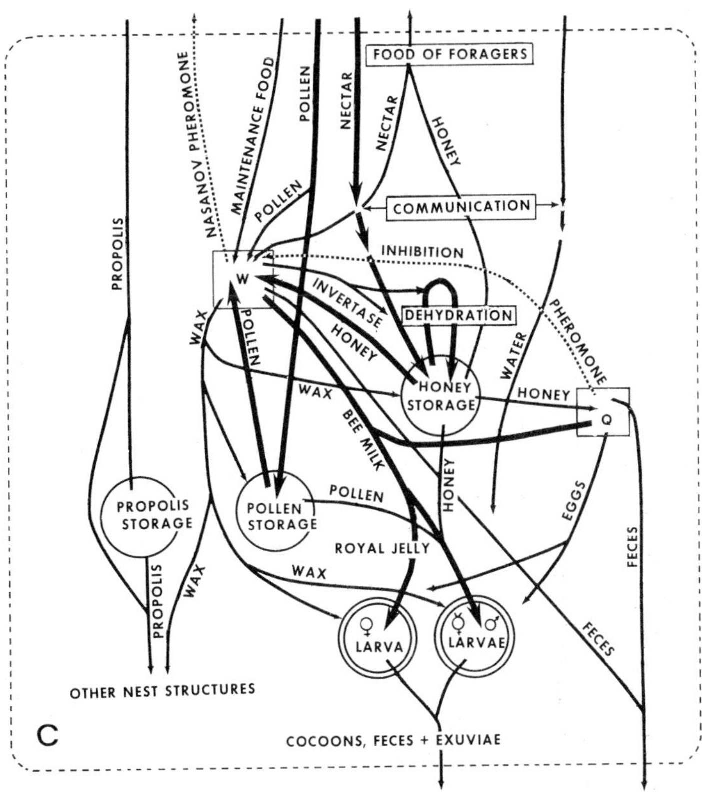

FIGURE 16.9. Diagrams showing the flow of major materials in colonies of highly social bees. Q = queen, W = worker. (*A*) A *Trigona* without trophic eggs. (*B*) A *Trigona* with trophic eggs. (*C*) *Apis. H-P gland* = hypopharyngeal gland. For further explanation see text. (Original drawings by Barry Siler.)

G. The flow of materials

Figures 16.7 to 16.9 are designed to suggest the flow of materials in colonies of certain bees, with comparable diagrams for solitary halictids and subsocial colonies (gyne nests) of allodapines and *Bombus* for comparison. The lines represent transfer of materials by bees from one place or individual to another. Some of the lines representing food transfer are thickened for ready comparison among the figures. "Maintenance food" means nectar and pollen brought in from outside the nest for the maintenance of certain adult bees. This line is included only because the bees concerned regularly or occasionally go out to flowers and return after feeding, but without scopal pollen loads and with indications of making such trips to satisfy hunger rather than for any social purpose. Frequently there may be no distinction between such food and that retained by the bee after ordinary pollen- and nectar-collecting trips (not necessarily separate trips) such as are indicated in Figure 16.7A. In a primitively eusocial halictid such as is represented by Figure 16.7B, the queen receives no food directly from ordinary collecting trips, since she makes no such trips, but gets it from cells and in some species from occasional "maintenance" trips.

Larval exuviae and feces are omitted in Figures 16.7A, 16.7B, and 16.8A to 8D, since they are not removed from the cells. In some halictids of the genus *Lasioglossum*, subgenus *Evylaeus*, they are taken from the cells but probably left in the nest.

As indicated in Figures 16.7C and 16.7D, allodapine larvae are free in the nest; cells are shown as double circles in all other forms. The pollen dump is the pollen mass sometimes found inside of allodapine nests.

In *Bombus* (Figure 16.8) honey storage first appears, and the flow of materials is further complicated by hypopharyngeal gland secretion, the use of pollen-wax for construction, and the re-use of pollen-wax in making cells. Once a *Bombus* colony reaches the eusocial stage in its development, pollen pots or pockets provide for pollen storage, and the diagrams are complicated not only by the presence of workers but by pollen from the pots contributing to the pollen-wax used for construction. Figure 16.8C is merely Figure 16.8B expanded by showing laying workers. Such expansion would also have been possible for halictid and allodapine colonies in which egg laying by workers occurs.

The line labeled "food of foragers" (Figures 16.8B to 16.9C) relates to honey taken from storage prior to foraging trips. To what extent such feeding occurs in *Bombus* is not obvious, but it is well known in highly eusocial bees. In the figures this line may appear to be associated with the label "dehydration." There is no relation; "dehydration" relates to the taking of honey from storage and dehydrating it, then returning it to storage.

The complexity of the diagrams for the highly eusocial bees (Figure 16.9) results largely from the increase in number of types of storage and from additional interactions involving workers. The increase in the latter comes from transfer of foods to other workers by returning foragers. This occurs principally with nectar, but pollen loads of returning foragers may also be nibbled by other bees. The diagrams show in only a feeble way the vast effect of food transfer among highly eusocial bees. This is because the transfer involves large numbers of individuals, pots, cells, and the like, whereas all of a given class are shown in the diagram as a single entry. Moreover, to maintain simplicity in the diagram, the line for nectar coming into the nest is broken only twice to show transfer of the food to other individuals, although the food dispersal is actually very wide. Nectar is transferred to so many individuals that this becomes the heart of a whole series of social interactions, as noted in section B above. Involved are such things as communication of food sources, development of distinctive colony odors, and bringing of water to cool the nest.

Accompanying the increasing complexity and intracolonial integration resulting from factors listed above is increasingly complex temporal division of labor. In primitively eusocial halictines the life of an adult worker is ordinarily divisible behaviorally into only two parts. In highly eusocial bees, however flexible and even reversible their behavioral ontogeny may be, the recognizably different parts of the adult life are more numerous (Chapter 12, B).

If we are ever to properly understand the workings of insect societies—why populations increase or decrease when they do, for example, and why colonies of different species attain different sizes—we must obtain quantitative data for at least some of the "flows" indicated in Figures 16.7 to 16.9. Such data are needed for colonies at various ontogenetic stages. As they stand, these diagrams have some heuristic value—they remind one of things—but they are largely indications of ignorance.

17 *The Control of Physical Conditions within Nests*

One of the principal attributes of social behavior is the ability of colonies of organisms to control the conditions under which they live better than related solitary organisms. Such social homeostasis involves virtually all of the functions of the colony, including food supply and defense against natural enemies, as well as control of the physical conditions within the nest, the topic of the present chapter. Emerson (1950) has argued that evolutionary progress generally consists of or can be described as improved homeostasis. It is, of course, easy to argue that social progress involves improved social homeostasis, for there is a direct relationship, on the average, between colonial size, complexity of colonial integration, and homeostatic control of conditions in the nest. At least one species, *Apis mellifera,* has reached the stage at which, while it is individually poikilothermic, it is socially nearly as homeothermic as are birds and mammals, and successful brood rearing is dependant upon this homeothermy.

By contrast, the allodapine bees including eggs, larvae, and pupae are peculiarly resistant to fluctuations in temperature and humidity: their nests in dry, dead, pithy stems, often in the sun, must be characterized by remarkable seasonal and diurnal fluctuations in these factors. However, some species of the same group, probably less tolerant, nest only in the shade (Michener, 1970b). Allodapine bees have evolved to greatly increase their tolerance to temperature and probably to humidity fluctuations. Over these factors their social homeostatic control is minimal.

A. Halictine bees

Solitary bees and those in small colonies have little potentiality for altering temperature and humidity. They must nest in suitable places and make cells that provide proper conditions for the young. In burrowing bees, the burrows go to a soil level which is characterized by suitable temperatures and humidities. In halictines this is often evidenced by the progressive deepening of the nests and lowering of the level of cell construction during the progressively hotter and drier weather of summer, especially during periods or years of especially dry or hot weather. For example, in *Lasioglossum imitatum* in Kansas the mean depth of the uppermost occupied cell drops from 7.5 cm in moist, cool, spring (April-May) soil to 41.5 cm in the hot, dry, August soil (Michener and Wille, 1961). Species such as *L. rhytidophorum,* living in constantly humid environments, such as certain tropical or subtropical areas (in this case, the southern Brazilian plateau), do not deepen the nests seasonally but construct cells at more or less the same level irrespective of the time of year (Michener and Lange, 1958b).

1. Nest burrows and chambers

There is no good evidence that parasocial or primitively eusocial halictine or allodapine bees modify the physical environment in the nest more than their solitary relatives are able to do. However, I have seen fanning at the nest entrance by *Augochlorella striata,* a ground-nesting, primitively eusocial halictid. Such fanning is suggestive of that of *Bombus* and the highly eusocial bees; an *Augochlorella* female, half out of the nest entrance, facing outward, fans her wings vigorously as though directing an air current down the hole. As these observations were made on particularly hot, wet days after rains, it was assumed that this activity had something to do with drying out the nest, although it is not clear how effectively an air current could be directed down the simple burrow into the nest, for there was no other obvious exit for the air. Similar behavior occurs after nests of *Augochlorella* are entered by the parasitic bee *Sphecodes* (Ordway, 1964); the fanning may be associated with clearing the nest of a *Sphecodes* odor. In any event, there is no obvious explanation for such fanning except control of some aspect of the nest environment.

A remarkable aspect of the nest architecture of various unrelated halictine bees, as well as of certain other solitary, semisocial, and primitively eusocial bees, is the construction (in the soil or occasionally in rotting wood) of clusters of cells, very close together, and surrounded by burrows or more commonly by an air space so that the clump of cells is supported only by pillars of substrate. The earthern walls between cells in some species are regularly less than 1 mm thick and may be as little as 0.25 mm thick. These structures are discussed in detail by Sakagami and Michener (1962); examples are shown in Figures 23.3 and 25.3, among others.

A noteworthy fact is that such relatively elaborate nest construction is not correlated with level of social attainment. As Sakagami and Michener (1962) pointed out, cell clusters surrounded by air spaces are found in solitary forms like *Augochlora pura* and *Halictus quadricinctus,* in semisocial forms like *Pseudoaugochloropsis,* in scarcely eusocial forms like *Augochlorella,* and in fully eusocial forms like *Lasioglossum duplex* and sometimes *L. malachurum.* The structure arose independently in most of these groups, as well as in various others, such as the Nomiinae. It seems that it must have an important function; it is not surprising that it is found in forms that exhibit other derived characters in various groups.

Such structures, made of earth, must be susceptible to destruction by excess water, since the thin earth walls, even though strengthened by the waxlike internal linings of the cells, quickly break down in water. Structures of this sort suggest a premium on the close association of young with one another or with adult bees in the nest, while at the same time maintaining each cell as a separate unit. The outside walls of the cluster are often almost as thin as the walls between cells, so that the shapes of the various cells can be easily seen as convexities on the outside of the cell cluster (Figure 23.3).

The function of such construction is not known. Relations to gas exchange, prevention of accumulation of excess moisture around the cells, and other functions have been suggested. However, Verhoeff's old suggestion (1897) that the space and thin walls permit brooding by the adult bees still seems interesting and possible. The importance of brooding to raise the temperature of *Bombus* cells is well known; other bees, even though smaller, might do the same thing. If this were so, one could expect cells in clusters surrounded by a chamber to be found more often in temperate than in tropical climates. This does not appear to be the case; such construction is found in tropical as well as cool climates and in moderately dry as well as very humid areas. But it must be remembered that temperatures in tropical soils are not necessarily higher than warm season temperatures in temperate soils, and may often be lower. Incubation could be important in both areas.

In support of the idea of temperature control as the *raison d'être* of the chamber and thin cell walls, the following two points seem worth noting: (1) in *Lasioglossum malachurum* it is only the spring nests, constructed at a season of low soil temperatures, that sometimes show the chamber (Chapter 25, B1); (2) in general it is the larger bees, whose temperature control should be most effective, that construct chambers. Thus it is the largest species of *Halictus* (*quadricinctus*) and of *Lasioglossum* subgenus *Evylaeus* (*malachurum, duplex, calceatum*) that are involved. There are, however, groups of bees which never make such nests regardless of the large size of some species (*Lasioglossum,* s. str.), and one group, the Augochlorini, which almost always makes such nests in spite of the small size of some species, such as those of *Augochlorella.*

2. Cells

The construction of halictine cells has to occur in substrate moist enough to be workable, a fact which has a relation to both humidity and temperature of the soil. However, the cell itself has the potentiality of controlling to some degree the humidity inside, as indicated by May (1970a, b, 1972). She showed that the smooth secreted cell lining of *Augochlora pura,* a solitary species, is made up of high molecular weight lipids with ester bonds. Although the cell closure is not sealed with similar material, humidity in the cell remains near saturation even when that in burrows or other spaces drops somewhat lower as the substrate dries out. High humidity in the cell is important for survival. For eggs, 84 percent survived at 95 percent relative humidity, 60 percent at 93 percent r.h., and 0 percent at 90 percent r.h. Larvae are more resistant to lower humidities, but nonetheless high humidities are important, for they absorb water from the atmosphere of the cell: thus the mature larvae before defecation can weigh 62 percent more than the original weight of the provisions plus egg. At any one time the saturated air volume of a cell could contain only about 1 percent of the amount of water absorbed by an *Augochlora* larva during its growth. It is therefore apparent that water must enter the cell during larval growth, and the lining may play a role in removing water from the substrate to maintain the inside air near saturation. No doubt the cell lining also excludes excess water when the substrate interstices are filled by rain.

B. Bumblebees

In both *Bombus* and the highly eusocial bees, control of the physical environment in the nests is well developed. Temperatures in *Bombus* nests vary less and are normally higher than those outside. Hasselrot (1960) has given an excellent historical account of studies of control of *Bombus* nest temperature, beginning with the investigations of Newport published in 1837. Hasselrot's own work was concerned with the temperature control through the season of *Bombus* activity, and utilized only daily maxima and minima. Figure 17.1 gives a sample of his results. Unfortunately, for control purposes he recorded outside air temperatures rather than temperatures in an unoccupied nest, but the difference probably would not have been great except for maxima on sunny days (the nest was in the sun), as indicated by the curves at the right of Figure 17.1, after the colony of bees had died out. It will be noted that before about June 28, while the colony contained less than five workers, minima varied considerably. From June 29 to July 25, with more numerous workers and an abundant honey supply, the temperature fluctuated but little, being in the 30–35° C range, in spite of environmental variation. Temperatures ordinarily vary least when worker populations are largest; such uniformity may be important in rearing the sexual brood. Thereafter, during the decline of the colony, when the sexual forms had used up the honey supply, temperatures, especially the minima, fluctuated greatly. Fye and Medler (1954) made nest temperature studies concerned with fluctuation during a single day. They are partly without appropriate controls, but the data for a small colony of *B. fervidus* in a wooden box have an appropriate control and clearly show the ability of 15 to 17 bees (queen and workers) to maintain a temperature on the brood comb (Figure 17.2) which, compared to that of comb in a similar but uninhabited box, is markedly elevated.

The first group of young, tended only by the lone gyne, is always arranged in a manner to facilitate warming by the gyne, there being a broad depression, known as an incubation groove, across the top of the cell or on top of the group of young; the body of the gyne nestles into this depression. Subsequent groups of young are not arranged in this way, the larger number of bees probably making a more general warming of the cells possible. The importance of the incubation groove is indicated by the early emergence of bees from the central, warmest part of the groove as compared to those in marginal positions. This fact has been noted for various species by several authors, most recently by Alford (1970).

The stimuli that lead the *Bombus* gyne to act "broody" should be investigated. A gyne that has been captured and confined may show no signs of broodiness for weeks. After she becomes broody, if she has as yet laid no eggs, she will sometimes brood strange objects in her nest, such as the edge of a tin can or places on the walls or floor of an artificial domicile. Ultimately, after laying, she broods the cell and the growing larvae.

Heat generation in *Bombus,* as described by Hasselrot (1960), is associated with bees which consume honey, exhibit lively respiratory movements of the abdomen, and press themselves down between the cells or groups of cocoons. Incubation activities are engaged in by gynes and workers of all ages and even by males during the short period that they are in the nest. At lower temperatures the bees vibrate their wing muscles and even run about vibrating their wings, thus presumably increasing their metabolic rates.

The temperature in the thoracic wing muscles of *Bombus* ranges from about 37° C to about 39°, whether the

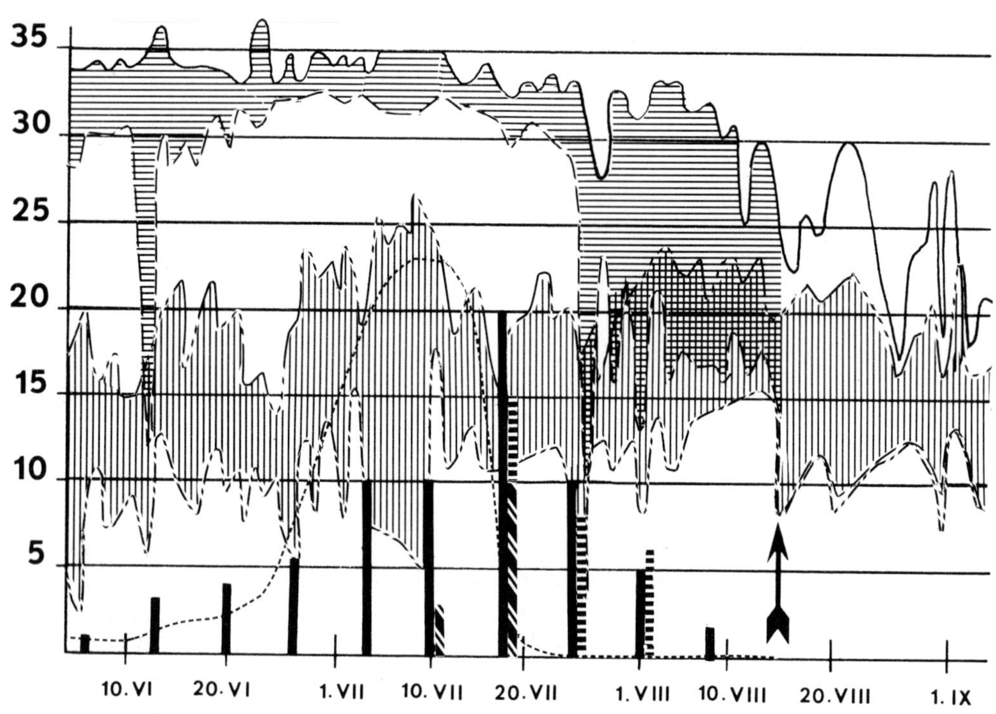

FIGURE 17.1. Seasonal changes in temperature in a nest of *Bombus terrestris* in Sweden. *Top, solid curve* represents daily maximum temperatures in the brood area; *broken line* represents daily minimum temperatures at the same place. *Horizontal shading* thus shows the temperature variation before August 15, when the colony died out. *Vertically shaded area* represents temperature fluctuation of the air outside the nest; after August 15 the minima correspond to the minima in the nest. *Dotted curve* represents the amount of honey stored. *Vertical bars* show the numbers of bees in the nest—*black,* workers; *obliquely lined,* males; *transversely lined,* young gynes. *Numbers* on the left show degrees C, units of the honey store (one filled cocoon of medium size = 1 unit), and tens of adult individuals. *Vertical arrow* indicates the time when the colony died out. (Modified from Hasselrot, 1960.)

bee is flying (even at low ambient temperatures, 2°–3° C) or warming its cells. In contrast, the thoracic temperature of a resting queen that is not "broody," or of any other inactive bumblebee, is near the ambient temperature (Heinrich, 1972a). The major contact of an incubating bee, at least a queen, is made by applying the undersurface of the abdomen to the brood cells. The leg bases prevent close contact of the thorax with brood cells. Therefore heat is transferred from its source in the thorax to the abdomen (Heinrich, 1927b). Such transfer does not occur in a dead bee, a fact which suggests that it is accomplished by blood flow. Heinrich, in contrast to some other observers, noted no wing movement associated with incubation.

Wojtowski (1963) reports average temperatures during the period of the nests' thermal stability for several species of *Bombus* as 32° C, or 33° for *B. terrestris.* In a colony of *B. terrestris* studied by Hasselrot, vibration of wing muscles began at night when the nest temperature was 32.2° and the outside temperature was 12.8°. Four hours later the nest temperature had dropped to 31.6°, the outside temperature to 12.0°; at this time four to six bees began running about and buzzing energetically. With unchanged outside temperature the comb temperature rose to 32.5°.

Excessive heat stimulates fanning, starting for *Bombus terrestris* and others at about 33° C. The bees stand on the tops of the cocoons or cells and fan their wings. At extremely high temperatures (36°–38.5°) Hasselrot says that the majority of the bees in the colony, including the

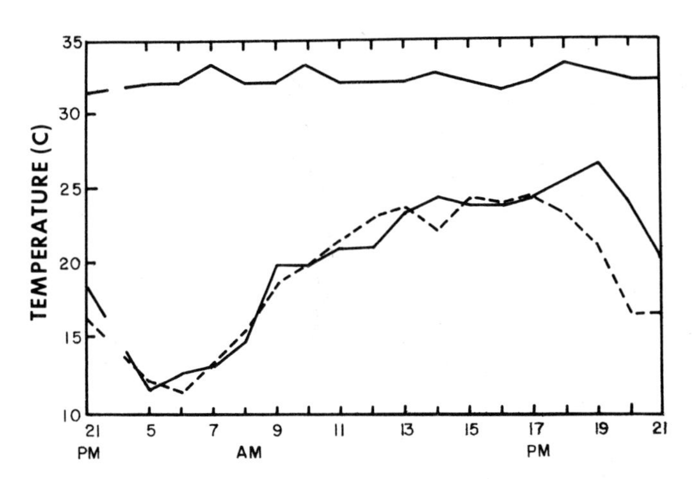

FIGURE 17.2. Graph showing temperature variations during one day in a box containing a nest and a colony of 15 to 17 *Bombus fervidus,* and in a box containing an abandoned nest but no bees. *Broken line* represents the air temperature; *upper solid line* represents the temperature at cocoons in the active nest; *lower solid line* represents the temperature at the remains of cocoons in the abandoned nest. (Modified from Fye and Medler, 1954.)

queen, were so located, often in long rows, and fanning vigorously. Sometimes they make holes in the roof of the nest in the middle of the day: thus heated air can escape. So far as known bumblebees, unlike honeybees, never carry water into the nest to increase the cooling effect of air currents.

Old literature on bumblebees is full of accounts of trumpeters, that is, bees that fan and thus make an audible sound. As early as 1700 Goedart described this activity and thought that the trumpeter was calling "his" companions to work. Plath (1934) reviews the history of such interpretations. A clearly demonstrated fact is that fanning is frequently heard for a time in early morning, sometimes also in the evening, and not necessarily in the hottest part of the day. It seems likely that there are cues other than temperature, possibly humidity or CO_2 concentration, that also lead to fanning.

C. *Apis dorsata*

Nests of the giant honeybee of southeast Asia are covered, essentially at all times, by a blanket of young bees not engaged in obvious activities (Morse and Laigo, 1969). Its thickness varies with the population, but on populous nests it is three or more bees thick; around one nest studied it was 3 to 7 cm thick. For a few days after departure of a swarm it was thinner and in some places incomplete. The bees cling to one another with their claws and are regularly oriented, heads up, backs out (that is, away from the comb), abdomens slanting slightly outward, and wings flat over their backs, as though to maximize the protection (Figures 17.3 and 30.5).

The blanket hangs from the nest support, free from the comb, leaving a crawl space for working bees between it and the comb. It is possible to lift the blanket away from the comb with a stick or paddle, but it quickly dissolves as the bees take flight. An area at the bottom of the comb, varying in size with weather and flight activity, is called the active area or mouth by Morse and Laigo. It is characterized by less regular orientation of the bees in the blanket, and it is here that most bees coming and going enter and leave through the blanket. The communicative dances take place on the outside of the blanket in this area.

The presence of a continuous thick blanket of bees surrounding the comb serves an important temperature control function, as well as keeping wind and rain off of the comb and brood. Morse and Laigo show that the temperature at the surface of the comb is maintained at about 31° C in spite of considerable fluctuation in temperature outside the blanket (Figure 17.4). Viswanathan (1950a) also notes the importance of a blanket of bees in temperature control of tropical species of *Apis.* As shown in Figure 30.7, *A. florea* combs are also covered with a blanket of workers.

The details of the temperature control mechanisms have not been studied, but the bees in the blanket are farther apart, permitting better ventilation, in warm weather than in cool. Moreover, *Apis dorsata* does sometimes collect water, and bees in the blanket have been seen fanning. Warming must be by metabolic activity.

The covering of the comb by a blanket seems to be an inefficient use of bees, but probably is a preadaptation that made it possible for *Apis mellifera* and *cerana* to invade dark cavities as nesting sites. In these situations, such a blanket not being needed, young bees can perform other functions. Morse and Laigo estimate that a minimum of 80 percent and in cool weather 90 to 95 percent of the bees of a colony of *A. dorsata* are in the blanket, and thus not active in foraging or brood care.

FIGURE 17.3. The blanket of bees covering a nest of *Apis dorsata,* the giant honeybee of Southeast Asia. (Photograph by R. A. Morse.)

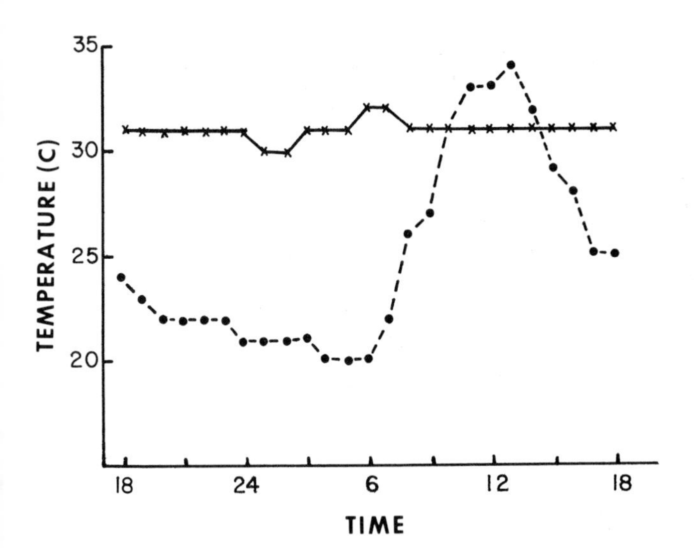

FIGURE 17.4. Comparison of the temperature at the surface of the comb of *Apis dorsata*, under the blanket of bees (*solid line*), with outside air temperature (*broken line*), during a 24-hour period. (From Morse and Laigo, 1969.)

D. *Apis mellifera*

In its ordinary nest sites the common honeybee's combs are not covered by blankets of bees like those of *Apis dorsata*. The behavior pattern exists in the species, however, although not often elicited. The rare exposed combs may be covered by similar protective blankets. Moreover, the bees that cover brood areas and the outer layer of bees around a swarm seem comparable to the blankets of *A. dorsata* and *florea*. For ordinary colonies of *A. mellifera* the walls of the nest cavity serve the protective function of the blanket.

Colonies of the common honeybee maintain the temperature of the brood near 35° C irrespective of the outside temperature (Ribbands, 1953; Steiner, 1947). Brood survives from about 32°–36° and the brood rearing area is always kept within this range, but usually between 34.5° and 35.5°, just below human body temperature (Lindauer, 1961).

1. Cooling

Cooling is accomplished by fanning, which brings in outside air not warmed by metabolitic activity and which also causes air circulation that increases the evaporation rate from moist surfaces—the bees themselves, their larvae, uncapped honey, etc. At times of extreme heat *Apis* brings water in quantities from outside the nest. This occurs when the outside air approaches the optimum brood temperature, say 34° C. The water is placed in drops within brood cells (Figure 17.5) or elsewhere in the nest. At the same time many workers in the brood area repeatedly expose films of water by proboscis movements, producing large water surfaces for evaporation. These movements are the same as those that are used, also for evaporation, in curing honey. The air currents produced by fanning pass over the water, causing evaporation from its surface and decreasing the temperature. In *Apis* the fanning is done by workers of all ages (Allen, 1965b), and the bees almost always orient themselves so that the air currents enter through the nest entrance and go out elsewhere, sometimes through another part of the entrance in a hive with a slit-shaped opening. The effectiveness of such cooling, especially if the air is dry, is amazing. Lindauer (1961) placed a hive of bees in full sun on a lava field in southern Italy where the outside temperature reached 70° C, but even then, as long as the bees had access to nearby water, the temperature in the hive remained at 35° and the combs did not melt.

Lindauer (1961) summarizes his studies and those of Kiechle (1961) on the interactions in *Apis* that lead foraging bees to bring water instead of nectar to a colony that is too hot. It should be noted that water must be brought in quickly in the event of a sudden temperature rise (as when the sun strikes a hive on a hot day) and that forag-

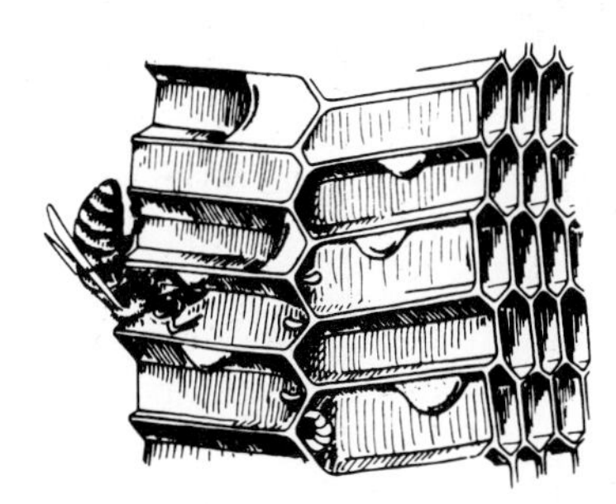

FIGURE 17.5. A sectional view of a bit of comb of *Apis mellifera*, showing drops of water (as well as eggs and a small larva in the bottoms of cells). When the hive is overheated, drops of water are placed in the brood cells by worker bees. Evaporation of the water, especially when the bees are fanning, cools the surrounding area. (From Lindauer, 1961, after O. W. Park.)

ing for water also must cease abruptly either when enough is at hand or the temperature falls, for, unlike nectar, water cannot be stored.

A bee that detects overheating does not simply fly out for water. The hottest part of the nest is usually the brood area, because of the density of adult bees and the metabolism of brood there. The adults there are mostly young, not experienced in foraging; experienced foragers are needed to find and bring back water, but they are usually in less hot parts of the nest or in the field.

2. Solution of the water problem

Lindauer (1961) shows that bees from the brood area are able to cause, indirectly, a change in behavior of foragers from food gathering to water gathering. It is, of course, young adults that accept much of the nectar from returning foragers. When the brood temperatures rise dangerously the nurse bees first deposit their own crop contents in and about the brood cells. This material contains about 60 percent water, and evaporation from it cools the nest. These bees then beg from other bees in the nest. Lindauer says that, from the center of overheating, solicitation, by stretching of the proboscis and tapping of antennae, spreads through the colony, and that the liquid reserves normally carried in the crops of bees are transferred to the heated area and evaporated. The receivers of nectar from foragers thus find themselves without much of their usual crop contents; the amount of water in their collective crops is reduced.

These bees now, instead of accepting sweet syrup with alacrity, accept most readily poorer food with more water. A forager with rich syrup is delayed in getting rid of her load. This behavior serves to increase the water content of the liquid being brought to the hive, and dancing comes to indicate the locations of more watery sources.

Ultimately either a pioneer among the foragers brings back water and finds it promptly accepted, or, more likely, a bee that is carrying water anyway has the same experience. (Some water carrying goes on most of the time. Water is used to dilute honey in preparation of brood food.) When delivery times are short (under 40 seconds), foragers even dance to indicate the location of water, and thus other bees are directed to and bring in water. This is a remarkable behavioral change for a creature that, during the rest of her life, seeks sugar solutions, the more concentrated the better, and communicates their locations only if they are sufficiently rich.

Foraging by bees is ordinarily facilitated by odors, and workers of *Apis* commonly mark rich but odorless food sources, like a dish of sugar syrup, with their own Nasanov gland odor. Free and Williams (1970) show that bees collecting water do the same thing, thus providing an odorless water supply with an attractant.

To turn off the water supply when the temperature drops, the receivers have only to delay accepting water or dilute syrup. When the delivery time of the material is short (20–60 seconds), collecting continues, and is commonly accompanied by dances recruiting more bees to the same source. But when the delivery time goes above 60 seconds, dances become rare and collecting decreases. With delivery times over three minutes, the forager constantly trying to get rid of her crop contents and being frequently refused by receiver bees, collecting practically ceases. Figure 17.6 shows the decrease in flights for water as delivery time increases, as well as the occurrence of dances only when delivery time is short.

3. Warming

Increasing the temperature is well known for races of *Apis mellifera* and *cerana* from temperate regions. In these bees the warming is a metabolic result of muscular activity of the bees (Lindauer, 1961). Roth (1965) says that the heat production is by microvibration of the thoracic flight muscles, and Esch (1964b and in Chauvin, 1968) has shown that the warming is associated with action potentials from the wing muscles even when no wing movement occurs. Roth found that heat production de-

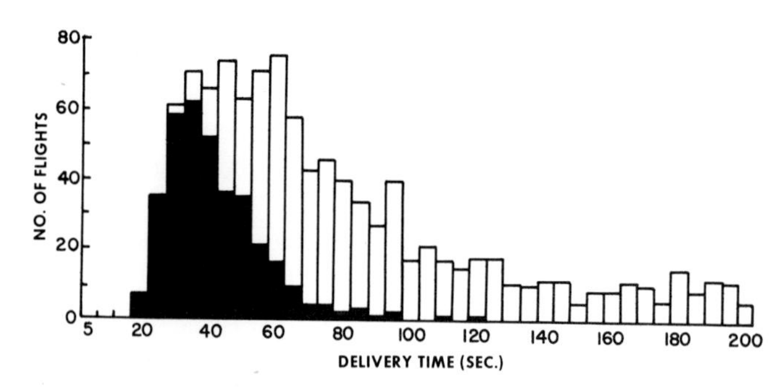

FIGURE 17.6. The decrease in the number of water-collecting flights of *Apis mellifera* as the time required to deliver water to receiving bees increases. *Black area* represents water-collecting flights that led to dances. Dances occur only when the delivery time is relatively short. (From Lindauer, 1961.)

creases the closer the ambient temperature is to the normal brood temperature. At 13° C one bee can produce 0.65 calories per minute for 20 minutes; bees in a group produce less, for example, a mean of 0.26 calories per minute per bee.

Apis not only maintains the brood at a relatively constant temperature near 35° C as indicated above, but also survives the winters by forming a cluster in each nest. A cluster consists of all the bees in a colony, forming a more or less spherical mass among the combs; in *A. mellifera* the cluster is subdivided by the combs into parts that have no direct contact with one another, but in *A. cerana* combs are partly removed in the cluster area so that the bees form a continuous mass (Sakagami, 1958b). As a colony of *A. mellifera* is cooled, clustering becomes evident at about 18° C, and a cluster is definitely formed at 13°, but small groups of bees may remain apart from it until about 0° is reached. During particularly cold periods the bees are tightly crowded in the cluster, while during warmer times they are more loosely placed (Figure 17.7). In the center of the cluster, temperatures are maintained that are much warmer than those out of doors, but they fluctuate considerably (Figure 17.8). The usual range is 20°–30°—impressively above the environmental temperatures on cold winter nights. A difference of 59° C (106° F) has been recorded between the environment and the center of a cluster on a cold night.

There are reports of constant rotation of individuals in the cluster, those on the outside being supposed to work their way to the center when they become chilled. Also, the cluster is said to move slowly among the combs, feeding on the stored honey and using this energy source as the basis for the muscular activity that produces the heat. Neither of these statements is accurate. There are glandular differences (described elsewhere) between outer and central bees of a cluster, showing that there cannot be a constant rotation of individuals. When the surface of the cluster reaches about 10° C, it seems likely that the outer layer of bees becomes more active and that the resultant movements stimulate the inner bees, so that more heat is produced (Himmer, see Ribbands, 1953). The result is that a drop in the outside temperature commonly causes a *rise* in the cluster temperature (Figure 17.8), as does any artificial disturbance of the cluster. If their bodies are cooled to about 8° bees chill, so that if surface bees reach this temperature they fall from the cluster and die. As to movement of the cluster, its center

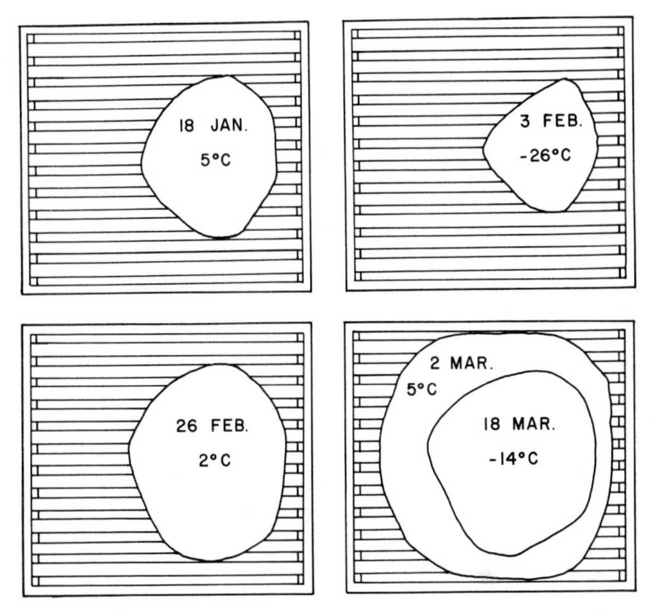

FIGURE 17.7. The relation between the size of the winter cluster of *Apis mellifera* in a hive, as seen from above, and the outside temperature. (From Ribbands, 1953, after H. F. Wilson and V. G. Milum.)

remains relatively fixed and nectar stored distant from that point must be obtained during warmer winter weather, when the cluster expands in size. A prolonged very cold period may exhaust food supplies available in the contracted cluster, with resultant death of the bees.

In late winter and early spring, brood rearing begins in many *Apis* colonies, and when this occurs the temperature of the brood area is, as at any other time of year, about 35° C.

E. Stingless bees

In the Meliponini cooling is also accomplished by fanning and resultant air currents, but the bees often orient themselves so that the air currents go out through the nest entrance. In *Melipona* and some *Trigona* the batumen plates (thick plates of mud or resin or a mixture of the two which close off the nest cavity from adjacent areas, for example, separating a relatively small nest cavity from the rest of a hollow trunk) contain multiple minute perforations, and fanning usually causes air to

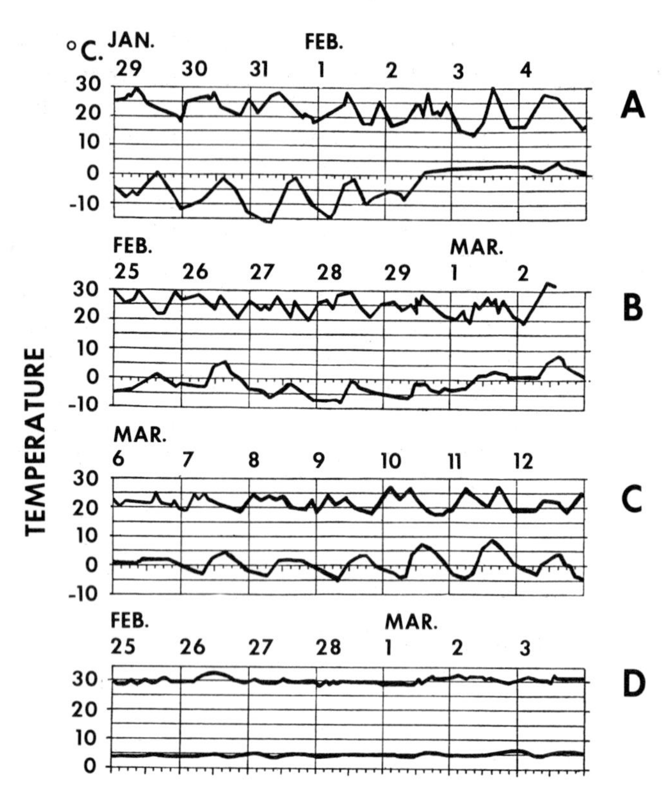

FIGURE 17.8. Parts of a record of winter air temperatures and cluster temperatures of *Apis mellifera.* (*A*) *to* (*C*) normal colonies in hives; (*D*) a colony in a hive in a cellar where the environmental temperature was nearly constant. For each record *upper curve* represents cluster temperature, *lower curve,* air temperature outside the hive. (From Ribbands, 1953, after A. Himmer.)

enter through these small holes (Nogueira-Neto, 1948). In *Dactylurina staudingeri,* an African meliponine that makes its nests in the open instead of in hollow logs or cavities in the ground, large numbers of temporary small openings, about a millimeter in diameter and lined with sticky resin, are made in the nest covering, supposedly to increase ventilation when the temperature rises (Darchen, 1966). Darchen says that the number of such openings varies with temperature, increasing during the day, and all being closed at night. His data show that for some nests the increase was slight or erratic, although for others additional ventilation seems reasonably convincing. Also, a large opening, similar to the entrance, is often made near the base of the nest and opposite the entrance. While the entrance proper is open each day, the second opening may be open, partially open, or closed, presumably depending on the temperature (or humidity?) inside.

As would be expected, Meliponini are also able to raise their nest temperatures, and in the state of São Paulo, approximately on the Tropic of Capricorn, temperatures in the brood chambers are almost always above the ambient temperature. To what extent this is simply the result of the metabolism of immature bees in a well-insulated space, and to what extent the temperature may be raised by activity specific to that end, is not known. Zucchi and Sakagami (1972) give temperature data for brood chambers of seven species of *Trigona* and *Melipona,* while Kerr and Laidlaw (1956) give such data for one of the species of *Melipona.* One nest of each species was used by Zucchi and Sakagami; all were in artificial hives except *T. spinipes,* the colony of which was in a nest moved to a site beside the laboratory. The insulation, colony sizes, and other factors therefore may not have been as in nature. The data indicate varying degrees of temperature elevation and control. In one 24-hour period during which temperatures were taken every half hour, ambient temperatures varied from 15.5° to 28° C. The brood temperature of the least well-regulated species, *T.* (*Trigonisca*) *mulleri,* varied almost as much, from 16.5° to 29.2°. This species normally lives in small colonies, often poorly insulated. In contrast, for the best-regulated species, the temperature was not only more uniform but higher, 34.1° to 36.0° C. This is *T.* (*Trigona*) *spinipes,* which lives in enormous, well-insulated colonies. The other species were intermediate both in mean temperature and in uniformity. The authors provide much more data, taken over a period of months, for *T. spinipes.* Because of seasonal temperature variation, the ambient varied from 8.2° to 30.3° C while the brood chamber varied only from 33.3° to 36.2° C. Such a species obviously has temperature uniformity approaching that of *Apis,* although some of the other species provide much less effective control.

F. Possible control of humidity and respiratory gases

Temperature and humidity control are closely interrelated, but there are relatively few data on the humidities inside of nests compared to those found outside.

Hasselrot (1960) and Wojtowski (1963) have shown that *Bombus* can maintain the relative humidity in the nest within certain limits (for instance, between 60 and 70 percent when it would otherwise have been saturated for long periods) by fanning and opening holes in the nest covering. In this case it is not clear whether the bees were responding to high humidities or to some other environmental factor (perhaps temperature) when opening holes and fanning. The relative humidity, of course, is closely related to temperature, as shown in Figure 17.9, and may be controlled merely as a byproduct of temperature control. But the control of humidity is presumably important in preventing fungal growth, even if it is merely a result of temperature control.

Apis nests are commonly not in places where humidity threatens to approach 100 percent if uncontrolled, as could easily happen in subterranean *Bombus* nests. Humidities from 20 to 80 percent have been recorded for *Apis* nests, but in Wohlgemuth's (1957) experimental hives, under normal circumstances the water vapor content of hive air was uniform and such that at the temperature of the brood area it contained 40 percent relative humidity. When the nest was being cooled by fanning about the same humidity was maintained by evaporation in the brood area, but it was lower elsewhere. When temperature was being raised in a cold environment, respiratory activity (breathing movements and the like) raised the humidity in the brood region to 50 to 70 percent, and water sometimes condensed in the outer parts of the hive. There seems to be no evidence of special humidity control (Simpson, in Chauvin, 1968; Stussi, 1967).

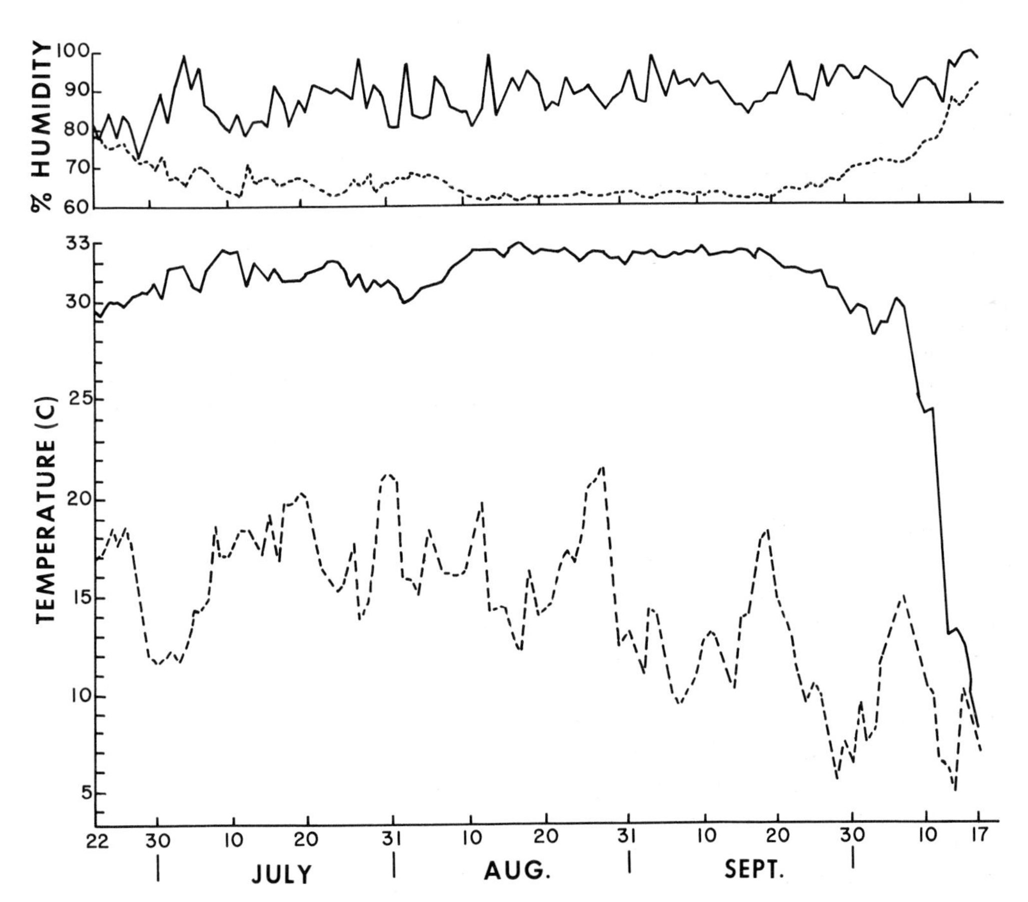

FIGURE 17.9. Temperature and relative humidity inside (*solid lines*) and outside (*broken lines*) a nest of *Bombus agrorum*. (From Wojtowski, 1963.)

Because evaporation from surfaces cools them and is related to the humidity at a given temperature, and because the drying power of air varies with the temperature at any given water content, it is likely that temperature and humidity controls are not readily compatible for bees. It seems that temperature is the factor of principal importance and is more or less accurately controlled in *Bombus,* meliponines, and *Apis,* and that humidity fluctuates as necessary in view of the temperatures and the atmospheric and substrate humidities. But experiments should be devised to test this view.

The respiratory gases in a large colony might influence bees; low oxygen tension or high carbon dioxide concentration could stimulate fanning. Stussi (1967) says that in *Apis* ventilation maintains oxygen concentrations at 15 percent or more, and that such a low figure does not reduce the metabolic rate of the bees.

G. Absconding

Of course there are occasions when no amount of activity practical for bees can control the conditions adequately at a given nest site. Because of inability to control physical conditions, or probably more often because of natural enemies or lack of food or water, colonies of highly eusocial bees may abscond. This means that the entire colony leaves its nesting site and goes to another. Thus it resembles swarming in that large numbers of bees are moving, but it does not involve division of a colony into two. In the genus *Apis* absconding is especially well known in *A. dorsata* and is common in all tropical species and races; it is rarest in European races of *A. mellifera.* In some areas of India *A. dorsata* migrates seasonally, so that at certain times of year whole colonies of thousands of these large *Apis* may be seen flying to new sites, having abandoned their combs at the old location. Migration to new sites, leaving locations with inadequate food or especially inadequate water, occurs also in *A. mellifera* in subsaharan Africa. Such behavior probably explains in part the rapid spread of the African race *A. mellifera adansonii* after it was introduced into Brazil in 1957.

In *Trigona* absconding is rare and resembles the swarming of that genus in that it is a gradual process, workers carrying building materials to a new and necessarily nearby site. When the bees finally leave, the queen is left behind to die alone (Portugal-Araújo, 1963), because she is unable to fly.

Some solitary bees also leave an unsuitable area to nest miles away (*Megachile brevis,* Michener, 1953a), although most are attached to their nesting sites and one has at least the impression that they do not readily leave them once nesting has begun.

Absconding is unknown in parasocial and primitively eusocial bees. So far as known, if conditions are bad, the colony dies out *in situ.* Presumably this is because among the bees in such colonies there are no mechanisms to integrate flight activities; there is nothing to cause simultaneous departure or to keep the colony together while moving to a new location.

18 *Defense*

This chapter concerns the abilities of bees to protect themselves from other organisms, whether parasites or predators, diseases, casual intruders, or other bees (even of the same species) that may rob or take over the nest. Material on parasitic and robber species of bees will also be found in Chapter 19.

Almost any bee will defend itself by stinging if it is captured, for example between the fingers, or if it is a stingless species, by biting with its mandibles or by releasing odors of various sorts, depending upon the kind of bee. Male bees do not possess stings and therefore are dependent on the last two modes of defense. These responses of individual bees occur not only at the nest, but also elsewhere, for example on flowers if the bees are molested there. They are the defense reactions of individuals and there is no social aspect to them (except that in some of the highly eusocial bees, as described below, the sting or defensive responses of one bee makes that of others more probable because of the liberation of pheromones promoting aggressiveness). From the standpoint of this chapter, on the other hand, defense is a social phenomenon and involves defense of colonies, or at any rate of nests.

A. Defense against unrelated animals

This section concerns defense of nests against other species, from vertebrates like bears or man to parasitic insects. A small nest entrance at which a defensive stand can be made is a usual feature of bee nests. Often the entrances are inconspicuous or seemingly hidden. Defense behavior and construction is otherwise so variable that there is little more to be said about it that is applicable to all or most social groups of bees.

1. Burrowing bees and allodapines

There are various accounts of solitary bees that live in aggregations, especially the European *Anthophora parietina,* attacking people who approach the aggregation or who get into the line of flight of the bees. Buttel-Reepen (1915) presents several such reports. Thorp (1969) describes how in California *A. edwardsii* buzzes around an observer in great numbers at a nest aggregation, perhaps discouraging trampling of nests, but does not sting. Nonetheless, solitary bees rarely defend nests against large enemies like vertebrates in any obvious or effective way, although exceptions are known (for example, *Lasioglossum leucozonium,* Knerer and Atwood, 1967). Parasites and predators are often ejected from the nests of solitary bees, but only on an irregular basis because such a bee must leave her nest unguarded during foraging trips. Although she may close a nest with soil from within in bad weather or when disturbed, she does not, in fact, give much attention to guarding the entrance even when she is in the nest. Nevertheless, parasites and other habitual intruders into solitary bee nests usually have unusually strong exoskeletons, protective spines or ridges over delicate parts of the body, and long stings, showing that defensive responses of the hosts are frequent and effective enough to lead to evolutionary responses by the intruders.

The elaborateness of nest entrances and of nest guarding behavior gives some insight into the importance of nest defense for parasocial and primitively eusocial bees. In the halictine bees the burrows, generally in the ground, often show peculiarities specific to the particular kind of bee. Some species place the entrance under a leaf, stone, or clump of vegetation. In others the tumulus of loose earth pushed up from beneath as the bees make tunnels and cells often closes the entrance, so that returning bees must work their way through this material to get to the burrow. This sort of loose closure is not consistently used by any halictid known to me, although such nest protection is of regular occurrence in some andrenid bees (Michener and Rettenmeyer, 1956). Sometimes the central core of the tumulus, surrounding and above the burrow mouth, is solidified by tamping and by a secretion so that, when the rest of the tumulus blows away or is washed away by rain, a turret remains over the original nest entrance. In other species a turret is built up from the ground surface, above the entrance. Such turrets can be constructed only when the soil is moist so that particles stick together, but when dry a turret is rather firm and can be picked up in the fingers as a unit. Turrets probably tend to deflect wandering mutillids and other wingless enemies or casual intruders from the nest entrances, although from above they seem quite conspicuous, with their shadows and central openings. Turrets must also deflect limited quantities of water flowing on the soil surface.

Nearly all halictine and allodapine nests are constricted at the entrances (Figure 18.1), generally to about the diameter of the heads of the inhabitants. Sakagami and Michener (1962) cite many examples among halictines. Figure 18.2 gives some idea of the method of constructing a constricted entrance and of the attention expended on it by many halictines. The similar constrictions of allodapine nest entrances are made of bits of pith, fiber, or frass excavated from the stems or wood in which the nests are made. Those that excavate their own nests in pith initially make the burrow with a narrow entrance which is only shaped and perfected by construction. Those that use pre-existing burrows made by beetles or other insects have to make the entire collarlike constriction.

In halictine bees, whether belonging to species that are typically communal, semisocial, or primitively eusocial, burrows of solitary individuals are generally unguarded even though the entrances may be constricted, while burrows of the same species inhabited by two or more females are guarded much of the time, usually more effectively the larger the colony. Such a statement is usually quite valid for any one species, but among species the relation between guarding and colony size occasionally fails, as shown especially by *Lasioglossum marginatum,* which makes the largest colonies of any halictid known but leaves the nest entrances unguarded. In some species guarding activity varies seasonally; thus the queen

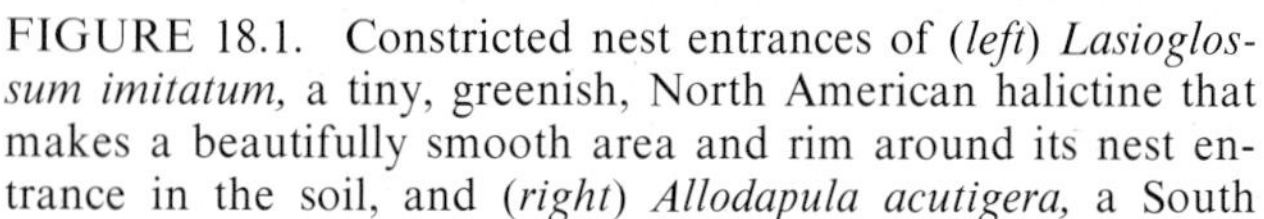

FIGURE 18.1. Constricted nest entrances of (*left*) *Lasioglossum imitatum,* a tiny, greenish, North American halictine that makes a beautifully smooth area and rim around its nest entrance in the soil, and (*right*) *Allodapula acutigera,* a South African allodapine bee, which in this case nested in a stem blackened by a brush fire, so that the collar of fresh, pale pith fragments that narrows the entrance is conspicuous.

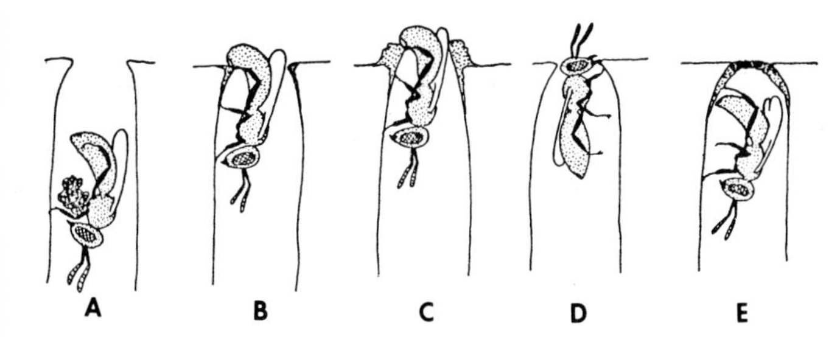

FIGURE 18.2. Sketches showing the method of construction or repair of the narrowed nest entrance of *Lasioglossum zephyrum*. (*A*) Guard bee carrying soil up to repair the entrance. (*B*) Tamping soil into place with the apex of the abdomen. (*C*) A continuation of the same processes has produced a small turret. (*D*) Nibbling and shaping the entrance. (*E*) A solitary gyne of the same species closing the nest with soil tamped into place below the entrance rim. (From Batra, 1964.)

may guard the nest during the spring semisocial phase, but during the subsequent eusocial phase guarding may be irregular or absent.

The guarding activity of halictids, as seen in most of the social species, involves a bee spending considerable time with her head plugging the nest entrance. Such a guard is not always the same individual, although on some occasions one bee may occupy this position for hours at a time. More frequently, the guard is replaced every few minutes, although it is common for one bee to hold the position most of the time in spite of frequent changes. Often there is one or more bees in the entrance burrow just below the guard; when the latter flies out or turns to go back into the nest, the next bee in line takes over as guard. It is often workers that act as guards. In some species, however, the guard is the queen of the colony; sometimes a particular "worker" with queenlike attributes shares the guarding of the nest with the queen (an example is *L. calceatum,* Bonelli, 1965b; see Chapter 12, A).

A shadow or other disturbance may cause a halictine guard to withdraw, and on hot days, when the soil surface reaches temperatures far above 40° C, the guards disappear out of sight down the holes. Nevertheless, disturbance with a small object such as a blade of grass or a bristle will usually cause the bee to bite at the offending object or, on further stimulation, to turn and block the nest entrance with the posterior dorsal part of the abdomen. Mutillid wasps and other parasites cause similar responses, and in this position, with the nest entrance blocked with the abdomen, halictid bees sometimes successfully keep mutillid parasites and doubtless other enemies out of the nests. Some much-studied species, however, such as *Lasioglossum duplex,* seem to lack the abdominal burrow-blocking response completely, although a close relative, *L. calceatum,* shows it readily.

The same defense behavior exists in the allodapine bees, although the nests are usually burrows in broken stems. In the allodapines the surfaces of the last three abdominal terga are somewhat flattened, and their pubescence is usually of a specialized sort. The flattening makes this part of the abdomen an effective plug for the nest entrance. These bees also produce from the mouth region (mandibular glands?) an unpleasant brown secretion. It has the odor of tenebrionid or carabid beetles. The bees emit this material either at the nest or, if captured or greatly disturbed, away from the nest. On one occasion in New South Wales I found females of *Exoneura bicolor* so commonly visiting flowers of certain bushes that when the flowers were beaten with a net, the odor could be perceived several feet away. The effectiveness of adult females of *Exoneura* in defending their nests is shown by a series of nests of *E. variabilis* which were moved to a new location for observation near Brisbane, Australia. Within a day those containing no adults were robbed of all immature stages by ants, while those containing adults as well as immature stages mostly survived for many days.

Repeated disturbance of a nest entrance often causes halictine bees to close the entrance with soil from inside, in the manner shown in Figure 18.2e. Rain can have the same effect. In some species the nests are closed in the heat of the day. A modification of this behavior reported by Knerer (1969a) is closing of nests from inside by a single stone pushed up into the entrance. Whether an entrance is closed by a stone, soil, or the head of the bee itself, the closure makes nests in soil less visible to people and probably to flying enemies like bombyliid flies. In addition, of course, closure presumably makes discovery by casual intruders and surface parasites like mutillids less likely. That such enemies are commonly diurnal is suggested by the fact that many halictines leave their burrows open and unguarded at night. However, *Lasioglossum duplex* closes its nests, usually between 1:00 and 2:00 P.M., and keeps them closed until the following morning (Sakagami and Hayashida, 1968), and

L. umbripenne does likewise except that closure often occurs at about 11:00 A.M. (Wille and Orozco, 1970).

Improved protection against mutillids and the like, enemies capable of digging around a guard bee, is probably provided by the tamped, hardened, and smoothed area around the nest entrance in *Lasioglossum imitatum,* and by the tamped burrow wall, hardened by a salivary secretion, in *L. malachurum.*

In *Lasioglossum zephyrum* more than one female from a nest may attack a mutillid, even if it is several centimeters from the entrance (Lin, 1964; Batra, 1965). Since forceps and needles that have been in contact with the mutillids are also attacked, it seems that the principal cue for such an attack must be chemical. Such behavior has not been seen in other halictids, where individuals, but not groups, attack natural enemies at the nest entrances but not away from them.

2. Bumblebees

Defense of nests against a large intruder by *Bombus* and *Apis* takes a well-known form, bees swarming out to sting. But in *Bombus* each bee finds its own way to the enemy; there is no concerted attack such as sometimes occurs with highly eusocial species. Chemical stimuli promoting aggressiveness, that is, alarm pheromones, are not known in parasocial or primitively eusocial bees, even *Bombus* (Maschwitz, 1964a, b), although brood and stored food of *Bombus* are sufficient to attract bears and skunks. The species of *Bombus* vary in their reactions. Many times nests can be opened without even a single bee flying or stinging.

Often when a *Bombus* nest is opened some of the bees take a distinctive defensive position instead of flying out at the enemy. They roll over nearly onto their backs, with one or two legs up, mandibles open, and stings partly out, ready to attack anything that touches them (Figure 28.7, QR). Such behavior is especially common in the species known as carder bees, whose nests of loose material on the surface of the ground are probably particularly subject to being opened by vertebrates. Small intruders that have entered the nest are stung and carried outside, or sometimes left inside and covered with wax or propolis, so that they become part of the wall or floor of the nest. Of course workers are sometimes killed in dealing with intruders, even small ones like other bees.

Bombus fervidus, according to Plath (1934), has an unusual way of dealing with dangerous small intruders, such as other bumblebees or *Psithyrus.* Each defender cautiously approaches the enemy and regurgitates honey onto it. Soon it is so wet and sticky that it retreats. Similar defense is known only in certain stingless bees.

One or more guards may often be seen inside the entrance of a *Bombus* nest. Such individuals often inspect returning bees and are the first to fly out and sting in the event of a disturbance. As with the halictine bees, defense of large colonies is more effective than that of small ones, not only because of the greater numbers of individuals but because of their aggressiveness when the colonies are large. In a small bumblebee nest there are usually no guards, although certain workers may be more aggressive than the rest (see Free, 1958). Strongly aggressive workers or guards in larger nests have ovaries more developed than those of less aggressive individuals (Chapter 12, A). Since guarding by workers has physiological correlates, it is not surprising that the same bees serve as guards day after day.

3. Stingless honeybees

In the Meliponini, or tropical stingless honeybees, there are no functional stings. The very notion of bees without stings seems anomalous, but in fact stings are lacking in all male bees, and in females of many species they are reduced and do not function in defense. This is not so surprising for solitary forms which do not store food or make nests containing many edible young. However, the Meliponini occur in large colonies, store abundant pollen and honey, and are highly attractive to many natural enemies, including man, for the honey of some species is of excellent quality. In spite of lacking stings, they are by no means defenseless. Some species swarm out of the nest to attack a large intruder more aggressively than do workers of *Apis.* However, instead of stinging, they bite, and can be very effective because they crawl into the eyes, nose, ears, or hair; their mere presence is a terrible nuisance and their ability to nip with the jaws makes almost any enemy retreat.

Species of one subgenus, *Trigona* (*Oxytrigona*), place the caustic secretion of the enlarged mandibular glands (Figure 18.3) (Kerr and Costa Cruz, 1961) on the skin of the intruder and then bite it into the flesh with the mandibles. The result is very painful and causes lesions (Figure 18.4) that last for many days and may cause almost permanent scars. (A secretion of the same glands

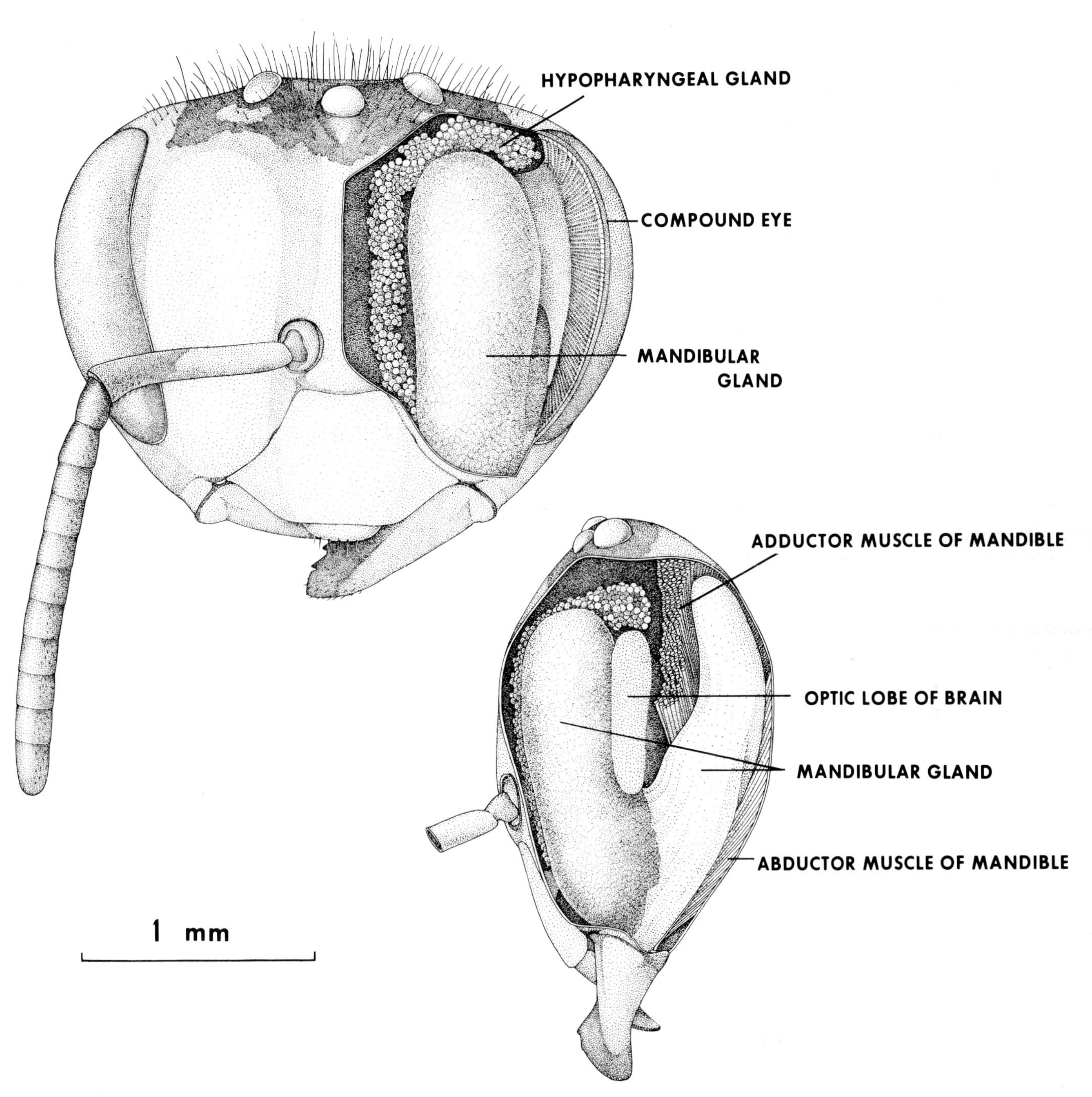

FIGURE 18.3. Dissections of the head of a worker of the fire bee, *Trigona* (*Oxytrigona*) *tataira,* so called because of the intensely burning secretion from the enlarged mandibular glands. (Original drawings by J. M. F. de Camargo.)

FIGURE 18.4. Lesions on the arm caused by bites of the fire bee, *Trigona* (*Oxytrigona*) *tataira*. The victim is Sr. Egidio (Lilo) Diaz, who has discovered and helped to open many a bee nest in Costa Rica. (Photograph by Alvaro Wille.)

is used by this species as well as by others for communicative odor spots leading to food sources, Chapter 15, C.)

Although *Oxytrigona* nests in hollow trunks, the most aggressive species generally are those with exposed nests—the more exposed the more aggressive. Thus in Africa the only species with a largely exposed nest (*Dactylurina staudingeri*) is the most aggressive African meliponine (Darchen, 1966), while in America species like *T.* (*Partamona*) *cupira,* with nests often partially exposed, are very unpleasant but less so than are those *Trigona* s. str. with fully exposed nests, such as *T. spinipes* and *corvina.*

The many species of meliponines which do not seem aggressive to us nonetheless seem to have effective defenses. I have watched a colony of *Trigona* (*Trigona*) *fulviventris,* a gentle species that does not attack human intruders, when a raid of army ants (*Eciton*) covered the ground and tree roots around the nest entrance. The ants would have swarmed into any other hole, but the *Trigona* nest was undisturbed. The ants went to the very rim but did not enter, presumably for some chemical reason; there was no contact between the bees and the ants.

Other forms, perhaps less effectively protected, nest in or beside nests of ants or termites, from which they presumably gain protection. Thus *Trigona* (*Scaura*) *latitarsis* always inhabits active nests of nasute termites; *T.* (*Paratrigona*) *peltata* inhabits nests of the neotropical spinning ant *Camponotus senex,* made in clumps of leaves (Wille and Michener, in press); and *T.* (*Trigona*) *compressa* inhibits nests of the ant *Crematogaster stolli,* in hollow tree trunks (Kempff, 1962). Kerr, Sakagami, Zucchi, Portugal-Araújo, and Camargo (1967) have reviewed the literature on such associations and illustrate a nest of *T.* (*Trigona*) *cilipes* in an *Azteca* ant nest (Figure 18.5); the same *Trigona* also inhabits termite nests.

The diverse aspects of the nest entrances of meliponine bees, illustrated in Figures 18.5, 18.6 and various figures in Chapter 29, serve for ready field identification of many species. For the bees, they may facilitate return to the nest. But it seems probable that they also have defense functions in making it less easy for small animals to find the entrances and get in.

Kerr and Lello (1962) list no less than 13 different means that meliponines use for defending their colonies, some of them overlapping the methods described above. Their list is as follows (wording of these authors slightly modified):

(1) Plastering of resins or gums, sometimes mixed with wax, upon the body of an invader. This system is used by *Trigona,* subgenera *Trigona, Plebeia, Tetragona,* and others.

(2) Robust mandibles sometimes reinforced with teeth able to cut an invader apart are characteristic of the genus *Melipona* and of *Trigona* subgenus *Trigona.*

(3) Unpleasant taste and smell. This defense is considered the most important by some authors. It is noted in *Trigona* of the subgenera *Trigona, Scaptotrigona, Geotrigona* (here regarded as a group of the subgenus *Tetragona*), and in some species of the genus *Melipona.*

(4) Massive attack to repel a man or large animal, penetrating nostrils and ears, biting at eyes and mouth, and the like. Such a method is used by *Scaptotrigona, Oxytrigona, Trigona, Partamona,* and several other subgenera of the genus *Trigona.*

(5) Large number of workers. Huge colonies are found

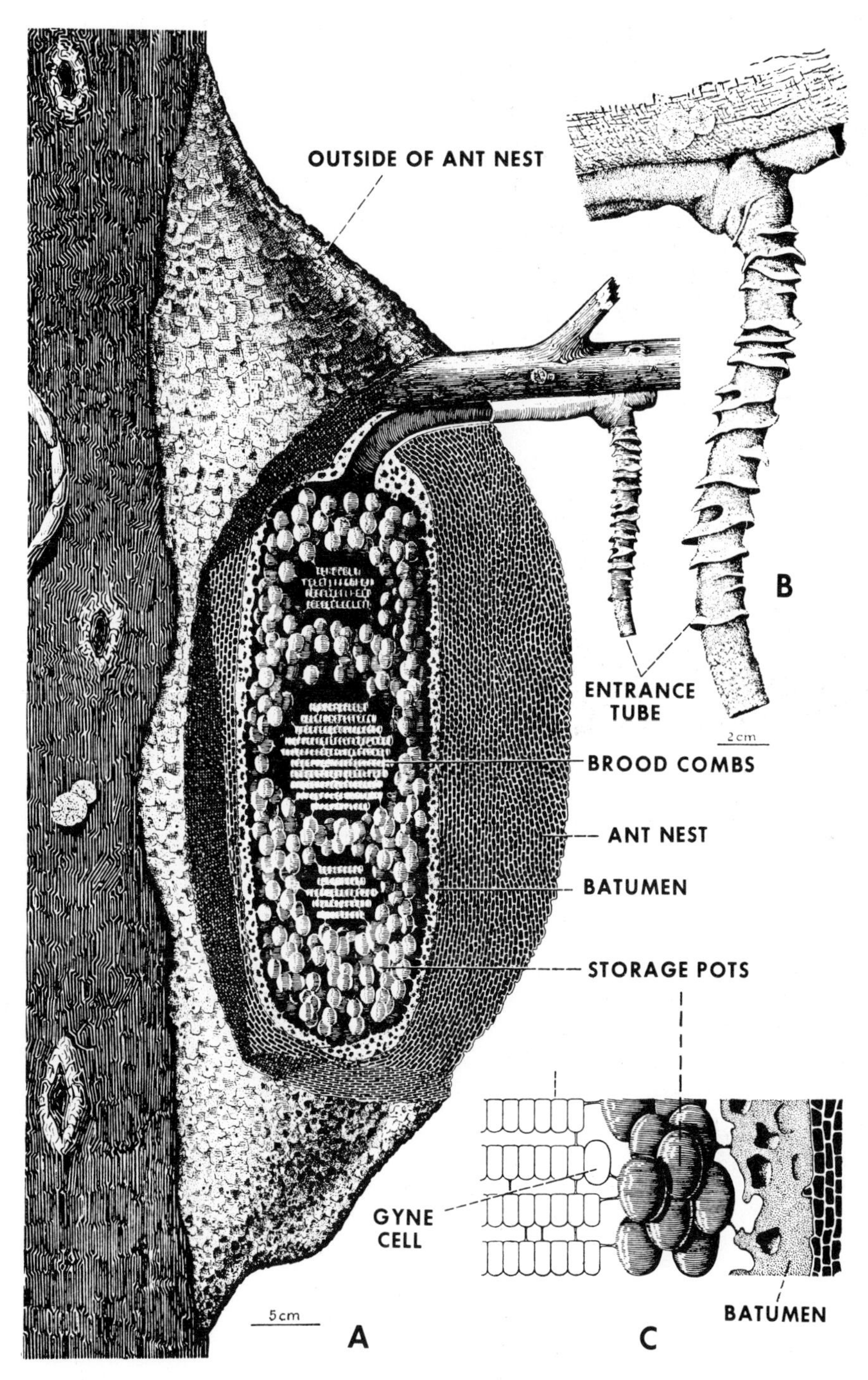

FIGURE 18.5. Nest of *Trigona* (*Trigona*) *cilipes* in the nest of an ant of the genus *Azteca*. (*A*) Sectional view; (*B*) the entrance tube; (*C*) detail of structure, showing a gyne cell. (From Kerr et al., 1967.)

FIGURE 18.6. Nest entrances of three species of Costa Rican stingless bees. *Above: Trigona jaty.* Note perforated walls of the entrance tube and the long abdomen of the bee in flight. (Photograph by C. W. Rettenmeyer.) *Below: Trigona mellaria* (*left*) and *T. testaceicornis perilampoides* (*right*). Both nests were in logs. Note the ring of guarding bees at the entrance of the latter. (Photographs by C. D. Michener.)

in many species such as *Trigona corvina, jaty, pectoralis, hyalinata,* and *cupira.* Large populations give these species special advantages in using defense methods 1 to 4.

(6) Constricted nest entrance to allow only one bee at a time to go in or out. Such an arrangement improves the defense system considerably and is used by almost all species of *Melipona* and some species of the genus *Trigona.*

(7) Closure of the entrance during the night (and in bad weather) with wax or cerumen. Such a system is used by many weak colonies and by all colonies of many of the smaller species.

(8) Blockage of the entrance with sticky wax or resins. According to Fiebrig (in Maidl, 1934) workers of *Lestrimelitta limao* place in the elaborated entrance tube small pieces or blocks of wax and resin to keep ants from getting in. When an attack is over the bees remove them. Maidl (1934) says that workers of *Trigona* (*Tetragona*) *canifrons* build a constantly renewed ring of sticky resin around the entrance. Some species of the subgenera *Plebeia* and *Paratrigona* have the same behavior.

(9) Playing dead. Some species of stingless bees such as *Trigona silvestrii* play dead when touched by a large enemy.

(10) Camouflage of the nest or nest entrance. This is done by several species, such as *Melipona quadrifasciata, Trigona silvestrii,* and *T. cupira.*

(11) Mimicry. Kerr (1951) cites nine cases of mimicry where the stingless bee resembles a bee or a wasp possessing a strong sting.

(12) Use of a caustic secretion. *Trigona* (*Oxytrigona*) *tataira* uses a mandibular gland product to inflict on the enemy, especially mammals, a terrible "burning" (see above and Kerr and Costa Cruz, 1961).

(13) Use of honey on invaders. *Trigona braunsi* defends itself against a robber bee, *Lestrimelitta cubiceps,* by pouring honey on the invaders (Chapter 19, C). (*Bombus fervidus* uses a similar method of defense.)

Kerr and Lello postulate that the effectiveness of methods of defense such as are described above made stings unnecessary and indeed a liability, because use of them so ofen results in death of the bee in the genus *Apis.* This need not be so, however, for there are other mechanisms for reducing mortality due to loss of the sting in *Apis;* in fact in *A. cerana* the sting is usually worked free and the bee flies away intact (Sakagami and Akahira, 1960). Most bees other than species of *Apis* have unbarbed stings, and bumblebees can sting two or three times in quick succession before they seemingly exhaust the venom supply (Free and Butler, 1959). It is more likely that the use of space and nutrients for sting development was selected against in view of other effective means of defense and the need for space in the abdomen for nectar transport and the production of trophic and male-producing eggs.

Some species of *Trigona,* especially those that use defense method number 4 above, behave as though influenced by alarm pheromones. As with *Apis* (see below), disturbance of a few workers is sufficient to cause hundreds or thousands to fly out of the nests and attack a person. The substances involved have been investigated for two species of the subgenus *Scaptotrigona, T. postica* and *T. tubiba* (Blum, 1970). In both species the trail-marking materials from the mandibular glands consist of a mixture including various methyl ketones. At least two of these, 2-heptanone and 2-nonanone, are attractants at moderate concentrations, as along an odor trail (Chapter 15, C), but at high concentrations are the releasers of attack behavior. (Some other constituents of the trail-marking material are attractants but do not release attack behavior at high concentrations.) Presumably the high concentrations needed to cause attacks are easily attained in a nest, while outside, where odor trails are used, such levels are not attained. This is another example of the multiple use, or parsimony in the use, of pheromones. In the *Lestrimellita limao* the alarm pheromone includes citral. Its other functions are enumerated in Chapter 19 (C).

The funnel-shaped nest entrances of *Trigona* of the subgenus *Partamona* lead into a cavity full of anastomosing rods (Figure 18.7). An intruder may well get lost in this maze, but a more likely function is as a resting place for hundreds of aggressive bees that stream out at the slightest disturbance, presumably stimulated by alarm pheromones from the first bee disturbed.

4. *Apis*

In honeybees certain workers, at about the age when they are becoming foragers, spend much time at the nest entrance serving as guards (Chapter 12, B2). These are the bees that are most active in inspecting incoming foragers and that are most prompt to attack and sting in the event of a disturbance.

Of course the most aggressive individuals often expose

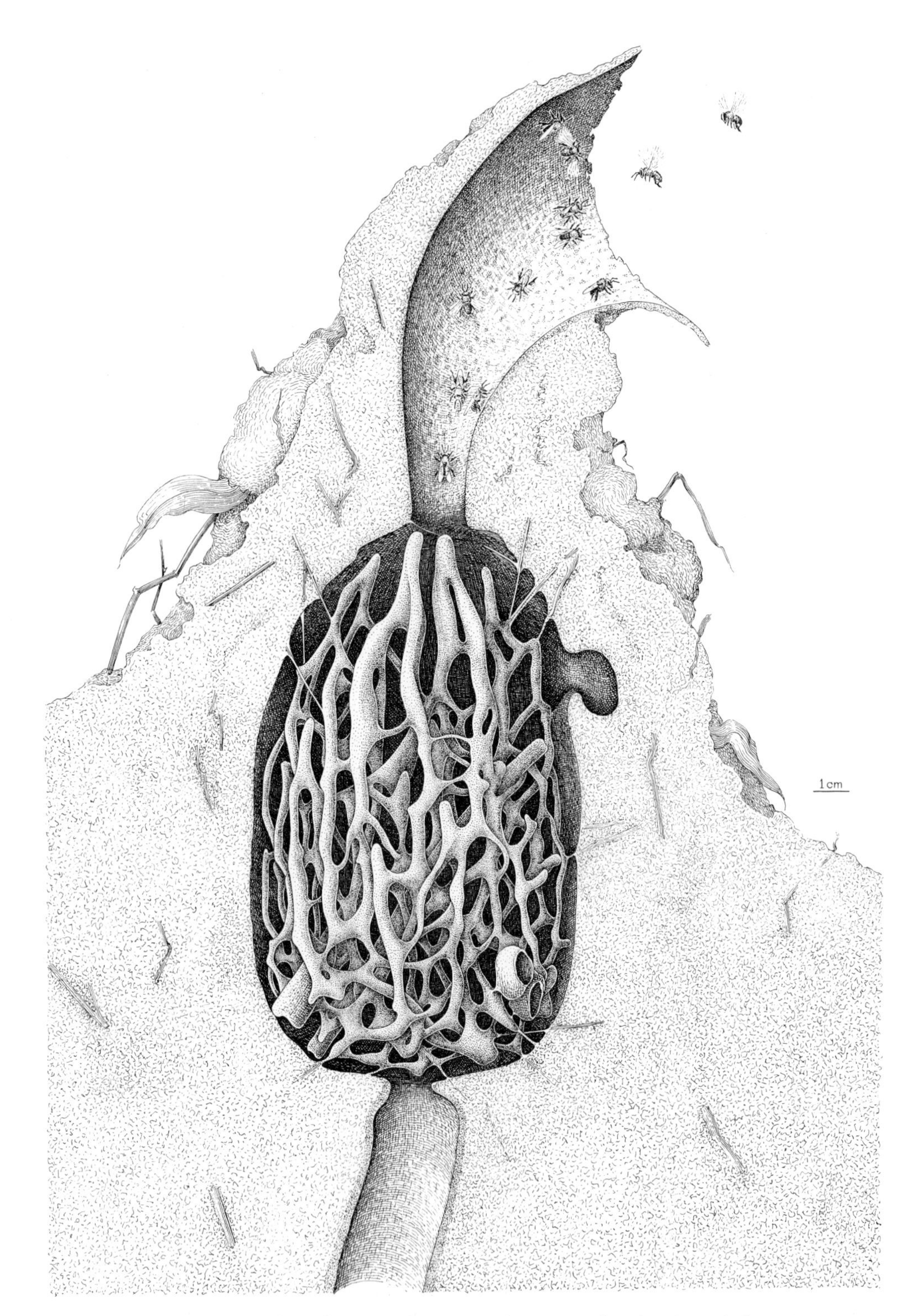

FIGURE 18.7. Nest entrance of *Trigona* (*Partamona*) *testacea*, showing funnel-shaped entrance and cavity within, in which anastomosing rods presumably provide resting space for the defense force of the colony. (From Camargo, 1970; original drawing provided by J. M. F. de Camargo.)

themselves to the danger of being killed by natural enemies, whatever they might be. Such individuals are thus altruistic in the sense that they defend the colony in spite of the danger to themselves. Mechanisms by which such altruistic behavior could evolve in spite of the earlier average mortality of individuals exhibiting such behavior are discussed in Chapter 20. A level of altruism involving not merely behavior but also structure is reached in workers of *Apis mellifera* where the barbed sting commonly pulls out after use, killing the bee.

As implied above, large colonies of any kind of bee and especially those which contain much brood and stored food must require particularly good defenses. For such colonies, it would seem advantageous if any individual discovering an intruder around the nest or entering it would not only attack the intruder herself but would simultaneously alert and promote the aggressiveness of other individuals in the colony. The result would be joint action very likely to chase away if not kill the intruder.

In all species of *Apis* an alarm pheromone is known, partly isopentyl acetate (Boch and Shearer, 1966; Morse, Shearer, Boch, and Benton, 1967; Free and Simpson, 1968). As early as 1814 Huber noticed that fresh stings or the odor of stings elicited attacks by workers. The substance that produces such behavior is liberated by the stinging apparatus of workers (Butler, 1967a; Maschwitz, 1964a, b), more specifically by a minutely setose membrane at the base of the sting. The excited bee raises the abdomen, opens the sting chamber, and exserts the sting; the pheromone is dispersed in the air by the fanning activity of the disturbed worker (Figure 18.8). The venom itself produces no alarm reaction in other bees. Another alarm substance, 2-heptanone, is released by the mandibular glands of workers in *Apis mellifera* (but not other species) that have reached the guarding and foraging ages (Boch and Shearer, 1967; Morse, Shearer, Boch, and Benton, 1967). It produces alarm reactions and provokes attacks by all species of *Apis,* whether they secrete it or not. For *A. mellifera* to produce alarm responses at the nest entrance, 2-heptanone is needed in concentrations 20 to 70 times as great as isopentyl acetate (Boch, Shearer, and Petrasovits, 1970).

Workers of *Apis,* stimulated by alarm substances, which they appear to detect with their antennae, run excitedly about, sometimes fly, and are ready to attack and sting any moving object. Even at subfreezing temperatures bees on the outside of a winter cluster protrude the sting and expose the sting chamber if disturbed. They are too cold to fly but the alarm substance liberated evidently stimulates warmer bees inside the cluster to come out and attack (Morse, 1966).

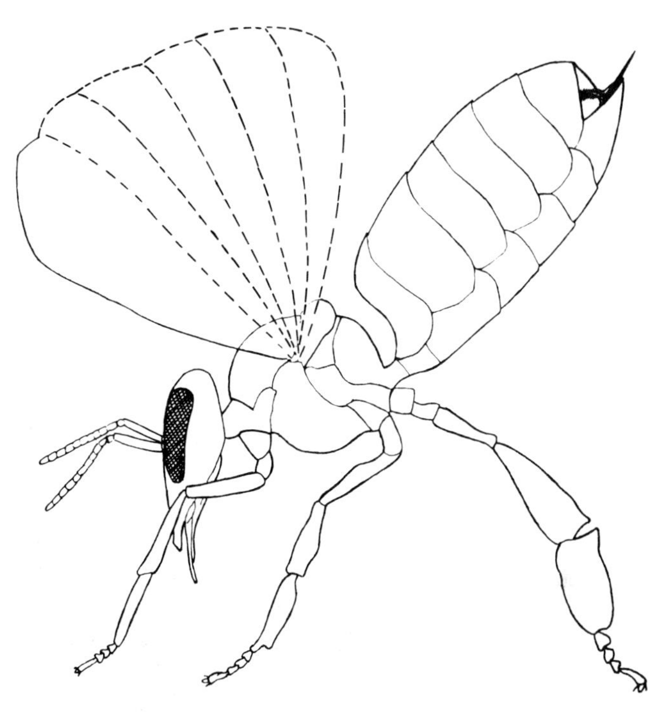

FIGURE 18.8. Sketch of a worker of *Apis mellifera* in the position in which she releases an alarm pheromone, isopentyl acetate, from the base of the sting apparatus. The abdomen is elevated, the sting exserted, and the air currents from the vibrating wings disperse the material. Compare with Figure 15.6. (From Maschwitz, 1964b.)

Like many other pheromones, alarm substances of *Apis* do not have their usual effect in all contexts. Morse, Shearer, Boch, and Benton (1967) report that the alarm pheromones do not release alarm behavior in bees visiting flowers.

Synchronized by alarm substances, the attacks that can be made by some stingless honeybees and by the species of *Apis* can be impressive indeed. Hundreds or thousands of bees may buzz about and at least hundreds may attack an intruder almost simultaneously. Morse and Laigo (1969) say of the giant honeybee *A. dorsata* of southern Asia that there is no question but that it "is the most ferocious stinging insect on earth. It is not unusual for 1,000 to 5,000 bees to leave a nest within a few seconds to attack an enemy."

Apis mellifera is responsible for more deaths annually in the United States than any other poisonous animal (including all snakes taken together). Most result from a few stings or even one, inflicted on hypersensitive persons. But massive attacks resulting from accidents or misjudgements can be dangerous to anyone. The highly aggressive Brazilian honeybee, derived in part from the African race *A. mellifera adansonii,* is now spreading rapidly in the Western Hemisphere; it will be discussed further in Chapter 30.

B. Defense against fungi and microorganisms

Mold fungi are a significant hazard for nearly all bees, except perhaps the highly eusocial forms; and for ground-nesting bees in cool, moist climates the impression that one gets, with no hard data, is that mold on food and larvae may be the principal mortality factor (see Batra, Batra, and Bohart, 1973).

Bees may reduce mold or bacterial attacks by the use of competing flora. For example, the food masses of *Anthophora* and *Amegilla* are full of yeast and always smell yeasty. Stored pollen of *Apis* is preserved in part by an abundance of *Lactobacillus.* The lactic acid produced by these organisms reduces attack by others. What bees do to assure an abundance of the appropriate organisms, or to assure conditions favorable to them, is not clear.

Many bees that make firm food masses have interesting devices for reducing the area of contact of the food mass with the brood cell walls. Presumably the areas of contact are the best places for mold to start. Usually the food mass is a ball which rests on the cell wall at only one place (Figures 24.7 and 24.9), but there are other interesting shapes in different bees which accomplish the same thing. The extreme is reached in the nonsocial allodapine genus *Halterapis,* in which the food mass never touches the substrate, being supported at first by the egg and later by the larva (Michener, 1971). Presumably the tendency to isolate food from the substrate culminates in the progressive feeders, since they do not leave food around long enough to get moldy. Yet even the bee milk of *Apis* has bacterial inhibiting properties, partly due to the fatty acids of the bee milk, perhaps partly to other materials (Lavie, in Chauvin, 1968).

Ground-nesting bees commonly line each brood cell with a waxlike layer derived at least in part from Dufour's gland in halictids (and presumably also in related forms, for example, andrenids). In panurgine bees the food ball is also covered with what appears to be the same material, or in some species of *Perdita* the food ball is covered but the cell wall is not. It seems very probable that this lining contains or consists of a mold inhibitor. May (1970a) points out that high molecular weight lipids and esters such as those of the cell lining of *Augochlora pura* have antimicrobial activity, although Bienyenu, Atchison, and Cross (1968) could not demonstrate such activity in the cell linings of *Nomia melanderi.* May suggests that they may have been stored too long to retain antimicrobial activity. A reason for believing in mold inhibition by cell linings is the repeated experience that, soon after being opened so that the food mass is exposed to the soil, a cell becomes moldy, while unopened cells mostly do not.

Of course the high osmotic pressure of material containing some 80 percent sugar, as does *Apis* honey, is inimical to microorganisms. Bacterial inhibition agents have been established, however, for pollen stored by meliponine bees (Gonnet, Lavie, and Nogueira-Neto, 1964) and *Apis* (Lavie, in Chauvin, 1968). Meliponine and *Apis* honeys also have an antibiotic content. For *Apis* White, Subers, and Schepartz (1963) found this to be due to hydrogen peroxide produced in the honey. Glucose oxidase, added to honey by worker bees, slowly breaks down glucose, releasing H_2O_2, a strong bactericidal agent. This reaction goes on principally in dilute honey, where the osmotic pressure gives inadequate protection. Lavie also reports other antibiotic substances in honey. It seems very likely that antibiotics are present in food masses in the brood cells of other bees, for example, halictids. One has to take it as rather remarkable that a nutrient mixture such as pollen mixed with nectar ever survives long in moist soil. It seems probable that there are inhibitory agents in the food itself as well as in the cell linings.

The sources of the inhibitory agents are not fully known. It seems certain that some of them are in the pollen and nectar and therefore are products of the plants. Lavie (in Chauvin, 1968) tabulates the antibiotic effects of pollen extracts from a number of plants; some have no such effect and others considerable effect. Less is known about antibiotics in nectar, but they exist in the nectar of some plants. This may be one of the remarkable aspects of the symbiotic or mutualistic relationship between plants and insects, for especially with plants whose flowers are ephemeral, the antibiotics are probably not directly important to the plants. Plants doubtless receive better pollination, however, if they protect their pollinating bee populations, and one can easily imagine that a plant population providing such protection would have an advantage over populations not doing so. Thus the basis exists for evolution of such protective agents in pollen and nectar.

Resins (propolis) used by honeybees are often strongly antibiotic because they contain terpines, which are both bactericidal and bacteriostatic. The synthesis of terpines presumably serves primarily the plants from which the resins are taken, but the materials are also important to the bees. According to Lavie (in Chauvin, 1968) the antifungal material in *Apis* hives is the propolis. Bees use propolis to fill all crevices, which in the warmth and humidity of a hive are potential sites for starting infections of fungi and bacteria. They also cover rough surfaces and bodies of animal that are too large for the bees to remove. Such a covering often attains a thickness of 2 mm or more, and effectively excludes such possible sources of infection from the hive. Meliponine bees exhibit similar behavior, providing physical and probably also chemical protection from molds or microorganisms.

Bees probably also produce their own inhibiting agents. Bienyenu, Atchison, and Cross (1968) found microbial inhibition by prepupae of a solitary halictid, *Nomia melanderi.* Lavie (in Chauvin, 1968) gives much data on an antibacterial substance found on the body surfaces of *Apis* workers, principally on the head and thorax. He finds an inhibiting substance that appears in the hypopharyngeal glands when workers are two days old, and he believes it is deposited by grooming on the body surface. This may be the same antibiotic present in bee milk. Incidentally, material from these three sources but not from other hive sources inhibits pollen germination.

Lavie (1960; in Chauvin, 1968) emphasizes the importance of antibiotics as social products. Without doubt they are important in the survival of colonies of social bees. However, I cannot believe that they will not be found to be important, also, to solitary forms.

A behavioral rather than chemical defense against microorganisms is known as part of *Apis* resistance to American foulbrood, *Bacillus larvae.* It is described in Chapter 7 (D5). Another behavior pattern that is beneficial in terms of nest sanitation is the removal of all movable loose objects. This includes dead bees as well as small intruders. Morse (1972) points out that a bee finding such an object carries it for some distance, usually toward the nest entrance, and drops it. Aother bee soon finds it and carries it farther. Ultimately a bee carries it out the entrance and if possible flies away with it for over 15 m before dropping it. The result is cleanliness around the nest and removal of possibly diseased bees.

C. Colony odor and defense

Bees sometimes enter and usurp the nests of other individuals or colonies. Usurpation is not uncommon for solitary, parasocial, and primitively eusocial bees. Nests that do not contain large stores of provisions are not attractive as foraging places for other bees, and it is the construct itself that is sometimes attractive and occasionally is taken over by an intruder. On the other hand, nests of *Bombus* and highly eusocial bees contain stored honey and pollen. Such nests may be usurped, but are more often the objects of robbing. Robber bees are individuals that enter nests of their own or sometimes of related species and take away food or building materials. Robbers are foraging in nests of their relatives instead of on flowers.

Solitary bees generally defend their nests against other bees, as shown, for example, by Stockhammer (1966) for *Augochlora pura.* Such defense is one reason that these species are solitary. As soon as a female reaches reproductive condition, she ejects any other individuals that may be in the nest with her, at least if they are also becoming reproductive, although newly emerged young females tolerate one another. In spite of the tendency of solitary bees to defend their nests, their burrows are sometimes taken over by other bees. For example, gynes of *Lasioglossum malachurum* in spring (ordinarily solitary at that time, later primitively eusocial) frequently usurp nests of their own species or even of related species (Plateaux-Quénu, 1960a). Communal bees may also defend their nests against others of the same species, but briefly and ineffectively. For example, a bee in the nest entrance of *Pseudagapostemon divaricatus* was often seen to bite at other individuals looking for nesting sites and attempting to enter the burrow. However, such strangers often did successfully enter and establish themselves, so that a colony developed (Michener and Lange, 1958c). Breakdown of the aggressiveness that individuals of solitary species show to one another must be one of the factors permitting colonial behavior.

Some primitively eusocial bees may have little or no ability to recognize and exclude from their nests individuals from other colonies of the same species. One can transfer adults of *Lasioglossum imitatum* into foreign nests without any fighting, and the foreign bees will stay there providing they have not yet begun foraging (Michener and Wille, 1961). If they have started foraging activities and are familiar with the outside environment, transferred bees will leave such a nest and return to their own. In *Allodape, Allodapula,* and *Braunsapis,* 90 percent of the nests in several artificially constituted groups of 50 nests each were destroyed, so as to liberate all adult bee. These bees, searching for their nests, often entered the 10 percent remaining nests, were never excluded, but left in a few minutes and ultimately disappeared from the vicinity (Michener, 1971). Apparently the intruders were not excluded by the usual inhabitants, but left of their own accord. In *L. versatum,* which nests in small dense aggregations, returning bees frequently enter the wrong nest entrances, sometimes repeatedly, and may even leave pollen loads in the wrong nests, perhaps in the cells where the pollen is used by the usual inhabitants (Michener, 1966c).

It is fairly clear that the halictine or allodapine recognizes its own nest and will usually leave other nests if it enters them, unless it is forcibly introduced while too young to have had experience outside. On the other hand, there is conflicting evidence on whether or not the inhabitants of a halictine nest are able to distinguish and exclude foreign bees, for example on the basis of a colony odor. It is true that guard bees often exclude foreign bees (including parasitic species). This is perhaps sometimes a matter of recognition of the hesitant manner in which a bee unfamiliar with the nest entrance must approach it. After any disturbance in the vicinity of the nest, the rightful inhabitants also may approach hesitantly, presumably because the local landmarks ordinarily used have been shifted about, and they may then be persistently excluded by the guard for a considerable period of time before they ultimately gain entrance (Aptel, 1931; Michener, personal observation). If there were a colony odor that could serve for identification, one would expect such returning bees to be quickly recognized and allowed to enter. On the other hand, various authors (such as Legewie, 1925a for *Lasioglossum malachurum*) clearly state that returning foreign females are excluded while nest mates are admitted, and moreover that males, regardless of origin, are admitted. Admission of such males does not fit the theory presented above, for they would have to approach the nest hesitantly. Studies now in progress in Kansas show that even in the scarcely eusocial *Lasioglossum zephyrum,* foreign females are excluded from mature nests. More work on the subject is needed.

Gynes of *Bombus* frequently enter the nests of their own or related species, either before workers have been produced or afterwards. Fighting results and either the intruder or the original gyne is killed; in the latter case the nest, provisions, and brood are usurped by the intruder. Species such as *B. terrestris* and *lapidarius* seem to have usurping or parasitic tendencies, and the gynes often enter nests of their own species. Sladen (1912) records a single nest of *B. terrestris,* one with a conspicuous entrance, in which were the bodies of 20 dead gynes.

Bumblebees also rob provisions from other colonies. Sometimes a robber can repeatedly enter a foreign nest, take honey from the honey pots, and escape without being attacked. At other times robbers are recognized and killed.

In *Bombus* and in the highly eusocial bees, the storage of considerable quantities of provisions makes a nest highly attractive to individuals from other colonies who find foraging there more worthwhile than in the field. It is difficult to introduce adults (workers) of these bees from one colony into the nest of another because they are often recognized by what is widely called a colony odor and killed. Free (1958) found that of anesthetized *B. agrorum* placed in their own colony, only 5 percent (1 individual) were attacked, whereas of those from another colony 90 percent were attacked. (Other species may be less efficient in recognizing foreigners; for example, in *B. ruderarius* comparable percentages were 10 and 29.) As there could be no behavioral differences in anesthetized bees, there must have been a difference in odor, that is, a colony odor. Since bumblebees hardly ever exchange food among adults, such an odor, to be uniform among the members of a colony, must be adsorbed onto the surfaces of the bees from the nesting or food materials. Workers of *B. agrorum* placed on cells and nest material, but not with bees, of a foreign colony for an hour or two and then returned, anesthetized, to their own colony were often attacked, as are foreign bees, indicating acquisition of a foreign odor. Similar results were obtained if bees were merely suspended in cages over a foreign comb. The acquisition of the foreign nest odor was incomplete, however, for such bees were also attacked in the foreign colony. Such experiments show the importance of acquired odors and suggest that they consist at least in part of volatile substances adsorbed onto the body surfaces. Probably combined odors of cells, food, and nesting sites and materials are involved. The possibility of inherited odors also exists. However, genetic segregation would seem to raise problems with inherited colony-specific odors.

For *Apis* there is an old theory that differential colony odors come from the Nasanov gland, but Renner (1960) has shown that its secretion is neither colony nor race specific. Nonetheless, honeybees often exclude foreign workers while admitting to their nest colony members. The colony odors probably actually come from odors of the nesting cavity and of the food supplies. Such odors are likely to be adsorbed onto the bees' bodies, but extensive food sharing may also tend to spread a uniform odor mixture. Thus thc individuals in a hive acquire an odor corresponding to the proportions of foods from different flowers brought in by the foragers of that colony and most likely different from the odors of neighboring colonies (Ribbands, 1953). If food content of a number of nearby hives is artificially made the same, widespread mutual robbing among the colonies occurs, showing that the ability of the guard bees to recognize individuals from foreign colonies is diminished. Renner (1960) showed that the odor does not have to be acquired by feeding or contact, but may be adsorbed, presumably onto the cuticle, from the air in the hive. He placed black treacle in hives in such a way that bees could not reach it. They nonetheless acquired a distinctive colony odor recognizable by other bees. It is very likely that colony odors of the same sort make possible the defense of colonies of Meliponini from robbers of their own species.

The contention of some authors that colony odors have a genetic basis seems to be outweighed by the evidence of odors acquired from the environment. However, it may well turn out that there is a genetic as well as an environmental component in these odors.

In spite of colony odors, guard bees, and alarm pheromones, highly eusocial bee^c often do drift to nearby colonies or rob colonies of their own and related species. In the Meliponini robbers take not only food but mud, propolis, and cerumen for construction purposes. Ribbands (1954) has shown that *Apis* workers from nearby colonies are readily accepted during times of abundant food, but are rejected and often killed when food is scarce and when robbing is therefore likely.

Nogueira-Neto (1949) notes that there can be both mild robbery, continuing sometimes for weeks, and exterminating robbery in the Meliponini. He recounts cases of mutual robbery, where two colonies rob one another, of a change in status of a colony from the robbed to the robber, and other interesting interactions. Selection would be expected to have improved defenses against robbery, since loss of food or building materials must in the long run adversely influence reproduction. On the other hand, selection may have improved robbing abilities, since nests of other colonies are such excellent food sources. Against this latter development is the likelihood of transmitting parasites or disease organisms from weak colonies—those most readily robbed and at the same time most likely to be suffering from infection—to the colonies of the robbers. A well-known illustration is the spread of European Foulbrood, a commercially important bacterial disease of *Apis,* by robbing from disease-weakened colonies.

As indicated in Chapter 19, it is possible to produce artificially mixed colonies containing workers of two or more related species. Such colonies have been made in *Bombus,* Meliponini, and *Apis.* Their success further supports the view that recognition odors are probably not inherited but are acquired from the colony and nest in which the bees live. Yet it is always more difficult to mix individuals of different species than those from different colonies of the same species. This indicates that inherited odors or behavior patterns play a role in the recognition of foreign bees, at least at the species level.

Species of bees that live as robbers or as social parasites will be dealt with in Chapter 19.

19 *Mixed Colonies: Parasitic and Robber Bees*

This chapter concerns special kinds of interactions between bees in which one species enters into the construct or social organization of another. Unrelated parasites and associates—protozoa, nematodes, mites, beetles, moths, flies, Strepsiptera, parasitic Hymenoptera—exist in some numbers, some kinds exclusively on social bees, but so far as known they do not enter into the social organization as do some inhabitants of nests of termites and ants. Both in species and in individuals they are much less numerous than are the associates of ants and termites. Even though these parasites and associates must influence colony size and population dynamics, they are not dealt with in this book. Too little is known of their influences to make a special section of them worthwhile, although some are considered in appropriate places in Part III.

Parasitic[1] bees whose relations with their hosts are brief and not especially dependent on the level of sociality of the host are numerous and of two principal types. First, there are those in which the parasitic bee opens a host cell, destroys the egg, replaces it with her own egg, and recloses the cell. For example, bees of the halictid genus *Sphecodes* enter the nests of both solitary and eusocial halictids (as well as solitary and communal andrenids), open the completed cells, and replace host eggs with their own. They then close the cells and either leave or close the nest and remain inside to deposit more eggs on subsequent days, depending at least in part on the number of cells available for parasitization (Knerer and Atwood, 1967; Ordway, 1964). Some species of *Sphecodes* accomplish parasitization without killing adult hosts (Ordway, 1964; Eickwort and Eickwort, 1972b), while others regularly do so. For example, Legewie (1925c) recorded the results of 76 attacks of *S. monilicornis* on colonies of *Lasioglossum malachurum.* The *Sphecodes* succeeded in 75 cases, and 283 *Lasioglossum* were killed. *S. pimpinellae* appears to chase away the hosts, *Augochlorella* spp., without killing them (Ordway, 1964). Replacement of the host egg is like replacement of conspecific eggs by females in halictine colonies such as those of *L. zephyrum* (Chapter 16, D1). It is possible that this type of parasitism arose in nests of colonial halictines and spread to solitary forms.

A second type of parasitism is the cuckoo-like relationship well known among various solitary bees. Such social parasites have arisen independently among various of the families and subfamilies. The Nomadinae and some tribes of the Anthophorinae are exclusively parasitic (see Chapter 3) as are various genera of Megachilidae, Euglossini, and so forth. Females of these parasites lay their eggs in incomplete or complete cells of the host. The young larva kills the egg or young larva of the host and eats the provisions in the cell. There are species of *Nomada* parasitic on communal species of *Andrena, Exomalopsis,* and others, but no such parasites are known to attack semisocial or eusocial hosts.

Sometimes a normally nest-making bee takes over and remains in the nest of another species. Knerer and Plateaux-Quénu (1967a) show that gynes of *Halictus scabiosae* frequently usurp nests of *L. nigripes,* merely utilizing the burrows, constructing their own cells, and filling in around the cell clusters of *nigripes* with loose earth. They report one cell of *nigripes,* however, which contained an

1. Parasitic bees have larvae which feed on the food placed in their cells by the hosts. Adults eat similar food in nests or take nectar from flowers. They may kill eggs, larvae, or adults of the host, depending on the kind of parasite, but they are not dependent on the hosts' bodies for food. Their way of life is more like that of a cuckoo than that of typical parasites. They are sometimes called social parasites, as are cuckoos, or more specifically, cleptoparasites.

egg believed because of its size to be *scabiosae;* if their interpretation is correct this is an interesting approach toward the behavior of *Sphecodes.*

The third type of parasitism found among bees, discussed in greater detail in section B below, occurs only with social hosts. The female of the parasitic species supplements or usually replaces the queen of the host.

A. Mixed colonies of nonparasitic bees

The relationships which are the principal subject of this section differ from those mentioned in previous paragraphs in that they involve more or less protracted cohabitation of two nonparasitic species of bees. The occurrence of two species together in certain communal groups, as an occasional event, has been described in Chapter 5 (E1). Most such records involve the perhaps almost accidental use of a nesting burrow by normally solitary bees of two species. Sometimes two communal species are involved. Sakagami and Michener (1962) list some examples.

Natural mixed colonies of *Bombus* species occur if a queen of one species invades the nest of another, destroys the original queen, and appropriates the brood, which is reared along with her own (Sladen, 1912; Plath, 1934; Free and Butler, 1959). Such mixed colonies are common, the best known cases being colonies of *B. lucorum* usurped by gynes of *B. terrestris* in Europe. In northern Japan nests of *B. schrencki* are usurped by *B. pseudobaicalensis* (Sakagami and Nishijima, 1973). Many mixed colonies of *Bombus* have been observed in Alberta by K. W. Richards (personal communication), who verified that they always involve mixtures of species in the same subgenus. Queens of different subgenera evidently do not so often enter one another's nests, and perhaps if such an intruder survives, she destroys the brood of the original nest owner.

Nogueira-Neto (1950) recorded a colony of *Trigona schrottkyi* in which there lived and worked for at least 13 days two workers of *T. mosquito* that must have got into the nest when disoriented about their own nearby nest.

Artificial groups of two species of *Bombus* (Munakata and Sakagami, 1958; Free and Butler, 1959), of Meliponini (Nogueira-Neto, 1950), and of *Apis* (Sakagami, 1959; Atwal and Sharma, 1968) have been made by various investigators, usually by putting brood of one species into the nest of another. Adults emerging from such brood are frequently accepted into the host colony, where they may live a normal life in full association with their "hosts." Or they may be attacked immediately, or after varying intervals. Interestingly, although some rather closely related species, such as certain species of *Melipona,* got along poorly together, the distantly related *Trigona postica* and *M. marginata* lived well together. When such mixed colonies of Meliponini contain large numbers of individuals of two species, the architecture and behavior may show attributes of each. For example, in a mixed colony of *T.* (*Nannotrigona*) *testaceicornis* and *T.* (*Plebeia*) *mosquito,* the former kept making its typical long entrance tube and the latter kept removing it. Or in two colonies of *T.* (*Plebeia*) *schrottkyi* mixed respectively with *T.* (*Tetragona*) *jaty* and *T.* (*Plebeia*) *mosquito,* the latter two species made involucra around the brood chambers, features not found in nests of *T. schrottkyi,* the workers of which insistently removed them. Such conflicting instincts are not the ones that often led to interspecific attacks in the mixed colonies; the reasons for attacks usually were not obvious and are likely to involve scent.

In two recent studies of mixed colonies of *Trigona* (Rezende, 1967; Oliveira and Fonseca, 1973), queens of *T. remota* and of *T. saiqui* were placed in dequeened colonies of a somewhat smaller species, *T. droryana.* All these are related species of the subgenus *Plebeia,* and the foreign queens were accepted by the *T. droryana* workers. In both cases the young produced at first had the size of the host, *T. droryana,* as they were reared in cells made and provisioned by that species. Later, the sizes of the new workers as well as the nest architecture changed to those typical for *T. remota* and *saiqui.*

B. Parasitic queen-replacing species

This section concerns parasites that enter the nests of the hosts and remain there, functionally replacing the host queen. In all such cases the host is a primitively eusocial bee, more or less closely related to the parasitic species. Also, in all cases the parasitic species lacks a worker caste. Since they make use of the workers of the host, the females of the parasite are not divided into castes; all function like queens. This is a clear case of loss of the caste system during the course of evolution of parasitic behavior, as occurred also during the evolution of some parasitic species of ants and of vespid wasps.

1. Bombini

As has been mentioned, gynes of *Bombus* often enter nests of their own or closely related species and kill the original queen. It could well be from such behavior that parasitization as practiced by the parasitic bumblebees arose. The arctic *Bombus hyperboreus* is an obligatory parasite of *B. polaris* at least in some areas (Milliron and Oliver, 1966; K. W. Richards, personal communication). Workers (small females) of *B. hyperboreus* are reported but are rare in the Eurasian arctic. Reports of workers may result, however, from misidentifications. Workerlike females are seemingly absent in the American arctic, where the relation with *B. polaris* has been investigated in northern Ellesmereland by K. W. Richards. Here overwintered females of *B. hyperboreus* enter nests of *B. polaris* and kill and replace the *polaris* queens. Females of *B. hyperboreus*, like ordinary species of *Bombus*, have fully developed corbiculae for carrying pollen, but they do not do so. They may visit flowers for nectar, but in two years of study by Richards were not seen collecting or carrying pollen. It is clear that at least in northern Ellesmereland, where these are the only two species of *Bombus, B. hyperboreus* is a *Psithyrus*-like parasite on *B. polaris*. The host and parasite are both members of a single subgenus, *Alpinobombus*.

A second workerless, obligately parasitic species of *Bombus* probably exists. It is *B. inexspectatus*, a rare species of the subgenus *Agrobombus* found in the mountains from Spain to Austria. Yarrow (1970) reports that over 10,000 bumblebees have been taken on flowers in areas where this species occurs without finding any *inexspectatus* that could convincingly be called workers. Of further interest, no specimens of this species showed wax secretion and none were carrying pollen loads. There is a reduction of the hairs at the lower end of the corbicula and of the auricle on the basitarsus which pushes pollen up into the corbicula. The host (assuming that the species is a parasite) remains uncertain, but is likely to be the closely similar and sympatric *B. (Agrobombus) ruderarius*.

The best-known queen-replacing parasitic bumblebees are more different from their hosts than those listed above; they constitute the genus *Psithyrus*. All the species of *Psithyrus* are social parasites in the nests of *Bombus;* for general accounts see Sladen (1912), Plath (1934), and Free and Butler (1959). The female *Psithyrus* have the size and form of queen *Bombus*, although with a somewhat more pointed abdomen. Species of *Psithyrus* may or may not resemble the host in color pattern. Much of the old literature emphasizing such a resemblance was based on presumptions rather than facts about the host species for various species of *Psithyrus*. Hobbs' publications show that at least some species are not host specific; a given *Psithyrus* species often could not resemble in color all its hosts. K. W. Richards in a recent, unpublished study in Alberta also found that the three species of *Psithyrus* there all enter *Bombus* nests regardless of species and subgenus. In other areas there are, however, *Psithyrus* species that are probably host specific and that may resemble their hosts closely in color.

There are unresolved differences of opinion as to whether the various species or groups now placed in *Psithyrus* evolved polyphyletically from different *Bombus* or, alternatively, arose from a common parasitic ancestor (Richards, 1927). The species of *Psithyrus* have many common characters differing from those of *Bombus*, including stronger cuticle, sharper mandibles, a stronger sting, and lack of the pollen-carrying corbiculae (Figure 19.1). But most such characters are related to parasitism and might have arisen independently among different parasitic forms. Similarities in male genitalia among species of *Psithyrus*, however, would not seem to be likely convergences. From this I would suspect that *Psithyrus* had a single origin from a *Bombus* ancestor.

The females of *Psithyrus* typically have more or less the size of *Bombus* queens, although they range down to worker size. The *Psithyrus* female, regardless of her size, seeks out and enters an established *Bombus* nest; therefore she leaves hibernation later than the queens of the host species. Ordinarily she enters, or is successful in, a colony of moderate size. If there are not enough host workers they will not be able to adequately care for the progeny of the *Psithyrus*, while if there are too many, they are likely to fight and kill or eject the invader even though she is strong and heavily armored.

The activities of a female *Psithyrus* when she first enters a host nest vary with the species of parasite and the species of host as well as the size of its colony. Sometimes the invading *Psithyrus* hides under the comb, avoiding contact with host bees for hours, presumably meanwhile acquiring the nest odor, before beginning to move about

FIGURE 19.1. Hind legs of a female of *Psithyrus rupestris,* a parasitic bumblebee, and a queen of *Bombus lapidarius,* a working bumblebee. (From Wilson, 1971, after Sladen, 1912.)

and make contact with individuals of the host colony. If molested, the *Psithyrus* may pull in her legs against her body and remain motionless until the disturbance fades away. On the other hand, the *Psithyrus* may act aggressively when she first enters the nest, grasping workers as though to sting them, but rarely doing so.

The *Bombus* may be rather passive, or alternatively, especially if the colony is large, may attack and ball the intruder, attempting to sting her. Establishment of a *Psithyrus* in colonies containing egg-laying workers is particularly difficult because of the aggressiveness of such workers. But the size, activity, armor, and strong sting of the *Psithyrus* usually enables her to kill many worker bees even if she is ultimately killed or ejected. Plath reported that in the fighting after a *P. laboriosus* entered a nest of *B. terricola,* the *Psithyrus* and 15 *Bombus* workers were killed.

A critical relation is that between the invading *Psithyrus* and the queen *Bombus.* Sometimes, as in Plath's colony of *B. affinis* parasitized by a *P. ashtoni,* the two large females live together without evident animosity. G. A. Hobbs states (personal communication) that he has no evidence of *Bombus* queens being killed by *Psithyrus* in his extensive studies in Alberta. In such cases the *Psithyrus* destroys eggs laid by the *Bombus* or even young larvae; Plath saw a *Psithyrus* destroy a clump of *Bombus* larvae. Production of *Bombus* thus is terminated even though the *Bombus* queen survives. Such behavior is apparently similar to that postulated for the parasitic allodapines discussed below. In other cases, for example, *B. lapidarius* and *terrestris* parasitized by *P. rupestris* and *vestalis,* respectively, Sladen reports that the intruder always kills the host queen. This may occur on the arrival of the *Psithyrus* but typically occurs sometime after, when she is about to start laying, and fighting is said to be often initiated by the *Bombus* queen. The fighting may be initially similar to the fighting that establishes social dominance among individuals of *Bombus;* it may start when the *Psithyrus* ovaries enlarge so that she becomes queenlike.

The *Psithyrus* female makes cells on top of cocoons and oviposits in them, penetrating the wall of the cell with the sting before each egg is laid, just as does a *Bombus* queen. Care and feeding of the young is by the *Bombus* workers; the methods are the same as for *Bombus.*

The ultimate result of parasitization of a *Bombus* nest by *Psithyrus* is production of a brood of male and female *Psithyrus* instead of the males and gynes of the *Bombus.* Sometimes the replacement is incomplete, but in general a parasitized *Bombus* nest can contribute little or nothing to the next generation of *Bombus.* The number of *Psithyrus* offspring produced depends in large part on the number of *Bombus* workers present as adults and immatures in the *Bombus* colony at the times of arrival of the *Psithyrus.* Plath records over 80 *P. ashtoni* from a colony of *B. affinis.* The young females of *Psithyrus* may remain in the host nest for a long time, but in autumn they disperse and hibernate in the ground like gynes of *Bombus.* In some instances maturation of the *Psithyrus* occurs well before the usual season for maturation of sexual forms of the host.

2. Ceratinini

Another group of bees with parasitic species and genera which replace the queens of the host is the *Allodape* group of Ceratinini. In the genus *Braunsapis* there are parasitic species, independently evolved, in Australia (Michener, 1961c), Malaya (Michener, 1966d), and Africa (Michener, 1971), and in Africa each of the genera *Allodape, Allodapula,* and *Macrogalea* contains a presumably parasitic species in addition to nonparasitic ones. These parasitic species, seven in all, are very similar to the hosts in appearance and are placed in the same genera as their hosts; usually the pollen-collecting scopas are reduced (Figure

19.2) and sometimes the mouthparts are shorter than in their nonparasitic congeners. Of these species *Macrogalea mombasae* and *Allodape greatheadi* from Africa are the most modified as parasites. Still more specialized parasites are placed in the separate genera *Inquilina* (parasitic on *Exoneura* in Australia, Michener, 1965b), *Eucondylops* (parasitic on *Allodapula* in Africa, Michener, 1971), and *Nasutapis* (parasitic on *Braunsapis* in Africa, Michener, 1971). The parasites given separate generic status are quite unusual bees, *Eucondylops* in particular being morphologically one of the most curious bees known; details of the parasitic forms from Africa are given by Michener (1970a).

Most females of the parasitic allodapines show in varying degrees several attributes that seem related to their parasitic mode of life. The parasites and their hosts are listed in Table 19.1, some of the convergent features of the parasites in Table 19.2. See also Figures 19.2 to 19.5.

None of the three parasitic genera has been taken on flowers in spite of extensive field work in areas where they are common. Apparently they feed only in the nests, on food brought in by the hosts. Reduction of eyes and wing veins is probably related to decreased flight activity. Thus they are much more highly modified parasites than is *Psithyrus.* The parasitic species of *Braunsapis, Allodapula, Macrogalea,* and presumably *Allodape,* however sometimes visit flowers and are in varying degrees less specialized parasites.

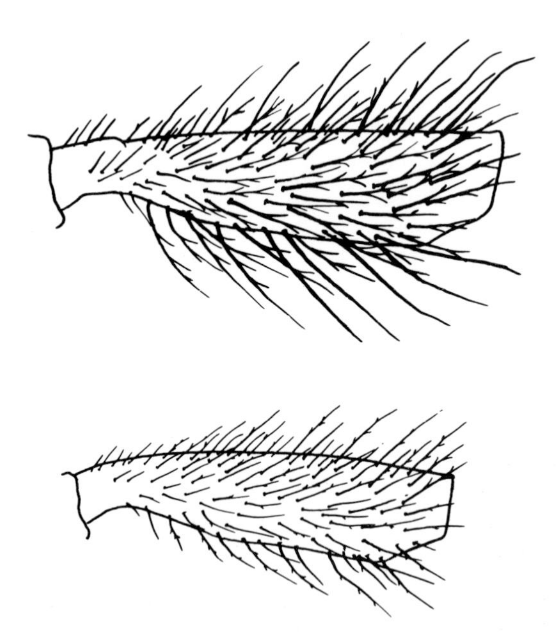

FIGURE 19.2. Hind tibiae of females of a typical *Braunsapis* (*above*) and of *B. breviceps,* a presumably parasitic species of the same genus. (From Michener, 1966d.)

The relations to the host are best known for *Inquilina* and *Nasutapis;* presumably all the other parasitic allodapines behave in a similar way. After the female parasite enters the host nest, production of young of the host ceases. The reason for this is not known. There may be an inhibition exercised by the parasite that prevents the host from laying eggs. Alternatively, and more probably, the parasite may eat or otherwise destroy the eggs of the host. The finding of occasional host eggs and hosts with enlarged ovaries in parasitized nests otherwise lacking young hosts supports the latter suggestion.

After entering a host nest the parasite soon begins to lay her own eggs, to judge by the scarcity of nests containing a parasite but no eggs or young. The resulting larvae are cared for, along with the older young of the host, by the female or females of the host species that are present in the nest. Ultimately, after the host larvae and pupae have all matured, all the young are those of the parasite. On maturity, the young adult parasites leave the nest; no colony of parasites develops. Occasionally one finds a nest that has been producing parasites, as

TABLE 19.2. Some convergent features of parasitic allodapine bees.

Feature	Species[a]
Face flattened or concave (in 9–11 with projections above or below concave area) (Figure 19.5)	6–11
Eyes reduced (shown, for example, by lengthened malar area) (Figure 19.5)	7–11
Mandibular dentition (♀) reduced (Figure 19.5)	7–11
Proboscis reduced (Figure 19.3)	5–11
Wing venation reduced (Figure 19.4)	10–11
Fore basitarsal hairs (♀) not or slightly bent	5?, 6–9
Legs robust (♀) (Figure 19.5)	8, 10, 11
Tibial spines on fore- and midlegs enlarged (♀) (Figure 19.5)	8, 10, 11
Brush at base of midfemur and apex of trochanter reduced or absent (♀)	6–11
Scopa reduced (Figure 19.2)	3–11

[a]Numbers refer to species as listed in Table 19.1.

TABLE 19.1. Parasitic allodapine bees and their hosts.

Species	Area	Host[a]	Reference
1. *Braunsapis associata*	Australia	* *Braunsapis unicolor*	Michener, 1961c
2. *Braunsapis praesumptiosa*	Australia	*Braunsapis simillima*	Michener, 1961c
3. *Allodapula guillarmodi*	S. Africa	*Allodapula* sp.	Michener, 1970a
4. *Braunsapis natalica*	S. Africa	* *Braunsapis grandiceps* and *facialis*	Michener, 1970a
5. *Braunsapis breviceps*	Malaya	*Braunsapis* sp.	Michener, 1966d
6. *Allodape greatheadi*	Uganda	*Allodape interrupta*	Michener, 1970a
7. *Macrogalea mombasae*	Kenya, Tanzania	* *Macrogalea candida*	Michener, 1970a
8. *Inquilina excavata*	Australia	* *Exoneura variabilis*	Michener, 1965b
9. *Nasutapis straussorum*	S. Africa	* *Braunsapis facialis*	Michener, 1970a
10. *Eucondylops konowi*	S. Africa	* *Allodapula variegata*	Michener, 1970a
11. *Eucondylops reducta*	S. Africa	* *Allodapula melanopus*	Michener, 1970a

Note: Species are listed in order of increasing degree of difference from nonparasitic relatives.
[a] Names marked with an asterisk represent hosts in whose nests the parasites have been found. The others are close relatives of the parasites and probable hosts.

shown, for example, by several pupae of the parasitic form, which has thereafter reverted to production of the host. This must be a result of the death or departure of the female parasite, under which circumstances eggs of the host species are laid and survive.

The extremely small sizes of allodapine colonies have been noted (Chapter 6, F). One might assume that such small groups would not be suitable for the successful intrusion of a queen-replacing parasitic bee. This assumption is an error; the point is that in colonial allodapines any adult female seems to care for whatever young are present. Thus even if there is only one adult host living in a nest, she cares for the young, including those of an intruder that may be present and that functions in the capacity of a queen.

3. Sphecodes (Microsphecodes) and Paralictus

The species of *Sphecodes* have been described above as halictids that enter nests of their hosts, killing or chasing away the adult hosts. But Eickwort and Eickwort (1972b) give an account of a Costa Rican species, *S.* (*Microsphecodes*) *kathleenae,* females of which enter nests of *Lasioglossum* (*Dialictus*) *umbripenne* either by passing the guard as would a *Lasioglossum* or by locating unguarded nests. Female *Sphecodes* were found in nests along with workers and queens of the host and are believed to locate the host cells, open them, eat the host eggs, deposit their own, and close the cells. More than

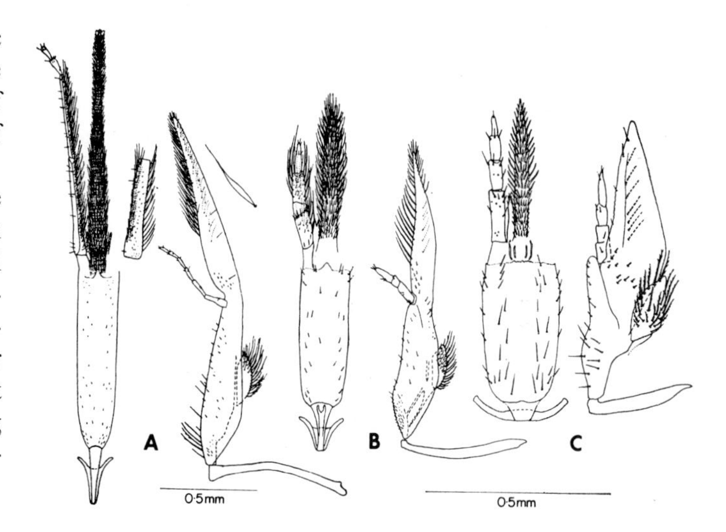

FIGURE 19.3. Proboscides (labia in ventral—that is, posterior—view, maxillae in lateral view) of females of allodapine bees. (*A*) *Allodapula melanopus* (with first segment of right labial palpus shown in side view and with an enlargement of a marginal galeal hair). This is a nonparasitic species. (*B*) *Nasutapis straussorum,* and (*C*) *Eucondylops reducta,* both parasitic species. Note that to accommodate the drawing to the space the proboscis of the *Allodapula* is much less enlarged than the others. The palpi of both parasites and the mentum and submentum of *Eucondylops* are especially reduced. The bodies of the three species are roughly the same size. (From Michener, 1970a.)

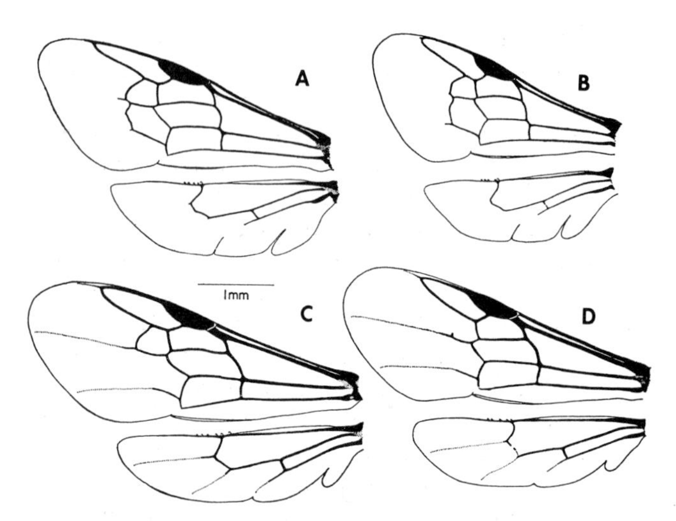

FIGURE 19.4. Fore and hind wings of females of allodapine bees. (*A*) *Braunsapis facialis;* (*B*) *Nasutapis straussorum;* (*C*) *Eucondylops konowi;* (*D*) *E. reducta.* The last three are parasitic species. (From Michener, 1970a.)

one *Sphecodes* was often found in a single nest. The situation differs from that of *Psithyrus* and the allodapines in that a parasite appears not to remain in the host nest for so long a period. The *Sphecodes* may remain long enough to be thought of as a partial queen replacement.

The Brazilian *Sphecodes* (*Microsphecodes*) *russeiclypeatus* was found in a *Lasioglossum* (*Dialictus*) nest and probably has similar habits. The several North American species of *Paralictus* are also parasites in *Lasioglossum* (*Dialictus*) nests and adults of host and parasite are regularly found together in the same nest. Probably they, too, have habits similar to those of *S. kathleenae.*

C. Robber species

A unique sort of relationship exists between some species of *Trigona* and bees of the genus *Lestrimelitta. Lestrimelitta* is closely related to and no doubt derived from *Trigona.* It has a small number of species in both Africa and tropical America. Probably they were independently derived from *Trigona* in the two continents, in which case the African forms can be put under the generic name *Cleptotrigona.* The species of *Lestrimelitta,* unlike the parasites discussed above, are highly eusocial, with queen and worker castes, but the workers, like queens, lack pollen-collecting and pollen-carrying apparatus and are never seen on flowers. Instead, they rob the nests of various species of *Trigona,* or, less commonly, *Melipona* or even *Apis.*

Since they have no pollen-carrying apparatus on the tibiae, workers of *Lestrimelitta* carry food back to their nest from nests of the host in a liquid form, mixed pollen and honey, in the extraordinarily enlarged crop. Firm pollen, in pollen pots, is made into a paste by addition of an acid oral secretion, and is then eaten. Other meliponines exhibit similar behavior when given firm lumps of pollen (Kerr, 1959), but do not show it on a grand scale

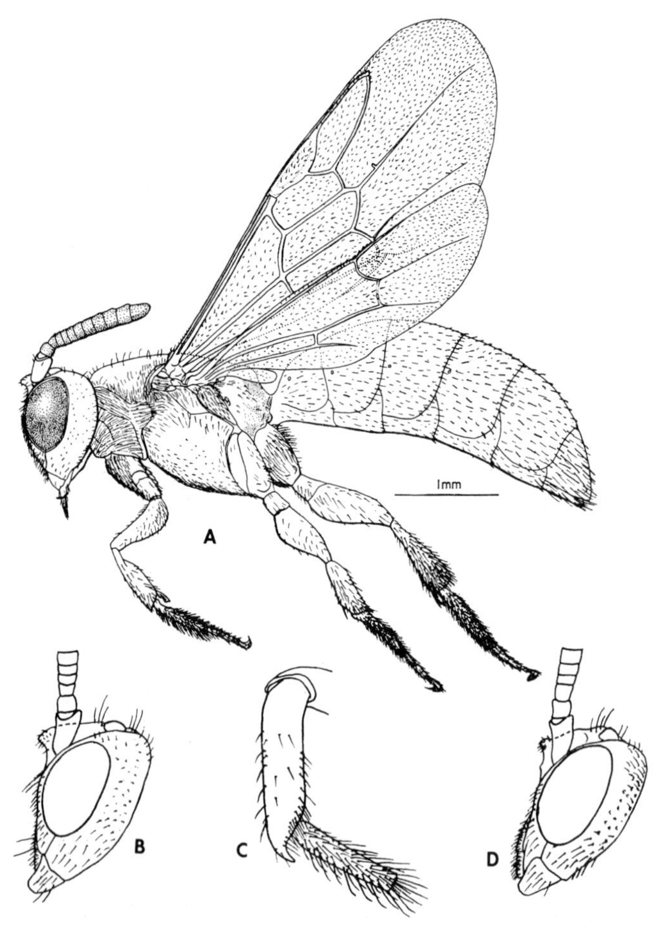

FIGURE 19.5. *Eucondylops,* a parasitic allodapine genus from South Africa. (*A*) *to* (*C*) *Eucondylops reducta:* (*A*) female; (*B*) head of female; (*C*) fore tibia and basitarsus of female, showing apical tibial spine. (*D*) *Eucondylops konowi,* head of female. (From Michener, 1970a.)

as does *Lestrimelitta.* The latter, however, is obviously using for its robbing an ability possessed also by related forms. The result of this method of food transport is that food is stored in the nest of *Lestrimelitta* in the form of a liquid mixture of pollen and honey, instead of these two materials being stored separately, as in other highly eusocial bees. Reports that honey of *Lestrimelitta* is often poisonous are perhaps due to the abundance of certain pollens in it.

The common neotropical species is *Lestrimelitta limao,* a shining black bee whose raiding behavior has been observed repeatedly. Attacks may often occur irregularly, but several authors have reported repeated attacks on a given *Trigona* nest, as though a particular robber colony returned to a known food source from time to time. For example, Sakagami and Laroca (1963) report a case in which a colony of *T. emerina* was attacked by *L. limao* at intervals of two or two and one half months for several years.

In the attacks on small and relatively timid species of *Trigona* described by Kerr (1951), Sakagami and Laroca (1963), and others, one or a few *Lestrimelitta limao* get into the host's nest, where they are attacked and may even be killed. The mandibular gland secretion of workers of *L. limao* contains citral (neral plus geranial) which has the strong lemon odor for which the species is named (Blum, 1966, 1970; Blum, Crewe, Kerr, Keith, Garrison, and Walker, 1970). This is the alarm pheromone of the species and of course is liberated by the *Lestrimelitta* workers that are attacked in the host's nest. This material has been adopted as the key element in the robbing attacks. It serves to disturb the host, disrupt its social organization, and probably to overpower its own alarm pheromones; adults of the host show escape responses and cower in groups in the recesses of the nest while plundering is going on. Species of stingless honeybees that are not susceptible to social disruption by citral are never or rarely attacked by *Lestrimelitta.* Presumably the robber bee is not able to exploit them efficiently. Citral also attracts more *Lestrimelitta* to the nest being attacked; Stejskal (1962) indicates that the route from the *Lestrimelitta* nest to the nest being robbed is marked by trail-marking odor spots having the same scent. Thus, as in the *Trigona* discussed in Chapter 18 (A3), the attracting trail-marking pheromone probably is the same substance or part of the same mixture which at higher concentrations is the alarm pheromone.

After the first few *Lestrimelitta* enter a host nest and are killed, dozens or hundreds arrive more or less all at once and enter unmolested. The robbers carry away not only food material but also cerumen (wax and resin). Workers of *L. limao* are always active at the entrance of a nest being robbed, guarding it and preventing the return of host workers which were foraging at the time of the initial attack. Therefore there is sometimes a considerable swarm of host workers, many of them carrying pollen on their hind legs, flying about the entrance and trying to enter. The black shiny *Lestrimelitta* lunge at them with open jaws to drive them off. Meanwhile, the *Lestrimelitta* modify the entrance tube of the nest and often carry away bits of the cerumen of which it is made. Immature stages of the host are not attacked, but the destruction of cells results in death of many young.

Sometimes a nest may be attacked several times with only short intervals between. It seems possible that such events may sometimes represent tentative moves toward establishing a *Lestrimelitta* nest, for *L. limao* establishes many and perhaps all of its nests by destroying a colony of some other stingless bee and taking over the nest and food reserves. Sakagami and Laroca (1963), for example, record cases where *L. limao* took over nests of *Melipona quadrifasciata.* In one case, a week after arrival of the *Lestrimellita,* most of the *Melipona* had disappeared but the gravid queen and about 200 workers, mostly still too young to fly, remained and from time to time attacked the *L. limao.* In spite of being much larger than the invaders, the *Melipona* were killed by them. The nest entrance had already been modified to resemble that of *L. limao.* A full colony of the later species had not moved in, however. Probably a site was being prepared for later arrival of more workers and a young queen and establishment of a new colony of *L. limao,* in the same manner that new colonies of *Trigona* are established (Chapter 13, A).

Sometimes *Lestrimellita limao* attacks weak colonies of *Apis.* I have seen such an attack in Belém, and Stejskal (1962) records attacks on mating nuclei (artificial small colonies) in Venezuela. In both cases the *Apis* were being killed in spite of the fact that they have stings and the *Lestrimelitta* do not, Stejskal says by a venomous liquid from the mandibles (that is, from the mandibular glands). Presumably this is the citral, which Blum (1970) remarks should be an excellent topical irritant and defense (or offense in this case) substance. Nogueira-Neto (1970b), however, emphasizes that the robbers, sometimes in large numbers, may kill host bees with their mandibles.

The behavior of the very small African *Lestrimelitta cubiceps* has been studied only by Portugal-Araújo (1958), who observed this bee in Angola robbing nests of the minute *Trigona* (*Hypotrigona*) *braunsi*. It may restrict its attacks to *T. braunsi, gribodoi,* and related forms. The very similar *T. araujoi* is apparently not attacked in Angola.

The following numbered paragraphs are quoted with certain modifications and deletions from Portugal-Araújo's chronological account of attacks of *Lestrimelitta cubiceps* on *Trigona braunsi:*

(1) Two or three *Lestrimelitta cubiceps* are seen flying close to the entrance of a hive of *Trigona braunsi.* They seem to be inspecting the base of the waxy entrance tube and its location.

(2) A great number of defenders come out, flying close to the entrance tube.

(3) The robbers increase their number and start to destroy the base of the entrance tube until it falls down. During this phase, one very interesting thing happens: some honey is poured by the workers of *Trigona braunsi* (not by the attackers) from the inside into the remains of the entrance tube. This closes the entrance for all the bees remaining outside, both attackers and defenders. Usually the defenders and guards are slain by the *Lestrimelitta.*

(4) The number of the attackers increases, reaching 40 to 60. A good number of these newcomers bring on their hind legs small pellets of a viscous propolis or cerumen, apparently from the stores in the *Lestrimelitta* nest. This material is put in small dabs around the entrance in an area 2 to 3 cm in radius. The role of the propolis is not clear; it may mark the entrance for approaching *Lestrimelitta* not familiar with the site. Interestingly, it is common to see workers from other colonies of *Trigona braunsi* collecting this propolis and carrying it home.

(5) Foragers of *Trigona braunsi* and some defenders stay on the wall of the hive, close to the entrance, and occasionally fight the attackers. The *Trigona* are usually killed in any such encounters. The honey that is poured in the entrance tube by the bees inside the colony is collected by the attackers.

(6) At the end of the day the attackers go home, the honey stops being poured, and the foragers of *Trigona* enter their nest. During the night the entrance tube is rebuilt.

(7) On the following morning the attackers come again to the *Trigona braunsi* nest. The guards make short runs and nervously beat their wings; some of them are killed by the *Lestrimelitta cubiceps.* Some of the attackers penetrate the nest. At the same moment the honey again appears in the entrance of the *Trigona* nest and usually the robbers that have entered the hive come out cleaning themselves from this sweet bath. This honey flows and drips to the ground at a rate of about one drop every 2 minutes; this may continue for 12, 24, or 36 hours, without stopping now, even at night.

(8) After a while the robbers seem no longer interested in collecting the honey that pours out and try, without success, to open the entrance more widely. Also, more bees from other hives of *Trigona braunsi* appear to suck the honey being poured out and to collect the propolis that is continuously transported by the *Lestrimelitta cubiceps* workers from their nest to the vicinity of the entrance of the one being robbed. In the evening the robbers return to their nest and the defenders stay in the robbed nest, grouped together on the front wall. This same situation may continue for one, two, or three days.

(9) After one to three days, without an indication that the situation will change, the honey stops pouring. The attacking *Lestrimelitta cubiceps* and the workers from the neighboring *Trigona braunsi* hives suck up all the honey close to the entrance. Now the attackers enter the nest without resistance being offered by the *Trigona.* Soon after, workers of *L. cubiceps* start coming out of the hive carrying propolis and cerumen on their legs, sometimes in such quantities that the legs are stuck together, while others have the abdomen swollen with honey or brood food. After a number of *L. cubiceps* have penetrated the colony, a group of robber guards is established in the entrance, as occurs also during attacks of the Neotropical *L. limao* on nests of *Trigona.*

(10) When the *Lestrimelitta cubiceps* workers finish the pillage, one can see in the robbed nest mutilated storage pots and larvae starving because their cells have been opened and brood food taken away. But there are still good quantities of stored pollen and honey left, along with a reduced population of adult *Trigona.*

No odor obvious to the observer, like that of *Lestrimelitta limao,* accompanies attacks of *L. cubiceps,* but the group action of the attacking bees suggests the existence of a pheromone that the bees can detect and to which they respond.

20 *The Evolution of Social Behavior in Bees*

One of the pastimes of persons studying social insects has been to speculate on the origin and evolution of social behavior. I do not underestimate such activity, first, because it is interesting, and second, because for certain types of evolution (such as kin selection and evolution of altruism) the social insects may be exemplars from which can be derived evolutionary ideas applicable to many other organisms. I have not avoided such speculations myself and will indulge in some here. (Earlier papers including such materials for bees and references to numerous other papers are by Lin and Michener, 1972; Michener, 1958, 1969a; and Sakagami and Michener, 1962.)

Bees of several groups have independently evolved eusocial behavior (Figure 2.2). Presumably the ancestors of such species passed through a series of stages, beginning with species never going beyond the solitary level and ending with those that develop eusocial colonies or, for highly eusocial forms, those that never live except in eusocial colonies. The present chapter indicates the prerequisites necessary for a species to enter upon this evolutionary pattern, some of the advantages and disadvantages of doing so, and the selective mechanisms that may operate to produce a nonreproductive caste.

A. Features associated with social evolution in bees

A number of attributes seem important to any hymenopteron that is to evolve into a eusocial species. Attributes 1 to 5 below appear to be prerequisites for the evolution of eusociality, while 6 and 7, if not prerequisites, at least must be needed before high social levels are attained.

(1) *The construction of a nest, in which young are reared, and to which the mother returns time and again.* Solitary bees and wasps qualify in this respect. Involved, of course, are (a) possession of physical features necessary to construct a nest and to carry food and building materials and (b) possession of nervous and sensory equipment to permit orientation and regular return to the nest.

(2) *Sex ratio subject to strong female bias.* The social organizations of Hymenoptera are based on females. Females make and provision nests; workers are all females. Production of large numbers of females as compared to males, at least at certain seasons, is an important potentiality of the sex determination mechanism if a female worker caste is to evolve. This potentiality exists because of the haplo-diploid sex determination, which also has had a number of other influences related to the evolution of castes, as detailed in Section D below.

(3) *Long reproductive life of adult females.* The mother must live at least long enough to associate with her adult offspring if a eusocial group is to develop. Long-lived adults in groups not having social behavior are well known in solitary Halictinae and especially in Xylocopinae and Euglossini. For example, females of *Xylocopa caffra* live almost until the brood matures (Skaife, 1952). Examination of the classification of the bees given in Chapter 3 indicates the positions of these three long-lived, nonsocial groups near eusocial ones in the Halictinae, the Xylocopinae, and the Apidae, respectively. Long reproductive or postreproductive life is unusual or absent in other nonsocial bees, that is in groups not listed above. Long prereproductive adult life, such as occurs in *Andrena,* is not related to sociality, but rather to hibernation or diapause.

(4) *Tolerance of other bees of the same species in the*

nest. This factor received strong emphasis by Legewie (1925a, b). Solitary species are often aggressive toward other members of the same species, the degree of intolerance decreasing with the sociality of the species (Batra, 1968). Tolerance could evolve along with other attributes of sociality, but since various parasocial colonies exist in taxonomic groups related to the eusocial bees (as well as in unrelated groups), a prediliction for group life may well be common to the antecedents of eusocial groups.

(5) *Ability to omit part of the stereotyped behavior pattern of cell construction and provisioning.* In typical solitary bees, there is commonly an essentially unvarying sequence of cell construction—provisioning—egg laying—closure, followed by construction of the next cell and repetition of the sequence. It is common to assume that such a sequence is invariable, but Legewie (1925a) gives a list of six solitary species (Megachilidae, Colletidae, Anthophoridae) which under unusual circumstances can omit parts of the sequence at least to the extent of provisioning cells made by other individuals. In the allodapine genera this sequence is broken down by the complete disappearance of cells, so that the young are reared in a common space. In some Halictini and Euglossini the sequence is broken down by taking advantage of preexisting cavities for nesting (as in *Augochlora pura,* Stockhammer, 1966), by reusing old cells (*Lasioglossum seabrai*—Sakagami and Moure, 1967; *Augochloropsis*—Michener and Lange, 1959; *Pseudaugochloropsis*—Michener and Kerfoot, 1967; *Eulaema*—Bennett, 1965; and *Euglossa*—Roberts and Dodson, 1967), or by starting to make new cells before the preceding ones are completed so that two or three cells are in different stages of construction or provisioning all the time. Synchronous work on more than one cell was recorded for *Halictus quadricinctus* (a solitary European species famous for its large and beautiful earthen combs found in the soil) by Verhoeff as early as 1897, and was considered by him as a significant prerequisite to social evolution. Such breakdown of the cell-making sequence is essential if bees are to make cells cooperatively. For example, if a bee must construct a cell before lining it, it would be impossible for two bees to work sequentially, one constructing it, the second lining it, because the second bee would be unable to provide a lining until it had done the constructing. In social bees one finds no such limitations. They seem able to omit any part of a behavioral sequence relating to construction and provisioning and are presumably stimulated by the nest and its contents to go on with appropriate activity (Chapter 7). What one would really like to know is whether in solitary bees there is an action sequence not or little controlled by sensory feedback, in contrast to a reaction sequence in which sensory stimuli control each activity to be performed, as is presumably the case in social bees. More work on solitary bees is needed to determine whether there is actually a difference in this respect between solitary and social forms.

There is considerable interspecific variability in the work sequence among eusocial halictines, as shown by the following comments on species of the subgenus *Evylaeus* of *Lasioglossum.* Gynes of *L. calceatum* follow the usual sequence of solitary bees (Plateaux-Quénu, 1964). In *L. marginatum* all the cells are made, then all are provisioned, and finally all the eggs are laid. In an indoors nest of *L. nigripes* three cells were made, then provisioned, then eggs laid, after which the bee went on to construct and provision a fourth cell (Plateaux-Quénu, 1965b).

(6) *Ability to interact socially or to communicate.* As social bees arose from solitary ones, the attributes of solitary bees that could take on socially significant functions would be important in the evolution of sociality. There are few totally new structures or functions involved with the evolution of sociality or of anything else. To elaborate this point in detail would involve much duplication and reiteration of the obvious. One example is perhaps worth citing, however. Mandibular glands, while commonly much smaller than in highly eusocial bees, are the only large cephalic glands in solitary and primitively eusocial bees. Their primitive function may have been secretion of defense substances, but probably in all families they produce volatile substances used either by males or females as sex attractants (Nedel, 1960). The glandular openings have valves which can be opened or closed, controlling the escape of the secretion. The glands thus have a communicative function even in various solitary bees. As demonstrated by highly eusocial bees, they had the potentiality for further evolution.

In workers of stingless bees the mandibular glands are enlarged, and the secretion has functions in defense recruitment (alarm), to which is added marking of food sources and, in some species, trail marking for recruitment to foraging activities, and in one subgenus (*Oxytrigona*) defense against large enemies, such as man. The glands have apparently been released from their original function in workers of Meliponini which do not mate. It is likely that they secrete a sex attractant in queens.

In *Apis* workers the mandibular glands secrete part of the food for larvae, and thus have nearly lost a communicative function, but they do produce a minor alarm pheromone. In queens the communicative function is elaborated; the glands are largest at the time of the nuptial flight (Nedel, 1960), but as is well known their two major products function not only to attract males but to assemble swarms of workers and, inside the hive, to limit production of queen cells and enlargement of worker ovaries.

Presence of mandibular glands has facilitated important parts of social evolution. Parsimony in the use of pheromones, as described elsewhere (Chapters 18 and 19), has enabled the bees to evolve more communication (both more messages and more targets) from one pair of glands and a few chemical products than would *a priori* have seemed likely.

(7) *Species and nest-mate recognition.* Section C below lists many features which have certain common threads that are related to social interactions or communication. One of the most obvious of such threads is the ability of individual social insects to recognize other individuals of their species or colony. Nonsocial insects often show the capacity to identify as conspecific other individuals of the same stage; among adults this is true especially if they are of the other sex. But often there is no evidence of ability to recognize individuals of other stages as conspecific. Individuals of predatory species, for example, may eat young of their species, even their own progeny. Social forms, however, usually have the ability to recognize their own species or at least colony mates in whatever stage they may be encountered. Eating of eggs and larvae by adults occurs but it is one of the social controls, not feeding for the sake of nutrition alone. Bees are otherwise not predaceous.

Progressive feeders obviously recognize the developmental stages of their young; eggs, young larvae, older larvae, and pupae are all treated quite differently. Conspecific adults may recognize one another and distinguish those from different colonies through adsorbed food or nest odors and food and pheromone interchange, as in *Apis.* Young queens recognize one another, prior to fighting, by sounds (piping). Dominance-subordinance interactions as in *Bombus* require recognition of groups or individuals, possibly by odor or behavior related to ovarian development. In some parasocial and primitively eusocial forms the abilities to recognize conspecific individuals and age groups presumably are little developed beyond those of the solitary ancestors. Recognition and discrimination of various stages, age groups, and castes become highly elaborated in forms such as *Apis.*

B. Origins of eusocial behavior in bees

1. The groups involved

Presumably the eusocial allodapines arose from subsocial allodapines in which lone adult females cared for their young but in which adults did not live together. Such species are well known (for example, *Exoneurella lawsoni* and *Allodape mucronata*). Larger colonies, when they occur, originate from just such subsocial colonies, which attain a primitively eusocial level.

In other eusocial bees, however, progressive feeding of larvae occurs only in *Bombus* and *Apis,* rather specialized forms, and presumably mass provisioning characterized the initial stages of sociality; hence most eusocial bees passed through no subsocial stage. For this reason and because of the various sorts of parasocial colonies known in halictines and other bees, Michener (1958) suggested that eusocial halictines probably arose from forms whose communal and semisocial colonies were groups of not necessarily related individuals. The main point was that eusocial bees did not arise from subsocial ancestors, as had been claimed by various earlier authors. But whether eusocial bees or some of them arose from communal and semisocial ancestors is still not settled. Communal and semisocial species occur both in the Augochlorini and Halictini, those in the former tribe (such as *Pseudaugochloropsis*) being semisocial, in the latter (*Pseudagapostemon,* for example) communal; temporarily semisocial groups of gynes also are common in the Halictini. Primitively eusocial groups exist in both tribes. The various eusocial groups might have arisen from communal or semisocial antecedents, but alternatively, daughters might merely remain in the maternal nest so that eusocial groups might arise from solitary antecedents without an intervening parasocial stage (Sakagami and Michener, 1962). Ontogenetically, eusocial groups of halictines arise from both solitary and parasocial groups. The important problem of how selection could operate to establish a nonreproductive worker caste will be discussed in section D.

There are no bases for speculating on the origin of the sociality in Apidae, at least until the Euglossini are well studied, for the parasocial apids are little known, and apids having eusocial behavior simpler than that of *Bombus* are probably extinct.

2. The number of origins

Figure 2.2 shows that eusocial behavior has arisen independently in several groups of bees. It appeared at least once in allodapines. However, even the most social allodapines are only temporarily eusocial in an ontogenetic sense, and it seems likely that there has sometimes been evolution toward subsocial life from eusocial ancestors, as well as the reverse. Obviously the parasitic species have lost the worker caste, and some nonparasitic ones may well have done likewise. Possible evidence is found in the female-biased sex ratios of some subsocial forms, like *Exoneurella lawsoni* (Chapter 8, B). Such preponderance of females may indicate descent from a eusocial ancestor in which the female-biased sex ratio was related to worker production. In summary, it is likely that there has been repeated development, probably loss, and redevelopment of eusocial behavior in allodapines, the social level vacillating according to changing environmental conditions that favor or discourage colonial life. It is not even possible to say whether certain species are eusocial or not. What fraction of the colonies must become eusocial, and for how long, to justify the term eusocial for the species?

In the Halictinae, eusocial behavior occurs in at least two genera (*Augochlora* and *Augochlorella*) of the Augochlorini and in the following groups of Halictini: *Halictus* (*Halictus*), *Halictus* (*Seladonia*), *Lasioglossum* (*Dialictus*), and *Lasioglossum* (*Evylaeus*). The only three species of *Augochlorella* whose life cycles have been studied are primitively eusocial. The related genus *Augochlora* contains both solitary and eusocial species. Moreover, in view of the level of sociality in *Augochlorella,* I would be surprised if there are no solitary species within the genus. In all the subgenera of Halictini listed there are solitary as well as eusocial species. There is no reason to believe that eusocial behavior arose only once in each of these subgenera, although it is quite possible that the rather distinctive sociality in Holarctic *Evylaeus,* with workers produced in discrete broods, did arise only once. In the other groups listed the situation is more as in the allodapines, with possibilities of evolution toward either increasing or decreasing sociality and with probabilities of multiple origins of eusociality. The strikingly different social levels of populations of the same species (for example, *Lasioglossum umbripenne*) in different habitats suggests social adjustments to local conditions and the possibility of evolution toward less as well as more elaborate social behavior.

In the Apidae, if one believes Figure 2.2, it appears that unless the Euglossini have lost sociality, eusocial behavior has arisen at least twice, once for the tribe Bombini and once for the subfamily Apinae. I think loss of sociality in Euglossini improbable, but it is possible, as shown by the loss of a worker caste in parasitic Bombini. The Meliponini and Apini have so many common features that they probably arose from a common eusocial ancestor. If the apid branch of Figure 2.2 is wrongly branched, then a single eusocial ancestor of the Bombini, Meliponini, and Apini might have existed.

I would emphasize that the oft-repeated question about the number of origins of eusocial behavior cannot be answered for any group in which sociality is in its initial stages, because sociality may have come and gone, possibly repeatedly, even in one phyletic line, and also because the lines of descent are not known in sufficient detail. It is not really the right question to ask when such groups are involved. However, the summary provided in Table 20.1 may have some value.

C. An analysis of social levels

In the evolution of organisms there is frequent discordance between rates of changes of the various features. It is common to speak of primitive organisms and of specialized ones, but this is unfortunate generalization, since virtually every species has a mixture of ancestral and derived characters. Even a form with many ancestral features almost surely has some derived ones to enable

TABLE 20.1. Number of origins of eusocial behavior in bees.

Allodapine bees	1 (probably many)
Halictinae	
Augochlorini	1 or 2
Halictini	4 (probably many)
Bombini	1 (possibly with Apinae)
Apinae	1

it to survive in today's world. A glance at some of the elaborately ornate solitary bees, for example, some males of *Megachile,* shows a multitude of what must be derived morphological characters that are not associated with social behavior. It should not be surprising, therefore, that the various aspects of social evolution and of nesting biology have not always evolved hand in hand, although in the highly eusocial bees, derived states of numerous such characters are associated.

Table 20.2 lists 28 variables or characteristics associated with social levels. This is not a comprehensive list; the items were selected partly on the basis of the possibility of articulating two or more alternative states. Some obvious candidates for inclusion, like the extent and complexity of pheromonal communication, were excluded because of lack of knowledge and of easily understood alternative states. Logical duplicates and items that would seem to duplicate information were excluded. For example, queen foraging is included, but the manner in which nests are established, by lone queens or by swarms, is excluded because it is so closely related to foraging. Finally, the brevity required in a summarization such as this leaves much to be said, mostly appearing elsewhere in this book, and requires various degrees of artificiality, particularly in assigning codes for the states of the variables.

The assumption in preparing Table 20.2 has been that evolution has been in the direction of increasing sociality. The character states are, where possible, listed in the order from ancestral to derived. However, evolutionary reversals may sometimes have occurred, as indicated in section B above and Chapter 5 (H), so that for certain forms an "ancestral" state may have been derived from a "derived" one.

Table 20.3 shows the distribution of the various character states from Table 20.2 among 18 representative kinds of bees. All other species whose biologies are well known would fall very near one or another of these species. The 18 species are arranged in a general way according to systematic groupings and increasing sociality. Species 1, 2, and 10 are solitary; the highly eusocial species are 15–18. In general the upper part of the table has low numbers, the lower part high numbers, showing that by and large the various features judged to be socially meaningful have evolved together. This could be partly a result of circular thinking, for some characteristics that clearly do not vary in parallel with the rest may have been excluded as not being socially meaningful. However, it seems that in a general way various socially meaningful characters have evolved together, despite the warning against this expectation in the first paragraph of Section C.

Sakagami and Michener (1962) emphasized the lack of relation between nest architecture and other aspects of social evolution in halictine bees. Efforts to show such a relation in *Lasioglossum,* subgenus *Evylaeus,* are unimpressive because, for example, one of the socially most derived species, *L. marginatum,* has ancestral type architecture. On the average, however, for most characters, ancestral character states are associated with ancestral states of most other socially meaningful characters, and derived states are associated with derived states of most other characters.

As shown in Table 5.1, the classification of social levels outlined in Chapter 5 and used throughout this book is based on only a few of the characteristics listed in Table 20.2. It is therefore not surprising that the various social levels—solitary, communal, semisocial, primitively eusocial, highly eusocial—for species included in Table 20.3 are not equally different from one another when all the attributes of Table 20.2 are considered.

A principal components analysis (Rohlf, 1968) based on the data in Table 20.3 was made with the objective of showing the relative similarities of the species listed in that table. Considering all the characteristics and weighting them equally, the species are divisible into four discrete clusters. Although the characteristics used were all biological attributes thought to be important in determining social level—none were morphological features used in bee taxonomy—these clusters all turned out to represent taxonomic units rather than social levels. Species 1–9 are Halictinae; 10–12, allodapines in the Ceratinini; 13–14, Bombinae; and 15–18, Apinae.

Figure 20.1 is constructed with the first three principal components as its dimensions and shows the positions of the 18 species with respect to these dimensions. The four clusters are easily recognized. The Apinae at the left, with *Apis* rather different from the Meliponini, are highly eusocial. The Bombinae in the center, as well as species 11 and 12 in the allodapines and 5–9 in the Halictinae, are primitively eusocial. Species 4 is semisocial, 3 is communal, and 1, 2, and 10 are solitary.

Clearly the taxonomic units are far more coherent than are the social levels. Solitary forms appear at the right-hand edges of the two right-hand clusters. Primitively eusocial forms are found in three clusters. Probably this

TABLE 20.2. Some characteristics (variables) related to social level.

Variables or characteristics	States
1. Maximum normal number of females per nest	(1) 1; (2) 2–19; (3) 20–199; (4) 200–1999; (5) $>$1999.
2. Sex ratio in early colony growth (that is, seasonal change in male production)	(1) Normal; (2) few males; (3) no males.
3. Frequency of reduced reproductivity in ♀♀	(1) None; (2) few ♀♀ to common; (3) all ♀♀ except queens with reduced reproductivity.
4. Queen (or lone ♀) foraging	(1) Pollen-collecting and carrying apparatus as in ☿[a]; (2) with such structures absent, must start colony with swarm of workers.
5. Social dimorphism	(1) ☿ absent or =♀; (2) ☿ mean size $<$♀; (3) ☿ usually or always discontinuously $<$♀; (4) ☿ smaller than and morphologically different from ♀.
6. Excess gynes	(1) No such thing; (2) killed by other gynes; (3) killed by workers.
7. Queen ovarian specialization	(1) Ovaries as in solitary relatives (solitary forms placed here); (2) ovaries $>$ than in solitary relatives; (3) ovaries $\gg$ than in solitary relatives; (4) ovaries $\gg$ than and with many more ovarioles than in solitary relatives.
8. Queen (or lone ♀) longevity	(1) Much $<$1 year; (2) about 1 year; (3) several years.
9. Worker fertilization	(1) Regularly fertilized; (2) $<$50% fertilized; (3) rarely fertilized; (4) never fertilized.
10. Worker laying of reproductive eggs	(1) Unfertilized bees with queenlike ovaries among the workers; (2) of regular occurrence, but workers always less productive than queens; (3) uncommon, only occasional worker laying; (4) absent in presence of queen.
11. Larval food	(1) Mostly honey and pollen; (2) major glandular component; (3) mostly of glandular origin.
12. Eggs and larvae closed in cells	(1) Yes; (2) no.
13. Progressive feeding of larvae	(1) Absent; (2) well developed.
14. Food storage for adults	(1) Absent; (2) well developed.
15. Food exchange among adults	(1) Absent; (2) via brood cells or brood food only; (3) via storage pots; (4) direct and uncommon; (5) direct and extensive.
16. Removal of wastes from larval quarters before pupation	(1) No; (2) yes.
17. Secretion of cell construction material	(1) None; (2) inner lining; (3) major constituent mixed with pollen or resin; (4) entirely secreted.
18. Re-use of cells or space where larvae are reared	(1) No; (2) frequent, but secreted part not re-used; (3) material re-used but cells reconstructed; (4) cells or space consistently re-used.
19. Stage of caste determination	(1) The adult, or a result of season; (2) late larva, or largely food quantity; (3) early larva, food quality; (4) the zygote.
20. Communication as to food sources	(1) Unknown; (2) social facilitation including odor; (3) odor trail and leading; (4) partial leading; (5) dance communication.
21. Defense at entrance	(1) Irregular and no obvious guarding; (2) constricted entrance and one guard; (3) larger entrance often with several guard bees.
22. Young in clusters or combs presumably permitting temperature or other environmental control	(1) No; (2) earthen cell walls thick; (3) earthen cell walls thin; (4) cell walls not of substrate material; (5) cell walls absent or several larvae per cell.
23. Temperature control (cooling)	(1) Absent or rare; (2) including fanning; (3) including fanning and bringing in water.

TABLE 20.2 (continued)

Variables or characteristics	States
24. Temperature control (warming)	(1) Absent; (2) individual or group incubation; (3) group warming of large area.
25. Alarm pheromones	(1) Unknown and probably absent; (2) present.
26. Survival through unfavorable seasons	(1) Fertilized females usually alone; (2) colony.
27. Response to emergency conditions such as starvation	(1) Inactivity, death; (2) absconding.
28. Generations of workers and queens	(1) The same; (2) usually different (not applicable to allodapines).

[a] ☿ = worker

implies that evolutionary changes from solitary to primitively eusocial are relatively easy—I have suggested that they might go in either direction. By contrast, a change to the level of the primitively eusocial *Bombus* or the highly eusocial Apinae is a major evolutionary development requiring changes in a variety of characters and probably taking a great deal of time. The immediate nonsocial or scarcely social ancestors of these two groups are extinct.

These findings confirm the view that the levels termed solitary to primitively eusocial are actually similar to one another in the totality of biological attributes. There is nothing wrong with providing separate terms for them, except that the terminology itself tends to lead the user to feel that the various named social levels are equally distinct, which they are not. There has been fine "splitting" among the initial stages of sociality compared to other levels.

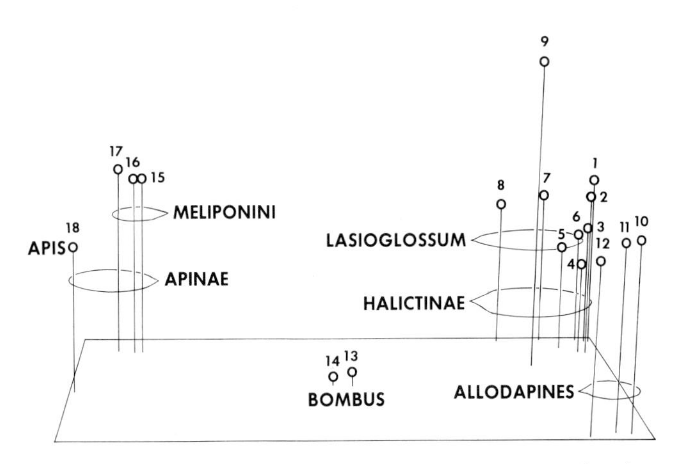

FIGURE 20.1. Three-dimensional scatter diagram showing the positions of the 18 species of bees listed in Table 20.3, with respect to the first three principal components extracted. The data used appear in Table 20.3; the characters or variables are listed in Table 20.2. (Original drawing by Barry Siler.)

It is noteworthy that within the halictine cluster, the genus *Lasioglossum* shows more social diversity than all the other Halictinae taken together, in spite of great morphological similarity. In fact, solitary species of the genus exist (Chapter 24), but were not included in the numerical study because none has been investigated in great detail. Such *Lasioglossum* would doubtless fall as far to the right in the figure as any other solitary halictine bee.

As to the meaning of the three principal components appearing as dimensions in Figure 20.1, the first (across the page) clearly relates to a number of the features generally regarded as socially significant. The right-to-left sequence of species is one of increasing social elaboration. This component (Factor I) explains 69 percent of the variance among species in Table 20.3. The heaviest loadings on Factor I are from the following characters, listed in order starting with the one with the highest loading, using the numbers for characters from Table 20.2: 24, temperature control (for warming); 11, larval food; 15, food exchange among adults; 14, food storage for adults; 25, alarm pheromones; and 1, maximum normal number of females per nest. Other characters loading nearly as heavily on this factor are 4, 5, 27, and 28, with 3, 6, 7, 20, and 23 only a little lower. Clearly these are some of the basic characters that are considered to vary with social level.

Factor II (front to rear in Figure 20.1) explains 15

TABLE 20.3. Distribution of the states of variables from Table 20.2 among various bees.[a]

Variables (from Table 20.2):	1	2	3	4	5	6	7	8	9	10	11	12	13	14	15	16	17	18	19	20	21	22	23	24	25	26	27	28
Most derived state:	5	3	3	2	4	3	4	3	4	4	3	2	2	2	4	2	4	4	4	5	3	5	3	3	2	2	2	2
Halictinae																												
1. Solitary *Lasioglossum*	1	1	1	1	1	1	1	2	N	N	1	1	1	1	1	1	2	1	N	N	1	1	1	1	1	1	1	N
2. *Paroxystoglossa*[b]	1	1	1	1	1	1	1	2	N	N	1	1	1	1	1	1	2	1	N	N	1	3	1	1	1	1	1	N
3. *Pseudagapostemon divaricatus*	2	1	1	1	1	1	1	1	N	N	1	1	1	1	1	1	2	1	N	1	2	1	1	1	1	1	1	N
4. *Augochloropsis sparsilis*	2	1	2	1	2	1	1	1	1	3	1	1	1	1	2	1	2	2	1	1	2	2	1	1	1	1	1	1
5. *Augochlorella striata*	2	2	2	1	2	1	1	1	2	3	1	1	1	1	2	1	2	1	1	1	2	3	1	1	1	1	1	2
6. *Lasioglossum zephyrum*	2	2	2	1	2	1	1	1	2	1	1	1	1	1	2	1	2	1	1	1	2	1	1	1	1	1	1	2
7. *Lasioglossum imitatum*	2	2	2	1	2	1	2	2	3	2	1	1	1	1	2	1	2	1	1	1	2	1	1	1	1	1	1	2
8. *Lasioglossum malachurum*	3	3	2	1	3	1	2	2	3	3	1	1	1	1	2	1	2	1	2	1	2	2	1	1	1	1	1	2
9. *Lasioglossum marginatum*	4	3	2	1	1	1	3	3	4	2	1	2	1	1	2	2	2	1	1	1	1	1	1	1	1	1	1	2
Allodapines																												
10. *Allodape mucronata*	1	1	1	1	1	1	1	2	N	N	1	2	2	1	1	2	1	4	N	1	2	5	1	1	1	1	1	N
11. *Allodapula acutigera*	2	1	2	1	1	1	1	2	2	3	1	2	2	1	4	2	1	4	1	1	2	5	1	1	1	1	1	N
12. *Braunsapis leptozonia*	2	1	2	1	2	1	1	2	2	3	1	2	2	1	4	2	1	4	1	1	2	5	2	1	1	1	1	N
Apidae																												
13. *Bombus hypnorum*	4	3	2	1	2	1	3	2	3	3	2	1	2	2	3	1	3	3	2	2	3	5	2	2	2	1	1	2
14. *Bombus terrestris*	4	3	2	1	3	1	3	2	3	3	2	1	2	2	3	1	3	3	3	2	3	5	2	2	2	1	1	2
15. *Trigona emerina*	5	3	3	2	4	3	3	3	4	3	3	1	1	2	5	1	3	3	2	2	3	4	2	3	2	2	2	2
16. *Trigona postica*	5	3	3	2	4	3	3	3	4	3	3	1	1	2	5	1	3	3	2	3	3	4	2	3	2	2	2	2
17. *Melipona quadrifasciata*	5	3	3	2	4	3	3	3	4	3	3	1	1	2	5	1	3	3	4	4	2	4	2	3	2	2	2	2
18. *Apis mellifera*	5	3	3	2	4	2	4	3	4	4	3	2	2	2	4	1	4	4	3	5	3	4	3	3	2	2	2	2

[a] N = no comparison, that is, not applicable, usually for logical reasons. For example, variable 9, worker fertilization, is not applicable to bees that lack a worker caste.
[b] A solitary augochlorine which makes very well-developed cell clusters (Michener and Lange, 1958e).

percent of the variance among the species listed in Table 20.3. The only character loading heavily on this factor is 13, progressive feeding of larvae. The next highest are 18, re-use of cells or space where larvae are reared, and 12, eggs and larvae in closed cells. These and other characters suggest that this factor tends to have to do with the nest construct and its functions. These characters together must be largely responsible for the forward positions of *Apis* and the allodapines in Figure 20.1.

Factor III (the vertical axis in Figure 20.1) explains 7 percent of the variance (making a total of 91 percent for the first three factors together). Although no character loads heavily on this factor, 8, queen longevity, is heavier than others. Characters 16 and 21 are next in order. No general biological meaning of this factor is evident.

This numerical study emphasizes the well-known fact that evolution is likely to involve many of the characteristics of organisms. In constructing a classification of social levels, we have selected a few attributes, and on the basis of these, we can show parallel or convergent evolution among different taxonomic groups. Thus communal behavior or primitively eusocial behavior, as defined in Chapter 5, have arisen independently in various taxonomic groups of bees. But when one considers more attributes, even though all are attributes believed to have to do with levels of social attainment, the parallelisms vanish and groupings that are strongly taxonomic emerge.

D. Selection for social attributes

The question arises how some of the social attributes discussed in preceding sections could evolve, or more specifically, how selection could operate to establish them. The problem of how selection could lead to the origin of a worker caste which has reduced reproductive ability and to its subsequent evolution has been considered by numerous evolutionists from Darwin to the present, most recently by Lin and Michener (1972), through whose paper the principal antecedent works can be located. The material below is largely derived from that paper.[1]

Much social behavior involves traits that are called altruistic. Sometimes altruism concerns an individual's own progeny, for example, when a mother bird's conspicuous activity draws a predator away from her young at some danger to herself, or when a subsocial insect guards her young, no doubt exposing herself to increased danger by so doing. Such activity is easy to explain, for in protecting her young, the mother is protecting descendants of her own genes and increasing her own fitness in an evolutionary sense. In studies of social behavior of insects, however, the altruistic traits discussed are those that enhance the reproductivity of relatives other than progeny, for example, of the mother (queen) at the expense of that of workers.

In the evolutionary sense the word altruism is in a way contradictory. Activity that promotes an individual's own fitness is called selfish, but promotion of the same genes via another individual is regarded as altruistic. A worker's activities that promote the survival and spread of her own genes, via her mother, are being considered here, as will be explained below. Altruism in the narrow sense, meaning promotion of welfare of others, without reward, is unknown (except as accidents or as a possibly aberrant phenomenon in man).

Social behavior requires evolution of mutual tolerance and social interactions of some sort among the individuals of a colony, and involves division of labor and castes, which in insects differ in reproductivity. Much social behavior (for instance, many business transactions, bartering, or exchanging in man, or common defense of a nest burrow in bees) is mutually beneficial, hence not altruistic but selfish in the sense that the participants both (or all) benefit. Such activity is here called mutualistic social behavior, to distinguish it from typical selfish nonsocial interactions in which, although one or more participants benefit, others lose. Thus social behavior can be divided into altruistic and mutualistic components. Much of the evolution of social attributes must have involved selection for both—compromises achieved through either simultaneous or alternating selection for altruistic and for mutualistic behavior patterns.

Basic questions are (1) whether social behavior and a worker caste can arise without altruism in semisocial or primitively eusocial groups, and (2) the allied matter of whether social behavior could originate among insects living as groups of unrelated individuals, or requires for its origin groups of closely related individuals, such as families. The traditional view has been that families are required, because insect societies are mostly eusocial family groups. In Hymenoptera most colonies consist of

1. Courtesy of *Quarterly Review of Biology,* Bentley Glass, editor.

a queen and her female offspring, the latter serving as workers. The existence of subsocial insects supports a familial origin, the subsocial units, which are families, being thought of as evolutionary antecedents to the well-known eusocial units.

A possible case of altruistic death of supernumerary gynes is described by Yadava and Smith (1971). In a study of the attacks of workers of *Apis* on gynes introduced into their colonies, these authors concluded that a gyne disturbed mechanically or by workers or by the presence of another gyne or queen secretes a stress pheromone which promotes attack upon herself by the workers. The postulated substance seems to be quite volatile, for its effect disappears in a few minutes. Tranquilized or anesthetized gynes are not attacked; presumably they do not liberate the material. How could the production and synthesis of such a substance evolve? In nature, excess gynes ordinarily are sisters, although sometimes having mother-daughter relationships. We might here have a case of altruism in which a gyne (*A*), in the presence of another, promotes her own death, thus assuring the success of the other gyne (*B*) and propagation of genes like those of *A*. Fights between the gynes may be dangerous to both (although such fights often occur) and thus to be avoided even at the expense of killing of one of the gynes by workers. Survival of both may lead to disruption of successful colonial cooperative behavior, and thus to reproductive failure of both.

1. Kin selection

The origin and evolution of social behavior are usually believed to depend upon kin selection. This is selection among kinships or families rather than among individuals, operating directly upon individuals without or with reduced reproductivity (workers). Workers, however, contribute to the welfare of the whole family, that is, the colony, and thus to the reproductivity of the queen who produces the sexual forms of the next generation. To the extent that members of the colony are genetically alike, the unit of selection is thus the colony, which for evolutionary purposes may be looked at as an individual. Such selection must favor the sexual forms that best represent the genetic backgrounds of the most effective workers. Usually this means selection for reproductives that are parents or siblings of effective workers. Hence familial (or other close) relationship would apparently be needed to start and carry on social evolution. Hamilton (1964, 1973) has treated such kin selection in detail, placing emphasis on the necessary proportion of genes identical by descent among siblings, between parents and offspring, or between other relatives, if altruistic traits are to arise. Altruistic traits arise, presumably, only through kin selection. Hamilton's views have been succinctly restated by Wilson (1966), Eberhard (1969), and others.

Joining or remaining with another female, as workers remain with queens, would be advantageous for a female (worker) that, as a result, becomes less reproductive, providing that as a result of the association more replicates of genes like hers were produced than she could produce if alone. Eberhard (1969) has rephrased Hamilton's basic expression as follows: Association is beneficial to a joiner if it meets the condition

$$\frac{P_{c+j} - P_c}{P_j} > \frac{1}{r},$$

in which P_{c+j} is the productivity of the colony with a joiner (worker), P_c is the productivity without the joiner, P_j is the productive capacity of the joiner if she remained alone instead of joining, and r is the coefficient of relationship between the joiner and the layer (queen). This coefficient represents the probable fraction of genes identical by descent. Ignoring sex chromosomes, r is $\frac{1}{2}$ for both parent-offspring and sibling-sibling relationships in ordinary diploid monogamous animals. In Hymenoptera, with male haploidy, r is $\frac{1}{2}$ for the mother-daughter relationship but $\frac{3}{4}$ for sister-sister. Therefore, assuming only one mating, selection for reduced reproductivity of a female hymenopteron that stays with her mother is understandable. If by so doing she enables the mother to produce more offspring, she is increasing the number of her own sisters, individuals sharing by descent $\frac{3}{4}$ of her genes, whereas if she mated and founded her own nest, she would produce daughters sharing by descent only $\frac{1}{2}$ of her genes. For a given number of young produced by the mother as a result of the daughter's joining, or alternatively by the daughter living alone, more replicates of the latter's genes would result from joining than from living alone. If these young contribute equally to subsequent generations, this situation should be one of the explanations for the several origins of eusocial behavior in Hymenoptera, as compared to only one such origin (for the termites) among all the other insects.

If the above reasoning is correct, selection for eusocial behavior in Hymenoptera is explicable but selection for semisocial behavior is difficult to explain. Even when the members of a semisocial colony are sisters, their joining is not convincingly explained by Hamilton's arithmetic unless there has been inbreeding in prior generations. When an individual joins such a unit and becomes a worker, she enhances reproductivity of her sister, thus increasing the production of her nieces. But the coefficient of relationship between the joiner and her nieces is $^3/_8$; the better strategy should be to live alone and produce offspring with a relationship of $^1/_2$.

Hamilton notes the diversity of factors that may be involved in joining. For example, the advantages of joining as an unmated worker will be increased by presence of an existing nest into which the insect can go, and by avoidance of the risks involved with courtship and mating. He has indicated how various types of behavior other than eusocial might be explained by kinship selection. To a major extent these explanations depend on "viscosity" of populations, that is, reduced vagility and resultant inbreeding so that the true relationships among individuals, considering past generations, are higher than the r values noted above, based only on the preceding generation.

The point of the kin selection theory is that one expects behavior to evolve in such a way as to make altruism more likely the closer the relationship among individuals concerned. The finding, detailed below, that distant relatives join to form colonies, requires the search for supplementary or alternative hypotheses only if joining is sufficiently frequent and joiners show altruistic behavior. Quantitative field data on frequencies and on levels of relationships (when neither mother-daughter nor sibling) are nonexistent, but reports of unrelated members in colonies are sufficiently numerous that hypotheses explaining social evolution among unrelated adults should be sought to supplement the theory of kin selection.

2. Indications that social groups frequently contain distantly related individuals

a. Semisocial colonies. Certain halictine species (*Augochloropsis sparsilis* and *Pseudaugochloropsis* spp.) make colonies which in the mature condition appear to be semisocial (Chapters 5, E3; 6, B2; and 22). The data are based on numerous nest censuses and examinations of ovarial conditions of the females, the generation to which the bees belong being judged by relative wear of wings and mandibles and knowledge of the life history. I do not believe there is any doubt about the semisocial nature of the colonies, although the data could be regarded as equivocal since the bees of no one colony have been followed as individuals through their lives.

Nests of these bees are often re-used, and inhabitants of old nests are likely to be sisters that have retained their association with the nest and hence with one another. New nests, however, are started by single individuals and if inhabited, as is sometimes the case, by two or more bees, the others joined subsequently and are not likely to be close relatives.

At least some of the nests occur in aggregations so that one might explain the worker caste as altruistic and its evolution as based on the viscosity of populations—a high rate of inbreeding and tendency of females to return to the vicinity where they were reared. Such an explanation for relationships among females much higher than the r determined from the preceding generation may be correct, but it should be noted that mating of these species does not occur around the nest sites. Presumably it occurs on flowers or elsewhere where individuals from various nest sites meet, with resultant outbreeding.

Numerous primitively eusocial halictine bees facultatively or regularly begin their colonies as semisocial units which subsequently become eusocial. In these cases, as with permanently semisocial species, a nest started by a single individual (an acceptor, or future queen) attracts joiners (auxiliaries), which become workers although they are presumably gynes (potential queens). Such joiners are not always sisters of the queen, although temporarily semisocial groups are probably often sisters that overwintered in the parental nest.

The occurrence of a semisocial stage in the ontogeny of numerous eusocial colonies is suggestive of recapitulation, if the suggested origin of eusocial behavior of some groups from semisocial antecedents is correct (Michener and Lange, 1958d; Michener, 1958). In some halictines the auxiliaries die or disappear about the time of the emergence of the first ordinary workers, leaving a eusocial colony. In *Halictus scabiosae,* as noted in Chapter 6 (D), the queen becomes increasingly aggressive before emergence of her first progeny and drives off her auxiliaries even though they may be collecting pollen for her nest. They then establish nests of their own (but with what productivity is not known). On the other hand, in *H. ligatus* foraging by auxiliaries continues for some time after appearance of daughter workers. Whatever the details, the colony soon (by summer in temperate climates where nesting begins in spring) becomes eusocial, matrifilial, and monogynous as though it had been started by a single gyne.

b. Drift from nest to nest. Workers in eusocial colonies drift from nest to nest to a degree not necessarily consistent with the idea of colonies as matrifilial units in which kin selection could readily be a dominant evolutionary force. Thus 19 percent of the workers of a population of *Lasioglossum duplex* joined colonies other than those in which they were born and marked (Sakagami and Hayashida, 1968). Such workers, when not reproductive, are not only not contributing to the propagation of their own (or their mother's) genotype, but are contributing to propagation of different and presumably competing genotypes. Both in the laboratory and in the field, workers of *Lasioglossum* (*Dialictus*) often enter foreign nests; in dense aggregations of *L. versatum* the same bee sometimes carries pollen into different nests (Michener, 1966c).

In some eusocial forms there are behavioral patterns which largely prevent transfer of workers from nest to nest (Chapter 18, C). However, such patterns may have evolved to prevent robbing rather than to prevent transfer, and the colonies do not always consist only of the queen and her offspring. In *Apis,* drifting of workers from one colony to another is sometimes considerable (up to 16.7 percent of marked workers, Sekiguchi and Sakagami, 1966) and may be enough to influence honey production in spite of the well-known colony defense which often results in exclusion or killing of foreign workers.

In the laboratory, numerous observations indicate the attractiveness of existing occupied halictid burrows as well as the tolerance of the original nest makers to joiners. For example, in one instance field-caught overwintered gynes of *Lasioglossum zephyrum* and *Augochlorella striata* were liberated in a bee room containing a number of nests of the same two species. The nests were in earth between glass plates (Figure 24.8) for observation purposes (Michener and Brothers, 1971) and thus the entrances were not in the kind of position where nest entrances of the species would ordinarily be found. Nonetheless, although some of the field-caught females excavated their own burrows, most that survived entered existing burrows of their respective species, forming semisocial colonies of two, or in one instance three individuals. Similar experiments with workers of *L. zephyrum* showed that they, too, entered artificial nests inhabited by lone workers, forming colonies of up to four bees.

In another series of experiments on *Lasioglossum zephyrum,* female pupae from the field, which normally would mostly have become workers, were placed in artificial nests in a room without males. In some cases only one pupa was placed in a nest, in others, several. When the pupae matured, some of the resulting adults established themselves in the artificial nests, made cells and reared young (all males). However, by no means all of the females remained in the burrows where they emerged. There was considerable interchange among different nests, as shown by observations of marked individuals; sometimes a nest that had originally been established with a single pupa in it later contained three or four bees. Although some individuals excavated new burrows, the attractiveness of existing burrows and the acceptance of "joiners" by those already there resulted in the establishment of semisocial colonies showing division of labor with respect to both foraging and ovarian development (Michener, Brothers, and Kamm, 1971a). To carry out observations on nests with minimal interchange of bees, we had to separate the nests far more widely than is usual in nest aggregations in the field. Field observations of this species and of other primitively social *Lasioglossum* that form aggregations of nests strongly support the view that in nature, also, there is much exchange of individuals among the nests. Such exchanges have been verified in the field with marked individuals for *L. versatum, L. duplex* (see above), and in a very few instances for *L. imitatum.* As Sakagami and Hayashida point out, some drifting of marked bees from nest to nest may be a by-product of handling and marking, but much of the laboratory data and some of the field data and impressions do not involve marked individuals.

The above examples, along with others from wasps and ants, indicate that inhabited nests or the colonies in them are highly attractive to other (joiner) individuals of the same species and that nest builders often are acceptors. This is probably the usual situation in nest aggregations of halictine bees and perhaps other groups. The resulting colonies are therefore not closed groups and often contain individuals not necessarily closely related to one another. Among the unrelated individuals there is usually division of labor such that some (workers) enhance the fitness of others (queens) while perhaps reducing their own reproductive effectiveness, that is, they may (or may not) be altruistic. The mere existence of such colonies does not show that kin selection is unimportant; much depends on the degree of inbreeding, about which there is virtually no information, and the frequency of movement from colony to colony. Hamilton (1973) shows the degree of inbreeding necessary to make altruism worthwhile in colonies of various sizes.

3. The reproductive system of workers

A worker has a functional reproductive system. It produces eggs in smaller numbers than that of a queen, and in some species produces them only under certain circumstances, for example, in the absence of a queen. Even though the worker caste probably dates back to the Eocene (Chapter 2, B), ovaries and the related connections and biochemical and physiological mechanisms, as well as reproductive behavior such as laying at the proper times and place, have persisted. By contrast, other useless organs (such as wings in worker ants) and mechanisms (mating responses and functional spermathecae in *Apis* workers) have been lost by workers. Comparative anatomy is full of examples of losses of structures and functions that are useless, such as the eyes of cave fish.

It is therefore safe to conclude that laying by workers has a reproductive function important enough that selection maintains the reproductive system and behavior. This fact does not mean that kin selection, producing altruism, has not been involved in the evolution of a worker caste, but it does mean that mutualistic social behavior or selfishness has been involved. Presumably there has been an interaction between altruistic and mutualistic evolution. The larger the productivity of the joiner or worker, the less relevant is the coefficient of relationship to an understanding of evolution of the worker caste. If kin selection were highly predominant over individual (selfish or mutualistic) selection, one would expect the reproductive system to be functionless and vestigial or absent in workers of the more highly eusocial forms.

4. Influence of natural enemies and other environmental factors

A major factor favoring colonial behavior, irrespective of the mode of origin of the colonies, is almost surely parasite and predator pressure. In ground-nesting groups like Halictinae, this means pressure of the natural enemies in aggregations of nests; scattered nests presumably experience less such pressure and the population is less likely to evolve social behavior.

Other advantages of social behavior in established and large colonies must include the ability to control the internal nest environment. This control is provided by the numerous individuals living together with the help of integrative mechanisms, such as food storage and temperature control. The highly eusocial way of life has the further advantage of permitting use of widely dispersed food supplies by a large biomass. Imagine a single animal weighing as much as an *Apis* colony trying to survive by sucking nectar from individual flowers. This advantage can be fully developed only with a system of communication about food sources and is doubtless facilitated by the existence of specialists (through caste and temporal division of labor) for various functions.

The above comments concern presumed advantages for established societies, but to return to the origin of social groups: what are the advantages of a colony of three or four bees over the solitary condition? The number of offspring per female (both castes) in colonies found and censused in the field decreases as the colony size increases (Michener, 1964a). Apparent exceptions to this rule have been found in *Bombus* (Michener, 1964a) and various *Braunsapis* (Michener, 1971), but it is applicable to a wide variety of social insects, including bees from *Augochloropsis sparsilis* to *Apis*. Even in facultatively colonial species, lone individuals produce more offspring per female than do females in colonies (Figures 20.2 and 20.3). In forms that produce workers, production of offspring is not synonymous with production of reproductives for the next generation, but the two figures are no doubt positively correlated; a colony containing many workers will on the average produce more reproductives than a colony with but few workers. Therefore the production of offspring (either sex or caste) is an indication of the success that the colony will have in contributing to the next generation.

From the rule mentioned above, one might suppose that evolution should reduce colony sizes and establish solitary behavior, since the smaller the group the greater the contribution per female to the next generation. The number of reproductives maturing and producing the next generation presumably really does not vary in the way indicated, for if it did, social groups would never have evolved. Control of environmental factors (including

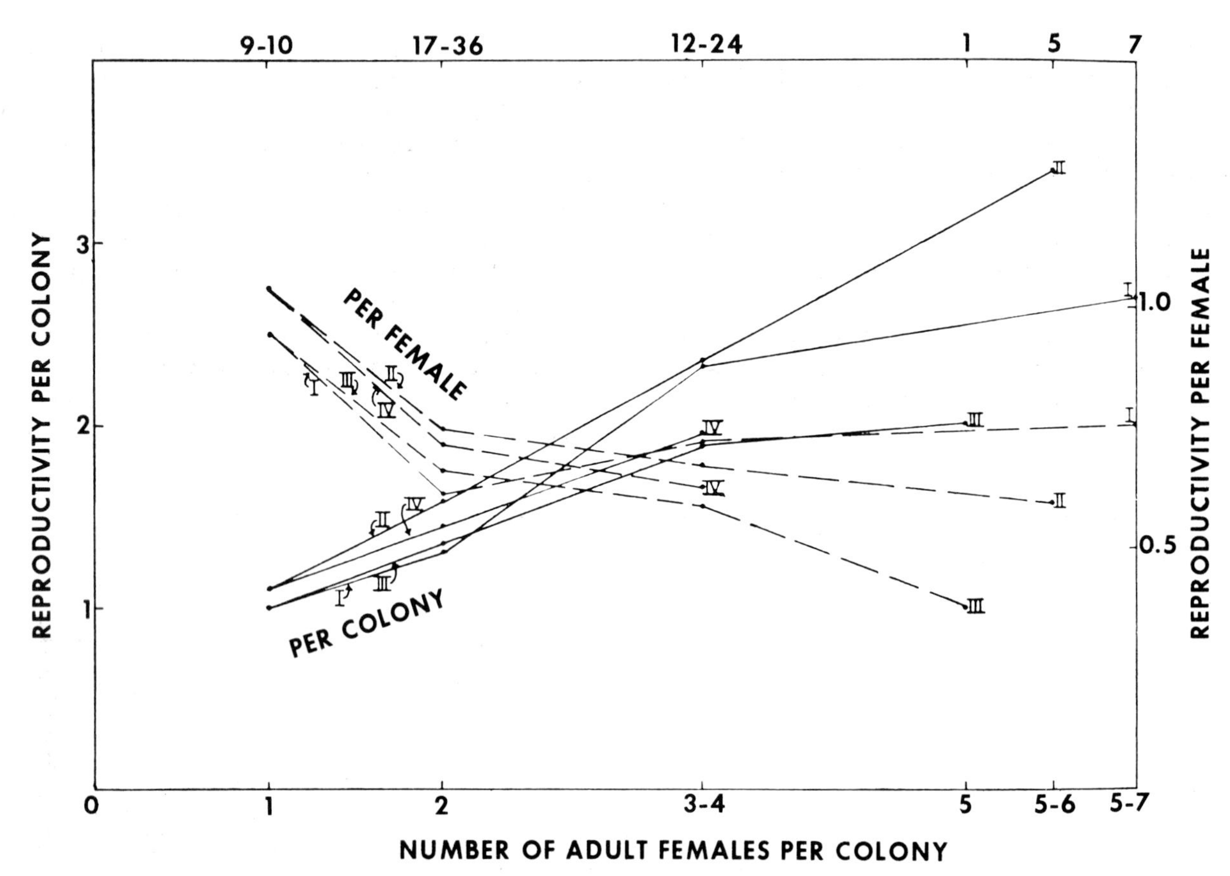

FIGURE 20.2. Mean reproductivity per colony (*solid lines*), based on the number of small larvae, eggs, and pollen balls in each colony, and mean reproductivity per female (*broken lines*), for nests of the semisocial halictine *Augochloropsis sparsilis* containing from one to seven females (queens plus workers). The numbers across the top of the graph indicate the number of nests upon which points in each vertical line are based. The different lines, marked by roman numerals differ in the way in which the numbers of bees in the nests were counted. *I* is based on total numbers of adult females. *II* differs in that unworn adult bees with slender ovaries ("inactive" females) were omitted from the nest censuses. *III* differs from *I* in that data were reduced to single queen equivalents; for example, if there were two queens in a nest, all census data were divided by two to obtain figures comparable to those from the more numerous monogynous colonies. *IV* is like *II* but reduced to single queen equivalents. (From Michener, 1964a.)

parasites and predators) should be better the larger the colony. It is unlikely that control of physical factors is appreciably better by a very small colony than by a lone individual. Natural enemies must therefore be the main factors removing colonies and lone individuals of parasocial and primitively eusocial forms from the census figures by exterminating them, presumably having more effect on small colonies and lone individuals of a species than on large, better-defended colonies. It therefore follows that the rule of inverse relation of colony size to individual productivity of colony members should be fallacious if applied to whole initial populations of parasocial and primitively eusocial forms; it is derived from census figures of surviving individuals or colonies and does not take into account those that have been killed. The applicability of the same rule to forms like *Apis* living in large colonies and even in hives cared for by man may result from entirely different factors, for example, reduced efficiency—getting in each other's way and the like—under crowded conditions. Parasite-predator pressure, in any event, can safely be considered one of the main factors leading to the success of social behavior (Lin, 1964; Michener, 1958).

Pressure by parasites or predators probably varies enormously and is likely to become devastating in some aggregations of nests. I have seen an aggregation of *Lasioglossum versatum* decimated by mutillids, and near Lawrence, Kansas, large aggregations of *L. zephyrum* have likewise faded away, apparently as a result of mutillids, myrmosids, and ants in different instances. Extinction may have occurred frequently for demes and species. It is perhaps significant that near Lawrence, Kansas, of several species of *Lasioglossum* (subgenus *Dialictus*) studied, the species that consistently forms the largest, densest aggregations (*zephyrum*) is the one with the best-elaborated defense behavior (Batra, 1965, and Chapter 18, A1).

Even the mere presence of one or more bees in a nest, while others forage, provides improved protection against intruders. In halictines guarding is uncommon when a nest houses only a single individual (Chapter 18, A1). During the periods when she is in the nest, even when apparently not occupied with other activities, such an individual does not spend much time at the nest entrance; perhaps the risk is excessive. However, when two or more individuals are in the nest, it is usual for one of them (not always the same one) to be at the entrance during

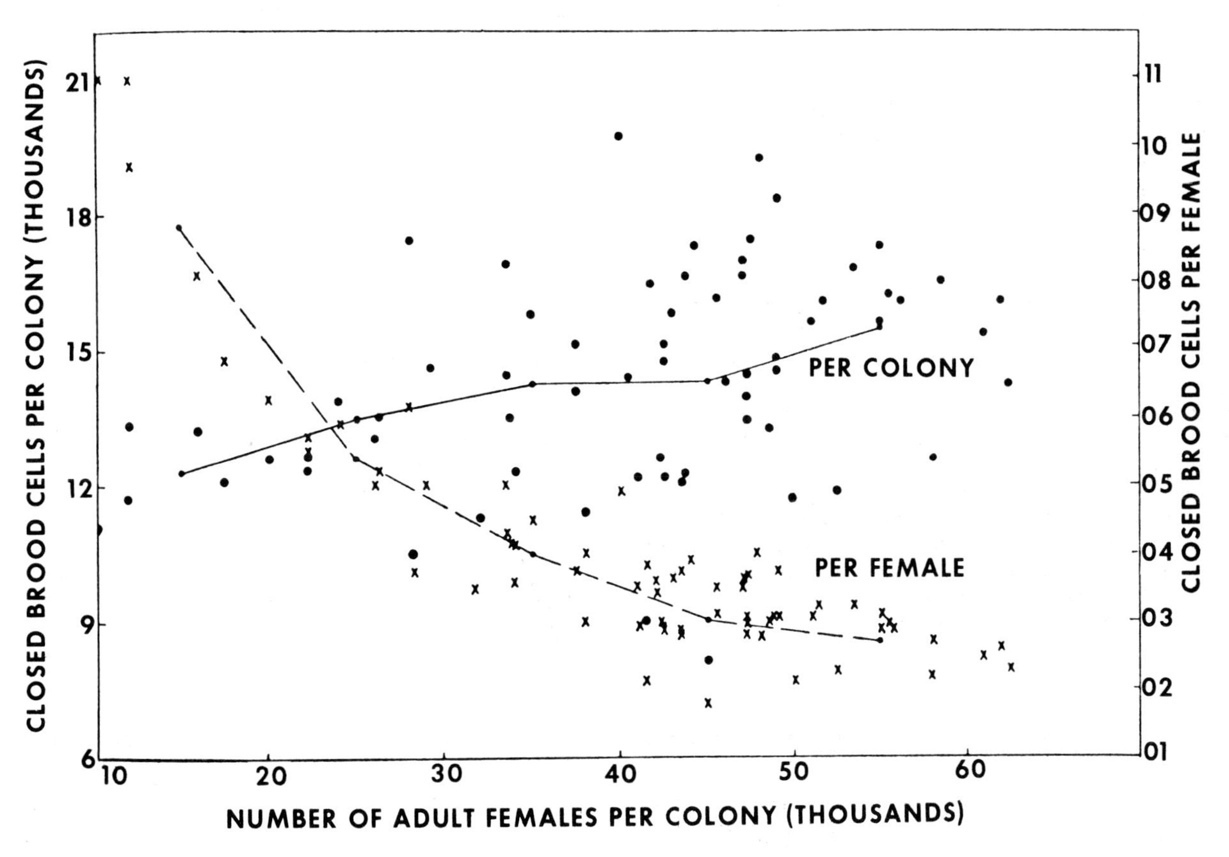

FIGURE 20.3. Reproductivity per colony (*dots*) based on numbers of closed cells (mature larvae and pupae), and reproductivity per female (*crosses*), in hives of *Apis mellifera* containing various numbers of females (workers and queen). Censuses of colonies were all made from June 23 to September 24, 1930. *Solid line* shows mean reproductivity of grouped colonies. *Broken line* shows mean reproductivities per female for grouped colonies. Colonies were grouped on the basis of population, each group having a range of 10,000. Means were not calculated for the largest colonies because, although they would continue the trends indicated, the number of such colonies was so small as to make such means of little value. (From Michener, 1964a; data from C. L. Farrar.)

the day and to exhibit defensive behavior, and there are various records of intruders (mutillids, ants) being thwarted by the guard (Batra, 1965; Lin, 1964).

5. Selection for joining and accepting behavior patterns

Evolution of joining and accepting behavior and the beginning of semisocial or eusocial colonies can be accounted for if we make the assumptions that (1) there is considerable variation in reproductivity of females even in solitary forms, (2) highly productive individuals are likely to make their nests first, before the less productive ones, thus increasing their advantage and (3) the percent mortality of offspring of all individuals in a colony is equal, and it is lower than that experienced when individuals are alone in their own nests. That these views may fit the facts is indicated as follows.

(1) The number of cells made by solitary females of halictine bees (*Lasioglossum* and *Halictus*), especially of the eusocial species, is highly variable. Even when not eliminated by a predator, lone females are often found that have stopped making cells and merely remain in the nest after completing only one or two, while others of the same species at the same time and place make eight or more. The variation is usually less in solitary species, but minimal producers often make less than one fourth as many cells as maximal producers in solitary species of *Lasioglossum* and *Halictus* and less than one half as many in solitary species of other halictid genera. Probably the low production is environmentally mediated. Otherwise it should have been improved by selection. The cells and protection given the young by low-producing females seem no better than by females that make more cells.

(2) Laboratory experiences with halictine bees (Michener, Brothers, and Kamm, 1971a) indicates that individual bees which establish nests early usually make more cells, while those that delay make but few. The paper cited concerns artificial colonies established in the laboratory in summer, but the same is true of the colonies established in spring by overwintered gynes. Bees that make nests early must at least start their own, they cannot be joiners, while those that become reproductive later can join if the earlier ones are acceptors. Thus a negative relation could be maintained between reproductivity and tendency to join. The same thing presumably happens in natural semisocial groups, whether permanent or temporary. Joiners seemingly would be unsuccessful if alone, or at least delayed and at a considerable disadvantage (also in *Polistes;* see Eberhard, 1969).

(3) The lowered mortality rate for offspring of females living in colonies would be due largely to improved defense, as indicated in the preceding section, and would be favorable to both low and high producers in a colony.

In nearly all eusocial insects, differences between queens and workers are environmentally, not genetically, determined, although the ability to produce such forms is of course provided for genetically. Selection for joining and accepting (worker and queen) behavior, hence colony formation, would occur in a population initially no more variable in reproductivity than that of solitary ancestors: that is, the joiners would initially be no less reproductive than the less successful females in a population of solitary bees. If such relatively unsuccessful bees are produced anyway and leave more offspring by joining, selection should favor factors that promote their doing so. Indeed selection would favor the accumulation of traits promoting mutual tolerance and cooperation, for both joiners and acceptors would be more productive in colonies than when alone. The colonies could be semisocial, the behavior mutualistic and without altruism, or with altruism if close relationship of colony members allowed kin selection to operate at the same time. Or the colonies could be eusocial, in which case it is more likely, although not inevitable, that kin selection and altruism would predominate.

A significant factor in the origin of joining behavior in some groups may be that joiners are in excellent positions to take over and become queens if the first queen dies. If the probability of success of a lone female is low, it may be good strategy for a female, especially if seasonally delayed, to join, survive, and await the possible demise of the queen.

6. Familial and semisocial evolution of eusociality

Once a joining class exists and is advantageous, selection would probably favor colonies that produce their own joiners. Such colonies would be more certain of getting joiners than if the latter came from other nests. After eusocial family groups exist, evolution of special queen and worker features, including reduction in worker productivity, become possible by kin selection.

Eusocial relations might arise through familial pathways in either of two ways. One is via subsocial forms. The other is essentially the same except that, because of the mass provisioning and the closed rearing cells of most bees, the growing young are not in contact with adults. On maturation, however, the young may remain with their nests or associates, including their mother. Sakagami and Michener (1962) suggested the latter as a possible mode of origin of eusociality in halictine bees, there being no subsocial forms in this group. In either case, survival of the mother until maturation of her offspring, a not uncommon occurrence in certain solitary bees, could result in a matrifilial or eusocial colony of adults.

Alternatively, the existence of semisocial colonies shows that in some societies castes have arisen among groups of individuals of about the same age. Such groups may consist of sisters or of not particularly closely related females (Michener and Lange, 1958d). From such semisocial groups, eusocial units could arise, as is seen in the ontogeny of numerous primitively eusocial halictine colonies.

For simplicity of expression the first route toward eusociality (either with or without true subsociality) is called the *familial route,* and the alternative is called the *semisocial route.*

The familial route seems almost certain for the allodapines in which each colony starts as a subsocial unit. There are various reasons for regarding the semisocial route as probable for at least some of the other groups of bees. It is likely that even within the Halictinae, some eusocial forms evolved via the familial route, others via the semisocial route. Aggregations should be particularly favorable places for establishing colonies. Aggregations contain a supply of occupied nests. The nests often contain acceptors, and an aggregation provides a good supply of potential joiners. The number of kinds of bees that nest in aggregations is enormous; thus there is ample potentiality for the establishment of colonial behavior. The very existence of semisociality, both temporary and permanent, suggests the semisocial route. In contrast, there are few subsocial bees (only allodapines and an early stage in the ontogeny of colonies of *Bombus*). In no bees do two generations of adults (parents and offspring) live and work together without caste differences. Such forms ought to exist somewhere as a step toward eusociality if evolution were largely via the familial route.

Semisocial joiners produce "instant colonies" and the resulting advantages are therefore available early in nest and colony development. The eusocial route provides adult joiners only after a considerable time for rearing young. The possibility exists that an initially eusocial organization might serve as a preadaptation for joining of auxiliaries early in the season when no daughter workers are present. This sequence, that is, early season semisociality derived evolutionarily from later season eusociality—the reverse of a suggestion made elsewhere—may have occurred in some cases. However, the young gyne fends for herself, without auxiliaries, in all the more elaborately social of the primitively eusocial forms (such as *Bombus*). Such forms are the ones which should have had time to evolve semisocial initial stages, had evolution been from eusocial to semisocial. Colonies pass through a temporary semisocial stage in species that have minimal caste differences and hence are presumed to be socially primitive. These forms have minimal caste differences not only in the semisocial but also in the eusocial phase. It is therefore probable that, from the viewpoint of sociality, the semisocial condition is primitive.

7. Joining and generation time

Given that joiners are reproductive even if only to a limited degree (see section 10d below), colony formation as a device that shortens generation time could be favored by individual selection operating on both joiners and acceptors. Such selection could operate both in semisocial and eusocial groups. Under certain laboratory conditions, lone individuals of *Lasioglossum zephyrum* make their first cells an average of 11.2 days after maturity. Colonies of two and of four females make their first cells 8.6 and 6.6 days after maturity of the first female (Michener, Brothers, and Kamm, 1971a). Probably both joiners and acceptors participate in the reproductive acceleration. In that case a joiner, even if its own productivity is reduced by joining, may in the long run produce more replicates of its genes by joining and laying a few eggs than by remaining alone, because of shortened generation time as well as for the reasons cited elsewhere.

8. Environmentally disfavored females as workers

While behavioral features such as mutual tolerance, accepting, and joining are likely to be under genetic control, caste differences in most highly eusocial insects are direct environmental effects. The same is likely for differences between acceptors and joiners in semisocial and primitively eusocial forms. Perhaps joiners in such species are those produced at less favorable seasons, when males also are relatively small (Michener and Wille, 1961; Batra, 1966b). For example, females maturing during periods of unfavorable weather or food shortage may be biased to become workers, or possibly they are provided by the mother for some other reason with below-average amounts of food. Reduced size, reduced reproductive potential, and reduced mating probability are likely to go together; these are workerlike attributes.

If a female has the potentiality to produce joiners which are not genetically mediated as such, it will be better to go ahead producing such individuals during unfavorable seasons than to produce nothing. A main feature of semisocial and eusocial species is the production of offspring through a long season, some of them much less reproductive than others. Production of individuals to become workers would presumably be accentuated by selection at the individual level for queens that reproduce during unfavorable seasons. The reduced productivity of the workers could be a resultant circumstance, not something engendered by kin selection.

The effect of the external environment might well be replaced evolutionarily by the queen's own modification of the environment in which her progeny are reared, so that as a result of the food, pheromones, or other nest conditions, she produces joiners (workers) during part of her life. Selection operating on queen productivity and her related ability to produce and control joiners could be a major factor in the evolution of eusocial groups. In this case the resultant workers are to be looked at as appendages of the queen; their altruism is irrelevant, but the group must be eusocial.

9. Evolution of caste differences

The fact that castes exist suggests that functional specialization is advantageous, and that selection has promoted mechanisms for producing division of labor by means of specialist individuals and for diminishing the number of intermediate individuals or intercastes which might not function usefully. The problem of intermediates is minimal when the castes are very similar to one another; when they are very different, intermediates are essentially nonfunctional, and selection against them must be strong.

Initially, females should be much alike and castes should intergrade completely. Castes would differ in the number and sexes of their progeny; those of queens would be more numerous and of both sexes, those of unmated workers, less numerous and all males. Mutual benefit from joint nest construction and defense can be looked at in part as benefit to the queen by workers. And it is likely that the more productive queen could sometimes lay in a cell provisioned by others and perhaps eat in the cell, thus making foraging trips less necessary for her, perhaps increasing her longevity. Such an organization, perhaps integrated by a dominance hierarchy, is as far as most semisocial and some eusocial colonies go.

Fertilized eusocial workers in *Lasioglossum zephyrum* (Batra, 1966b) are not recognizably different from semisocial workers (auxiliaries), except that they are associated with their mother instead of with sisters or others of their own generation; neither type of worker is distinguishable from queens except by degree of ovarian development and a slight average size difference. Among eusocial halictines, series such as *Lasioglossum zephyrum, L. versatum, L. imitatum,* and *L. malachurum* support the views explained above on the evolution of caste differences. The species in this series (only the first three are close relatives) show progressively increasing differences in size and in ovarian development between castes, decreasing frequency of worker mating, increasing queen longevity, and decreasing spring and early summer male production. It is presumably a series showing increasing efficiency of the two castes in their different roles.

10. The influence of male haploidy

The outstanding genetic fact about Hymenoptera is that males, unlike females, typically develop from unfertilized eggs and are haploids. This has an effect on the evolution of the parthenogenetic worker caste, but it also has more general influences.

Fully diploid genetic systems contain large numbers of rare recessive genes. There is much phenetically hidden genetic variability potentially available for subsequent evolution. But a beneficial mutant often is at first recessive, and there is a high probability that it will be eliminated before it evolves dominance or becomes fixed. This must be a weak link in the evolution of adaptations.

In a male haploid system, on the other hand, all recombinations and genes expressed in males act as dominants; if beneficial they are favored, if deleterious, they are selected against. Because of lack of meiosis, sperm cells are all identical in genotype to the male that produced them. For these reasons beneficial genes are likely to be widely and promptly spread and their fixation is more likely than in a diploid system.

a. Genetic variability. Male haploidy results in reduction of variability. This statement applies especially to highly deleterious alleles, although balanced polymorphisms may be as common in haplo-diploid species as in fully diploid ones (Crozier, 1970). At any one fertilization, only three fourths as many chromosomes are involved as in a fully diploid system.

Diploid organisms carry numerous highly deleterious recessives. Inbreeding results in some of them becoming homozygous, commonly with disastrous results to the individuals and populations concerned. Because of haploid males, Hymenoptera lose genes that are highly disadvantageous to males, and therefore should be partly immune to the deleterious effects of inbreeding.

Immunity to the effects of inbreeding presumably permits fixed nesting sites, aggregations, and the resultant conditions favorable for establishment of social behavior. The small number of queens (often only one) found in most colonies means that the genetically effective population is small even though the total population may be large. The result is increased inbreeding, which might sometimes be prohibitive in fully diploid forms.

There is a selective advantage for a female to return to nest at the site where she was reared: it is a site that was used successfully once and hence is likely to provide proper conditions again (Lin, 1964). Long-lasting nest aggregations may lead to build-up of parasites and predators, then to defensive evolution against them, and thus set the stage for social evolution, since social evolution is in part a response to the need for defense. Although some aggregations start with almost simultaneous arrival of many females (Chapter 6, A), others start from a single nest, with establishment of others nearby, built by descendants of the first arrival. In such cases at least, not only inbreeding but genetic drift must lead to reduced variability. The close relationship of the members of a population probably plays a significant role in social evolution by permitting kin selection more effective than that found in strongly outbreeding groups.

Another mechanism that limits variability, the tendency of Hymenoptera to mate once and store sperm, may be made more limiting by the identity of all sperms received at one mating.

In spite of male haploidy and the resultant reduced variability, there are obvious needs for variability if Hymenoptera are to adjust to locally and temporally varying conditions or to evolve at all. Not surprisingly, there exist various mechanisms which increase variability. There is no evidence that Hymenoptera actually suffer from reduced genetic variability. Potentially, however, it is reduced. The following can be interpreted as countermeasures—devices that increase variability: (1) multiple insemination (*Apis* and some other bees); (2) multiple queens, either simultaneously as in polygynous forms or successively (as in *Apis*); (3) queen, worker, or male transference or joining; (4) worker laying of male-producing eggs and sometimes of female eggs; (5) worker mating (as in some primitively eusocial bees); and (6) common mating places where young gynes and males from different colonies in a considerable area congregate. Most of these mechanisms occur in other groups as well, but their frequency in Hymenoptera is impressive and may be related to maintaining genetic variability.

Activities that promote increased variability at first seem to contradict the view that close relationships are necessary for the evolution of altruism in the initial stages of social evolution. But the activities listed above are noted primarily in highly eusocial colonies where kin selection among colonies is probably effective. The actual breeding structure must be a compromise, which doubtless has varied with changing conditions, sometimes favoring enhancement of altruistic features, at other times favoring outbreeding and increased variability. That outbreeding is common is suggested by the quick appearance of natural interracial hybrids when two or more subspecies of honeybees are kept in the same area.

b. Relative plasticity of males and females. The question arises as to why female Hymenoptera have so frequently developed two castes while males have not. (Males of a few species are dimorphic, but related behavioral differences suggestive of the female castes are unknown.) As indicated earlier, the close male-daughter relationship compared to the male-sister relationship might be responsible for lack of male altruism and castes, although morphological and behavioral attributes typical of males are probably more important. Another factor, however, is probably the haploid condition of males.

Diploids, with twice as many genes, may be more able to develop plasticity, so that different environmental conditions trigger different lines of development. As Williams (1966) notes, a facultative adaptation implies possession of instructions for two or more alternative somatic states (for example, worker and queen), together with sensing and control mechanisms to adjust the response to the environment. Haploid organisms are presumably less likely to possess such flexibility.

c. Sex ratios and determination. The haplo-diploid sex determination mechanism has an important influence on sex ratios, as it allows them to evolve freely, very far from one-to-one. Hamilton's (1964) views are that sex ratios in Hymenoptera should never be male biased. They sometimes are, for example, *Pseudagapostemon divaricatus,* 60 percent, and *Lasioglossum marginatum,* 76 percent males (Chapter 8, B). However, most ratios are female biased. But if, for social species, one considers only reproductives, as is fully justifiable for forms like *Apis* whose workers lay only or principally in emergencies, sex ratios are almost always strongly male biased. This is more consistent with outbreeding than with inbreeding and does not support the idea that kin selection operates beyond the immediate family units because of viscosity of populations. It seems likely that the evolution of sex ratio depends more on the natural history of the species than on the greater genetic contribution of females than of males to succeeding generations.

The sex-determining mechanism with one or two *x* loci producing females, sometimes only when heterozygous and never when hemizygous (haploid) (Chapter 8, A), is of course closely related to male haploidy. Inbreeding, such as would result from high viscosity (Hamilton, 1964) in a population, is not compatible with the sex determination system of *Apis.* In *Melipona, Meliponula, Trigona,* and perhaps most bees, the sex determination system could tolerate more inbreeding than that of *Apis* (Chapter 8, A1), and for at least the meliponine bees is associated with queens that mate only once.

d. The parthenogenetic worker caste. In almost all social Hymenoptera, as already suggested, some or many of the workers lay eggs. We are commonly not dealing with a nonreproductive caste, but with individuals that are in varying degrees less productive than queens, and in some cases are productive only under certain circumstances. In semisocial colonies lack of mating sometimes goes with joining or workerlike behavior. In eusocial colonies nearly all workers are unmated, although there are exceptions among primitively eusocial species.

In one sense laying workers are competing with the queen. However, at least in primitively eusocial forms like Halictinae, they appear also to work for the queen—guarding, provisioning cells in which she lays, and so on. In *Bombus,* once their ovaries enlarge, workers probably do not work for the queen except that they become aggressive and probably are important in defense of the nest. And in all species in which queen-produced males are scarce, there should be a premium on laying workers as sources of males. Or male production may be looked at from the queen's viewpoint: Given a limited productivity, she will transmit more genes to subsequent generations by devoting that productivity to diploid females rather than haploid males. Perhaps she cannot control the laying potentials of her workers; in this case laying mostly female eggs will be good strategy, letting the workers provide the males.

The importance of workers as male-producers has been discussed in Chapter 8 (A3). For some halictines, *Bombus,* and *Trigona,* eggs laid by workers seem to be the main source of males. Data on the origins of males in normal colonies are difficult to obtain and many more observations are needed to properly evaluate the importance of workers as male-producers. A queen doubtless produces more reproductive progeny than does any one worker, but since males commonly mate more than once, at least in primitively eusocial halictines, a worker's genes may be widely distributed by her male offspring.

Also important in considering male production by workers is the genetic nature of the males produced. Males and females are generally very different both morphologically and behaviorally. Many of the attributes of the two sexes must be controlled by different genes, and selection will operate very differently on the two sexes. For example, in most familiar temperate climate forms, like *Bombus* and halictines, females must survive the winter while males do not. The haplo-diploid sex determination mechanism is such that a male's relation to his "sons" is 0; his chromosomes go entirely to daughters. This means that selection operating on males of a population cannot influence males of the next generation, but only those of the one after that.

In primitively eusocial forms in which workers probably produce most of the males, however, males can contribute genetically through their parthenogenetic daughters (workers) to the next males that are produced. In other words, a male receives both maternal (queen's) and paternal (queen's mate's) genes via a worker mother. Thus selection operating on one group of males (for example, those active in the autumn of one year) will have an influence on the next such group (those active the following autumn). As a result of meiosis, males produced by workers also have the advantage of genetic recombinations, to some degree making up for the loss of variability that results from male haploidy. Workers thus prove to be signficant in improving the variability among male offspring. Such matters may have been significant in promoting, in Hymenoptera, the origin of a worker caste that does not mate, and in the establishment of eusocial behavior. The advantages of workers producing male offspring and of the queens producing these males via daughters (workers) rather than directly, must outweigh the disadvantage to the queen of such males sharing on the average only one fourth rather than one half of her genes.

11. Conclusions

A whole series of factors acting jointly or alternately is responsible for the several origins of eusociality in bees and other Hymenoptera, compared with only one origin in all the other insects. A key factor is the haplo-diploid genetic system, associated, of course, with brood care and ability to make and return to the nest, that is, the first of the prerequisites listed in section A above. Most of the other factors are operative only because of that system.

The following are particularly important results or concomitants of the haplo-diploid system: (1) partial immunity to inbreeding resulting from loss of lethals, allowing long-term aggregations and colonies and effective kin selection, (2) sperm storage by females, absence of a king in social groups, and the control of the sex of offspring, permitting the extraordinary sex ratios found in highly eusocial Hymenoptera, (3) the importance of male-producing workers in at least some primitively eusocial and highly eusocial forms, and (4) the closer relation of sisters to one another than to their daughters, an arrangement that encourages effective kin selection. Male-producing workers increase genetic recombinations and permit retention of the advantages of a haplo-diploid genetic system, while also permitting some advantages of the ordinary fully diploid system. The loss of variability due to male haploidy may be partly responsible for the evolution of outbreeding devices, including acceptance of joiners. The joiners are ordinarily workers which, as male producers, will increase variability, but may in error be related species, resulting in the potential for evolution of social parasites. Immunity to inbreeding permits aggregations, in which pressure from natural enemies is likely to promote development of colonies.

In the semisocial and most primitive eusocial Hymenoptera, queens and workers differ little more than do females of related solitary species. The most important factor favoring mutual tolerance and colony formation is need for defense. If both the more and the less reproductive individuals profit from the association, selection will favor colonies and result in mutualistic social behavior. The more reproductive individuals tend to start nests first and to be joined by delayed reproducers, which are often workerlike. Thus semisocial colonies of very similar castes can arise without kin selection or altruism. Such mutualistic social behavior is a preadaptation to eusocial evolution and provides immediate improved protection for the nest. Joiners will be more certain to arrive, although much delayed, if they are progeny of the colony or lone founder. This leads to eusocial colonies among which kin selection can operate. Eusocial behavior need not but probably often does arise from semisocial behavior, as suggested (1) by the frequency of nest aggregations providing numerous acceptors and joiners of the same generation and (2) by the ontogeny of many bee and wasp colonies. There is no certainty that kin selection and altruism will arise with eusociality, but probably they usually do; they may also arise in semisocial groups if, as a result of inbreeding, the individuals are related closely enough.

Long life, ability to produce female offspring under suboptimum conditions so that they will become joiners, or ability to dominate other females with the same result, and other queenlike features should be favored by selection operating on queens regardless of the origin of the society. These are among the most important aspects of social evolution.

Part III Natural History

Chapters 21 to 30 consist of general accounts of the life history, nest architecture, and other attributes for each of the groups of social bees. Such accounts are necessary to provide a well-rounded understanding of the social bees, since information on such matters is largely excluded from Part II. It must be emphasized that for many of the social attributes, the reader of Part III must refer to Part II. The index and contents will guide him to appropriate pages. It would have been inappropriate to describe these features in detail in both places.

Communal and quasisocial bees, as well as nest aggregations, are not treated in Part III. Chapters 5 and 6 contain the only material in this book on these matters.

21 *The Orchid Bees (Euglossini)*

The Euglossini are the closest relatives of the bumblebees (Bombini) and like them are large, robust insects. They vary in size from little larger than a honeybee to larger than the largest queen *Bombus.* In color they are sometimes black with yellow, whitish, or black hairs, sometimes forming conspicuous patterns, but the body is commonly at least slightly bluish or greenish. The most colorful forms, however, belong in the genera *Euglossa, Exaerete,* and *Euplusia,* whose species are mostly green, blue, purple, red, or brassy—the most magnificently brilliant of bees. As the tribal name, Euglossini, suggests, they all have long tongues; in the subgenus *Glossura* of *Euglossa* the tongue extends well beyond the apex of the abdomen even when folded. Figure 21.1 shows a male of a large species of *Eulaema.*

The orchid bees are restricted to the American tropics (southernmost Texas to Argentina), where they are most numerous and diversified in the forested areas. They are noted for the fact that males of many or all species are attracted to one or more species of orchids as well as to other flowers, and that they are the pollinating agents for many kinds of orchids (Pijl and Dodson, 1966). This habit, plus their large size and elegant colors, qualify them for their names—orchid bees. The females, however, rarely visit orchid flowers but go to a variety of other kinds of flowers for food.

The question of what the males are actually doing at the flowers has caused much curiosity. Many orchids produce no nectar. Orchid pollen is in pollinia and hence useless to bees. The male euglossines do not get food from the flowers. They rub certain specialized areas of the flower with their tufted front tarsi and then in turn rub the material into hairy grooves on the upper margins of the enlarged hind tibiae. Fine tubes lead from the groove to zones of specialized cells within the tibia (Cruz Landim, Costa Cruz, Stort, and Kitajima, 1965; Sakagami, 1965). One theory is that the male bee thus gathers substances from orchid flowers (and elsewhere, including fungus-infested wood) for storage in the spongy tissue of the hind tibiae, and that these substances have pheromone-like effects related to mating (Vogel, 1966). But female bees are not attracted to the substances in the orchid flowers and other places attractive to males. Hence, if there is any relation between sex pheromones and the substances collected by the male bees, those substances must by precursors of the pheromones and not the pheromones themselves.

Different species of orchids are attractive to different species of bees. Since the male euglossines are the pollinators of many orchid species, they are part of the isolating mechanism that prevents or reduces hybridization between some species of orchids. C. H. Dodson, R. L. Dressler, and their associates have identified several of the attractive substances produced by orchids, as follows: 1, 8-cineole, benzyl acetate, eugenol, methyl salicylate, methyl cinnamate (Dodson, Dressler, Hills, Adams, and Williams, 1969). These compounds in pure form are highly attractive to males of various Euglossini.

The tribe Euglossini consists of the following genera:

Euplusia (and its rare relative, perhaps only subgenerically distinct, *Eufriesea*). Nonsocial, although sometimes nests are aggregated. Cells made of resin and pieces of bark.

Euglossa. Contains both solitary and colonial species. Cells made of resin.

Eulaema. Large to very large species, perhaps all colonial. Cells made of mud or excrement.

Exaerete. Parasites in the nests of *Euplusia* and

FIGURE 21.1. A large male *Eulaema* removing with its front tarsi 1,8-cineole placed on a leaf as an attractant. (Photograph by C. W. Rettenmeyer.)

Eulaema. Very likely consists of two independent groups, one derived from *Euplusia*, the other from *Eulaema*, according to J. S. Moure (personal communication).

Aglae. Parasites in the nests of *Eulaema*.

The Euglossini are the only Apidae that are not eusocial. Known aspects of the nesting biology for the whole group are brought together by Zucchi, Sakagami, and Camargo (1969). Some species are solitary, each female making her own cells in a natural cavity in the soil, in a tree trunk, in nests of termites, wasps, or ants, or less commonly in small resin nests constructed by the bees on stems. *Euplusia* cells are usually in crevices in rock, under dead bark, in basements, in thatch, and the like.

It has long been known that more than one female of some species of *Euglossa* and perhaps all species of *Eulaema* commonly live together. Because the Euglossini are in the Apidae, the possibility of primitively social conditions in such nests has often been discussed, but because the nests are hard to find, there is little known about them. Only rarely has a nest population been preserved for dissection, much less, observed to see if there is any division of labor.

Nests consist of groups of oval, usually vertically oriented, contiguous cells. Superficially such a group resembles a cluster of *Bombus* cocoons. In the genus *Euglossa* the walls of the nest cavity are usually lined with resin; no nest linings are recorded for *Eulaema.* At least in some species of *Euglossa,* the nest cavity is enlarged by the bees after it has been nearly filled with cells. When a cell is empty due to emergence, *Euglossa* may either destroy it and re-use the building material (Dodson, 1966) or re-use the cell (Roberts and Dodson, 1967). The cells may also be re-used in the case of *Eulaema* (Bennett, 1965; Zucchi, Sakagami, and Camargo, 1969).

Sometimes when two females nest together there are two clumps of cells, probably each the work of one female (Dodson, 1966; Sakagami, Laroca, and Moure, 1967). Thus a communal colony probably exists. But sometimes there are more bees in a single nest cavity and the cells form a single mass. For example, Dodson records a nest of *Eulaema cingulata* in Ecuador containing 25 females and 386 cells. A small clump of cells of a similar species is shown in Figure 21.2. Adult males are not found in the nests. They evidently leave soon after emergence.

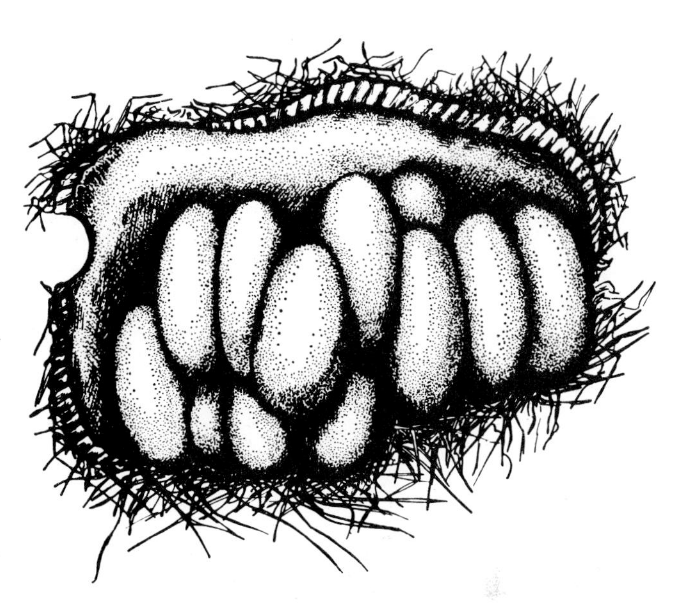

FIGURE 21.3. An opened nest of *Euglossa ignita,* from among orchid and fern roots, Iquitos, Peru. The nest cavity is largely lined with resin, except for the entrance hole, half of which is shown at the left. (From Dodson, 1966.)

FIGURE 21.2. A clump of cells of *Eulaema terminata* from Trinidad, West Indies. One cell at the bottom has been broken open. (Photograph by F. D. Bennett.)

Zucchi, Sakagami, and Camargo (1969), after reviewing all published nest descriptions, conclude that in the nonsolitary species of *Euglossa* and *Eulaema* the pattern of colony development is as follows: Lone females establish nests. Of the daughters, some leave the nest, presumably establishing their own, but two to several stay, jointly using the same cavity and making separate or merging clumps of cells. Some daughters of this and later generations may also stay, so that large masses of cells, many of them empty, may be found in a single nest, along with a communal colony of adults.

Hints that the relations among nest inhabitants are sometimes other than communal are found in reports of more bees working than cells being constructed or provisioned, so that one cell seems to be built and provisioned by more than one bee (Dodson, 1966; Roberts and Dodson, 1967). The latter authors dissected the populations of females from a few nests of *Euglossa* in Peru and Costa Rica (Figure 21.3). The maximum number of adults in one of these nests was eight; no known species of *Euglossa* has such large colonies as do some species of *Eulaema.* All eight females in a nest of *Euglossa imperialis* had well-developed ovaries, were fertilized, and presumably all laid eggs in spite of cooperative work on cells. This sort of relation is quasisocial.

Zucchi, Sakagami, and Camargo (1969), however, in studying a nest of *Eulaema nigrita* inhabited at the time by three females, noted that each female, to a large degree, builds and provisions her own cells, finishing one before beginning another. While one female is away, other females may do limited work on her cell, smoothing the interior or building up the collar, but they are usually chased away when the owner returns. Likewise, others may provide limited additions to the provisions in a cell, although the owner does most of the necessary foraging and lays the egg. Thus each cell is primarily the work of one bee, but limited help from others is tolerated or at least not successfully prevented.

These authors report no dissections; there is no information on ovarial conditions and fertilization. They do, however, report a curious dimorphism among females. Most were larger and were active for a few days in bringing in fecal material to a storage place for construction material in the nest, but did not forage for pollen, construct cells, or oviposit, and left the nest when young. The three resident females mentioned above were smaller, brought in little fecal material, were active in foraging for larval food, constructed cells, oviposited in them, and remained in the parental nest. Such a dichotomy, if found to be characteristic of the species rather than of only the one nest, represents a sort of division of labor without parallel among the other social insects.

An interesting parallel to other Apidae by which *Eulaema* differs from solitary bees is that building material, both feces and resin, are brought into the nest and placed in a single pile. Later, bees building cells take construction material from this pile.

The Euglossini should be further studied. Among its species will probably be found a number of interesting behavioral attributes that may help to explain behavior in *Bombus* and perhaps other Apidae.

22 *Semisocial Halictinae*

This chapter concerns those halictine bees whose colonies are semisocial when maximally developed. It does not concern the various other species of halictines that pass through a semisocial stage during colonial ontogeny but ultimately become primitively eusocial. Both of the genera treated here are members of the large American tribe Augochlorini. The first species, *Augochloropsis sparsilis,* is known only from southern Brazil, and has been studied in the equable temperate climate of Curitiba, on the southern Brazilian plateau, at an altitude of about 900 m (Michener and Lange, 1959). *Augochloropsis* is a very large genus; other species whose nests have been studied in detail are solitary or facultatively communal, except for *A. ignita,* which seems to have an unfertilized worker caste and may be semisocial or eusocial; data are entirely inadequate to decide (Michener and Lange, 1959). The second group treated consists of certain species of *Pseudaugochloropsis.* This is also a large genus, probably with considerable diversity in behavior, but the only three species that have been studied behaviorally are quite similar in their nests and life cycles so far as is known. One is *Pseudaugochloropsis graminea,* which has an enormous range, from northern Argentina to southern Texas. It has been studied both in southern Brazil and in Costa Rica (Michener and Lange, 1958a; Michener and Kerfoot, 1967; Sakagami and Moure, 1967). The others are *Pseudaugochloropsis nigerrima* and *P. costaricensis,* which have been studied only in Costa Rica (Michener and Kerfoot, 1967); the first is known only from that country, while the second is more widespread in central America.

All the species listed above are large Halictinae. They are mostly brilliantly green, although *Pseudaugochloropsis nigerrima* has an all-black morph which occurs in nests with green individuals, and *P. costaricensis* is bluish or greenish black with a brassy abdomen.

A. Augochloropsis sparsilis

1. Nests

Nests of this species were found in aggregations in more or less vertical roadside banks of decomposed gneiss or similar firm subsoil. Each nest consists of a burrow extending into the bank more or less horizontally to one or sometimes two or three clusters of adjacent cells which are nearly vertical (Figure 22.1). The cells open into a horizontal expansion of the burrow. The central cells of a cluster are those first made; the cluster is expanded by peripheral enlargement of the space into which the cells open and the excavation of new cells around the older ones. The cells are made by digging vertically, down into the substrate which remains unmodified among them. The short vertical burrows are converted into cells by addition of a layer of finely divided earth brought to the cell from elsewhere in the nest. With this earth, the cell is shaped and a beautifully smooth, earthen lining developed. Subsequently a thin, shining, light-brown, secreted lining having a waxy appearance is applied to the earthen surface to finish the cell. In the finished cell clusters there are almost always some marginal burrows not shaped into cells (Figure 22.2), and at certain seasons, principally in fall and winter, several of these are typically prolonged downward and then medially toward the center of the cluster but beneath the cells. Hibernation commonly occurs in these burrows, curled under the cells,

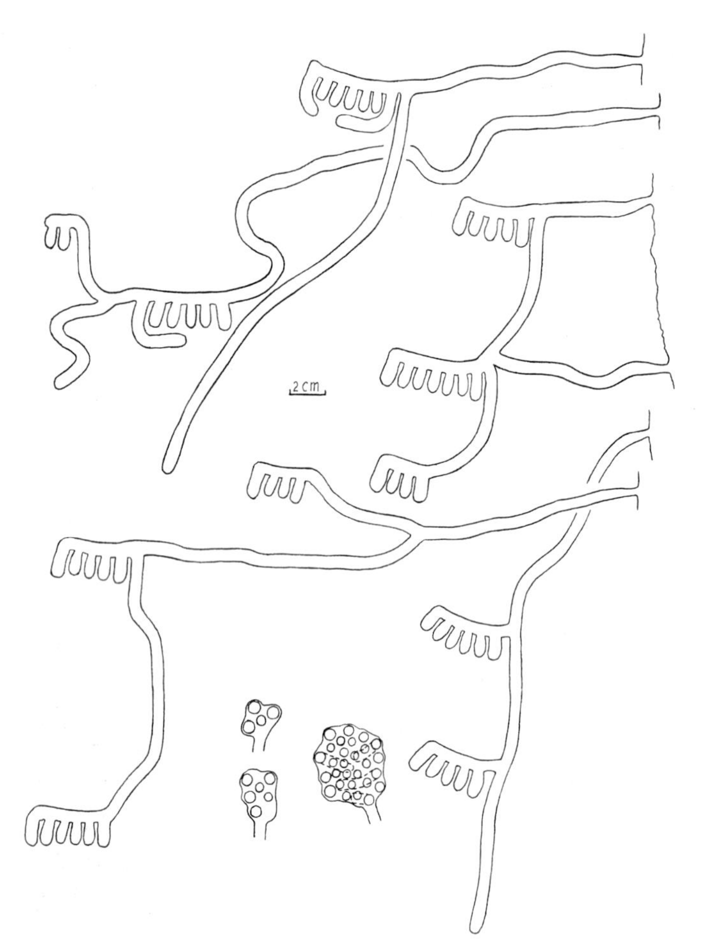

FIGURE 22.1. Diagrams of nests of *Augochloropsis sparsilis* from an earthen bank in southern Brazil. The upper two nests are in overwintering condition, with marginal burrows curving under the cell cluster. The other nests are in summer condition. Each group of cells is a cluster in the floor of a horizontal, more or less disc-shaped expansion of the burrow. *Below* are top views of three cell clusters, showing shaped and completed cells in the centers and roughed out larger lateral cells or burrows, which in the smaller clusters will be made into complete cells by a smooth lining of soil. *Broken lines* in larger cluster show some of the marginal burrows extended under the cells, as in winter. (From Michener and Lange, 1959.)

although bees may also be found resting in the cells. The number of cells in a cluster varies from 2 to 31, with a mean of 13.5 (43 clusters studied). The cells are commonly re-used, much of the old secreted lining usually being scraped off before a new one is applied. Early season activity in an old cluster of cells results in one or two newly refurbished cells in the center of the cluster. Gradually refurbishing spreads peripherally among the old cells.

One side of each cell is slightly flatter than the others, as usual in halictine cells. This surface is the homologue of the lower surface of horizontal cells of other genera of halictines. It is to this flatter side that the somewhat squarish, firm pollen mass is attached. A long curved egg is laid upon it. The pollen mass provides adequate food for the entire larval growth, the cell itself being firmly sealed (Figure 22.2) from shortly after egg laying until emergence of the new young adult.

2. Life cycle

Augochloropsis sparsilis, at least in the vicinity of Curitiba, overwinters as fertilized adult females in the nest. No males or immature forms were found among specimens examined in mid-September (early spring) before nesting activities had begun. Most of the females at this time were unworn and fresh in appearance, although two out of thirty were considered to have slightly worn mandibles and one had a nick in the margin of one wing. These spring bees, then, are a generation produced the preceding autumn, bees which had doubtless flown about and which had certainly mated but had returned to nests, probably usually to those in which they were reared, to overwinter in cells or in burrows beneath the cells.

During late September these bees begin activities, and earth is seen falling down the bank from the nest entrances, indicating some excavation within the nests. In early October cell provisioning begins and progresses slowly, the first eggs being found in mid-October, the first resulting adults at the end of December. Most of the bees that are active during the spring have disappeared by the end of January, a month marked by general inactivity, although a few bees can be seen carrying pollen from time to time. Most of the nesting activities of the second generation occur during February, and in March and subsequent autumnal months, little or no pollen collecting and egg laying take place. There are thus two generations, one emerging as adults in the autumn from cells provisioned mostly during February, surviving the winter as fertilized females, and provisioning cells the following spring, largely in November; the second maturing largely in January and provisioning cells in February. There is no evidence that adult females of different generations coexist. Both generations consist of both sexes, about 43 percent of the pupae developing into males. Figure 22.3 summarizes some of the data concerning the life history of this species.

The nests occur in rather dense aggregations, so that the entrance of one nest is usually only a few centimeters from entrances of others. Moreover, most nests contain more than one female bee. During winter hibernation the nest populations are higher than at any other season, up to 15 bees in one nest, although lone individuals are also found. During the summer the most populous nest found contained five females, many contained but one, and the average number of females per nest during half-month periods in November and February, the months of maximum cell-provisioning and egg-laying activity, ranged

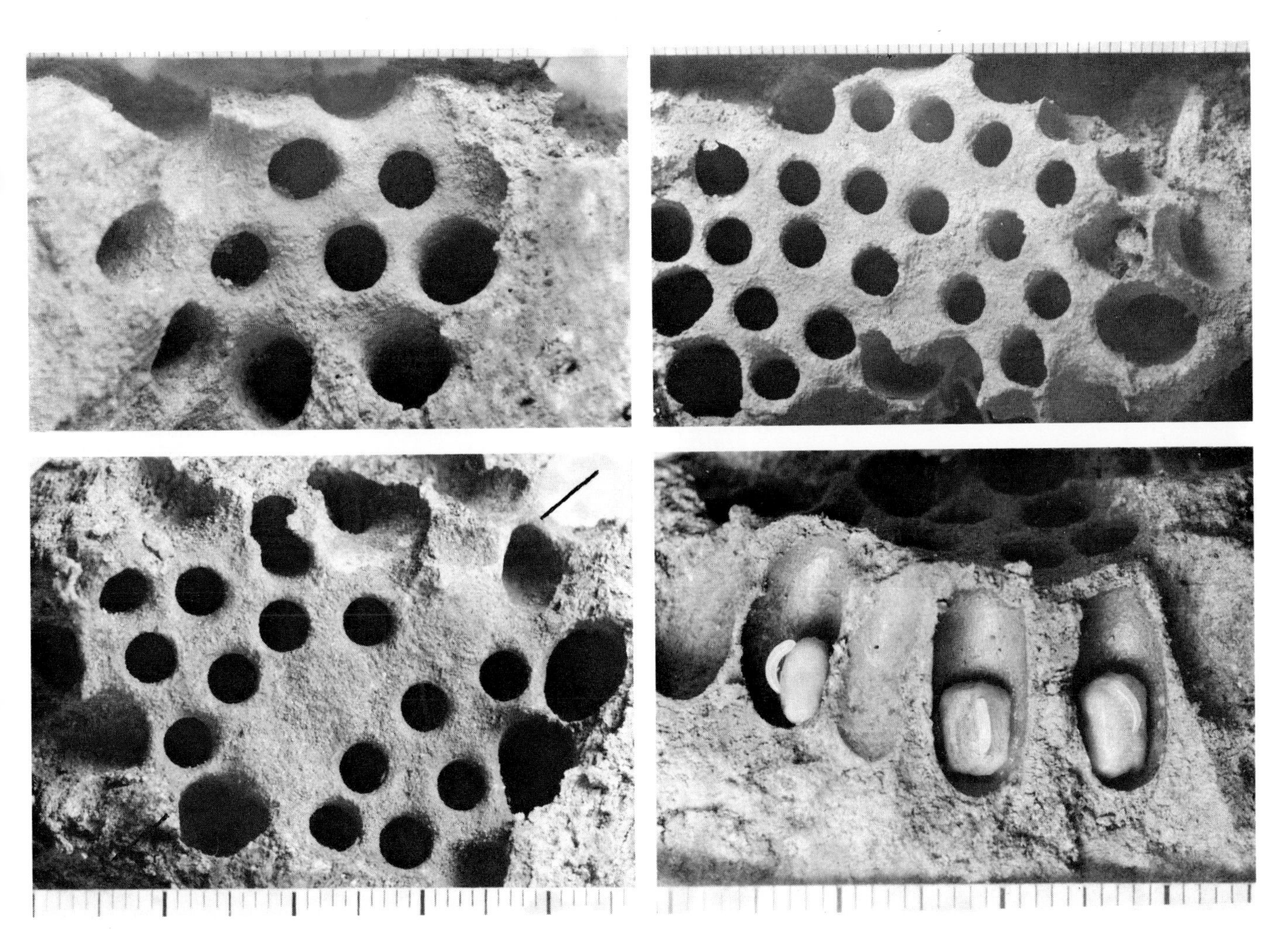

FIGURE 22.2. Top views and a sectional view of cell clusters of *Augochloropsis sparsilis. Above:* Small and large clusters seen from above, in overwintering condition, showing the large marginal burrows (some broken open at tops of photographs), which commonly curve beneath the central part of the cluster. *Below:* A cluster being re-used in the spring. *At left,* three closed cells seen from above among the open ones. *At right,* these cells are exposed and seen from side by opening the cluster approximately along the axis indicated at left. All scales are in millimeters. (From Michener and Lange, 1959; photographs by J. S. Moure.)

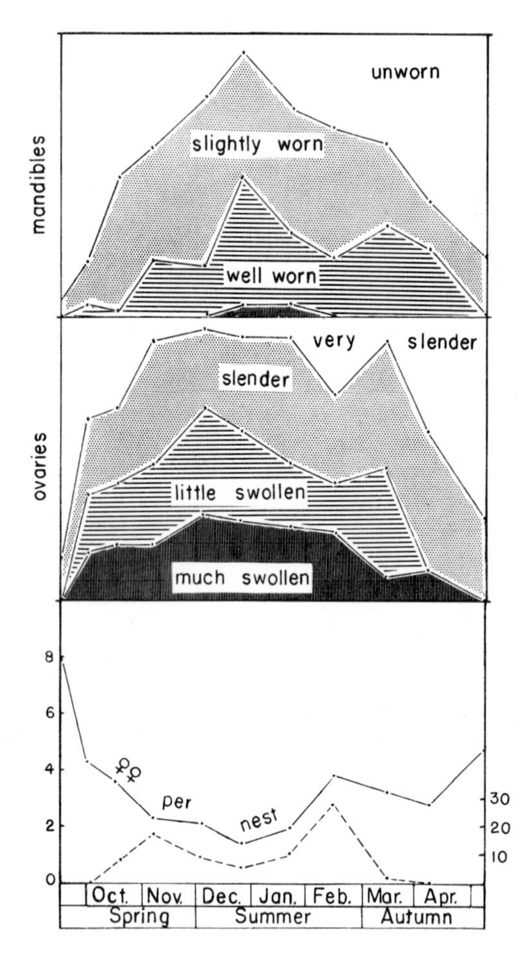

FIGURE 22.3. Nest statistics and conditions of female occupants of nests of *Augochloropsis sparsilis. Top: white and shaded zones* vary in width according to the percentage of individuals having mandibles classified as unworn, slightly worn, well worn, and much worn. The four zones total 100 percent in vertical distance at any date. *Middle: white and shaded zones* show fluctuations in percentages of females in four groups, classified according to size of ovaries. (For both these sets of percentages, the number of individuals examined was more than 20 for each date, with the exception of early October, when the number in the "middle" group was 11, and April, when the numbers in the "top" and "middle" groups were 13 and 12, respectively.) *Bottom: solid line* shows the average numbers of females per nest, plotted on the scale shown at the left (the number of nests examined, for each date, was 4 to 11, and trends were supported by data from many other nests); *broken line* shows the percentage of cells containing eggs, plotted on the scale shown at the right (the number of cells examined, for each date, was more than 49, except for April, when the number was 25). (From Michener and Lange, 1958d.)

from 2.3 to 4.0. In all probability the individuals in a given nest are commonly sisters born in and re-using the cells of the nest in which they live. We know, however, that individuals unlikely to be close relatives may be found in new nests, and doubtless sometimes associate in old ones.

The females in a colony in the active parts of the year when cells are being provisioned usually consist of a mixture of tattered and relatively fresh individuals. At least until the end of December, when young adults start to emerge, such variation indicates diversity in the activities of the overwintered nest occupants. As indicated in Chapter 9 (A), one female among 86 examined during the early summer and perhaps 15 or 20 percent of the females during the latter part of the summer were unfertilized, had slender ovaries, and their mandibles and wings became worn, while fertilized individuals believed to be the same age were mostly fresh looking. These occasional unfertilized workers, considerably less than one per nest, do not give a full explanation of the division of labor in this species, for among fertilized individuals there is also division of labor. This is shown by the small ovaries of fertilized pollen collectors compared to those of egg layers. Bees with large ovaries were rarely taken outside of the nests.

During the cell-provisioning season, nearly every nest contained one female with enlarged ovaries. Some nests contained two or three pollen-collecting individuals, but none contained more than one cell being provisioned at a given time. This means that the various pollen collectors were cooperatively provisioning a single cell in which another bee, an egg layer, would lay. Fertilized pollen collectors with slender ovaries may develop considerably worn mandibles and wings, indicating that they are functioning over a considerable period of time as workers; the evidence suggests that such bees die without becoming ovarially enlarged and reproductive. During much of the year some relatively unworn, fertilized females with slender ovaries can be found in the nests; such bees seem to be inactive for long periods and perhaps replace the fertilized workers as the latter die off.

Although the differences are not statistically significant, measurements of pollen collectors suggest that they are on the average smaller than egg layers. Since such size differences correlated with differences in activities are well known in other bees, it is quite likely that they are real, although more data are needed to establish this.

B. *Pseudaugochloropsis*

1. *Nests*

Nests of *Pseudaugochloropsis costaricensis* are scattered in sloping banks, while those of *P. graminea* and *nigerrima* are often in vertical or even overhanging banks and are commonly in aggregations. The nests consist of burrows, which in *P. costaricensis* slope downward steeply, but in the other species are more or less horizontal; in either case the burrows lead to cavities in the soil, often large and irregular, in which are found the clusters of cells. There is usually only one cluster per nest, but as in *Augochloropsis sparsilis,* there are occasionally two or three (Figure 22.4). Each cluster is supported by earthen pillars or by rootlets in the center of its cavity, so that it is almost entirely surrounded by an air space. The orientation of the cells varies considerably; typically they are more or less vertical, but sometimes they slant or are even horizontal. The cells are constructed from earth obtained elsewhere in the nest, probably from the walls of the cavity; they are not dug into undisturbed earth as in the case of *A. sparsilis.* In spite of the apparently different method of construction, the cells are similar in shape and lining to those of *A. sparsilis* and one may reasonably consider them and the cluster to be the homologue of the earthen linings applied to the insides of the excavated cells of the *Augochloropsis.* The provisions form a more rounded mass than in *Augochloropsis;* otherwise

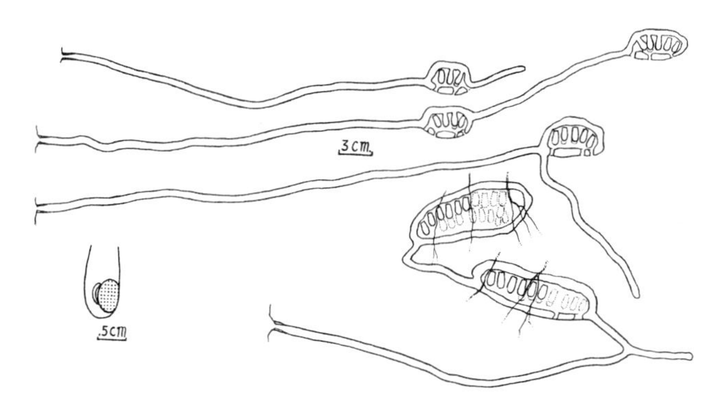

FIGURE 22.4. Diagrams of nests of *Pseudaugochloropsis graminea* from earth banks in southern Brazil. The cell clusters may be supported in their cavities by earthen pillars or by rootlets (*lower right*). *Dotted* cells are earth-filled and abandoned. *Lower left:* a single cell with provisions (*dotted*) and an egg. (From Michener and Lange, 1958a.)

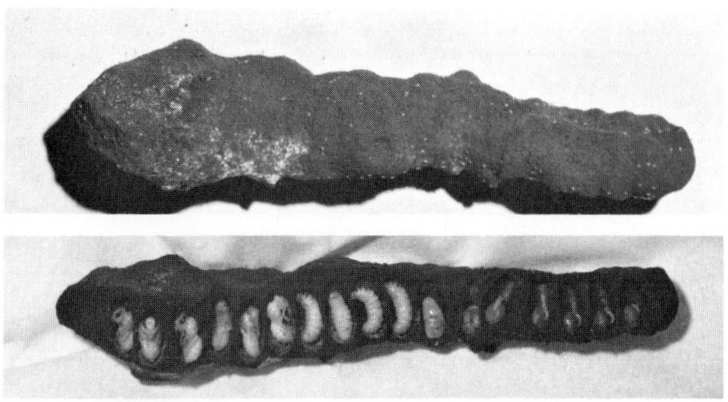

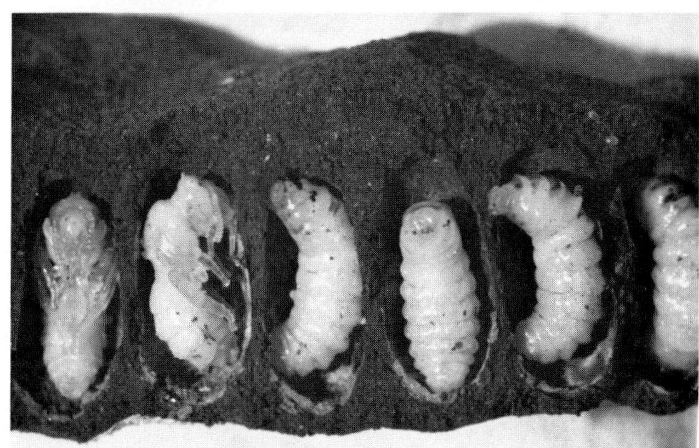

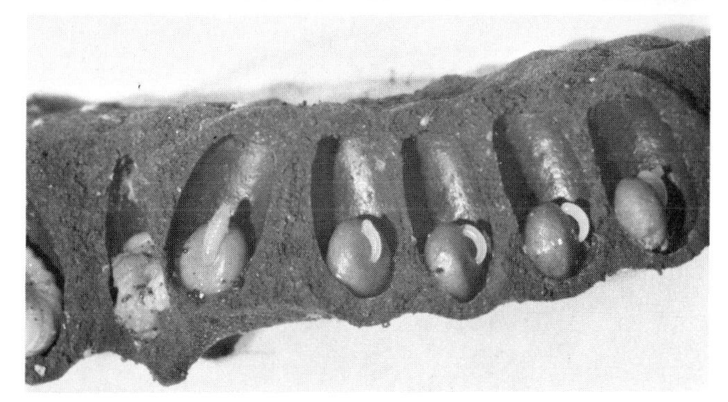

FIGURE 22.5. *Above:* An elongate cell cluster of *Pseudaugochloropsis nigerrima* from Costa Rica, intact above, opened for examination *below.* Enlarged sections show pupae and prepupae as well as a small larva and eggs on pollen masses. (From Michener and Kerfoot, 1967.)

they are similar to those of that genus (Figures 22.4 and 22.5). When cells are re-used, earth is often added to the inside before reshaping and application of the new secreted lining. The result is that an old cluster, after much re-use, tends to become thickened, with much earth below the occupied cells (Figures 22.6 and 22.7). The thickening in some parts of the left side of the cluster shown in Figures 22.5 and 22.8 probably has a different origin, for there was no evidence of re-use of these cells.

2. Life cycle

In Costa Rica *Pseudaugochloropsis graminea* appears to be active in nest provisioning throughout the year, and in such latitudes the other species probably are also. At Curitiba on the southern Brazilian plateau, *P. graminea* ceases nesting activity during the winter months, although the adults visit flowers occasionally for nectar on warm days. As in *Augochloropsis sparsilis,* males are produced in smaller numbers than females.

About two thirds of the nests of all three species studied in Costa Rica contained only a small cell cluster (one to six completed cells) and a single adult female (Figure 22.6). Such females behave as solitary bees, carrying on all the activities necessary to construct and provision cells and lay eggs in them. Re-use of cells in such small clusters shows that the same female, or probably another, may produce another small brood. Sometimes a single female is found in association with a large cell cluster, but in this case she only makes use of a few cells, the rest being unused.

Figure 22.9 shows the distribution of total cell numbers among various nests. Most of the nests have small num-

FIGURE 22.7. A cell cluster of *Pseudaugochloropsis graminea* from southern Brazil, showing evidence of re-use of certain cells, along with thickening of the lower part of the structure. (From Michener and Lange, 1958a.)

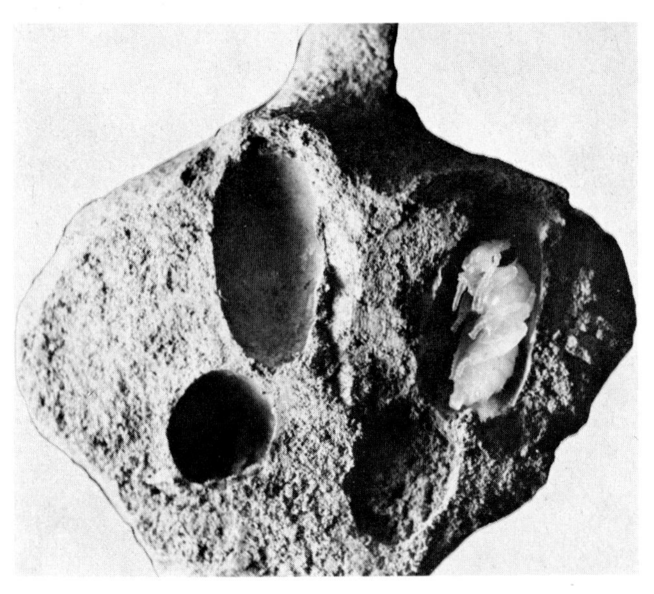

FIGURE 22.6. A small cell cluster of *Pseudaugochloropsis graminea* from southern Brazil. It has an unusual large pillar at the top as well as supporting pillars beneath. This nest contained only a single adult female. The cluster was being re-used and in the process thickened, as shown by the abandoned cell at the lower right. The cell at the lower left was subhorizontal. (From Michener and Lange, 1958a.)

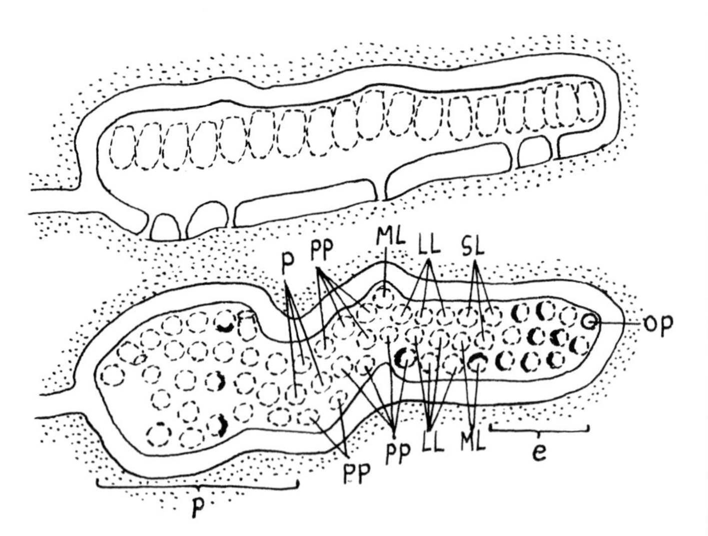

FIGURE 22.8. Sectional and top views of the cell cluster of *Pseudaugochloropsis nigerrima* shown in Figure 22.5. Cell contents are indicated as follows: *op,* open, empty; *e,* eggs; *sl,* small larvae; *ml,* medium-sized larvae; *ll,* large larvae; *pp,* prepupae; *p,* pupae. (From Michener and Kerfoot, 1967.)

bers of cells as described above for those occupied and constructed by solitary individuals. A few nests of all three species, however, have more cells (Figure 22.8), up to 70. Interestingly, nests of 61 to 70 cells are as common as those of 16 to 20 cells. Presumably this means that once a nest starts growing, with a colony of bees living in it instead of a lone individual, it has a good chance of survival and, furthermore, of growth. Perhaps this is because of the better defense which a colony provides. The number of adult females in colonies ranges as high as seven.

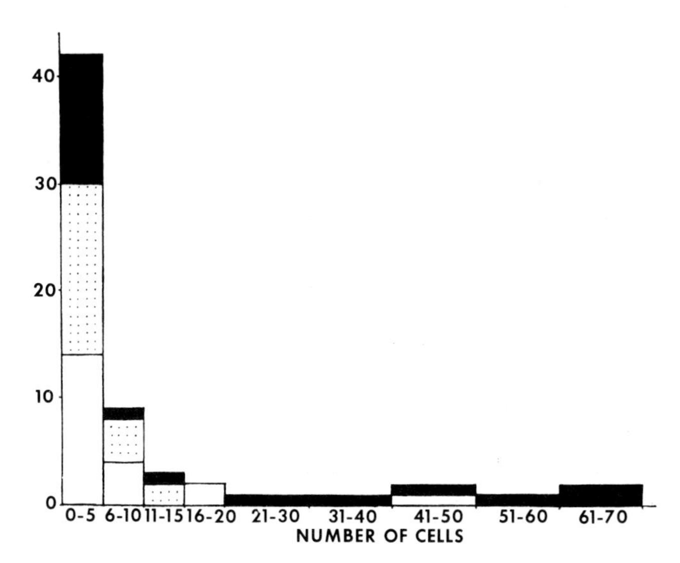

FIGURE 22.9. Histogram showing numbers of nests of *Pseudaugochloropsis nigerrima* (black), *graminea* (stippled), and *costaricensis* (white) having various numbers of cells. (From a study by Michener and Kerfoot, 1967, in Costa Rica.)

Due to the rather small number of nests of each species studied, and the escape of some bees, analysis of castes and of probable division of labor is rather unsatisfactory. However, considerable diversity of activities among the individuals in nests containing colonies of females is evident. There are fertilized bees with large ovaries, of which there may be one, two, or three in a colony. Curiously enough, there are also unfertilized bees with large ovaries, although such individuals are much less common and not more than one in a colony has ever been found. There are also unfertilized bees with slender ovaries and with worn mandibles or wings, showing that they are not recently emerged young adults but are instead workers. From one to three such individuals may be found in a colony. The egg layers and workers in a colony are more or less comparably worn and probably are ordinarily sisters or other individuals of about the same age, not mothers and daughters. There is no evidence that the lone egg layers that establish nests survive to act as queens of colonies, as is so common in colonies of comparable size among primitively eusocial bees.

As with *Augochloropsis sparsilis,* the mean size of workers is less than that of egg layers; this was noted for all three species in Costa Rica and for *Pseudaugochloropsis graminea* in Brazil.

23 *Primitively Eusocial Behavior in the Augochlorini*

The subfamily Halictinae is divisible into two tribes, the Halictini and the Augochlorini. Most of the eusocial groups have arisen in the former, but the semisocial species of *Augochloropsis* and *Pseudaugochloropsis* discussed in the preceding chapter belong in the Augochlorini. So also do the primitively eusocial species of *Augochlorella* and *Augochlora.* Three species of *Augochlorella* occur in colonies which attain a eusocial stage; such organization is otherwise not known in the Augochlorini, except in some of the species of *Augochlora.*

A. *Augochlorella*

Information on *Augochlorella* is available through the papers of Ordway (1964, 1965, 1966), based on data obtained in Kansas for two species, *A. striata* and *A. persimilis,* with limited additional information on two other species, *A. michaelis* from Brazil (Michener and Lange, 1958a; Sakagami and Moure, 1967) and *A. edentata* from Costa Rica (Eickwort and Eickwort, 1973a). All four species are small, 5 to 8 mm long and brilliantly metallic blue-green, green, or brassy green. It is unfortunate for studies of social behavior that, through most of the range of the species in the United States, two or more species exist, species sufficiently similar that not all females are distinguishable and even males often present problems of identification. Studies of social organization are based on females; such factors as size are difficult to treat since the species as well as the castes differ in mean size. Since the two species studied by Ordway nest and fly together, her studies of seasonally and socially related size variation are at times equivocal, because females of intermediate stature sometimes had to be excluded as unidentifiable.

1. Nests

The North American species nest in flat ground or in sloping banks of earth, usually in partly bare areas. The nests often form loose aggregations but also often are isolated. The nest burrow is irregularly vertical; at the surface it is slightly constricted. In *Augochlorella persimilis,* the smaller species, the entrance is usually only a neatly rounded and tamped hole at the surface of the soil, but in *A. striata,* if excavation has been done in moist weather when soil particles will stick together, the burrow is often continued upward as an earthen turret, sometimes as much as 30 mm high, with the entrance at its top (Figure 23.1).

At a depth of several centimeters, sometimes at the bottom of the main burrow but commonly above the bottom, the bee enlarges the burrow to make a chamber, in the center of which it builds of earth a subhorizontal cell of the usual halictine type, lined with smooth waxlike material. The cell is supported by columns or masses of loose earth (Figure 23.2). As the chamber is further enlarged, new cells are built beside the first, and the process is continued until a comb or cluster is formed of adjacent earthen cells separated from one another and from the surrounding chamber by earth walls less than 1 mm thick (Figures 23.2 and 23.3). Such a cell cluster completed by a single bee may have five or six cells, often less; but

FIGURE 23.1. Nest entrance of *Augochlorella striata.* This turret was about 10 mm high. (Original drawing by C. D. Michener.)

colonies make larger clusters, up to at least 53. In moist woodland soil the same chamber as well as cells themselves may be refurbished and re-used during the summer; in drier places or in dry summers new cavities and cells are made at lower, moist levels as the soil dries, and the upper cavities are abandoned and filled with loose earth from deeper in the nest. As a result, by late summer two or three clusters are sometimes found along a single vertical burrow, only the deepest being in use.

The provisions for the larva in each cell are packed into a slightly flattened ball (Figure 23.3), as in most *Lasioglossum.* The egg, larval feeding, and indeed everything inside of the cell are also as in *Lasioglossum* and most other halictines (see Chapter 24). Cells are closed with an earth plug immediately after oviposition, and there is no evidence that they are opened until emergence of the adult bee. Cells are ordinarily built and provisioned one at a time, all the pollen collectors in the colony cooperating to provision the cell. Provisioning is a slow process; I know of few bees so exasperatingly unhurried about their activities. A very few trips in the morning seem to be the extent of the outside activities of most bees. Table 23.1 indicates the outside activities for a day for a colony of three bees.

2. Life cycle

Augochlorella in North America overwinters as unworn adult fertilized gynes in the nests, or more accurately, in burrows dug downward in autumn from the lowest parts of their nests. Overwintering bees have slender ovaries; virtually all are fertilized.

In spring (April in Kansas) the bees emerge and establish nests. A new nest may be dug by a single gyne in a new location, or the old site may be used by one to five bees. In the latter case old cells and chambers are not rebuilt but new ones are constructed. Two or more egg layers are found in about 30 percent of the spring nests and, as explained elsewhere, one or more auxiliaries with slender ovaries are found in many nests. Figure 23.4 shows the relative frequencies of the various ovarian forms in spring as well as summer and autumn. In the

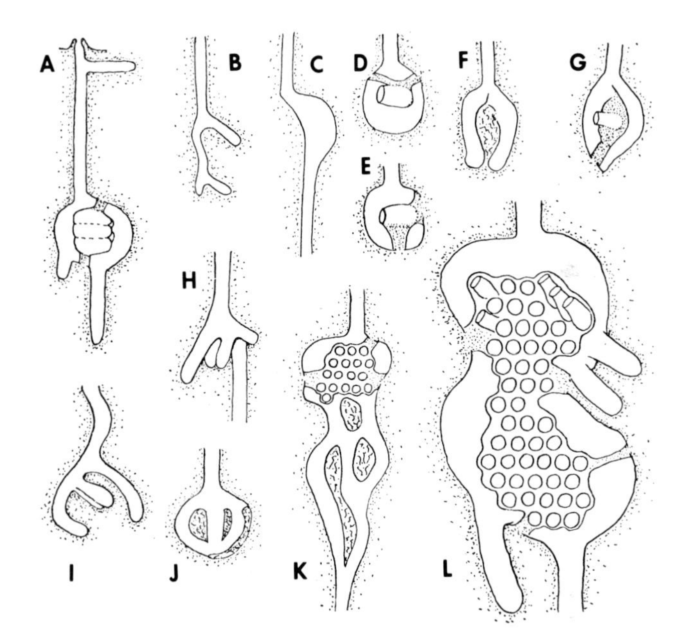

FIGURE 23.2. Diagrams of nests of *Augochlorella striata. White areas* indicate burrows or cavities. *Stippled areas* inside the nests indicate earthen supports connecting cell clusters to outside walls of cavities. *Irregular lines* inside nests indicate unexcavated earth around which connecting burrows have been made. (*A*) Complete nest of *Augochlorella striata,* showing short turret, constricted entrance, side burrow, main burrow, cluster cavity and cell cluster, a short blind burrow at the lower left, and blind terminal burrow. (*B*) The end of a hibernation burrow, showing hibernacula. (*C*) The enlargement of the main burrow to form a cavity in which a cluster of cells will be built. (*D*) A single constructed cell supported in cavity. (*E*) A single cell constructed on top of loose earth. (*F*) Construction of cluster cavity by branched burrows, leaving unworked earth in the center. (*G*) Construction of cell into center of loose earth. (*H*), (*I*) Cluster cavity formation by extensions of branch burrows. (*J*) Enlargement of cluster cavity by extension of branch burrows. (*K*) Cell construction before completion of a cavity. (*L*) Large cell cluster with constricted center between older and newer parts. Short side burrows often indicate regions for new expansion. (From Ordway, 1966.)

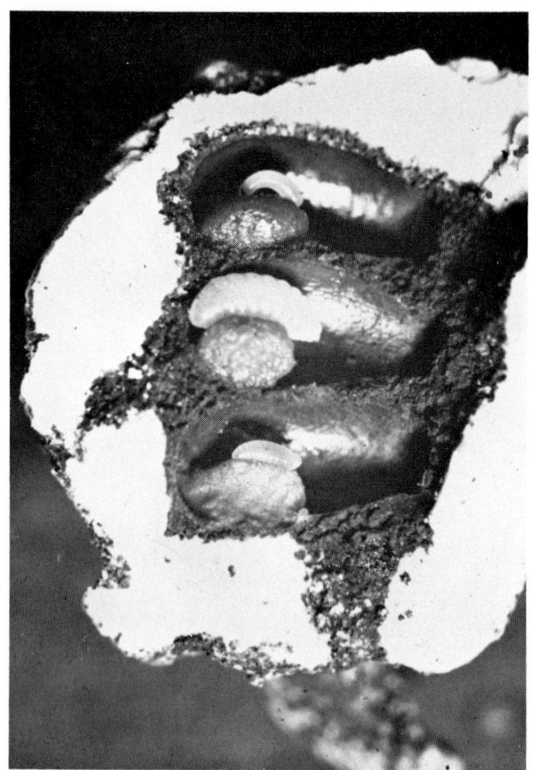

FIGURE 23.3. Cell clusters of *Augochlorella striata. Left:* A nest that had been filled with plaster of Paris (*white*), then opened. The cell cluster was completely surrounded by plaster of Paris except for the small earth pillars (one below, one at middle left, one at upper right). The thin earthen walls of the cells and their waxlike linings protect the cell contents—pollen masses with, *from top to bottom,* egg, middle-sized larva, and small larva. The scale at the left is in millimeters. *Right:* A cell cluster (upper cell broken and exposing a dark male pupa) removed from its cavity and showing how cells are recognizable from the outside as separate convexities, due to their very thin walls. (Photographs by C. W. Rettenmeyer and Ellen Ordway.)

spring nests, all the gynes forage, both queens and auxiliaries, and mean sizes of these two groups do not differ significantly.

Some new nests are started during May and early June, probably as a result of breaking up of polygynous colonies. Queens may stop laying after the first three or four eggs and become temporarily inactive, as noted for most halictines, but in many colonies there is continuing production of eggs from May through August. Males are produced throughout this period, but in lower percentages in spring and early summer than in late summer.

By June lone gynes are producing a few worker females and the spring semisocial groups are dissolved through dispersal and death, so that most nests are inhabited by small, primitively eusocial colonies. The mean nest population is down slightly from May, however, to slightly less than two adult bees per nest. At this season and on through the summer, colonies almost always contain only one queen; she does not collect pollen, although she sometimes goes out, presumably to feed, and she is on the average larger than her workers. Queens are sometimes replaced; to judge by the frequency of little-worn queens such replacement may perhaps be normal. Workers are presumably shorter lived than queens. In spite of continued cell provisioning and egg laying, the total adult nest population rises only to a mean of about three by August and the number of cells to about 11 (Figure 23.5). Slow colonial growth must be due in part to mortality of both queens and workers as well as to production of males which disappear. Somewhat higher colony sizes for *Augochlorella persimilis* (Ordway, 1966) may be real, but may reflect sampling errors; the number of nests of this species censused in late summer was small.

The Brazilian *Augochlorella michaelis* probably makes larger cell clusters on the average. Only nine nests have been found, and adults escaped from some when they were being studied. Some nests contained only one adult and a few cells—evidently relatively new nests. One nest, however, contained two large fertilized females with developed ovaries and three smaller unfertilized females with slender ovaries. This seemed to be a nest with two queens and three workers. Another nest contained four

TABLE 23.1. Activity taking place at the entrance of a nest of *Augochlorella persimilis*, 7 June 1957.

5:45	No bee seen at entrance
6:15	Guard appears (identity unknown)
6:15–6:55	W appears intermittently at entrance
6:56	Y guards intermittently
8:12	W guards
8:30	* Y leaves, W guards
9:04	Y returns with pollen, W guards
9:07	* Y leaves
9:30	W guards, R seen below
9:36	* W leaves
10:34	R guards
10:36	Y tries to return (but cannot get past R who has abdomen in entrance after being disturbed by observer); presumably she returned soon after without being noticed.
11:10	W guards (had returned without being noticed)
11:20	R guards
11:40	R still guarding
—	Nest not observed
12:30–16:45	Y guarding intermittently all afternoon, no other activity
—	Nest not observed
19:30	Y guarding, W in main burrow just below

SOURCE: Modified from Ordway, 1966.
Note: Bees were marked as follows: R = red (ovaries much enlarged, mated, mandibles slightly worn, *queen*); W = white (ovaries slender except for one enlarged ovariole, mated, unworn); Y = yellow (ovaries slender, mated, slightly worn). Guards could usually be identified only by disturbing them or waiting until they moved spontaneously so as to show the color marks on them; hence the periodic notations of which bee was the guard. Only three trips from the nest (marked *) were made during the day.

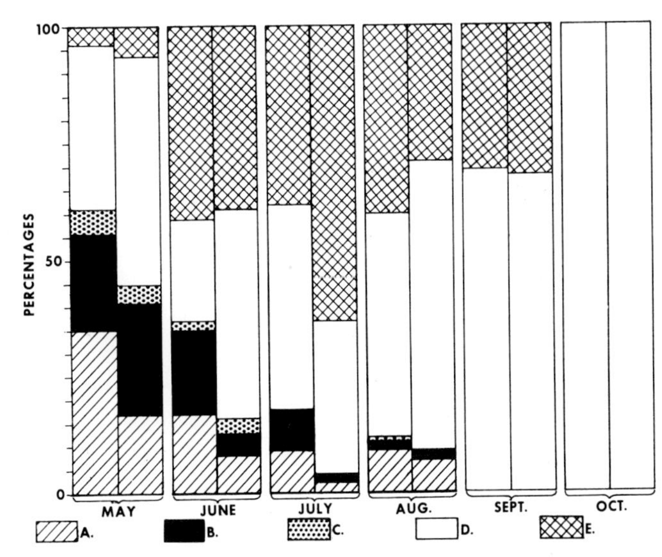

FIGURE 23.4. Percentages of females of *Augochlorella striata* (*left-hand columns*) and *A. persimilis* (*right-hand columns*) belonging to ovarian groups (*A*) to (*E*) during each month of the nesting season. Both field-caught bees and specimens from nests are included. Groups (*A*) and (*B*) are queens with larger and smaller but functional ovaries, respectively. Group (*C*) consists of fertilized bees with slender ovaries except for one large egg, more or less of a size to be laid. (*D*) and (*E*) consist of bees with slender ovaries, fertilized and unfertilized respectively. Newly emerged females all belong to group (*E*). The ovarian types are approximately like those of *Lasioglossum imitatum* shown in Figure 9-7. (From Ordway, 1965.)

bees with enlarged ovaries, but only one of them was fertilized. These data suggest larger colonies with more egg layers than in the two North American species, and imply an intriguing diversity of social habits in the genus *Augochlorella*, such as is found in various genera of Halictini.

In Costa Rica *Augochlorella edentata* has only been studied in July and August, during which time more than half the nests contained only one female, the rest two to five. The colonies seemed to be semisocial in that there was no evidence from wear that the bees belonged to two generations. Each colony consisted of a mated queen and unmated workers which averaged slightly smaller than queens. Interpretation of these observations would be premature; this could be a semisocial species, but it is likely that a eusocial stage follows that described above.

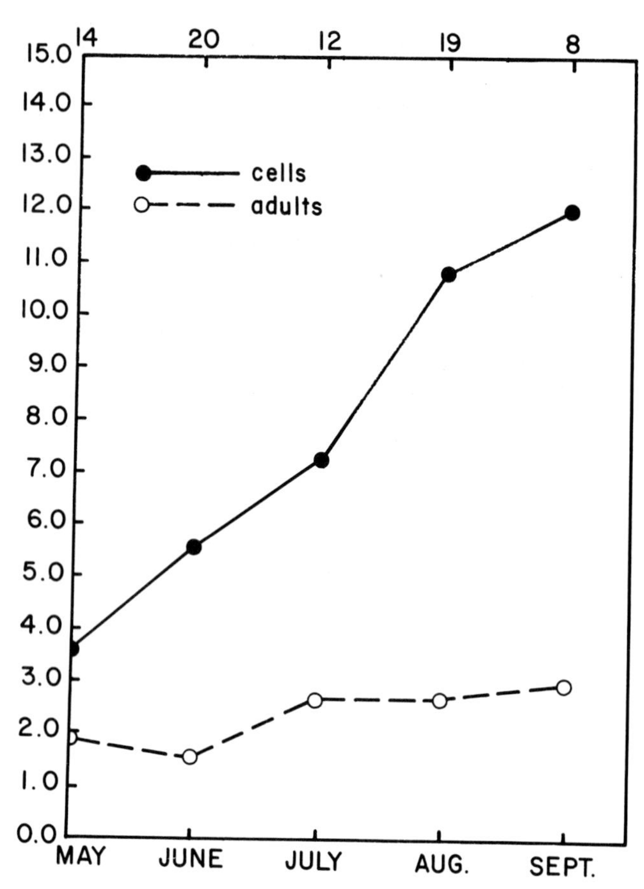

FIGURE 23.5. Mean numbers of cells and of adult females of *Augochlorella striata* per nest in various months. The number of nests examined per month is shown at the top. (From Ordway, 1966.)

B. *Augochlora*

This genus contains for the most part brilliantly green bees, morphologically very similar to *Augochlorella*, which exhibit considerable behavioral and social diversity. The best-known species is the North American *Augochlora pura*, which nests in rotting wood and is solitary (Stockhammer, 1966). Probably its close relatives in the tropics, which also nest in rotting wood, are also solitary (Eickwort and Eickwort, 1973b; Sakagami and Moure, 1967). But there is a tropical American subgenus called *Oxystoglosella* which nests in the soil and contains some species which are eusocial. None has been properly studied through the year; Eickwort and Eickwort (1972a) provide data on *A. nominata* from Costa Rica and summarize the published information on *A. semiramis* from southern Brazil. Data for each species are limited to one or two months out of the year. For other species even fewer data are available.

Both *Augochlora nominata* and *semiramis* nest in flat ground, making more or less vertical burrows, constricted

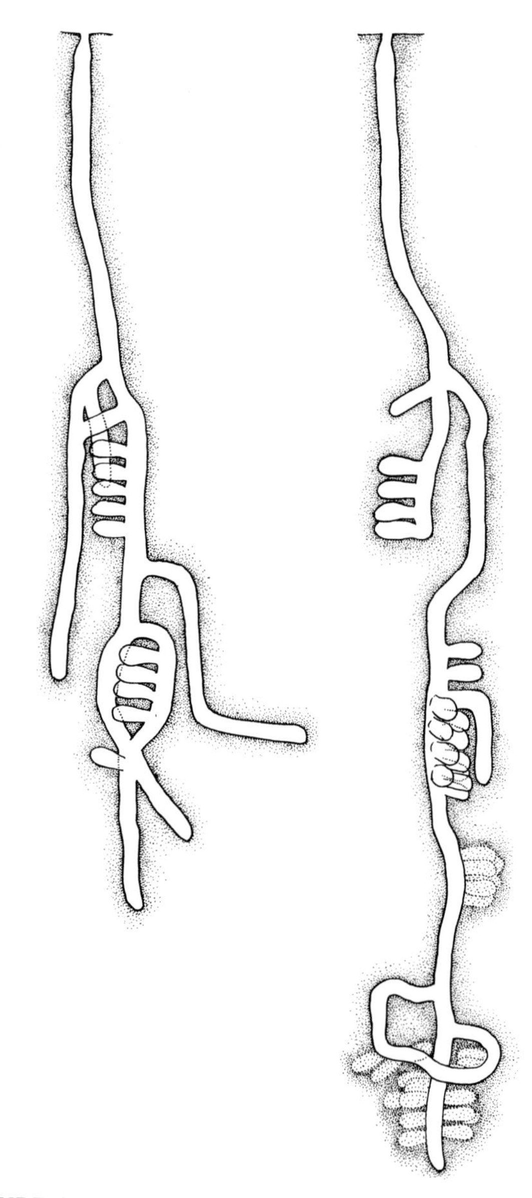

FIGURE 23.6. Nests of *Augochlora semiramis*. (From Sakagami and Michener, 1962, after Michener and Lange, 1958a.)

at the entrances, with subhorizontal cells extending out from the burrow much as in the common species of *Lasioglossum* subgenus *Dialictus,* described below (Chapter 24). In *A. nominata* the cells are rather scattered, but in *A. semiramis* they tend to be clustered and a branch burrow is often built behind them (Figure 23.6), suggesting an early stage in the evolution of the cell cluster in a chamber, a common feature in augochlorine architecture.

Perhaps seasonally, but at least not at all times or not in all localities, colonies of each species consist of a large, macrocephalic queen, and a few smaller workers with ordinary-sized heads, slender ovaries, and empty spermathecae. The Eickworts suggest that the large queens are the forms that ordinarily found new nests. In size they are almost discontinuous with workers. But at other seasons or locations the bees of the same species are all of the smaller-headed form, the largest individuals being mated queens. In such cases queens and workers overlap broadly in size, as in *Augochlorella* and most *Dialictus.* Such a colony might exist after the death of the large-headed founder, and might characterize a later stage in the history of most colonies. Males are produced in all broods, so that mating of replacement queens is possible.

In *Augochlora nominata* an average of two and a maximum of five workers lived with the large-headed queens. Thus the colonies are typically small, like those of *Augochlorella.*

24 *Primitively Eusocial Behavior in the Subgenus* Dialictus

The subgenus *Dialictus* of the genus *Lasioglossum* consists of small, greenish-black bees which are extremely numerous in individuals, with hundreds of species in the Western Hemisphere; a relatively few species occur in Eurasia. Although they are morphologically monotonous, the few species already investigated show great variation in levels of social development, from solitary to primitively eusocial. The principal life-history and behavioral studies of species in the subgenus are all on eusocial forms: *L. zephyrum* (Batra, 1964, 1966b), *rhytidophorum* (Michener and Lange, 1958b), *imitatum* (Michener and Wille, 1961), *versatum* (Michener, 1966b), and *umbripenne* (Wille and Orozco, 1970; Eickwort and Eickwort, 1971).

The behavioral diversity found among these and other, less well-known species can be outlined as follows (modified from Wille and Orozco, 1970):

Group A

Social behavior absent. All females are fertilized and have enlarged ovaries. *Lasioglossum opacum*, from Brazil (Michener and Lange, 1958b), belongs to this group. *L. herbstiellum*, from Chile, may also belong here (Claude-Joseph, 1926).

Group B

Eusocial behavior weakly developed and castes not well defined; all intergradations between queen and workers common. Fertilized and unfertilized females with fully enlarged ovaries are common. These females, regardless of size, are potentially able to become queens (they do so when alone); however, the smaller ones, when in a colony, tend to become workers. Queens are frequently replaced. Males are therefore produced throughout the reproductive season, as in group A. This category includes *Lasioglossum rhytidophorum*, from Brazil (Michener and Lange, 1958b); *L. zephyrum*, studied in Kansas (Batra, 1964, 1966b); probably *L. seabrai* and *L. guaruvae*, from Brazil (Michener and Seabra, 1959); and possibly *L. coeruleum* (Stockhammer, 1967), studied in Kansas, which nests in decaying logs in the eastern and central United States.

Group C

Established eusocial behavior, with more or less distinct castes differentiated externally by mean size. Fertilized workers, as well as workers with moderately enlarged ovaries, are common and therefore a clear line does not exist between queens and workers, although intermediates are not as common as in Group B. Queens are probably replaced during the life of the colony. A few males are produced with the first progeny of a colony, but males become abundant only later in the summer. Colony multiplication is perhaps by fission: new colonies are probably often established by opening new entrances to the surface from the overwintering nests. *Lasioglossum versatum*, studied in Kansas, belongs to this group (Michener, 1966b).

Group D

Established eusocial behavior and fairly distinct caste structure. Although the queens average slightly larger than the workers, a frequency curve based on sizes of all females is not bimodal (also true of groups B and C).

As in group C, the smallest workers are smaller than the smallest queens and the largest queens are larger than the largest workers. Workers with some ovarial enlargement are found in smaller numbers than in group C; very few fertilized workers can be found. There is thus evidence of a more conspicuous physiological and behavioral differentiation between the castes than in the previously mentioned groups. Queens commonly live for about a year and regularly survive for the life of the colony. Males are very scarce among the first progeny of a colony. This group includes *Lasioglossum imitatum* as studied in Kansas (Michener and Wille, 1961), and probably *L. umbripenne* (Turrialba population, Eickwort and Eickwort, 1971). The Ontario population of *L. imitatum* studied by Knerer and Plateaux-Quénu (1967d) probably belongs in group B.

Group E

Well-established eusocial behavior with very distinct castes. There is a definite difference in size between the queens and the workers, with no intermediates. The gyne-cell dimensions and food mass (size and shape) differ from those provided for rearing of workers and males. Workers with enlarged ovaries are common, and occasionally fertilized workers are also found. The queen survives for about a year, including the full life of her colony. *Lasioglossum umbripenne* (Damitas population, Wille and Orozco, 1970) is the only known representative of this group.

The kinds of differences listed above, especially when presented in the sequence of increasing social complexity, hint at steps in social evolution. Yet the adaptations, of course, are to the particular conditions under which the bees live, and evolution can undoubtedly produce either an increase or a decrease in social attributes. Probable social adjustments to local conditions are best shown by two populations of *Lasioglossum umbripenne* studied in Costa Rica by Wille and Orozco (1970) and Eickwort and Eickwort (1971). The locations are only about 65 km apart, but the climate and vegetation are quite different. Turrialba is on the Atlantic slope, with more or less year-round rainfall, while Damitas is on the Pacific slope, with a distinct dry season, although the area is forested. The bees from the two sites have been carefully studied by G. C. Eickwort and appear morphologically alike. If they should prove to be different species, they are at least extremely similar.

Yet in life cycle and social level the two populations of *Lasioglossum umbripenne* are more different than are some clearly distinct species of the subgenus. Table 24.1 enumerates the differences, with comparable features for two closely related temperate species. Particularly noteworthy, from the social viewpoint, are the large, long-lived queens in the usually monogynous Damitas colonies, the queens and most males being produced after worker production ceases. Thus, socially the Damitas population is the most specialized *Dialictus.* The contrast with the Turrialba population is impressive.

The only *Dialictus* that has been studied in some detail and judged to be solitary is *Lasioglossum* (*D.*) *opacus* from southern Brazil (Michener and Lange, 1958b). The judgement as to its solitary status is based on females taken from flowers and dissected at various times during the season of nesting activity. Essentially all of those except young unworn individuals were fertilized and had enlarged ovaries. This is in contrast to the other species which have been investigated, all of which have a worker caste which can readily be recognized among the individuals taken on flowers because of wing and mandibular wear combined with slender ovaries and, usually, lack of fertilization. Only five nests of *L.* (*D.*) *opacus* have been found, but each contained only a single adult female (except for one unworn and presumably recently emerged individual), thus supporting the surmise based upon individuals taken from flowers.

A. Nesting

The nests of nearly all species of *Dialictus* are similar in basic pattern, whether found in flat ground, in vertical banks, or in rotting wood. (For possible deviations from this pattern, which require further study to verify the systematic positions and check the identification of the bees concerned, see Eickwort, 1969b.) *Lasioglossum coeruleum* has a modified nesting pattern associated with its occupancy of limited and often irregular areas of rotting wood (Stockhammer, 1967). To simplify expression, and because the great majority of species nest in the ground, words such as earth and soil are used for the nesting substrate in the following account, even though a few species nest in rotting wood.

The nest entrance is smaller than the diameter of the rest of the main burrow. If it is artificially enlarged, the bees narrow it again by bringing earth up from the bur-

TABLE 24.1. Differences in social structure and nesting biology among forms of the *Lasioglossum versatum* group.

L. umbripenne (Turrialba)[a]	*L. umbripenne* (Damitas)[b]	*L. versatum* (Kansas)[c]	*L. zephyrum* (Kansas)[d]
Old nests typically with multiple entrances	All nests with single entrances	All nests with single entrances	All nests with single entrances
Active during wet season and probably throughout year	Inactive during wet season	Inactive during winter	Inactive during winter
Nests re-used	Colonies annual, nests not re-used	Colonies annual, nests re-used by hibernated queens	Colonies annual, nests sometimes re-used by hibernated queens
Queens may not live through year, may be replaced by daughters	Foundress queen lives through year; if she dies, not replaced	Overwintered queens may be replaced by daughters	Overwintered queen regularly replaced by daughters before maturity of colony
Older colonies often with several queens	Colonies usually monogynous, 10% with 2 queens	Early colonies with several foundress queens, older colonies with 1 to 4 queens	Early colonies usually monogynous, occasionally with 2 queens, replaced in older colonies by several daughters
Gynes apparently produced at same time as workers	Gynes produced after worker production stops	Most gynes produced at end of season, some may emerge earlier	Gynes produced at same time as workers
Numerous males produced at same time as workers	Most males produced after worker production stops	Most males produced near end of season, some earlier	Some males present throughout season, percentage increasing to peak at end of season
Queens only statistically larger than workers, largest bee in nest is queen	Queens much larger than workers with no overlap in sizes	Queens only statistically larger than workers, largest bee in nest is usually queen	Castes indistinct, queens statistically larger than workers but largest bee in nest may be a worker
Gyne cells and provisions similar to workers'	Gyne cells larger than workers', with larger rectangular provisions	Gyne cells and provisions similar to workers'	Gyne cells and provisions similar to workers'

Notes: Both localities for *L. umbripenne* are in Costa Rica.
[a] From Eickwort and Eickwort, 1971. This entire table is modified from one provided by these authors.
[b] From Wille and Orozco, 1970.
[c] From Michener, 1966b.
[d] From Batra, 1966b.

row, placing it about the entrance, and tamping it into position. Such activity occurs only when the soil is somewhat moist, for dry earth will not cling but is granular and falls into the burrow. The constructed entrance approximately fits both the head of the bee and the posterior dorsal surface of its abdomen; the significance of the constriction in the defense of the nest against natural enemies such as mutillid wasps has been discussed in Chapter 18 (A1).

Each nest consists of a main burrow, which, except in very young nests, has a few major branches. These branches are not laterals leading to individual cells, but usually continue in the same general direction as the main burrow, and should be regarded as multiple main burrows. The cells themselves are excavated horizontally into the soil, more or less at right angles to the main burrows, and without laterals or with only short laterals leading to them (Figures 24.1, 24.2, and so on). They are scattered

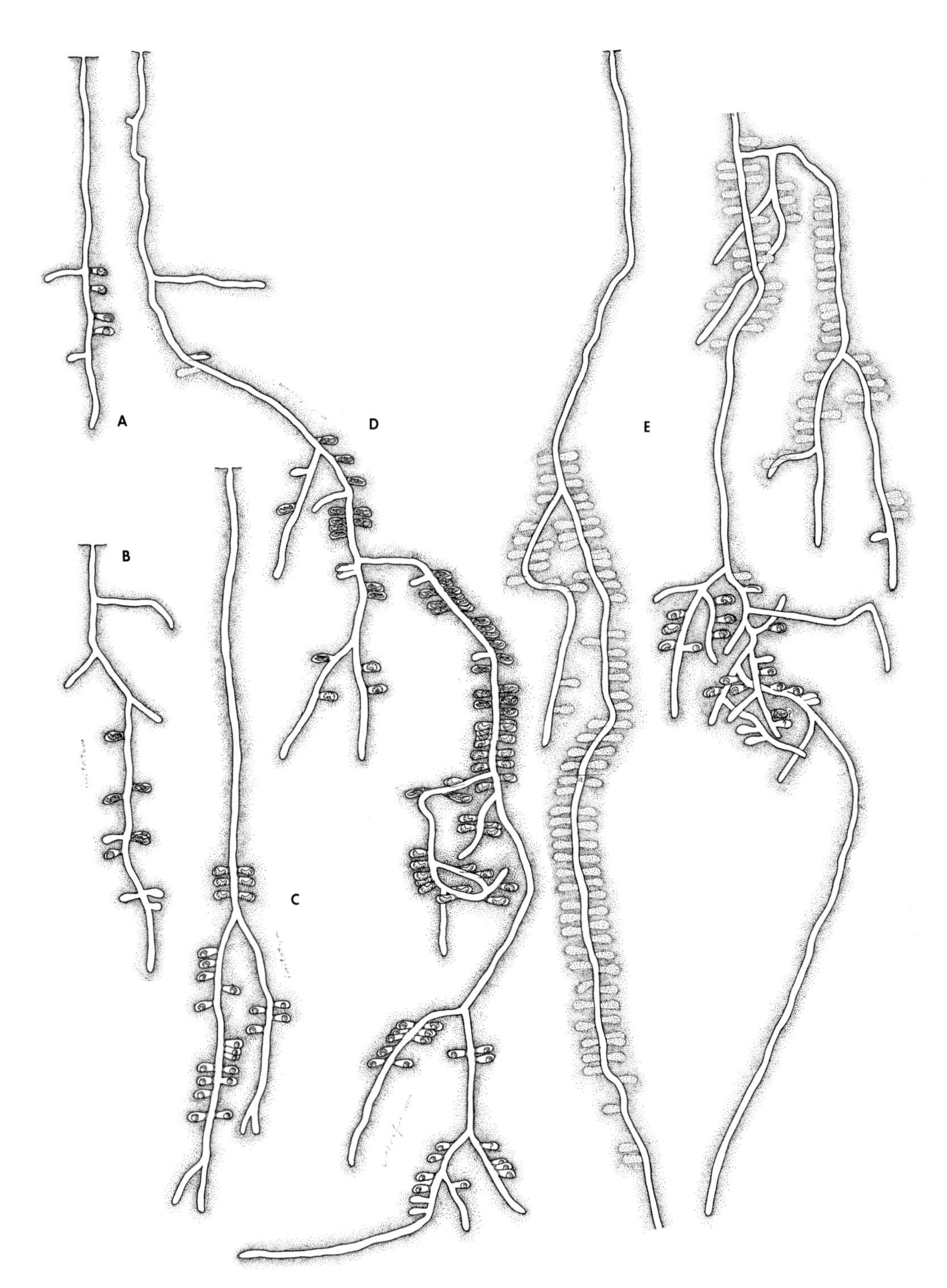

FIGURE 24.1. Diagrams of nests of *Lasioglossum* (*Dialictus*) *imitatum* in level ground in Kansas. (*A*) A nest (opened on May 6) with only a gyne and cells (oldest above, youngest below). (*B*) A nest (June 3) with a queen and two workers, and four cells with pupae; new cells made by the workers can be seen from the level of the lowermost pupa downward. (*C*) A nest (July 16) with a queen, two workers, and a fourth bee (worker?) which escaped. (*D*) A nest (June 26) with a queen and ten workers. (*E*) A nest (August 4) with two queens, 20 workers, and two females of doubtful caste; to save space the diagram is divided. (From Sakagami and Michener, 1962, after Michener and Wille, 1961.)

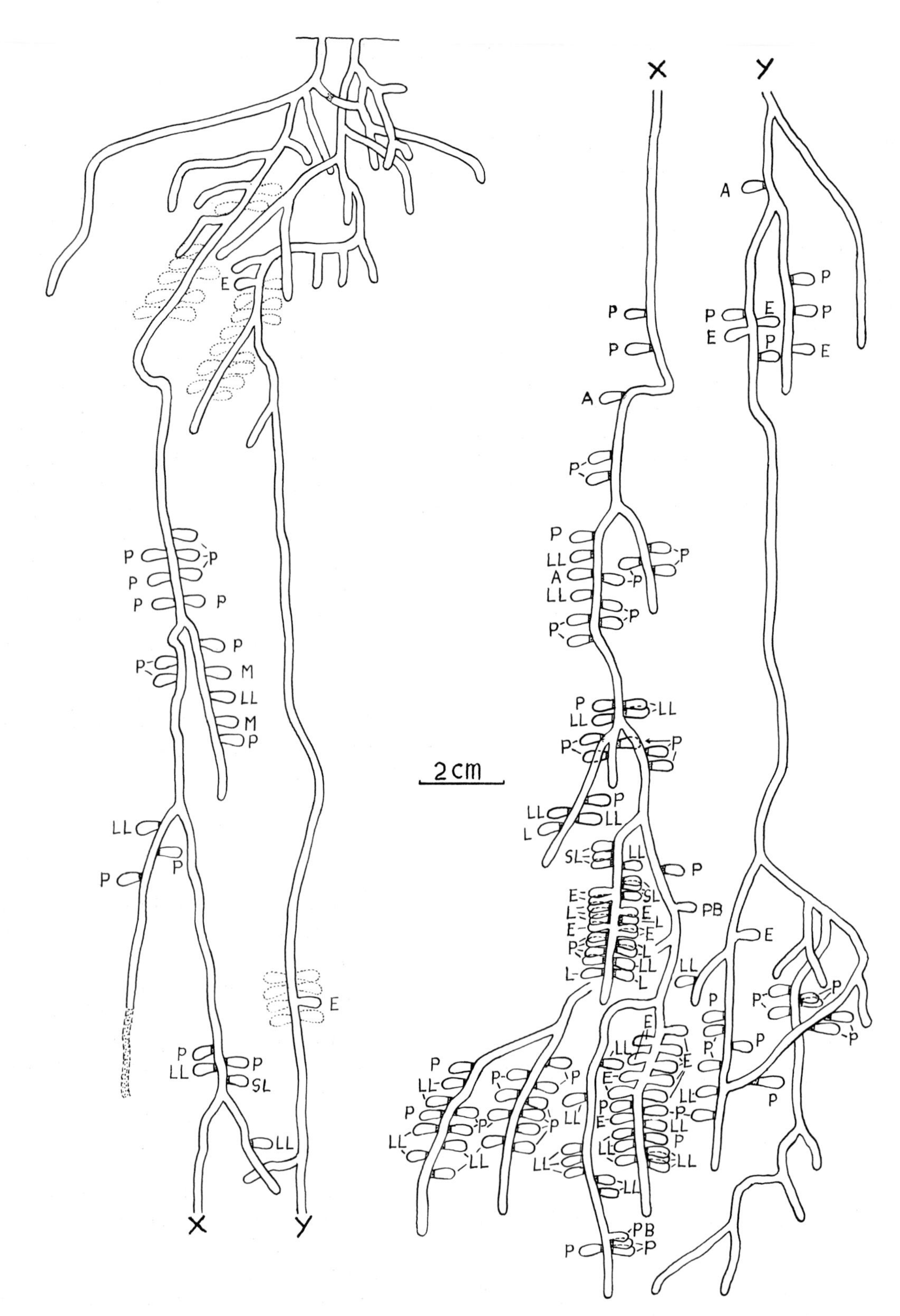

FIGURE 24.2. Diagrams of nests of *Lasioglossum* (*Dialictus*) *versatum*, to show some of the deeper and more intricate nests of the subgenus. Symbols beside cells are: *PB*, pollen ball but no egg or larva; *e*, egg; *SL*, small larva; *L*, moderate-sized larva; *LL*, large larva or prepupa; *P*, pupa (female); *M*, cell contents moldy; *E*, empty. *Dotted cells* were earth-filled. The right-hand burrows are continuations from points *x* and *y* on the left. (From Michener, 1966b.)

or sometimes clustered, especially where suitable substrate is limited, as sometimes in wood-nesting species.

Excavation of burrows in *Lasioglossum* (*Dialictus*) *zephyrum*, the species whose activity has been best observed in indoor bee rooms, thanks to glass-walled observation boxes (Batra, 1964), is mainly by females up to five days old. Figure 24.3 shows the development of a single nest of a related species, *L. versatum*, and the way in which, at least in our artificial nest boxes, burrows are built, filled, and new ones made, without obvious reason. Excavation in *L. zephyrum* as well as in its relatives is done largely—but by no means entirely—at night, the soil being loosened with the mandibles and pushed under the thorax with the front tarsi. After accumulation of a small pile, the bee picks it up under her thorax with her middle legs, sometimes curling the abdomen somewhat to help, and moves off backwards. As at other times when in burrows, she supports herself not with the tarsi but by pressing the outer surfaces of the legs outwards against the burrow walls, the basitibial plate and an apical hind tibial spine presumably facilitating such locomotion (Chapter 1, B1c). The loose earth may be carried only a short distance, later being picked up by other bees. Work extending a burrow may (or may not) then be continued, either by the same or by another bee. The soil goes into old, empty cells, into disused branch burrows, or to the nest entrance where it forms a tumulus of loose material on the surface of the ground. In artificial nest boxes of glass containing rather loose substrate, much of the excavated soil was tamped into the interstices in the burrow walls. Smoothing and tamping of the burrow walls is a common activity, apparently of all the bees in a colony. The bees bite at the surface, probably at projections or loose bits of soil, and then tamp any loosened particles into the wall by curling the abdomen and repeatedly striking the soil with the upper surface of the abdominal apex. As in other burrowing bees, the apex of the abdomen of the female is provided with a pygidial plate, a usually hairless, heavily sclerotized projection which presumably makes the tamping blows more effective or less damaging to the insect.

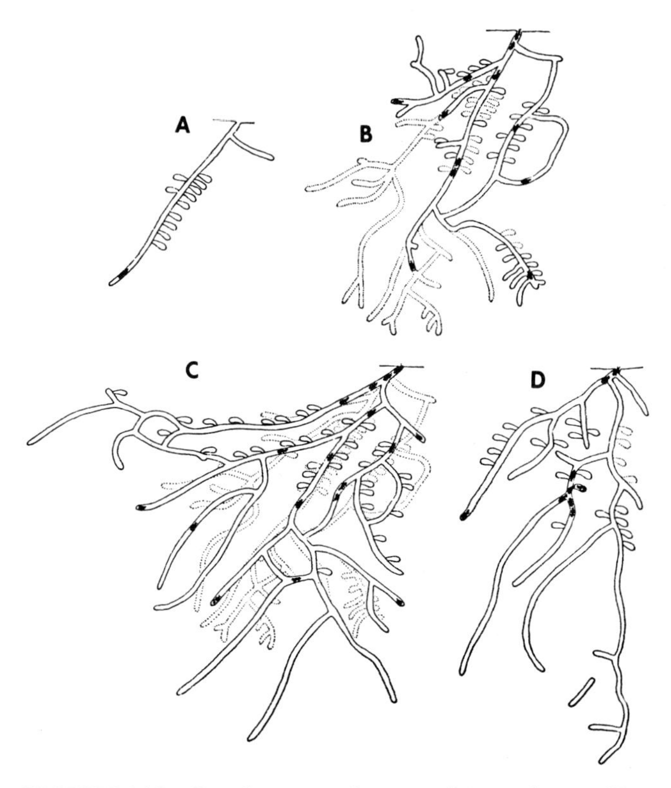

FIGURE 24.3. Development of a nest of *Lasioglossum* (*Dialictus*) *versatum* made in soil between two sheets of plexiglas in the insectary. *Solid lines* indicate inhabited cells or burrows, *stippled lines* indicate abandoned cells and burrows that have been filled with earth (largely omitted in *A*). (*A*) May 5, nest made by solitary female; (*B*) June 15; (*C*) July 28; (*D*) November 28, 1966. (From Batra, 1968.)

Probably all halictines also, at times, push loose excavated soil backward through the burrows with the somewhat curled apico-dorsal part of the abdomen. Some, or perhaps all, push soil with the face as well.

The result of soil transport along burrows and frequent tamping of it into the walls is that, as in most halictines, burrows of *Lasioglossum zephyrum* commonly become lined with soil from elsewhere, usually from deeper in the nest burrow system. In places where nests penetrate differently colored layers of soil, the result is a contrastingly colored burrow lining.

Depths of nest burrows vary widely. Some species, like *Lasioglossum rhytidophorum*, which inhabits rather uniformly moist soil, typically make burrows less than 10 cm deep (Figure 24.4), while most mature nests of *L. umbripenne* are over a meter deep.

The details of cell construction are probably more or less the same for all halictines. They have been directly observed in observation nests with glass walls by Batra (1964, 1968) for species of *Dialictus* and are therefore described here. In *Lasioglossum zephyrum* in the labora-

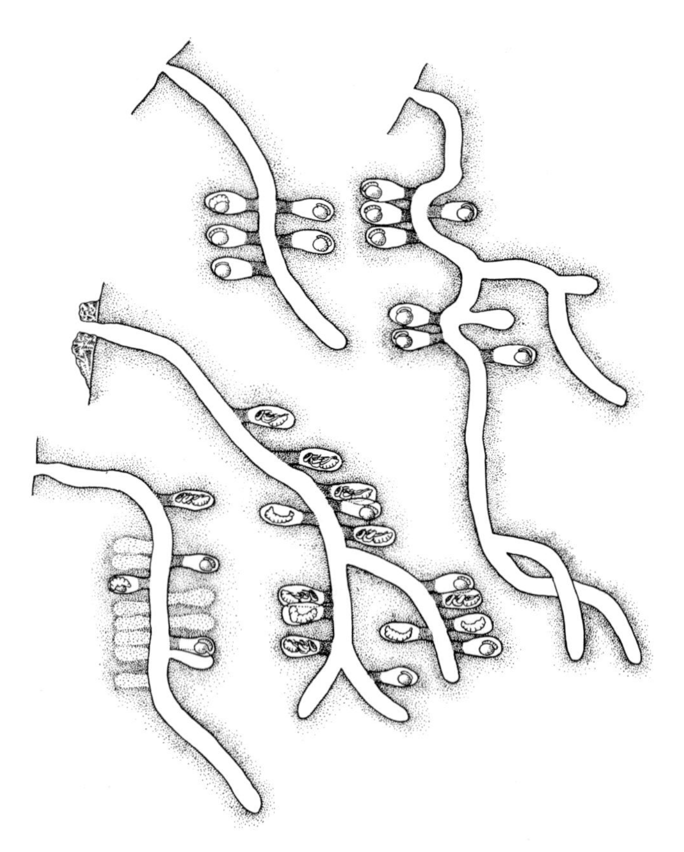

FIGURE 24.4. Diagrams of nests of *Lasioglossum* (*Dialictus*) *rhytidophorum*, which occur in earth banks in the southern Brazilian plateau. They rarely attain a depth of 10 cm. (From Sakagami and Michener, 1962, after Michener and Lange, 1958b.)

tory, cells are usually made from 11:00 P.M. to 7:00 A.M. along new burrows or ones with heavy traffic. Cells are also made at night in the field by *L. umbripenne*. As in most other halictines, the cells are initially roughed out somewhat larger than the finished size, excavated as for burrows, the loose soil dumped into the burrow. Sometimes two individuals of *L. zephyrum* work alternately excavating a cell, and in a related species (*L. versatum*) several females were seen working on the same cell. After the cell is roughed out, earth from elsewhere in the nest is brought in and tamped into place to give the cell its typical form (Figure 24.5), with a constricted entrance and with the lower surface slightly flatter than the other surfaces. This cell wall of earth is made beautifully smooth on its inner surface; one female *zephyrum* spent three hours tamping inside a cell, mostly at the rounded end.

The smooth earth cell wall is then coated with a thin secreted lining of waxlike material. Batra (1968) reports that lining material appears to come from two sources, the apex of the abdomen (Dufour's gland), and the mouthparts (perhaps salivary glands), and is spread by both front and hind tarsi, but especially by the paintbrushlike penicilli of the latter. Early in the deposition of this material, before the mouthparts have come into play, some of the posterior secretion soaks into the earthen cell wall, producing a dark zone and presumably making it firmer. May (1970a) interprets the matter differently, believing that the dark zone is due to moisture. She shows that for *Augochlora pura* most of the secreted cell lining originates in Dufour's gland, although she does not exclude the possibility of an important, possibly catalytic, addition from the mouthparts, for even fresh material of the cell lining appears to be an oxidation product of the material in Dufour's gland rather than the same substances. As discussed in Chapter 20 (A), two to

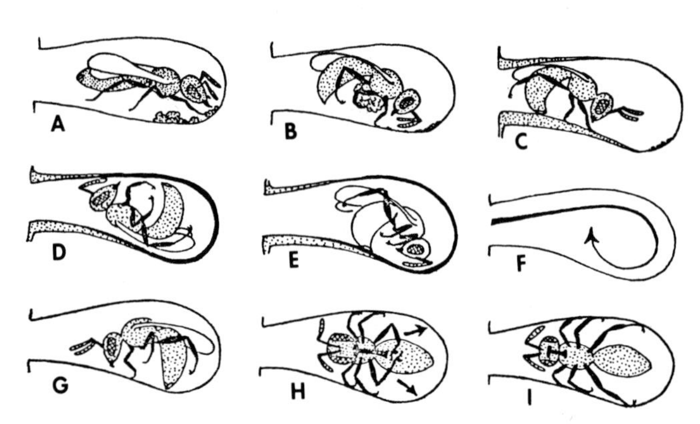

FIGURE 24.5. Sketches showing methods of cell construction by *Lasioglossum* (*Dialictus*) *zephyrum*. Different individuals are indicated by different patterns of shading. (*A*) Cell excavation; (*B*) and (*C*) excavated soil is carried back and used to form the narrow neck of the cell; (*D*) and (*E*) soil taken from burrows is used to line cell—here it is being packed down with the pygidial plate; (*F*) diagram of route commonly taken by bees while working in cell; (*G*) bee rubbing with apex of abdomen, presumably depositing secretion of Dufour's gland; (*H*) and (*I*) sequence in use of tongue and penicilli in spreading waxlike cell lining, at least part of which probably comes from Dufour's gland. (From Batra, 1964.)

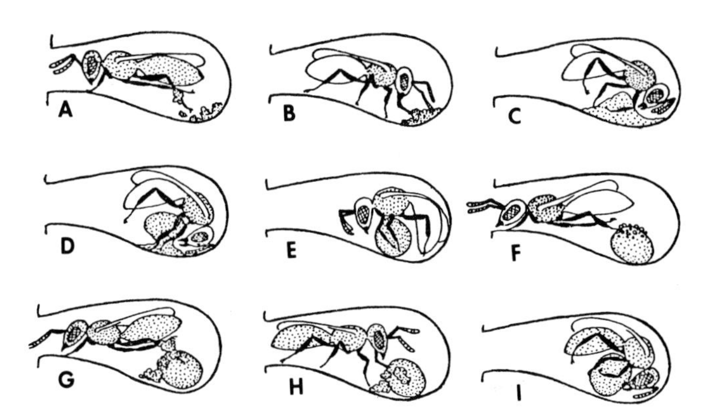

FIGURE 24.6. Sketches showing the provisioning of a cell by *Lasioglossum* (*Dialictus*) *zephyrum*. Different individuals are marked by different patterns of shading. (*A*) and (*B*) Deposition of first loads of pollen and nectar; (*C*) to (*E*) steps in initial formation of the pollen ball; (*F*) bee cleaning herself before leaving cell; (*G*) and (*H*) deposition of additional pollen and nectar loads; (*I*) perfection of a nearly complete pollen ball. (From Batra, 1964.)

several cells are often in various stages of construction or provisioning in nests of *Dialictus*.

As soon as the cells are completed, the bees provision them with a firm and rather dry mixture of pollen and nectar, which is formed into a smooth ball, somewhat flattened in the vertical axis. In *Lasioglossum zephyrum* up to six bees were seen provisioning a single cell, although of course when there is only one bee in the nest, she does all the work herself. Six to eight pollen loads suffice for one cell in the laboratory for *L. zephyrum*, while five is the regular number in the field for *L. umbripenne*. A bee (*L. zephyrum*) carrying pollen enters the cell, turns around, and brushes herself for one to two minutes, using movements that look like ordinary cleaning activity for hind legs and abdomen (Figure 24.6). She regurgitates nectar either before or after depositing the pollen. The loose pollen and nectar may be made into a miniature ball after the first load of food is put in a cell, or may be left unshaped until later. Shaping is done by any bees—foragers, egg layers, or others—and different bees may take part in the activity at different times in the same cell. The loose but moist pollen is brushed into a mass with the front and mid tarsi, the bee supporting herself with the hind tibiae pressed against the walls of the cell. The pollen is now packed together and the mass repeatedly rotated; in the process it becomes firm and quite round. If the ball was firmed up prematurely, additional loads of pollen are added in the same way. Finally the fully formed ball is left.

Workers show a certain "constancy" for a given cell in bringing in provisions, but if another bee is in it, they go on to another cell and leave their pollen and nectar there; their orientation seems poor, and sometimes a bee leaves provisions in different cells after different foraging trips for no obvious reason. Perhaps because of some sense of the proper proportions of pollen and nectar, a bee may put pollen in one cell, nectar in another. If a cell has received only pollen, several loads of nectar are brought to it. Sometimes a forager returns with pollen after all cells have been completed and closed; it then wanders about and ultimately gets rid of the pollen in the end of a burrow.

An egg is laid on the food mass, usually soon after it is complete, but at least within a day or two. If not, the cell is often abandoned and filled with soil. The egg-laying bee in *Lasioglossum zephyrum* again rotates and smooths the pollen, this going on for 8 to 78 minutes, even though the ball appears already perfect. She then deposits her large curved egg on top of it (Figure 24.7). The egg layer then closes the cell with earth, which she tamps into place until there is no visible evidence of the position of the cell entrance. However, the location is apparently recognizable to the bees, at least on some occasions, for, as observed by Batra and discussed in Chapters 7 (D1) and 16 (D1), the cell is sometimes opened later and the egg eaten and replaced or the contents of the cell may be merely "inspected."

For the observations discussed above on halictine behavior in the nest, we have been largely dependent on Batra's painstaking laboratory studies, made in Kansas. More recently this and other types of work with ground-nesting bees have been made easier by a simpler and cheaper type of study nest, illustrated in Figure 24.8. It works well at least for various halictines. Open nests, like the one illustrated, can be placed in a bee flight room. Alternatively, as discovered by Dwight R. Kamm, the bees can be kept in the nest by closing the entrance with a vial in which they are fed pollen and honey-water. Such nests can be moved readily, placed in incubators, and studied under a binocular microscope, opening the way for much further observational and experimental work.

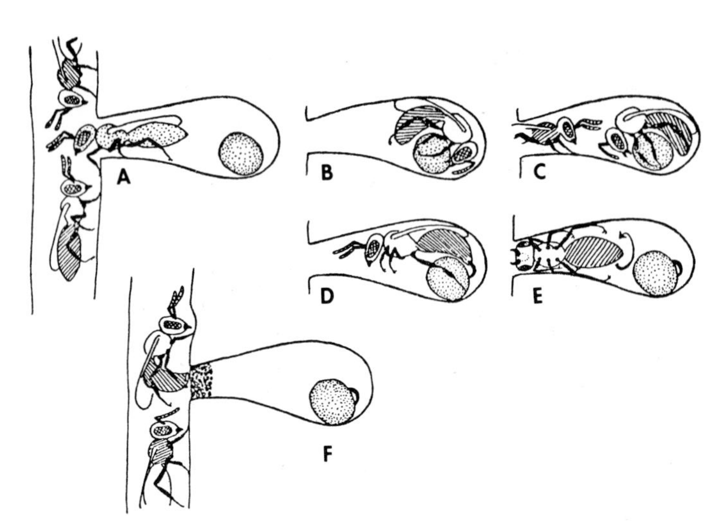

FIGURE 24.7. Sketches showing oviposition by *Lasioglossum* (*Dialictus*) *zephyrum*. Different individuals are marked by different patterns of shading. (*A*) Egg layers waiting in burrow as worker leaves cell; (*B*) and (*C*) egg layer pushing pollen ball back and forth in cell as she smoothes it, another bee inspecting cell; (*D*) oviposition; (*E*) egg layer rotating as she nibbles at entrance of cell before leaving; (*F*) egg layer plugs cell with soil, tamping it into place with the apex of the abdomen. (From Batra, 1964.)

As in all halictines, the larva, on hatching from the egg, feeds forward and downward and ultimately backward around the ball of pollen (see Figure 24.9). During this feeding period the larva, at first rather straight, becomes strongly curled. Before it reaches the stage of defecation, it always orients with its head towards the cell entrance and lies on its back. In this position, when it defecates, the feces are placed on the roof of the cell near the closed end. The larva therefore enters the prepupal stage, in which it acquires a nearly straight body form, with the feces clinging to the roof of the cell out of contact with the insect itself. The pupa occupies the same position; there is no cocoon surrounding it. The adult, on emergence, breaks down the loose plug of earth at the cell entrance and escapes into the main burrow of the nest.

After a bee has emerged from a cell, it is soon filled with earth. Except for a report of re-use of cells by *Lasioglossum seabrai* (Sakagami and Moure, 1967), there is no evidence that cells are re-used by *Dialictus,* although new cells may be constructed in the same general area, frequently cutting through old cells that have been earth-filled.

B. Life history

There are probably species of *Dialictus* which are active throughout the year in cell provisioning and egg laying. *Lasioglossum umbripenne* at Turrialba may be an example of such a form. However, all species that have been studied in detail and throughout the year pass through an unfavorable season as fertilized females in hibernation

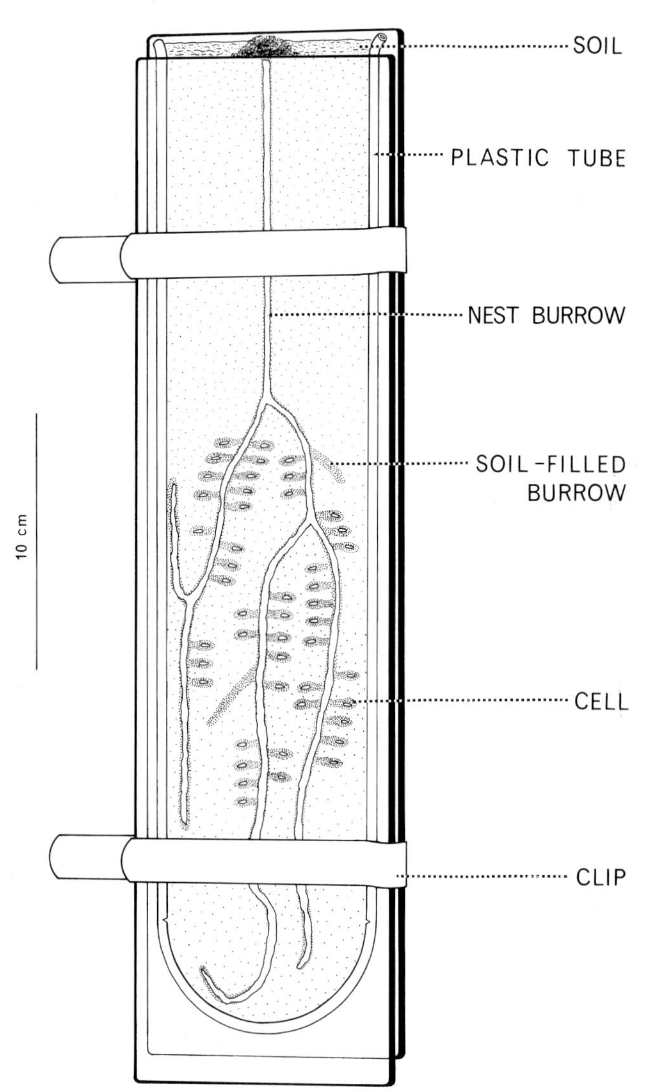

FIGURE 24.8. Nest of *Lasioglossum zephyrum* in earth between two sheets of glass separated by slightly less than the normal cell diameter. It is in such nests that observations of behavior of bees in their cells and burrows have been made. (From Michener and Brothers, 1971.)

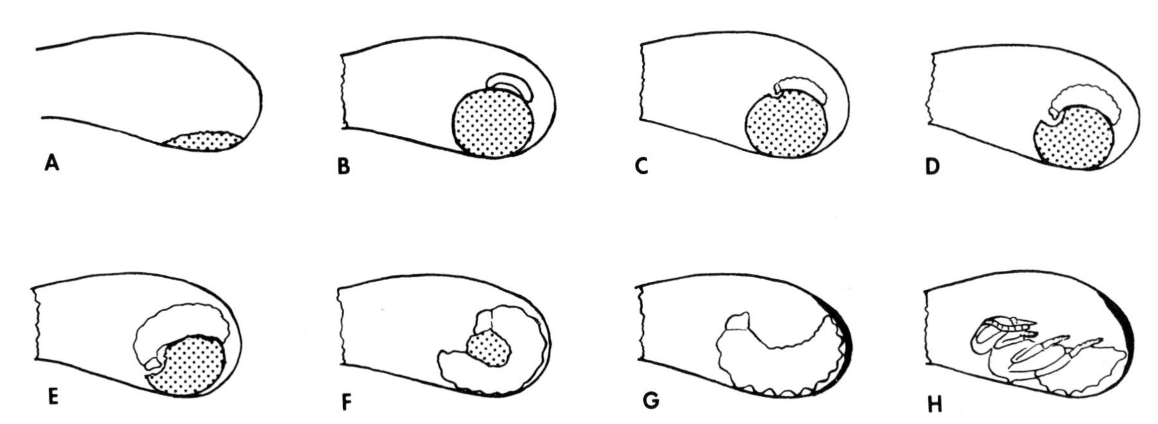

FIGURE 24-9. Growth of immature *Lasioglossum imitatum.* (*A*) Cell being provisioned; (*B*) complete pollen ball (*stippled*) with egg; (*C*) to (*F*) stages in larval feeding and growth; (*G*) prepupa; (*H*) pupa. In (*G*) and (*H*) the feces (*black*) can be seen on the upper distal part of the cell wall. (From Michener and Wille, 1961.)

or estivation. In the tropics such a period may be either the dry season or the wet season, depending on the species and area concerned; in temperate and subtropical regions it is the winter.

In the following explanation of the seasonal cycle, for simplicity of expression and because most of the species that have been studied in any detail are found either in the United States or in southern Brazil where there is a distinct winter, the account is written for species living in such areas. Most of what is said, however, is applicable to tropical forms as well, if in place of "winter" one substitutes "inactive season" or "unfavorable season." Other terms concerning the time of year (for example, spring), together with the word hibernation, must be altered accordingly.

In most species fertilized gynes, probably produced in the nest and therefore in most cases sisters, remain in the nest through the winter, going to the deepest part of it and frequently sealing themselves in by closing off the burrows behind them with soil. Some species do not or do not always hibernate in this way, however. For example, three different aggregations of *Lasioglossum rohweri* near Lawrence, Kansas, were quite suddenly totally abandoned by young gynes in the autumn. (The observations were made in different years.) The workers and males had mostly died by this time; the gynes were quite abundant and then, in a space of a very few days, vanished. A possible explanation is that the gynes leave their nest aggregations in the autumn and hibernate elsewhere. None of the aggregations mentioned above was reestablished the following season, but other nest aggregations of the same species in the same area have been re-used for several years. It is possible that the gynes sometimes return to the same place for nesting in the spring or that, like *L.* (*Evylaeus*) *malachurum,* they sometimes leave, at other times remain in parental nests through the winter. Or there may be some other explanation for the disappearance of the gynes in the three aggregations from which they vanished. *L. umbripenne* in Costa Rica disappears completely from its nesting sites in the inactive season, in this case the wet season, May through December, in a tropical forest near Damitas (Wille and Orozco, 1970).

In spring the overwintered fertilized gynes emerge from the hibernation quarters. Large numbers of them disperse from their overwintering nests and dig new burrows, in the same vicinity or elsewhere. In some species these are widely scattered; in others there are obvious aggregations of nests. Nearly all nests started by lone gynes continue to be inhabited by a single female until her progeny are produced. However this is not always the case. Sometimes a nest started by a single bee will later contain two or three, indicating that gynes looking for nesting sites sometimes enter burrows already established (see Chapter 20, D2b).

In certain species, such as *Lasioglossum imitatum,* nearly all nests that contain more than one gyne in the spring result from two to several overwintered gynes that, instead of dispersing, remain in the nest where they were probably reared and in which they presumably overwintered. This explanation is based on the observation that nests occupied by single gynes in the spring are nearly all new burrows showing no remnants of old cells or earlier occupancy of any kind, while nests occupied by two or more gynes are all old burrows around which one can discover remnants of cells made during the previous year. The lower parts of old burrows are closed off and the upper parts utilized by the groups of occupants as their spring nests. Such groups are sometimes communal, each bee with enlarged ovaries and presumably laying eggs. We do not know whether there is any cooperative cell construction or provisioning in such cases. Other seemingly similar groups of the same species are clearly semisocial, some of the gynes being nonreproductive and for practical purposes workers.

Lasioglossum versatum apparently consistently overwinters in considerable groups of young gynes, which then remain together in the nest in the spring. As indicated in Chapter 6 (A), new nests are apparently formed in most instances by bees digging up from different branches of the old nests and reaching the surface at different points, so that in the spring several nests may replace one of the preceding year, but all the new ones are in close proximity. The result is rather large semisocial or communal (actually intermediate) spring colonies forming small, dense aggregations.

After nest establishment in the spring, lone females of *Dialictus* species produce up to about six cells, although with surprising frequency the number is markedly less, down to one. Groups of females produce correspondingly more cells. After the cells have been constructed, provisioned, and the eggs laid, the females become inactive

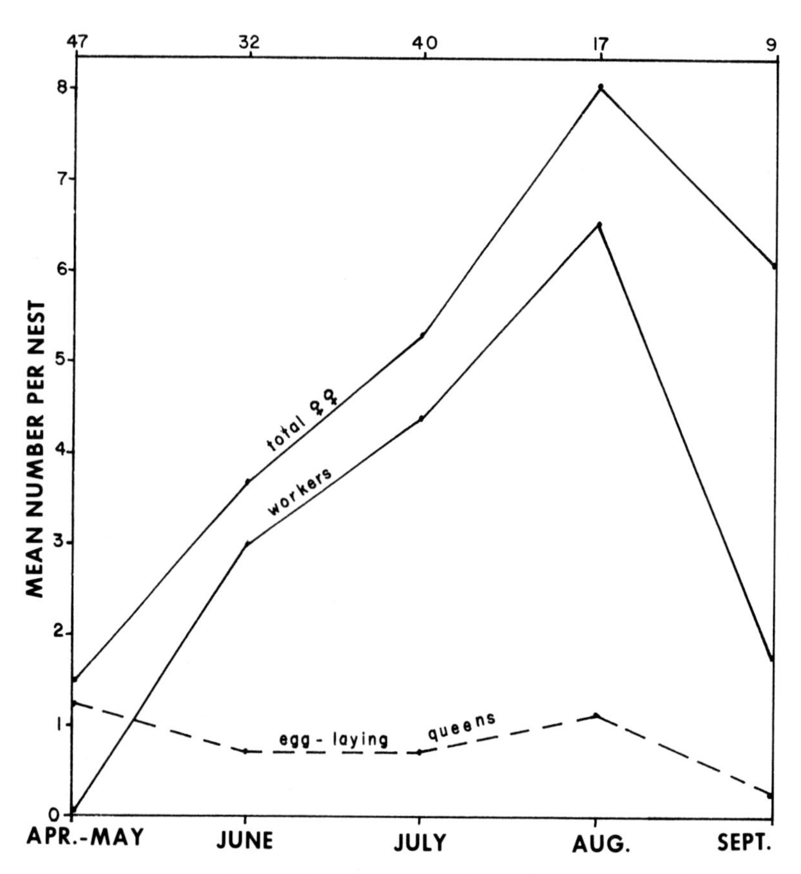

FIGURE 24.10. Graph showing the mean numbers of females of *Lasioglossum* (*Dialictus*) *imitatum* per nest at various times of year at Lawrence, Kansas. Numbers across the top represent numbers of nests censused at different times. (From Michener and Wille, 1961.)

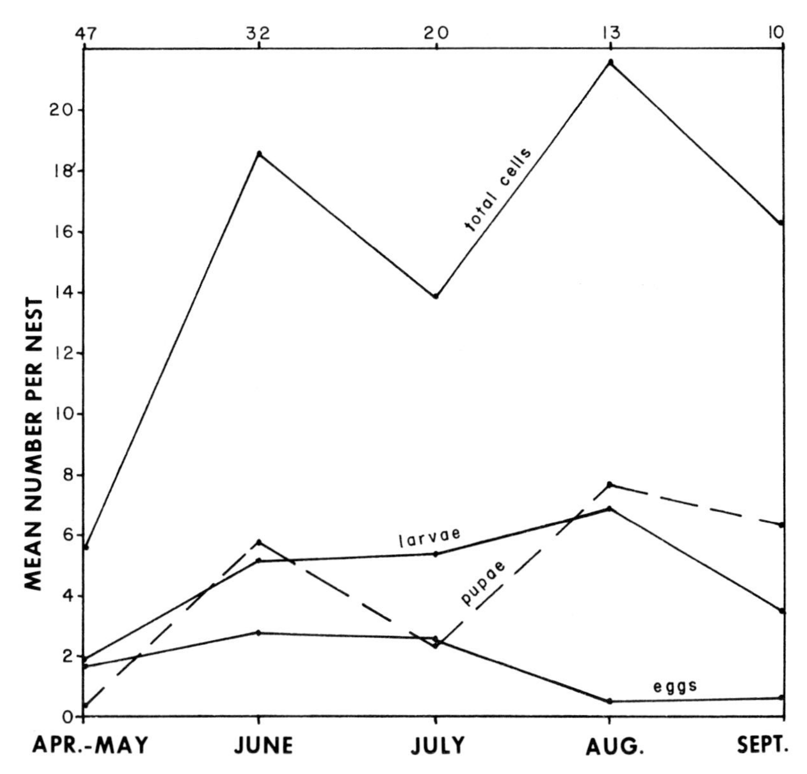

FIGURE 24.11. Graph showing the mean numbers of cells and of immature stages of *Lasioglossum* (*Dialictus*) *imitatum* per nest at various times of year at Lawrence, Kansas. Numbers across the top represent numbers of nests censused at different times. (From Michener and Wille, 1961.)

for a time, resting in their burrows, the entrances of which often become plugged as a result of wind and rain. By this time early spring nests containing communal or semisocial groups have usually experienced considerable reduction in population, perhaps as a result of the death of auxiliaries, perhaps as a result of departure of gynes to establish new nests of their own, or for other reasons, so that by the time of the emergence of the first brood of young, most nests are monogynous even if they started polygynously.

After the maturation of the first young a colony exists, and cell construction and foraging activities are resumed. There is no evidence of subsequent periods of inactivity on the part of the queen; as far as we know she lays eggs more or less regularly thereafter, until near the end of her life. The growth of nest populations of adults and of brood during the season for a Kansas population of *Lasioglossum imitatum* is shown in Figures 24.10 and 24.11. The total population does not grow as rapidly as these figures suggest, however, for while some colonies grow, many die out entirely. These figures take into account only those that survive.

At any time of summer in almost any species of *Dialictus,* an occasional colony will contain more than one gyne. This may be the usual state of affairs in *versatum,* although more investigation will be necessary to establish this. Also, in *zephyrum,* where intergrades between workers and gynes are extraordinarily common, it is frequently difficult to decide just how many gynes are present in a nest; but the number of bees that lay at least some eggs is commonly above one.

In some species, such as *Lasioglossum zephyrum* and probably *L. rhytidophorum,* overwintered queens die during the summer. For *L. zephyrum,* which in Kansas starts its first cells in early to mid-April, the overwintered queens die in June. Daughters and perhaps granddaughters of the overwintered queens serve as queens thereafter, until autumn. In certain other species, such as

L. umbripenne (Damitas population) and probably *L. imitatum,* the overwintered queens live through the active season and are the principal egg layers. In such forms, only a few males are produced early in the season. Regardless of the longevity of the queens, most species of *Dialictus* live in colonies of a primitively eusocial type from early summer until autumn.

In autumn the younger females become overwintering gynes, and males are also produced in large numbers (Figure 8.1). The males often become quite conspicuous in species that nest in aggregations, for during the daytime they fly about the nests in swarms, pouncing on females of their own or other species or even on unrelated insects of approximately the right size and flight behavior. A dense aggregation of nests can sometimes be found at this season because of the activity of the males around it; although they are small insects, the buzzing may sometimes be heard several meters away. Soon after, the males and workers die and the mated gynes become inactive.

The durations of the various stages vary with temperature conditions. Recent determinations by Wille and Orozco (1970) for *Lasioglossum umbripenne* in warm Costa Rican field conditions are as follows: egg, 52–55 hours; larva, 4–5 days; prepupa, 4 days; pupa, 11 days; adult queen, 1 year; adult worker, 30 days; adult male, 14 days. These figures are probably not greatly different from those found in summer for various temperate species.

25 *Primitively Eusocial Behavior in the Subgenus* Evylaeus

Evylaeus is another large subgenus of *Lasioglossum*. It consists of small to rather large bees (Figure 25.1) without greenish reflections, and is most abundant in the Eurasian continent, although there are numerous North American species and some species which at least for the present are included in it and range all the way to southernmost Africa. Many of the smaller species have nests constructed like those of *Dialictus*. Indeed, there is no justifiable distinction between *Dialictus* and the smaller *Evylaeus*, for there are species that can be either black or greenish and there are little species groups of obviously closely related forms, some of which are black and others greenish. There is only the historical tradition that the black forms should be put in one subgenus, the greenish ones in the other. There are even persons who place the forms in different genera on the basis of this single character—color. Obviously this is systematics at its worst and will certainly be changed when the group is revised. From the standpoint of this book, however, this problem turns out to be of little importance, because the species of *Evylaeus* whose behavior has been studied in detail and which are eusocial are rather large and differ from *Dialictus* considerably, not only in their behavior but usually also in nest form and in appearance. The species which, together with close relatives, are the topic of this chapter, are *L.* (*E.*) *duplex* from Japan, *calceatum* from Europe and Japan, and *malachurum* and *marginatum* from Europe.

In all the larger *Evylaeus* there are two or more discrete broods of progeny produced during the life of a queen. The discreteness of those broods, as distinguished from more or less continuous production of young, has been considered a specialized feature. It is true that the *Evylaeus* that exhibit discrete broods also have other features that make them legitimately regarded as behaviorally specialized halictines, but there is nothing in the existence of separate broods, itself, to justify "specialized" status. A female of a solitary halictid species produces one "brood" and then dies. Modification of this life cycle, with sociality, to provide for resumption of laying may take the form of either continuous laying (most social halictines) or of separated broods (certain social *Evylaeus*). Either could develop after the interruption in laying that follows the initial brood, which is the equivalent of the production of a solitary female.

A precaution in studies of broods is that in an area with a short summer season almost any eusocial species may appear to have two broods, while in a region with a longer summer the same species may show long-continued laying after resumption. Perhaps this is the explanation of the reported differences between Kansas and Ontario populations of *Lasioglossum* (*Dialictus*) *rohweri*. This species is said to have two broods in Ontario (Knerer and Plateaux-Quénu, 1967d), while in Kansas, after resuming laying, it continues laying, just as do most other *Dialictus*. It may be that in an area of short summers, potentially continued laying is cut short by autumnal conditions, so that only a limited second brood is produced.

A. Lasioglossum duplex and its relatives

This species has been studied in detail and expertly in Hokkaido, Japan, by Sakagami and Hayashida (1958, 1960, 1961, 1968); it is probably known in more detail

FIGURE 25.1. A female *Lasioglossum,* subgenus *Evylaeus,* resting on a leaf. Note the pollen load, largely on the hind femora although the tibiae, propodeum, and underside of the abdomen also carry some pollen. (Photograph by E. S. Ross.)

than any other social halictid. Moreover, its close relative *Lasioglossum calceatum* has been studied in Europe and recently also in Japan. Nests consist of burrows which occur in aggregations in flat or gently sloping soil; the studies of *L. duplex* here reported were made primarily in the botanical gardens at the University of Hokkaido.

1. Nests

A new nest in the spring extends downward 5 to 10 or even 15 cm, in the summer 10 to 16 or even 30 cm; the burrow is 5 to 6 mm in diameter. The burrow entrance is constricted to 3.5 to 4 mm in diameter and there is no turret, although of course there is often a tumulus. In spring the entrance is closed with dirt, except from about 10:00 A.M. to 1:00 P.M., varying with the weather. The cells are in clusters but do not open directly off the main burrow, as in *Lasioglossum malachurum.* Instead there is a lateroid burrow extending for a few millimeters to 1 cm (rarely as much as 3 cm) to one side from the main burrow, and opening into the cluster cavity. The cell cluster is a comblike structure with approximately horizontal cells that open on a nearly vertical surface. These cells are formed by excavation, not by building, although they are extremely close together. The usual number of cells is 3 to 8 in the spring nests and 4 to 18 in the summer, but a maximum of 30 cells was found in one cluster. The oldest cells are usually at the bottom of the cluster, the cluster being enlarged upwards; it is not surprising that the lateroid usually enters the lower part of the cluster cavity. The cavity is excavated around the cells, and when complete precludes further excavation of any cells in that cluster. It is most remarkable in that earth is removed from the sides and backs of the cells until one can recognize the cells by convexities of the sides and back of the cluster, the earthen walls at the backs of the cells often being only 1 mm thick. Similarly thin-walled clusters are made by various other, not necessarily social, halictids, but are usually constructed with earth brought to the cells and do not contain walls of unworked earth left by excavation (Chapter 17, A1).

The cells are similar to those of other halictids although rather elongate; they are very smooth-walled, slightly flatter on the under surfaces than elsewhere (thus bilaterally symmetrical), lined with a waxlike material. The provisions consist of a pollen ball; the egg is laid on top of the ball as in related species. The cells are not necessarily dug, provisioned, and sealed in sequence, one completed and sealed before the next is constructed; as in many halictines, this sequence is broken down and at least four (in the spring, when only one bee is in the nest) of five (in the summer when workers are present) cells can be in different stages of construction and provisioning at the same time. The cell closure consists of loose earth which can fall out of the cell entrances when a cluster is being manipulated for study. It has been noted that the cell closure is not necessarily constructed immediately after egg laying.

2. Life cycle

Soon after disappearance of the snow, overwintered females start nesting, usually as solitary individuals. Occasionally more than one coexist in a single nest, but this is rare and probably usually temporary; spring activity normally is like that of a solitary bee.

Eggs of the first brood are produced, for the most part, during the second two-thirds of May. Ovarian enlargement, as well as the regression following laying, are shown in Figure 25.2. Emergence of the resulting adults occurs in the latter part of June and through much of July. These adults are workers, except for about 10 percent males (plus 100 percent males in rare nests probably prepared by unfertilized overwintered gynes). The workers average slightly smaller than the overwintered gynes and they remain in the parental nests, participating in the work during the summer, assisting the mother or queen, who may survive until the end of summer. Workers usually do not mate and their ovaries remain slender, but a small number experience ovarian enlargement and perhaps participate in oviposition. The ovaries never attain the size of those of queens, but may contain a full-sized egg.

Although the average number of cells produced for the first brood is five, the mortality of gynes (hence failure to produce colonies) is so high that the total number of first-brood workers produced in an area is no higher than the total number of gynes which originated nests in the area earlier in the spring.

During the summer a second and sometimes a third cell cluster and cavity are constructed along the burrow, usually deeper than the one from which the first brood emerged (Figure 25.3). Egg laying for the second brood begins in early July, and the resulting adults leave their cells after the first week in August. The spring cell cluster and cavity are filled with soil during the construction of

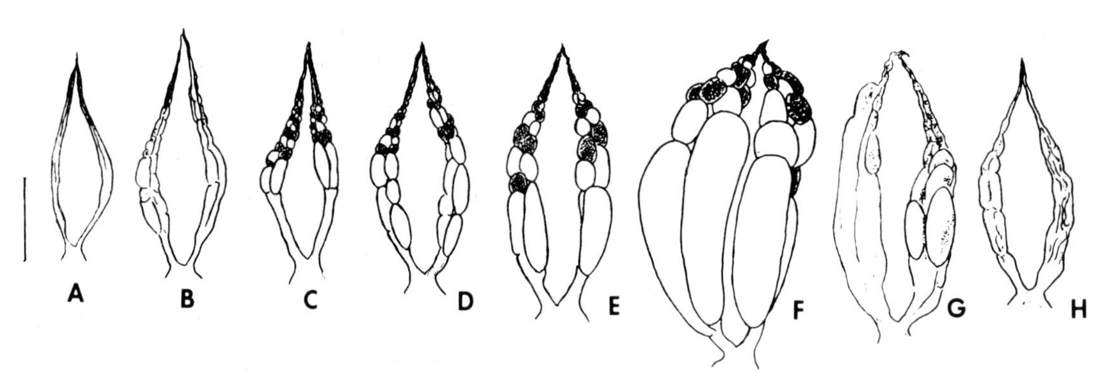

FIGURE 25.2. Ovarian development and regression in overwintered gynes of *Lasioglossum* (*Evylaeus*) *duplex*. The bees come out of hibernation with ovaries as in (*A*). (*B*) to (*F*) show stages in ovarial enlargement. When regression begins the large oocytes become irregular and yellowish (*G*). Regressed ovaries (*H*) differ from enlarging ones (*B*) in the large ovarian calyces. Ovarian enlargement is repeated for the second group of eggs. The scale line at the left represents 1 mm. (From Sakagami and Hayashida, 1961.)

the second cluster. When a third cluster is constructed, it is roughly synchronous with the second; there is no third brood of young produced. The sex ratio in the second brood is approximately 1 to 1 (Chapter 8, C).

The young females of the second brood mate and are gynes which overwinter, dispersing and excavating new nests in the spring. The old queens and the workers, as well as the males, die in autumn.

3. Lasioglossum calceatum and others

The nesting behavior of *Lasioglossum calceatum* in Europe has been studied over a longer period than that of any other halictid, the first account being that of Walckenaer in 1802, and one of the most extensive being that of Fabre (1915). These authors did not understand the life cycle and social organization, largely because they did not examine ovaries and spermathecae of females, and hence had difficulty in recognizing the castes and discovering the longevity of queens. The most recent interpretation of the life cycle in Europe is by Bonelli (1965b, 1968), and Sakagami has recently studied the species in the mountains of northern Japan (personal communication). Except in details of measurements, the nest is about like that of *L. duplex.* However, Bonelli says, and his drawings show, that the cell clusters are built rather than formed by excavation. He says the bees first construct a cavity, then build the cluster of cells in it (from the bottom upward), whereas in *L. duplex* the cells are first excavated, then the cavity formed around them. If Bonelli is right for *L. calceatum* and Sakagami and Hayashida for *L. duplex,* they seem to have shown a remarkable difference in construction methods leading to very similar nests in closely related species. But Plateaux-Quénu (1964) reports that in the laboratory cells of *L. calceatum* were excavated, the cells not being constructed in a cavity; that is, her observations are in agreement with those on *L. duplex.* Moreover, Sakagami (personal communication) does not emphasize a difference in cell construction between the two species in Japan. A possibility is that the two construction methods are not as different as they seem and that the same species may use each under different edaphic or other conditions. We know that some halictids do excavate their cells while others construct them, but this is a verbal oversimplification, for all *construct* the inner layer of substrate material on which the waxlike lining is deposited. Perhaps it is of no fundamental importance whether this layer is constructed inside an excavated cell or as an independent structure in a larger cavity.

The other noteworthy feature described by Bonelli (1968) for *Lasioglossum calceatum* is the overlapping of generations, not known in other temperate-climate halictines. The observations of *L. calceatum* in Japan do not suggest the same thing. Bonelli reports that, in addition to the principal cycle, which is like that of *L. duplex,* there is another, as follows: One female of the first (June) brood (Bonelli says the first to emerge), the same size as her contemporary worker sisters, becomes a guard instead of foraging, and remains in the nest. While foragers may sometimes guard the nest entrance, the queen or the guard is at the entrance most of the time. The guard, instead of dying in a short time, as do the foragers, survives with the queen and is present and guarding during the emergence of the second brood in August. Bonelli says that such a guard outlives the queen, who dies in autumn, and survives the winter in the nest with the fertilized gynes of the second brood; presumably the guard mates in August. In spring some of the gynes disperse and form new nests, but the guard remains with one or two other females in the old nest. The usual semisocial arrangement develops, the guard becoming the queen, the others becoming auxiliaries which die once the cells of the first brood are provisioned and closed. The guard (now the queen) remains through construction of a second cell cluster by the first brood of workers and dies at about the time of emergence of the second (bisexual) brood in August, at an age of 13 or 14 months.

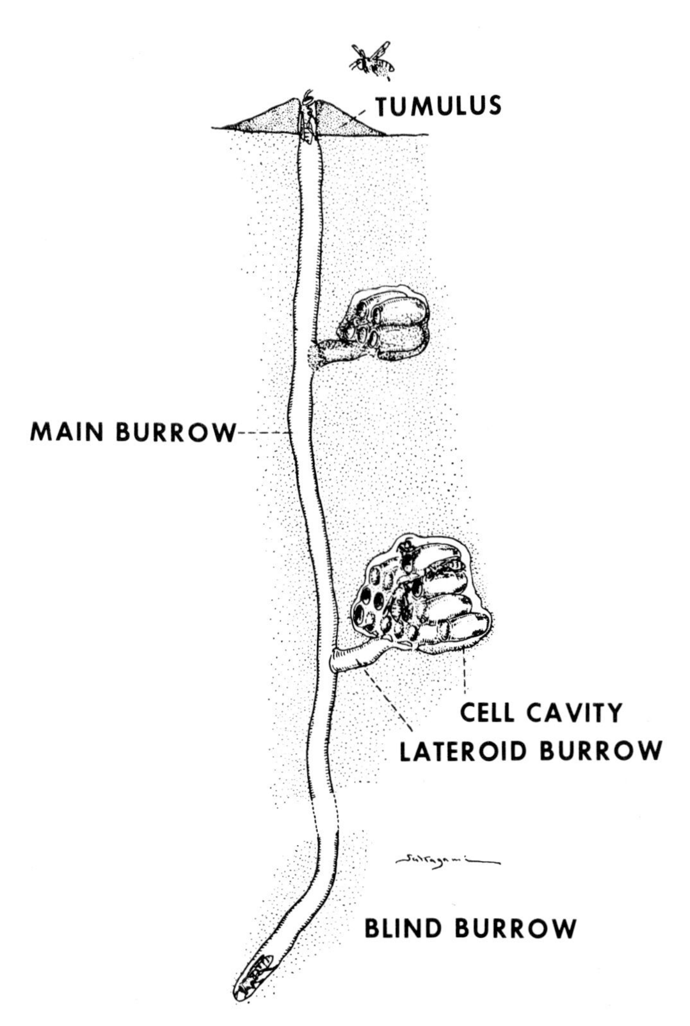

FIGURE 25.3. Nest of *Lasioglossum* (*Evylaeus*) *duplex* in summer, showing (*above*) the spring cell cavity, largely dirt-filled, and (*below*) the summer one, in use. (From Sakagami and Hayashida, 1960.)

Unfortunately, Bonelli gives little information on the bases for the statements in the 1968 paper; one does not know if he had marked bees known to be guards one summer and queens the next, but presumably he did not. One reason for suspecting that the long survival of guards still needs verification is that while the guards are worker-sized in the first summer and smaller than the queens, he speaks as though they were queen-sized and larger than workers the next summer. There is no doubt, however, of the presence of several females in some spring nests of *L. calceatum* (Vleugel, 1961).

Other species that make nests and have life cycles similar to those of *Lasioglossum duplex* are *L.* (*Evylaeus*) *nigripes* and *L.* (*E.*) *interruptum.* For a brief account of *L. nigripes* as studied in captivity, see Plateaux-Quénu (1965a, b). A more extensive account has been published by Knerer and Plateaux-Quénu (1970). For a brief account of *L. interruptum* see Grozdanić and Mualica (1968). Fragmentary information on *L.* (*E.*) *cinctipes* from Ontario and *pauxillum* from Europe suggests that they also have similar life cycles.

B. *Lasioglossum malachurum*

This species has been studied most intensively by Legewie (1925a[1]), Noll (1931), and Stöckhert (1923) in Germany and the bulk of the material presented below is derived from their works, although Bonelli (1948) in Italy and others cited below have contributed important material. Its most noteworthy differences from the species discussed in section A are the existence of two or three

1. Legewie's work is marred by failure to recognize the longevity of queens; he therefore thought that unfertilized females (now known to be workers) parthenogenetically produced females of the next generation, a viewpoint shared by Fabre, Armbruster, and other earlier writers. Nonetheless he presented much useful information.

worker broods before the autumnal brood of reproductives, and the distinctness of the castes, workers and gynes all being distinguishable by size (Chapter 9, B1a).

1. Nests

The nests are burrows into the packed soil of paths and roads, where they commonly are found in aggregations that are rather dense and that are known to survive for many years (Chapter 6, A). The burrows are approximately vertical, except that in deep nests the lower parts may be slanting or even horizontal or circuitous. They are constricted at the surface as usual in halictines; Bonelli describes in detail the ring of tamped soil 7 to 8 mm in diameter surrounding the entrance hole, which is about 2.5 mm in diameter in summer when the colony is active. In young nests the main burrow is unbranched and without elaborated entrance construction; in older nests there may be a few main branches (Figure 25.4). The diameter of the burrow is initially 4 to 5 mm, but late in the season, probably due to heavy traffic of bees, which tends to loosen particles from the furrow wall, the average diameter is somewhat greater. Bonelli (1948) notes that in nests found in Italy, the upper several centimeters of the burrows are lined with earth and saliva (Figure 25.4), but Legewie says that secreted material binds the loose earth only where the burrow passes through the tumulus. As stated in the preceding chapter, earth linings of burrows, perhaps accidentally produced, are common in halictids, and their visibility depends on local soil conditions. Such conditions, or differences in bee behavior, may determine whether burrows grow larger with use or are lined and thus maintained at about the same size. But the thick, hard wall of the upper part of the burrow, presumably consolidated with saliva, is a characteristic and unusual feature of the nests of *Lasioglossum malachurum* and presumably is important in defending them against natural enemies.

The cells, which are of the usual shape for halictines, with the lower surface somewhat less curved than the upper, project more or less at right angles from the main burrow, almost always without appreciable lateral burrows leading to them. That is, the cells open directly from the main burrow, as they do in most *Dialictus* and in *Lasioglossum* (*Evylaeus*) *marginatum,* discussed in section C below. Legewie has done an excellent job of surmising how the cells are constructed from the information gathered as he dug large numbers of nests; his conclusions agree in general with those of Batra, who actually observed *Dialictus* constructing their cells in observation nests. The cell lining, the somewhat flattened ball of provisions, the egg, and the manner of feeding by the larva are also as in *Dialictus.*

A striking feature of the nest of *Lasioglossum malachurum* is that the cells are made in dense clusters, so that adjacent cell surfaces are often only 1 or 2 mm apart. Their walls are presumably of worked earth, as in most if not all halictines, and are impregnated with a secretion and tamped firm by the bees. Seemingly, the cells are so close together that little unworked and untamped earth remains within the cluster area. It is therefore possible to free the whole cluster from the surrounding soil rather readily, although in most cases the bees themselves do not excavate any space around the cluster.

Spring nests of *Lasioglossum malachurum,* however, sometimes have a partial chamber excavated around the cell cluster (Figure 25.4). Such excavation was noted, seemingly, in all nests observed by Bott (1937) and Grozdanić and Vasić (1970), but in aggregations studied by Legewie, Stöckhert, and Bonelli a cavity around the cells was not found; Noll reports it occasionally. Knerer and Plateaux-Quénu (1967b) say that the nest founder excavates a cavity on one side of her burrow, then fills the cavity with loose earth in which she excavates the cells, forming the initial cell cluster. After all the cells are complete and provisioned and the nest closed, the bee excavates a cavity around the cluster, leaving the cells supported only by earth pillars. These authors do not indicate the number of nests that they studied, but their account would hardly explain the diverse observations of other authors; probably there is local variation, possibly dependent on edaphic factors, in the nesting behavior of this bee. The cells in a cluster are ordinarily constructed from the top downward, that is, the uppermost cells are the oldest; but there are occasional deviations.

Cell closure is another feature that may vary regionally. Knerer and Plateaux-Quénu (1967b) give the best recent treatment of the subject. It is based on a study made in southern France. They say that the cells are in general open until the larva is fully fed, but are than promptly closed, and remain so through the pupal period. Sometimes, however, cells are closed after egg laying, only to be reopened at about the time of hatching. Thus cells, in these authors' experience, are always open during the period of larval growth and closed thereafter until

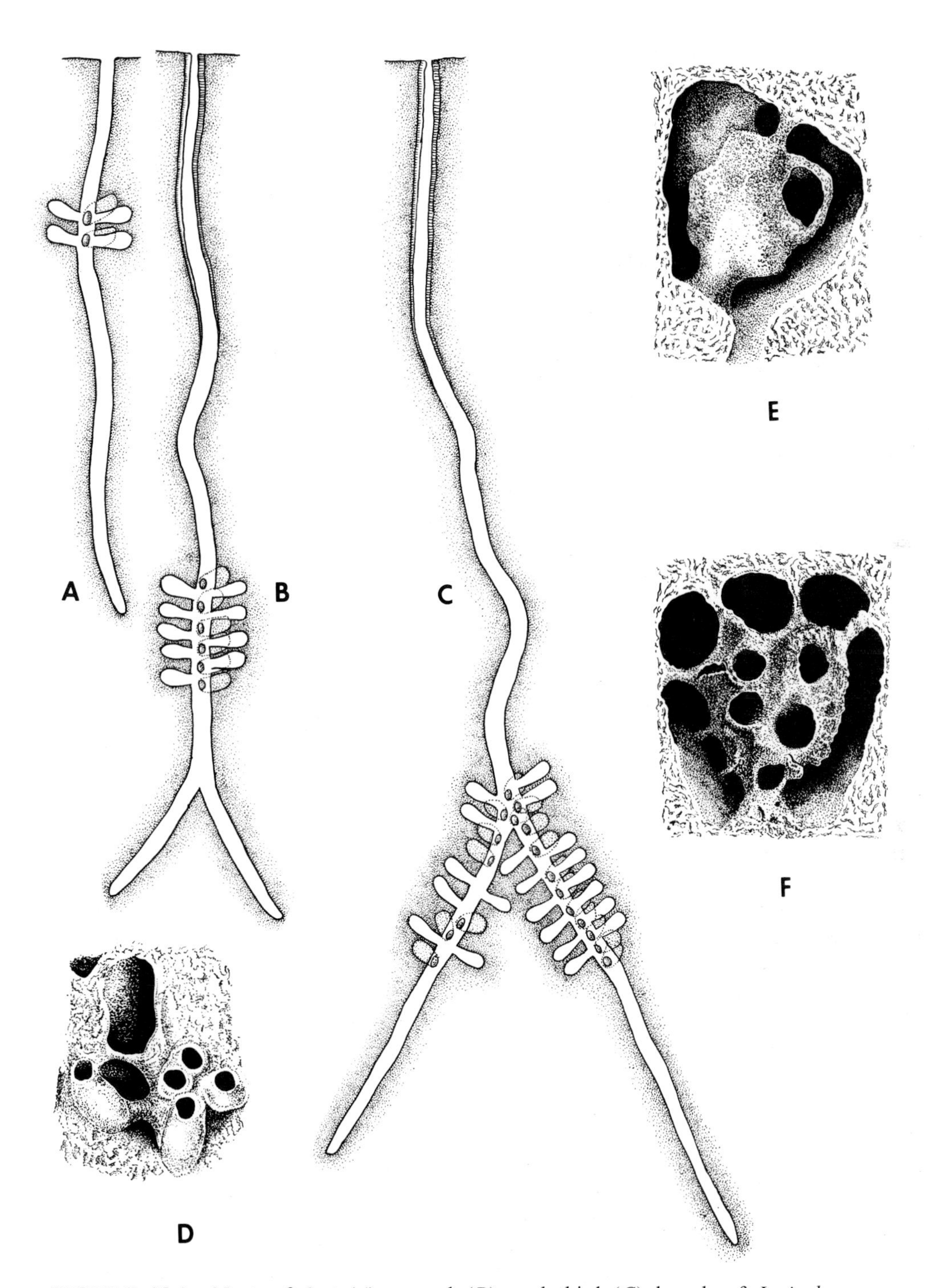

FIGURE 25.4. Nests of first (*A*), second (*B*), and third (*C*) broods of *Lasioglossum* (*Evylaeus*) *malachurum.* (From Sakagami and Michener, 1962, after Bonelli, 1948.) The cell cluster of the first brood may resemble those of later broods in that it lacks a cavity around the cells, as in *A*, or may have a partial cavity around the cells, *D* (after Noll, 1931), or a complete cavity, *E* and *F* (after Bott, 1937).

emergence of the adult. It is possible that some differing accounts of others result from insufficient or inadequately analyzed data. Legewie, however, working in Baden, reports the cells to be continuously open, even those containing pupae, and supports the statement by noting that in such cells larval feces are only in rare cases present (Chapter 16, E). Thus in his population adults appear to have access to cells at a time when they are closed in the French population. Stöckhert, in Bavaria, states that he found all cells containing eggs, larvae, or pupae closed. Grozdanić and Vasić (1970) say that in some spring nests in Yugoslavia the cells are always open; in others they are closed; while in still others both open and closed cells are found. For summer nests these authors say that the cells are all open during the entire brood development, although exuviae and larval feces are left in the cells. One can only conclude that there may be regional variation in cell closure, as in various other features. (For example, Stöckhert's bees had noticeably smaller colonies than those of Legewie and of Knerer and Plateaux-Quénu.)

2. Life cycle

According to Legewie, the gynes hibernate in unknown locations away from the nests (his excavations of the nesting area in late fall revealed no bees), a conclusion supported by Bonelli (1948); but Stöckhert reports gynes hibernating together in their mother's nests in the old burrows, in empty cells or in special hibernation burrows up to 5 cm in length branching off the main nest burrow. Grozdanić (1966) says that young gynes leave their nests in the autumn if weather is fine and warm, but that otherwise most of them winter in the nests and disperse in the spring.

In the spring (April in Germany; February to March in Yugoslavia) there usually remains only a single female in the nest, the others having dispersed and started new burrows, each for herself. Occasionally, however, two or three females remain in the same old burrow (Stöckhert, 1923; Grozdanić and Vasić, 1970). Krunić (1959), however, notes that after single females found their nests, they may be joined by others within a very few days. In several cases Stöckhert found two gynes in the same nest together with their workers in summer, showing that occasionally the communal or semisocial groups of gynes stay together and produce offspring together. Usually, however, any polygynous spring nests are monogynous by the time the workers appear in summer, due to the death or displacement of the excess females.

Knerer (1969c) emphasizes that in strong aggregations the number of females greatly exceeds the number of nests established in early spring. The excess females fly about, trying to usurp existing nests and sometimes succeeding in doing so. He removed the pollen-collecting gyne from each of ten nests. Sometimes within ten minutes and in all cases by the following day another bee had taken the nest and was collecting pollen. Removal of these bees resulted in similarly prompt replacement. Further evidence that the bees flying about the colony had the potentiality of establishing nests was found by taking some of them to the laboratory, where over 90 percent of them made nests and produced a brood of workers. Apparently density of nests or bees in the aggregation site in some way inhibited nest-making activities of other gynes. In April, when most of the original group of nests were closed, the occupants having provisioned their first batches of cells, the supernumerary females were suddenly released from the supposed inhibition and many new nest burrows were excavated. Non-nesting gynes still remained, however, and many of them presumably never were able to nest. Knerer interprets this as a population control device, limiting excessive increase and at the same time providing a reserve to replace losses at the season when lone gynes forage and are perhaps most subject to predation or other dangers.

Spring nests extend for 20 to 25 cm into the earth. The cluster of cells is usually made at a depth of 12 to 15 cm from the surface. Bonelli (1948), however, reports cells in spring only 4 to 6 cm from the surface in Italy, while Knerer and Plateaux-Quénu (1967b) say 7 to 9 cm in France; such variations often result from local soil and moisture conditions, but of course may also reflect differences in the bees. The spring cluster typically contains 6 to 7 cells, although Legewie found one with 11. After completing a cell, the bee flies out to collect pollen, and in the course of several consecutive flights brings in a supply adequate for provisions of that cell. Sometimes a new cell is started before the preceding one is completed and provisioned.

By early May (Germany) all of the cells to be constructed by the overwintered gynes have been finished, and these bees then remain in the nests, neither digging nor flying. The first brood of young, consisting only of females which are workers, is active in the latter part of May and through June. For many years the workers went under the specific name *longulus,* because they are smaller than and somewhat differently sculptured from the queens. These workers cooperate in the construction of new cells and in provisioning them. Contrary to the situation in the spring when the overwintered gynes were provisioning their cells, there is now a guard at the nest entrance most of the time. She is one of the workers, not the same one at all times; or, according to Stöckhert, she is more often the queen. The new cells are generally constructed somewhat deeper than the original cluster, the cells of which are now filled with earth. The number of cells in the new cluster is usually 19 to 20, with a maximum of 26 for the German records, but Grozdanić and Vasić mention unusual colonies with 30 or more workers, so that at least that many cells must be constructed on occasion. The queen lays an egg in each cell after it has been provisioned. After provisioning is complete, outside activity ceases and the bees remain in the nest. The workers of the first brood gradually die off, although some of them are still alive at the time of emergence of the second brood of workers.

The adults of the second brood, also workers, are active from early July to mid-August making and provisioning a third cluster of cells, usually still deeper in the soil, and containing 60 to 65 cells in general, with a maximum of 97 found in one case. Stöckhert reports an occasional overwintered queen still flying, even collecting pollen, but the majority are in the nests, commonly guarding entrances. (Aptel's report of 1931 of what amounts to communal behavior in August nests containing only 2 to 7 bees indicates either extraordinary diversity in behavior of this species or errors in observation or identification.)

The young produced in the third cluster, instead of being all females, are males and females (gynes) in approximately equal numbers, although sometimes there is a slight excess of males. At least the females result from eggs laid by the old queen. The males emerge first, from cells that tend to be in the upper part of the cell cluster and are the earliest cells constructed. The females come from lower and later cells. The males leave the nests and fly about during the daytime, feeding on flowers. At night they return to nest burrows, probably without regard to the nest from which they originated, although alien females are said to be excluded by the guard at the same season. Mating is said by Legewie to sometimes occur in the nests, but it also occurs outside. The behavior of the young gynes is markedly different from that of workers, for they appear lazy whether about the nests or on flowers.

The old queens, by autumn often unable to fly and so badly worn as to be almost unrecognizable, die first (although Stöckhert found one still alive on October 5; she was presumably 13 months old). Soon after, the workers die. Males spend a great deal of time on the autumnal flowers and ultimately also die. Only the fertilized young gynes survive the winter.

It is not surprising that the life cycle as reported by Bonelli (1948) for Bologna differs from that reported for central Germany by Legewie, Noll, and Stöckhert. Activity starts earlier, and there are three broods of workers instead of two before the bisexual brood appears in August. The Yugoslavian workers do not report an extra brood, however. At the northern edge of the range of the species, in northern Germany, the life cycle is also reported to be different. Because workers have not been collected there, the species is supposed to be solitary, "as is the case with the arctic bumblebees" (Stöckhert, 1933). This proves not to be the case with arctic bumblebees, and nesting populations of *Lasioglossum malachurum* need to be studied, or at least season-long collections and dissections of bees are needed to verify the reported lack of a worker caste. The evidence is inadequate; at best it is suggestive. Misidentifications may explain the reports.

C. *Lasioglossum marginatum*

This species has been studied primarily in France by Plateaux-Quénu (1959, 1960b, 1962, 1972), but Grozdanić (1966 and previously) and Vasić (1966) in Yugoslavia have also published information on certain aspects of its behavior. It forms the largest colonies of any halictine bee and has the most remarkable life cycle of any known species.

1. Nests

The nests (Figure 25.5) consist of burrows which are made in aggregations in hard, bare soil, such as that of paths or roads. The study by Plateaux-Quénu was made in an aggregation of some thousand nests which extended in a band about 80 cm wide along the shoulder of a road for some 100 m in the region of Les Eyzies in France. Like so many fine bee sites, this was destroyed by road construction.

A new nest consists of a nearly vertical burrow extending to a depth of 35 to 45 cm, with cells excavated into the soil laterally. The cells are connected directly to the

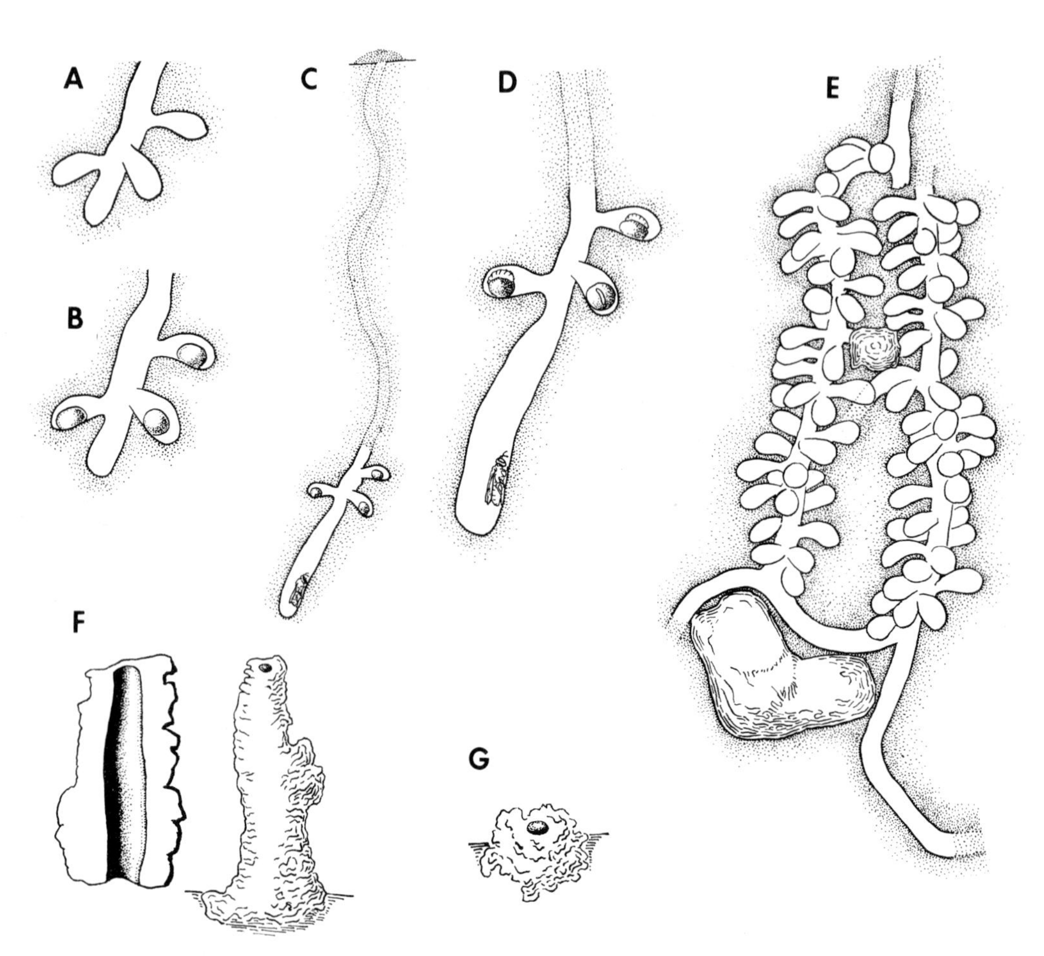

FIGURE 25.5. Nests of *Lasioglossum* (*Evylaeus*) *marginatum*. (*A*) to (*D*) First-year nest made by a lone gyne, showing synchronized construction of cells (*A*) and provisioning (*B*). In *C* the upper part of the nest is closed with earth and in *D* the eggs have been laid and two larvae are growing. (*E*) Part of an old, complicated nest. (*F*), (*G*) Turrets of old nests (one in sectional view) and a young nest, respectively. (From Sakagami and Michener, 1962, after Plateaux-Quénu, 1959.)

main burrow, and the nest structure is thus essentially like that of *Dialictus*. Older nests become enormously complex because of branching main burrows and numerous cells, but the basic structure is the same, like that of the branching older nests of *Dialictus*. As Vasić emphasizes, the turning and twisting of the burrows form a labyrinth; if they continued more or less straight they would extend to depths where soil textures would probably be unsuitable.

The sequence of cell construction, provisioning, and egg laying is unique, so far as known. the bees first construct all the cells, before provisioning any of them. Then they provision all the cells. Then the bees cease foraging activities as well as other trips to the outside and remain exclusively in the nest, the upper part of which, from the surface down nearly to the uppermost cells, is filled with earth. During the ensuing month the gyne or queen lays eggs in all of the cells. The cells are left open, not closed as in most halictines, but nothing is known of the attention, if any, given to the brood by the adult bees, except that the feces are removed from the cells (Knerer and Plateaux-Quénu, 1966c). The bees do not forage during the period of larval development but remain in the closed nest.

2. Life cycle

The life cycle of a colony requires five or occasionally six years, the life span of the queen (workers live 12 to 15 months). Fertilized females produced in a mature colony remain in their nest through the winter. In late March or April these gynes emerge to the surface and visit flowers for food, sometimes returning to the old nest, sometimes establishing new ones. By the end of April, according to Plateaux-Quénu, all are establishing new nests. Grozdanić (1966), however, reports that some nests are established by two or three females and speculates that such colonies may be important in developing the huge populations ultimately found in some of the old nests.

The number of cells in a newly established nest ranges from two to six. The nest is closed during May and will remain so until the following spring. Larval development is unusually slow for a halictine, requiring about two months from egg to adult. All the adults of the first brood of young, which mature in August, are females. They remain in their cells and the gyne remains in the burrow through the autumn and winter.

The following spring, toward the end of March, the daughters from the first brood open the nest, visit flowers, and provision a new batch of cells. These bees are workers. The nest is closed after the cells are provisioned, as in the first year, and the queen begins to lay about a month later. The number of cells constructed during the second year ranges from 6 to 17.

During the third, fourth, and occasionally fifth years, similar developments occur, the colony becoming larger each year. Egg laying is by the queen alone, who loses her power of flight, develops larger ovaries each spring, and never leaves the nest. If we can judge from *Dialictus*, she feeds from food placed in the cells by workers. The

TABLE 25.1. Diagram of the development of a colony of *Lasioglossum marginatum*.

1st year	2nd year	3rd year	4th year	5th year	6th year
G: Makes cells, provisions cells, lays 2–6 eggs.	No digging, no provisioning, lays 6–18 eggs.	No digging, no provisioning, lays 18–54 eggs.	No digging, no provisioning, lays 54–162 eggs.	No digging, no provisioning, lays 162–486 eggs. Often dies.	No digging, no provisioning, lays 486–1458[b] eggs. Dies.
W: 2–6 workers.	Dig (or restore) and provision 6–18 cells.	Dig (or restore) and provision 18–54 cells.	Dig (or restore) and provision 54–162 cells.	Dig (or restore) and provision (and lay in?) 162–486 cells.	Dig (or restore) and provision and lay in 486–1458[b] cells.
	6–18 workers.	18–54 workers.	54–162 workers.	162–486 workers.	
				OR	
R:				162–486 males + gynes	486–1459[b] males + gynes
				Fertilized gynes winter together, start new nests in spring.	Fertilized gynes winter together, start new nests in spring.
P: 2–6	6–18	18–54	54–162	162–486	486–1459[b]
A: 2,3,3,3,3,3, 3,4,4,4,6	6,8,9,10,12,17	15 (a bad year), 25,31,39,39	50 (a bad year), 58,81,104,108, 111,130	199,206,225,315, 319,411[a]	500,676[a]

SOURCE: Modified from Plateaux-Quénu, 1962.
Note: G = gyne; W = workers; R = new reproductives (males + young gynes); P = theoretical number of progeny, based on a threefold annual increase after the first year; A = actual numbers of progeny found in individual nests.
[a] Some additional actual numbers of sexual forms found in nests are shown in the last column of Table 25.2.
[b] These large numbers are theoretical; actual census data are in the last rows of this table and the last column of Table 25.2.

number of cells produced in the third year ranges from 15 to 39, in the fourth from 54 to 162. As before, the brood of young produced each year consists entirely of workers whose ovaries do not enlarge and who do not become fertilized.

When there are workers present in the nest, and when the nest is open in the spring, the burrow is prolonged above the surface of the ground in the form of a turret consisting of earth excavated from the nests. The height of the turret varies with the number of workers, therefore becoming higher each year with colony growth. The entrance is not guarded.

The final year, usually the fifth but sometimes the sixth, begins just like the preceding years except that, because the nest is very populous, the turret is very high. The cells constructed during the final year, which range in number from 199 to 411, according to Plateaux-Quénu's observations (but Grozdanić has recorded numbers as high as 897), produce both males and females, in contrast to preceding years. Males are usually more numerous than females, but as indicated below and in Table 25.2 there are strange variations in the sex ratio. Although in preceding years the workers do not show ovarian development, in the final year certain workers do show such development and, because they are unfertilized, are probably responsible for production of many of the male-producing eggs. Contrary to previous years, when the nests are only opened for one brief period, during the final year they are opened twice, in spring as usual for cell provisioning, and in September for the escape of the males. The males fly about, feed on flowers, swarm over the nesting site, entering any open (final-year) nests and mating within, and die before winter, as does the old queen. Females do not leave their nests at this season; in fact the two sexes are never on the wing at the same time. Therefore only females produced in final-year nests become fertilized and become the gynes which winter in the parental nest and will disperse and establish new nests the following year.

TABLE 25.2. Numbers of adult males and females in old nests of *Lasioglossum marginatum* opened in September in Yugoslavia and France.

Nest no.	♂♂	♀♀	Total
1	95	70	165
2	340	128	468
3	658	189	897
4	—	157	157
5	—	441	441
6	410	—	410
7	715	—	715
8	244	68	312
9	238	81	319

SOURCES: Numbers 1–7, data from Grozdanić, 1966; numbers 8–9, data from Plateaux-Quénu, 1959.

Table 25.1 summarizes the above account, and shows the theoretical numbers of cells that could be produced each year, together with (across the bottom) actual numbers of bees produced. The theoretical numbers are based on the observation that nest populations triple annually. (Note that censuses involve destruction of nests and colonies; no one colony was censused annually.) Therefore the population produced in a given year should be $3P(t-1)$, in which P is the number of workers produced the first year and t is the colony age in years.

The above account is based almost entirely on the studies of Plateaux-Quénu. Grozdanić (1966) raises some questions. He finds the annually increasing fecundity of queens hard to understand and asks the meaning of the diversity in types of nests shown in Table 25.2. Numbers 1 to 3 and 8 and 9 represent the type of nests described above. Nests 4 and 5 raise the question as to whether a single queen could lay so many eggs. He speculates that some females of the previous season may mate and remain in their nest to augment the output of eggs in the final year of the colony, but he reports no dissections or other studies to verify existence of such individuals.

26 *Primitively Eusocial Behavior in the Genus* Halictus

Halictus is a close relative of *Lasioglossum,* but differs from it in several morphological characters. Like *Lasioglossum* it contains biologically very diverse species. There has been less detailed study than in the case of *Lasioglossum,* and this chapter is therefore very brief. The genus contains mostly large, nonmetallic species in the subgenus *Halictus* proper, and smaller, slightly metallic greenish species of the subgenus *Seladonia.*

There is one famous European species of the subgenus *Halictus, H. quadricinctus,* that makes large clusters or combs of cells, excavated, not built, and supported by pillars in a cavity in the soil. Most *Halictus,* however, make nests whose structure is basically like those of *Lasioglossum* (*Dialictus*), that is, with isolated subhorizontal cells radiating into the soil around the main burrows. In some cases the nests are relatively small, but old nests of *Halictus* (*Seladonia*) *lutescens* in Costa Rica reach depths up to at least 140 cm. One excavated during the wet season when, although there was some flight activity, cells apparently were not being constructed, contained 342 adult female bees (Wille and Michener, 1971).

The subgenus *Seladonia* contains solitary species like *H. tumulorum* (Sakagami and Fukushima, 1961). It also contains eusocial species like *aerarius* (Sakagami and Fukushima, 1961), *confusus* (Dolphin, 1966), *hesperius* and *lutescens* (Wille and Michener, 1971), and *jucundus* (Michener, 1969b) in Eurasia, North and Central America and Africa. In no case is the life cycle well worked out, but it seems that it is similar to that of eusocial species of *Lasioglossum* (*Dialictus*). *H. aerarius* is noteworthy for great cephalic dimorphism in females, large-headed ones being gynes, small-headed ones being commonly but not always workers (Chapter 9, B1). In the other species listed such striking macrocephaly is not known, but it has been recognized among field-collected specimens of a few other species, especially *H. lanei* from Brazil.

In the subgenus *Halictus* proper there are also solitary species. *H. quadricinctus,* the large cluster-making European species mentioned above, is the best-known example. Apparent caste differences are described by Sakagami and Wain (1966) for *H. latisignatus* in India, although nest populations have not been investigated. Quénu (1957) records a queen and three smaller, unfertilized workers in a nest of *H. scabiosae* in France, while Batra (1966c) reported, for the same species in Switzerland, a picture suggestive of *Lasioglossum* (*Dialictus*) *zephyrum,* or with even more complete confusion as to caste. For example, seven of eight pollen collectors in different nests were fertilized, and two had well-developed ovaries. Nest populations ranged from one to six, with only one nest inhabited by a single female. A total of only seven nests was studied. Obviously the species needs further investigation. Caste differences and a colony organization similar to that of various eusocial *Dialictus* have been attained by several species: *H. scabiosae, maculatus, ligatus, rubicundus, parallelus,* and *sajoi.* In all cases the data are few, but in several of these species nearly all individuals of the first brood are females (workers) (Knerer and Plateaux-Quénu, 1967d; for a recent paper, Grozdanić, 1971).

Certain *Halictus* are reported to produce offspring in distinct broods, as do some species of *Evylaeus* in the genus *Lasioglossum.* According to Bonelli (1965a), overwintered females of *H. sexcinctus* start their nests in the ground individually. In late June and early July a brood almost exclusively of females emerges; they remain, unfertilized, in the mother's nest, are of the same size as the mother, and at first experience some ovarian development and perhaps lay some eggs. Both these young females and the mother provision cells, then the workers die. The mother remains alive while the bisexual autumnal brood matures; the brood emerges in August, and after mating the males as well as the old queen die. Even in this form, in which castes seem to differ only in longevity, fertilization, and degree of ovarian development and not in foraging behavior, there is a phenetic discontinuity between the castes as well as a temporal discontinuity between production of workers and of queens. More information is needed, however, to verify all this and to determine whether some *Halictus* regularly produce isolated broods of offspring. If so, this is an interesting parallelism to the larger species of *Evylaeus.*

27 *Allodapine Bees*

The allodapine bees are most abundant and diversified in Africa south of the Sahara. A few species occur in Asia Minor, but they are more abundant in southern and southeastern Asia and still more so in Australia, where species are numerous but not nearly so diversified as in Africa. Like their relatives in the genus *Ceratina,* they usually nest in dead, dry, pithy stems but they may also occupy burrows of beetles in stems or wood and cavities in galls or thorns. There is an unverified report by an excellent observer, the late Dr. H. Brauns of Willowmore, South Africa, of their also going into holes in the ground; but this is certainly not their usual habit. The most recent and extensive treatment of allodapine bees is that of Michener (1971), from which much of the following is extracted. The older papers on the group are summarized by Sakagami, 1960a.

A. Nests

The nest is a simple burrow. Nests are usually started where breaks in dead stems or twigs expose natural hollows, hollows made by borers, or pithy centers in which the bees can excavate. Less commonly, the bees enter unbroken stems through openings made by borers. It is often difficult to decide, for a given nest, whether the bees refurbished or enlarged a pre-existing burrow or excavated a new one. Many species probably never excavate their own burrows while others regularly do so.

Since dead, broken stems are the principal nesting places, various factors affecting the abundance of such sites are important in influencing populations of allodapine bees. Regular cutting of stems by people over long periods in the same area often leads to the buildup of good populations in parks and gardens, especially in tropical Africa. Movement and feeding of large animals, especially elephants, has the same effect. While fires burn dry stems, they also make more dead ones and may even burn them off so that subsequent breaking is not necessary. (These and other factors relating to the local abundance of allodapine bees are considered by Michener, 1970b.)

A most unusual feature of allodapine nests is that there are no cells. Young are reared together in the burrow, which is commonly deepened (except in certain genera, such as *Halterapis* and *Allodapula*) at the same time that young are being reared, the pith fragments being carried past the young to the nest entrance. Nest entrances are regularly constricted, either by being excavated in that way or by being narrowed by a constructed collar made of bits of material from the burrow such as pith or feces of burrowing beetles. Only in the Australian *Exoneurella* is such a constriction not evident; it is weak, however, in the African genus *Allodape.* As in halictine bees, if the constriction is removed, it is soon reconstructed.

B. Solitary species

The South African genus *Halterapis* contains solitary species whose habits differ from those of *Ceratina,* which makes individual cells for its young by partitioning the burrow, principally in the lack of partitions. The female lays an egg which is attached by its posterior end to the wall of the burrow and which projects outward into the burrow. Upon the egg she then places the loads of pollen, presumably mixed with nectar, as a firm mass, until there are provisions adequate for the entire larval growth. A most remarkable feature is that when smooth and shaped, these provisions nearly fill the lumen of the burrow yet do not touch its walls at any point, being supported exclusively by the egg. Without constructing a partition, the bee then lays another egg nearer the nest entrance and provides provisions for the larva that is to hatch from it in the same way. A series of eggs and developing young soon occupy the stem, much as in a *Ceratina* nest except for the lack of partitions. When the young become prepupae or pupae the mother may go past them to the lower part of the nest and lay at least one more egg, which is provided with provisions like the others.

The mother is still alive when her daughters mature, and one or two young adult females may be found in the burrow with the mother. There is no evidence that the daughters remain in the nest or that a colony of adults is formed. The number of nests studied, however, is small, and many details of the behavior of species of this genus remain to be discovered.

In one other allodapine, *Compsomelissa stigmoides* in Kenya, food may be present in mid-larval life in rather large quantities, as in *Halterapis,* but both younger and older larvae appear to be fed progressively.

C. Subsocial species

All other allodapines have progressively fed larvae which are regularly in contact with one another as well as with the adults in the nests; the latter bring in food and usually carry out feces. In some species a nest almost never contains more than one working adult. These are the subsocial species. *Exoneurella lawsoni* of eastern Australia (Michener, 1964c) has been studied in dead pithy stems of roadside weeds. As in most of the allodapines, the young are arranged from oldest near the nest entrance to youngest at the bottom, except that small larvae and eggs may be mixed in a cluster at the bottom. The moderate-sized and large larvae have structural features that support them in the burrow (Figure 27.1). The somewhat pointed apex of the abdomen and the dorsal projections on the body, combined with bristles on the body, are important. The apex of the abdomen and the dorsal swelling or projection (on the first body segment in middle-sized larvae and on the eighth segment in large larvae) are pressed against opposite sides of the burrow to support the larva, while the head extends upward. The underside of the body is somewhat concave and is directed toward the nest entrance. Pollen paste is placed on the underside of the larva by the adult bee.

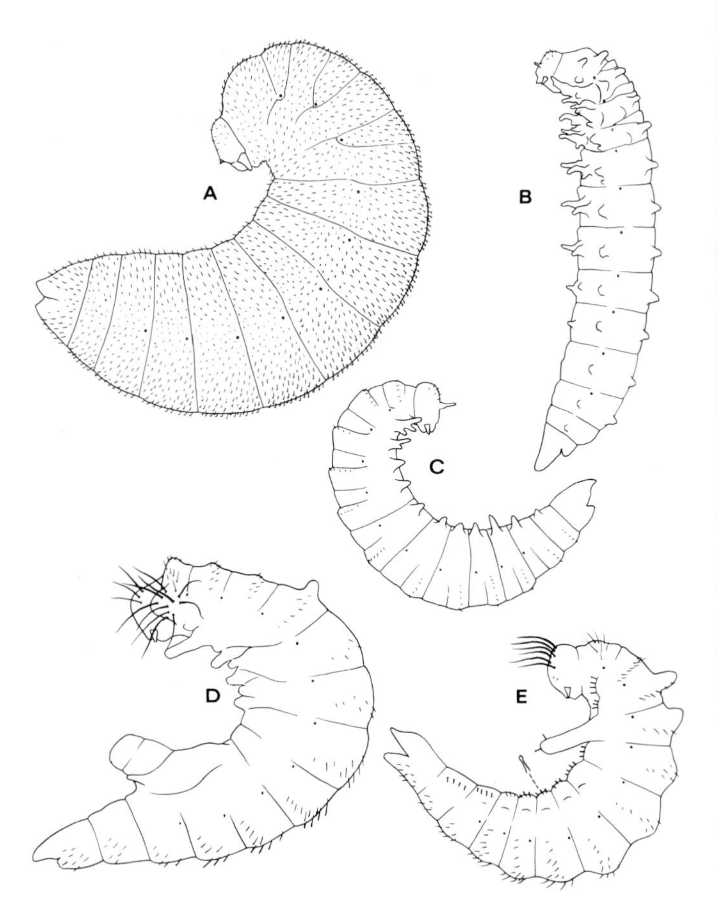

FIGURE 27.1. Larvae of allodapine bees: (*A*) *Macrogalea candida,* (*B*) *Allodapula melanopus,* (*C*) *Exoneura obscuripes,* (*D*) *Allodape mucronata,* (*E*) *Braunsapis foveata.* (Original drawings by C. D. Michener.)

Winter is passed by young adult males and unfertilized females in the nests, along with very small numbers of larvae. Mating occurs in the spring and the females disperse, only one remaining in each original nest, the others making new nests. The females soon start laying eggs, and continue to do so, so that by early summer the number of young of all ages per nest averages 11.5 and reaches a maximum of 28. These are subsocial associations, consisting of one female and her progeny. By late December (nearly mid-summer) numerous adults are maturing from eggs laid in the spring, and most of the overwintered females have died. In most nests at this season, a female, presumably a daughter of the overwintered spring mother, cares for the larvae and continues to lay. The other young females remain in the nests while they are still pallid callows, but soon after they have become black they leave and establish new nests. Young of all ages are present in the nests and the maturation of adults continues through the summer as does establishment of new nests. By late February, however, egg laying has ceased in most nests and by mid-March nearly all old adults are dead; the young adults are ready to pass the winter.

As can be seen from the above account, *Exoneurella lawsoni* is an essentially subsocial bee, although as suggested by Tables 6.3 and 9.2 and shown by Table 27.1, a small percentage of the nests contain two mature adults. *Allodape mucronata,* a South African species whose life cycle is known in considerable detail, has a life history similar to that of the *Exoneurella* although differing in some features; for example, the overwintering females nearly all mate in the autumn and relatively few adult males survive the winter.

D. Colonies of adults

As indicated by data contained in Tables 6.3, 9.2, and 27.1, there is every intergradation from essentially purely subsocial forms to allodapines which in some of their nests have semisocial or primitively eusocial colonies. Social organization among allodapines involves whatever mature females are in the nest—mother and daughters or a group of sisters. The young cared for may be their own offspring, their younger brothers and sisters, or perhaps sometimes nieces and nephews. Nests are started by lone females which raise young subsocially; when the young mature a primitively eusocial or ultimately a semisocial colony may exist.

The frequencies of nests containing two or more active mature females are indicated for several species in Table 27.1. When young are maturing, a more or less continuous process during much of the active season, there are commonly young adult females in the nests. Judging by the continuous establishment of new nests in most species, some such young females must leave their parental nests, but some remain to become working cohabitants in the

TABLE 27.1. Percentages of nests believed to have been inhabited for some time in the active season by two or more mature adult female allodapine bees.

Species	Percent of all nests	Percent of nests with eggs or young	No. of nests	Account of life cycle
Exoneurella lawsoni	6	8	64	Michener, 1964c
Allodape mucronata	6	7	106	Michener, 1971
A. panurgoides and *ceratinoides*	4	6	105	Michener, 1971
A. panurgoides and *ceratinoides*	9	15	34	Michener, 1971
Allodapula dichroa	3	4	150	Michener, 1971
A. variegata	7	10	55	Michener, 1971
A. acutigera	10	13	187	Michener, 1971
A. melanopus and *turneri*	20	23	103	Michener, 1971
Braunsapis facialis	34	43	166	Michener, 1971
B. leptozonia (Natal)	38	44	73	Michener, 1971
B. foveata and *leptozonia* (Cameroon)	28	40	90	Michener, 1968a
B. simillima and *unicolor*	29	36	17	Michener, 1962
Exoneura variabilis	27	30	109	Michener, 1965b
E. variabilis[a]	32	36	109	Michener, 1965b

Note: Exoneura, Exoneurella, and *Braunsapis simillima* and *unicolor* are Australian forms; all others are African.
[a] Percentages in this line are based on a broader interpretation of "mature female" than for those in the preceding line (see text).

nest where they matured. There is, therefore, every intergradation among the females between young adults, unworn, unmated, with slender ovaries and soft, pallid integument (tenerals or callows), and old adults with worn wings, often mated, often with enlarged ovaries, and with fully darkened integument. The percentages given in Table 27.1, therefore, sometimes represent decisions that would be difficult to defend as to which females in each nest were young adults that were likely to leave and probably were not yet active in foraging and care of young and which, on the contrary, represented mature adults, working in the nest, and so far as is known likely to stay there. For some species observations of foraging helped in making such decisions. For species that attain mature coloration late, so that acquisition of such coloration coincides with the age when most females leave their natal nests, nearly all females with such coloration were considered as probable permanent residents. For other species which attain mature coloration earlier, judgments had to be based on dissections and examination of wing margins, and even then were quite arbitrary.

The percentages of nests containing two or more fully mature females is low for species in the upper half of Table 27.1, considerably higher although under 50 percent in the lower half. The low figure, even for the most social species, results from two facts: (1) at any season at least in tropical areas some of the nests are newly established and hence contain only the founding female, commonly with her eggs and larvae; and (2) some of the old nests are abandoned by all but one of the mature adult females, sometimes the old founder, sometimes a daughter (probably when the founder has died). The only nests represented in the percentage columns in Table 27.1 are those in which young females mature in a nest and stay there, either with their mother or (if she dies) as two or more sisters. The number of such mature adult females in a nest is most often two but sometimes as high as six; it is among them that the possibility of division of labor has been investigated.

Occasionally there are two similar mature females in the same nest; for example, both may be mated, with enlarged ovaries, and with worn wing margins. No polymorphism is involved in such instances; both bees are apparently working and laying. In most nests inhabited by two or more mature adult females, however, dissection shows that there is one with enlarged ovaries that is nearly always mated and usually more worn, while others usually have more slender ovaries, are usually not worn or are less worn, and are frequently unmated. The former is the principal egg layer and may be called the queen, while the latter usually lay fewer eggs, sometimes apparently none, and are often workers. A noncommital expression, supernumerary females, is used for all such individuals in Table 9.2 because the word "worker" does not always seem appropriate. For example, in the four nests of *Exoneurella lawsoni* containing two mature females each, these females had ovaries of nearly the same size. "Worker," however, is used below in discussing the more fully social species (lower part of Table 9.2).

From Table 9.2 it can be seen that in some species (upper part of the table) most supernumerary females are mated (except for the five of *Allodape mucronata* dissected). On the other hand, in *Braunsapis* and *Exoneura* only 5 to 25 percent of such females are mated.

Data from certain species are given, as examples, in Tables 27.2 and 27.3 to show the variable relations among three attributes of each female in nests containing two or more mature females. These attributes are (1) ovarian development, (2) prior mating, and (3) prior activity as judged by wing wear.

Symbols used in Tables 27.2 and 27.3 are as follows:

A. All ovarioles, or the maximum number for the species, enlarged, and one with an egg more or less of a size ready to lay.

B. All ovarioles, or the maximum for the species, considerably enlarged but no oocyte approaching the size of a mature egg.

C. Ovaries moderately enlarged (about halfway between E and B).

D. Only one or two ovarioles enlarged, sometimes one oocyte greatly enlarged and egg-sized, but the other ovarioles slender.

E. Ovaries slender but not pedunculate.

F. Ovaries slender and pedunculate, that is, posterior halves or two thirds mere slender stems containing no oocytes.

The condition of each female represented in these tables is indicated by three symbols: (1) a letter showing the ovarian development; (2) a symbol indicating the content of the spermatheca, + for sperm-filled, − for empty; and (3) a number indicating the total number of nicks in the apical margins of the two forewings. Thus a bee described as A − 7 has much enlarged ovaries, is unfertilized, and is somewhat worn, having seven nicks in the forewing margins.

Of 187 nests of *Allodapula acutigera* examined, only

TABLE 27.2. Conditions of females in spring colonies of *Allodapula acutigera* from Cape Province, South Africa.

Nest No.	Condition of ♀♀	Nest No.	Condition of ♀♀
160	A + 105, C – 0	69	A + 166, C + 0, E + 0
550	A + 78, F + 3	113	A + 15, C + 1, E – 0
317	A + 23, F – 0	144	B + 87, C + 0, E + 5
17	B + 142, E + 0	7	F + 100, F + 0
60	B + 97, E + 0	18	C + 2, D + 0, E + 92
584	B + 96, C + 2	13	B + 6, E – 0, F + 0
560	B + 85, E + 3	73	B + 0, F + 0, F – 0
413	B + 54, C + 0	182	C + 0, C + 0, E + 3
557	B + 52, E + 8	100	C + 2, E + 0

SOURCE: Michener, 1971.
Note: Each colony contained two to three fully colored females, information on which is separated by commas. For explanation of symbols see text.

18 contained supernumerary females. As shown in Table 9.2, over 80 percent of these were mated, and as shown in Table 27.2, most had ovaries somewhat slender (classes C to F) and some had slight to considerable wing wear. As another example, out of 109 nests of *Braunsapis facialis* from Natal, 34 percent contained two or more mature bees (Table 27.1). Bees from 28 such nests were dissected, with results shown in Table 27.3. Most of the supernumerary bees were not mated; most but not all had somewhat slender ovaries, and some showed wing wear. Some idea of the contents of a *Braunsapis* colony can be obtained from Figure 27.2.

The importance of the workers in productivity of allodapines is suggested by the last column of Table 27.4.

It is quite possible that in all species or at least those where supernumerary females are of regular rather than "accidental" occurrence, productivity is increased by the presence of such individuals. The magnitude of the increase is indicated in Table 27.4. In every instance except *Allodape mucronata,* the difference in mean numbers of young in nests with one mature adult female from that in nests with two or more is significant at less than the .01 level. Even a single worker often results in doubling to tripling productivity, either because of its activity in

TABLE 27.3. Conditions of females in colonies of *Braunsapis facialis* from Natal, South Africa.

Nest no.	Condition of ♀♀	Nest no.	Condition of ♀♀
813	A + 16, C – 0	1212	B + 2, C – 0
1045	B + 14, B – 1	1235	B + 0, B – 1
1071	B + 19, C – 1	1239	A + 1, B – 2
1104	B + 8, C – 0	1457	A + 2, C – 3, F – 0
1125	A + 9, B – 1	1458	B + 3, B – 0
1231	A + 17, E – 0	1487	C + 5, E – 3
1242	A + 33, C – 0	1042	B + 2, B – 0, E – 16
1248	A + 23, C – 1, E – 0	1477	A + 1, C – 13, E – 0
1511	A + 15, C – 8	819	A + 18, B + 25, F – 0
1526	B + 5, B – 0, C – 0	1115	A + 22, A + 1
1044	B + 0, C – 1	1123	A + 7, B + 0
1045	B + 2, B – 0, C – 0	1217	A + 7, C + 0
1105	B + 0, E – 0	1227	B + 23, C + 0
1118	A + 1, B – 3, B – 0	1464	A + 6, B + 24

SOURCE: Michener, 1971.
Note: Each colony contained two to three fully colored females, information on which is separated by commas. Symbols are explained in text.

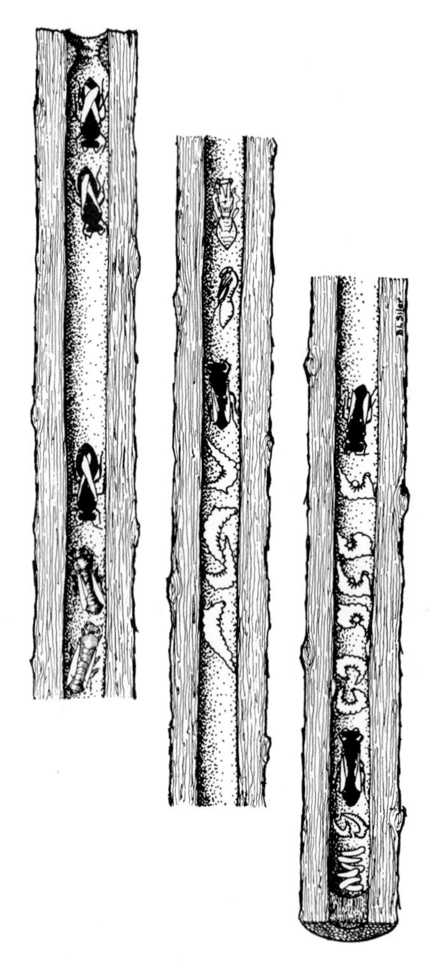

FIGURE 27-2. A nest of *Braunsapis,* the most widespread allodapine genus. The nest was in a dry, dead, pithy stem. *Upper left,* constricted nest entrance; *lower right,* the bottom of the nest. In the nest are several adult bees (the callows are pale) and immature stages in order from oldest (pupae) above to youngest (eggs) at the bottom. (Modified from Sakagami, 1960a, after Iwata, 1938.)

bringing in food and caring for the young or because of its egg-laying activity (even though normally less than that of the queen) or because of some interaction. It would seem that selection should favor colonial behavior; perhaps the small sizes of the colonies are related to the limited space available in the stems and twigs utilized as nesting sites.

Reports by Rayment (1951, and see Sakagami, 1960a) of direct exchange of liquid food among adult *Exoneura* have been verified by observations on *Braunsapis* (Michener, 1972a), although it seems to be a rather occasional event always involving a young adult. Similar behavior has been reported even for solitary Xylocopinae (Rau, 1933, for *Xylocopa virginica*). Such feeding may be important for the well-being of young adults, but returning foragers do not normally transfer nectar to receiver bees as in highly eusocial colonies.

E. Life cycles of social species

Among the better-studied, more or less social (that is, not merely subsocial) allodapines are several *Exoneura* from Australia (Michener, 1965b); various species of *Braunsapis* from Australia, southern Asia, and Africa (Michener, 1962, 1971); and several species of *Allodapula* and *Allodape* and one species each of *Macrogalea* and *Compsomelissa* from Africa (Michener, 1971).

In *Braunsapis,* as in all other allodapines, new nests are established by single females. They lay eggs at intervals in the bottom of the nest, sometimes loose, sometimes attached to the bottom of the burrow. Because the eggs are laid at relatively long intervals, young of all ages are ultimately found in the nest, but only a few of each stage. The larvae are fed progressively, as in *Exoneurella,* and

TABLE 27.4. Mean numbers of eggs and young in nests of South African allodapine bees containing one mature adult female or two or more such females.

Species		*Mean numbers of eggs and young if—* 1♀	2+♀♀
Allodape mucronata		6.2	8.3
Allodapula acutigera			
Spring, Sept.–Nov.		2.5	8.0
Summer, Feb.		1.6	4.1
A. melanopus and *turneri*			
Eggs and small larvae	(Oct.)	2.0	3.2
	(Nov.)	0.5	3.1
Braunsapis facialis			
Eggs and small larvae		0.4	1.9
Total young		1.8	5.9
B. leptozonia			
Eggs and small larvae		1.2	6.4
Total young		2.9	10.8

SOURCE: Michener, 1971.

TABLE 27.5. Statistics on eggs and young of *Braunsapis leptozonia* from Natal, South Africa.

	Eggs	Small larvae	Medium larvae	Large larvae	Prepupae	Pupae
Percent of eggs and young in each stage	45.4	8.4	9.8	15.6	6.8	14.0
Max. no. of each stage in one nest	12	6	6	7	3	4
Mean no. of each stage for all nests	3.1	0.6	0.7	1.1	0.5	1.0

SOURCE: Michener, 1971.
Note: Data were from 73 summer nests containing a total of 500 eggs and young.

as in that genus they are arranged consistently from youngest below to oldest near the nest entrance, the older larvae supporting themselves ventral side up in the same way as those of *Exoneurella,* although their dorsal tubercles are on different segments. As in *Exoneurella,* larvae eat food placed on their ventral surfaces, curling the head region over to reach the food mass. Prepupae and pupae are found in that sequence above the larvae and, just as in *Exoneurella,* must be passed by any bee bringing food to the younger stages. Table 27.5 gives an idea of the immature stages in nests of one African species. Initially such a colony is subsocial, although it has the potential of attaining other social levels because of association of adults.

Young adult females of *Braunsapis* emerging in the nest remain there for some time, so that there are often two, three, or more adult females in a nest. Unfertilized females do most of the pollen collecting, although they constitute only about half of the total population of the females, for example, in the case of Australian species (Michener, 1962).

Essentially the same type of life cycle is found in the Australian genus *Exoneura.* For *E. variabilis* it has been established that new nests may be made at any time during the active season by lone females (Michener, 1965b). However, apparently the majority of young adult females do not leave the parental nest but remain there. Some replace the original egg-laying female, who dies soon after laying two sometimes incompletely separated, batches of eggs. In *E. variabilis* and most other species of the genus the eggs are laid in small batches, rather than singly as in *Braunsapis.* Larvae hatching from the second batch of eggs are usually reared by one or more of the adult daughters from the first batch.

Eggs of most *Exoneura* are unattached, and the larvae remain in a cluster at the bottom of the nest until they grow large enough to maintain a position above the bottom. Larvae of other species hatching from eggs attached to the burrow wall may retain their position on the nest walls for a time and may be fed pollen in this position. They do not permanently occupy a position in a cluster above the bottom of the nest as in *Allodapula.* The young are kept more or less in age groups, with the oldest near the entrance, but there is considerably less regularity to this sequence than in *Exoneurella, Braunsapis,* and *Allodape.* Table 27.6 gives an idea of the varying frequencies of all stages through the season of activity in a species of *Exoneura.*

The South African genus *Allodapula* is *Exoneura*-like in most of its social attributes; the striking difference is in the manner in which the larvae live. In *Allodapula* and also in a few species of *Exoneura* the eggs are attached by their posterior ends to the nest wall, and project into the lumen of the burrow (Figure 27.3), typically not near the bottom of the nest. Eggs of some species of *Allodapula* (*A. acutigera, dichroa*) form a circle around the nest burrow; in other species they form a patch on one wall of the burrow (*A. melanopus, turneri*) or a slightly spiral but essentially lengthwise row (*A. variegata, rozeni*). As in *Exoneura,* the eggs of one such group are laid more or less at the same time (as a guess, one every day or so) and hatch at roughly the same time.

The larvae at first maintain their positions on the nest wall by leaving their posterior ends in the chorions or egg shells which are stuck to the burrow wall. At this stage the larvae project into the lumen of the burrow like the eggs, and are fed from a single (or occasionally more) common food mass placed on them by the mother bee (Figure 27.4), the food being supported by the egg remnants and the bodies of the larvae themselves. In most species, as the larvae grow they leave the egg chorions, but the chorion is relatively tough for a bee egg, and can

TABLE 27.6. *Exoneura variabilis* in Queensland, Australia. Average number of individuals per nest in various developmental stages on different dates.

Date:	X-3	XI-9	XII-8	XII-18	XII-27	I-11	II-15	II-25	III-22
No. of nests:	9	11	44	35	23	7	31	13	48
Eggs	—	3.5	2.9	3.2	4.0	1.0	0.5	—	—
Larvae	1.5	0.7	2.9	2.4	3.7	3.2	2.1	0.1	0.2
Pupae	—	—	2.0	1.4	1.4	1.7	0.7	0.1	0.2
Total immatures	1.5	4.2	7.8	7.1	9.0	6.0	3.2	0.2	0.4
Adult ♀♀ (total)	1.9	1.1	2.0	2.1	2.3	1.9	1.6	1.3	1.5
Callows	—	—	0.5	—	0.4	0.4	0.1	0.2	0.1
Fertilized	0.4	1.0	1.2	0.9	1.2	0.6	0.9	0.6	1.0
Worn	1.0	0.7	0.8	0.7	1.0	0.3	0.5	0.5	0.6
With enlarged ovaries	0.2	0.7	0.7	0.9	1.1	0.3	0.2	—	—
Workerlike	0.8	—	0.3	1.1	0.6	0.7	0.7	0.5	0.4
Adult ♂♂	—	—	0.5	0.1	0.2	0.1	0.2	0.1	0.2

SOURCE: Michener, 1965b.

still support the food mass to which small larvae cling and upon which they feed.

As the larvae increase in size, the group or clump of them becomes large enough to support itself in the burrow (Figure 27.4), which it more or less fills, by pressure against the walls. At this stage food is provided as a common food mass pressed down among the larvae. The older larvae, unlike those of the other genera, are relatively straight and covered with large tubercles (Figure 27.1), and the axes of their bodies are parallel to that of the burrow. The dorsal surfaces of the larvae are pressed against the burrow walls, the lateral surfaces against one another, and the ventral surfaces against the food mass, which is supported by the ventral tubercles and large lateral "pseudopods" of the larvae. They feed by bending their heads down to eat from the upper end of the food mass.

The adult bees are unable to remove feces from beneath the mass of larvae; therefore, unlike genera such as *Braunsapis* and *Allodape,* the larval feces accumulate in the lower part of the nest until the clump of larvae with its food mass breaks down. However, *Allodapula* larvae do not defecate until later in life than those of most other allodapine genera.

Breakdown of the clump of larvae occurs when the larvae cease feeding and enter the prepupal stage. At this time the larvae are distributed end-to-end up and down the burrow, just as in other allodapine bees; the pupae are similarly oriented. At about the same time the egg layer in the nest usually produces another batch of eggs, and these may be hatched and small larvae may be present in the nest before the older brood has reached adulthood.

In all allodapine bees except *Allodapula,* in which there is a large common food mass, and *Halterapis,* in which each larva is mass provisioned, the number of larvae having pollen supplies upon which to feed at any one time is small. For example, in one group of nests of *Exoneura variabilis* only 8 larvae out of 74 had food available when the nests were opened. Rayment (1951) showed that in 17 nests of *E. richardsoni* containing 209 larvae, only 6 nests contained food, and these had a total of 23 pollen masses. The fact that most larvae most of the time have no food supply must be related to the slow development of allodapine bees. This applies to all stages. Rayment showed that at the temperatures of late winter in Sydney, Australia, the egg stage of certain *Exoneura* lasted between two and three weeks, while Skaife (1953) reports that at the same season near Cape Town eggs of *Allodape mucronata* require four to six weeks to hatch. Rayment thought that the development from egg to adult of certain *Exoneura* required from early July (mid-winter) to early December (early summer); for *Allodape mucronata* Skaife indicates 14 to 15 weeks from egg hatching to maturity of adults. These are extraordinarily long growth periods for bees. Rayment (1951) also showed that both larvae and adults can survive up to 90 days closed in a container without food, and Erickson and Rayment

(1951) described survival of adults over 100 days at 10° C. Thus survival of both larvae and adults during periods of drought when flowers are scarce and during the winter is not surprising, although the principal stage of overwintering in temperate areas for most species is as fertilized adult females.

A high percentage of the females collecting pollen on flowers are unfertilized and have slender ovaries. All such bees were originally taken to be workers, although many of them were unworn (Michener, 1963b); but it has now become clear that at least in tropical Africa (Cameroon) the species of *Braunsapis* often collect pollen and carry it to their nests whether they have larvae to feed or not (Michener, 1968a). Thus if one opens nests at the right time of day it is not unusual to find that young nests, not even containing eggs and inhabited by lone young females, all contain a small patch of pollen on the wall of the burrow. In older nests such pollen or some of it is later used to feed larvae (Chapter 16, A), but adult bees eat it in nests without larvae.

Male allodapines are produced throughout the reproductive season. Except in *Exoneurella* in winter, the males disperse from the nests and presumably soon mate and die, so that adult nest populations are predominantly female. Rayment's (1951) claim that males guard nests and constitute a significant part of the society has not been verified.

Mating probably occurs away from the nests in all species except perhaps *Exoneurella.* In *Exoneura hamulata* and probably the rest of the section of *Exoneura* in which the males have very large eyes, males form swarms in certain places, and presumably the females passing near these swarms are fertilized. The bees seem to fly continuously, like the individuals in a swarm of midges. Males of *Braunsapis nitida* and *plebeia* in New Guinea were found to fly over certain sorts of leafy vegetation in large numbers throughout the warm period of the day. I could not find an objective way of recognizing such areas, but nonetheless I could frequently locate a suitable spot and find the bees there. The locations were not weedy but were vine covered, with leaves so oriented that they tended to cover the entire vine surface, allowing little light to pass beneath. The bees flew in the sun just over the leafy surface.

FIGURE 27.3. *Left:* Eggs of *Allodapula rozeni* in a row in the nest in a hollow stem. *Right:* Eggs and newly hatched larvae of *Allodapula dichroa* in a circle in a stem broken to expose them. (Photographs by C. D. Michener.)

FIGURE 27.4. *Upper left:* Eggs of *Allodapula dichroa* in a circle in a hollow-stem nest. *Upper right:* Small pollen mass supported in hollow stem by eggs and young larvae of *Allodapula acutigera. Below:* Middle-sized and large larvae of *A. acutigera* supporting and feeding on a large common food mass. (Photographs by C. D. Michener.)

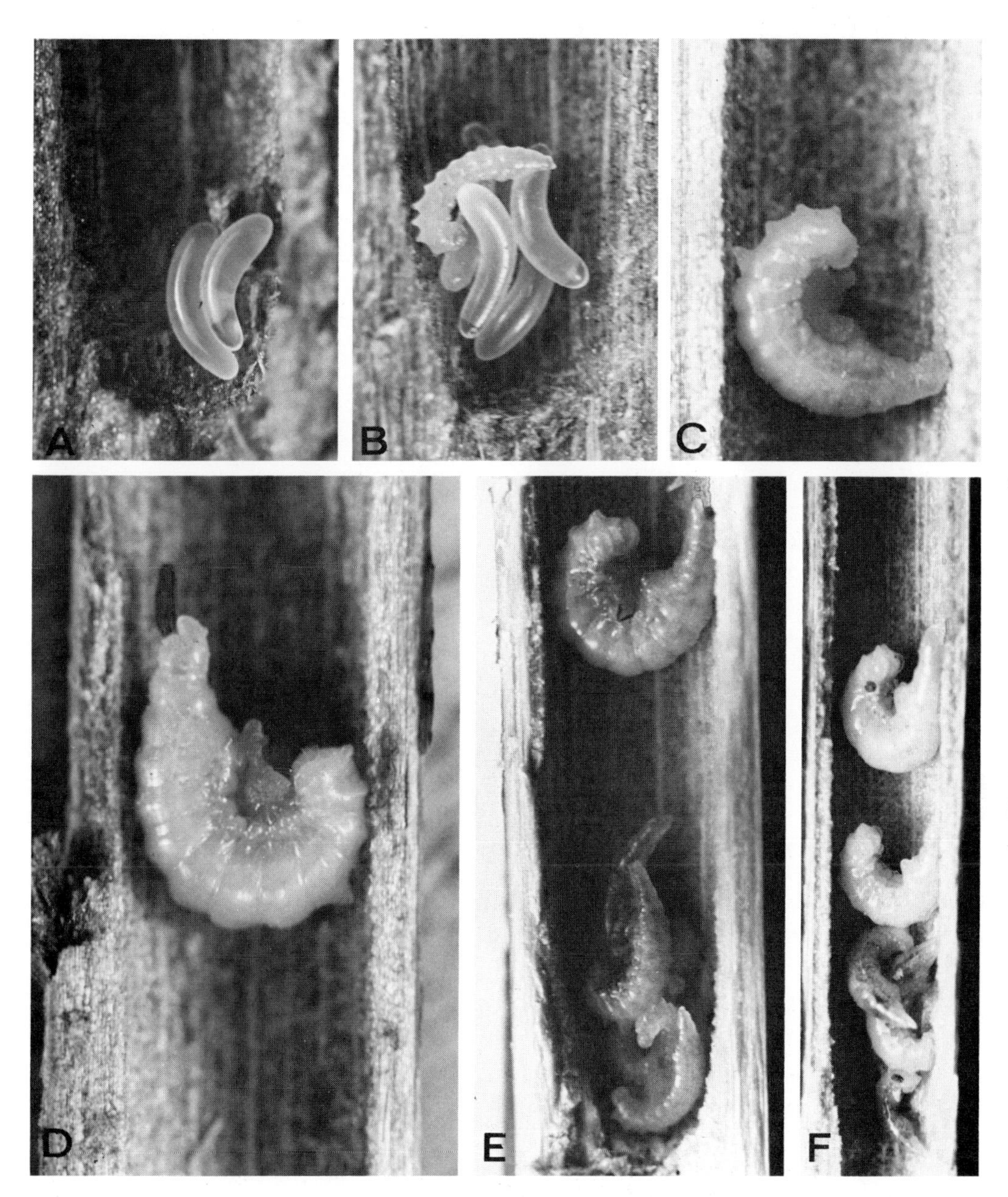

FIGURE 27.5. Portions of nests of *Allodape mucronata*. (*A*) and (*B*) Eggs and a young larva in bottoms of two nests. (*C*) and (*D*) Middle-sized and large larvae eating, the latter also defecating. (*E*) and (*F*) Groups of larvae. This species is essentially subsocial, but the appearance and behavior of the larvae are similar to those of many more social species of *Braunsapis*. (From Michener, 1971.)

F. Larvae

Larvae of *Exoneura, Braunsapis,* and *Allodape* commonly have ventrolateral projections, sometimes short, but sometimes certain of them are long and even branched. Frequently they are eversible, so that when the larva is in a resting condition they are relatively short but when the sensory organs of the head are stimulated by touch, the projections are everted, apparently by blood pressure, and may more than double their lengths. Alternatively, some species of *Braunsapis* which do not have such projections have rather long ventrolateral or ventral hairs. Certain species which have neither projections nor hairs have strong ventrolateral ridges, so that the venter of the body is concave, as in *Exoneurella.* In some *Allodape* there is a ventral midabdominal projection that can be everted to push food toward the mouth. Presumably all these devices (Figures 27.1, 27.5, and 27.6) have to do with holding and controlling the food masses, which in these genera are provided individually for each larva. They are characteristic only of the large (last stage) larvae, although much shorter projections may be seen on the penultimate stage in some species, especially the larger ones. Presumably the small food masses needed by the small larvae can be maintained, even on a convex venter, simply by their viscosity, whereas larger larvae might run the risk of losing a large food mass in the absence of specialized devices for holding it (Figures 27.7 and 27.8).

It is interesting that these lobes, hairs, and so on, for manipulating food masses appear to have originated independently in the different groups of allodapines, for they are often on different segments, even in different species of *Braunsapis,* and therefore are not homologous to one another. The same is true of the devices which support the larvae in the burrows, for example the dorsal tubercles, bristles on the body, the hooked hairs of *Macrogalea,* and the strong apex of the abdomen. The larvae

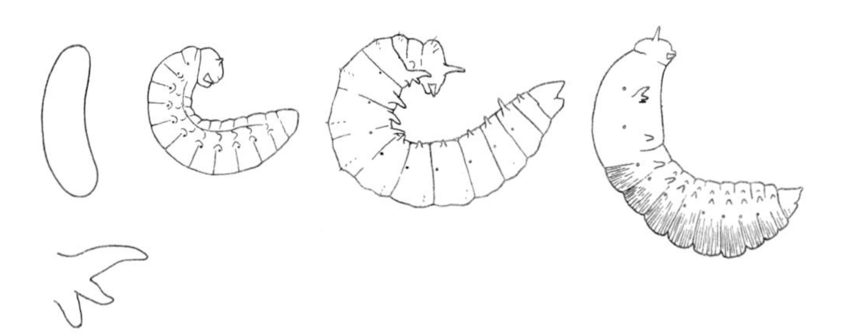

FIGURE 27.6. Egg, young and mature larva, and prepupa of the Australian *Exoneura hamulata,* with an enlargement of the thoracic appendage of a large larva. (From Michener, 1965b.)

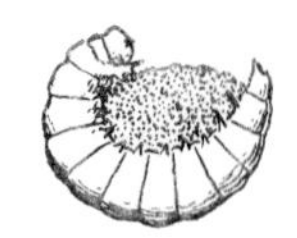

FIGURE 27-7. Larvae of *Exoneura variabilis* from Australia feeding on pollen masses. *At left,* a small larva eating a pollen mass which has been largely consumed. Others are moderate-sized and rather large larvae. (From Michener, 1965b.)

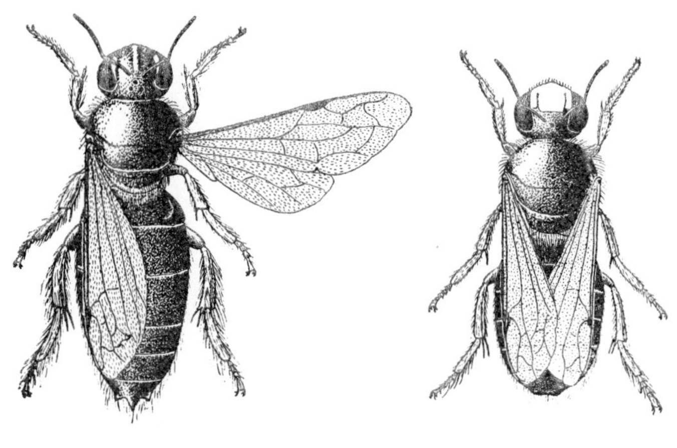
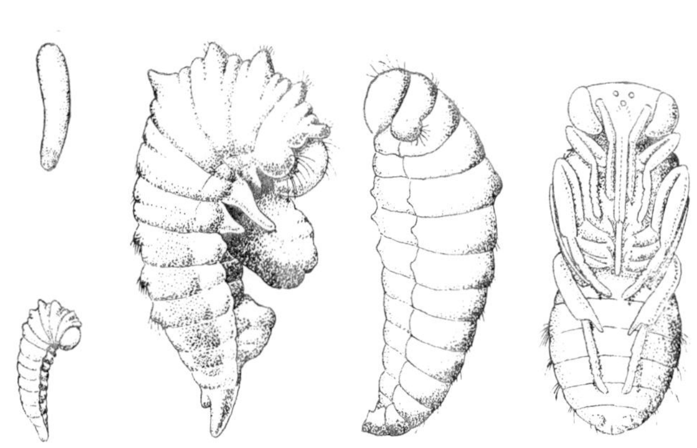

FIGURE 27.8. The subsocial species *Allodape mucronata* from South Africa. *Above,* adult female (*left*) and male (*right*). *Below,* egg, small larva, large larva with pollen mass, prepupa, and pupa. (From Skaife, 1953.)

also possess probable sensory devices more elaborate than those of ordinary bee larvae, which spend their entire lives closed up in cells. Allodapine larvae have relatively long antennae, or alternatively, long, specialized hairs on either the head or the prothorax, or alternatively (as in *Exoneura baculifera* and its relatives) a median frontal projection on the head. These are probably all tactile devices which evolved with the cell-less nests, where perception of and responses to activities of various associates are important. Like the supporting and food-manipulating devices, these structures appear to have arisen independently in different groups of allodapines, for they are by no means homologous to one another in the different groups.

For bees, the larval habitat of the allodapines is unique because, except in *Halterapis,* the larvae are constantly coming in contact with one another, with eggs, with pupae, and with adult bees, which move the young around from place to place in the nest. Larvae must support themselves in their burrows, hold food masses, and probably respond properly to adults to get food. It is therefore not surprising that there has been an adaptive radiation in allodapine larvae. The various groups are so different that they would probably be placed in different families, were the adults unknown. In general appearance there is more diversity among the allodapines than there is among all the other kinds of bee larvae considered together. By contrast, nonallodapine bee larvae are for the most part a monotonous lot, a scarcely surprising circumstance since nearly all of them are reared each in a separate cell containing an adequate supply of food; all the larva has to do is eat the food upon which it is lying, and before pupation orient itself with the head toward the cell entrance. (In some species it must also spin a cocoon, and forms which do so have better-developed antennae and palpi than those which do not.)

28 *Bumblebees*

Except for the species of *Psithyrus,* a parasitic genus, all the familiar, large, hairy bumblebees belong to the genus *Bombus.* This is a genus of a few hundred species which is found throughout the North Temperate region. It is particularly abundant in the cooler parts of this region and, in fact, extends far into the Arctic, to northern Ellesmere Island, only about 880 km from the North Pole. There are a few species and rather few individuals in the high Arctic, but they are the only bees to be found there. To the south, in both hemispheres bumblebees are scarce in warm arid climates like those of the Mediterranean and the southwestern United States, and absent in deserts; they are also scarce or absent in tropical areas. In the Western Hemisphere, however, the genus extends nearly as far south as there is land, the one Chilean species, *B. dahlbomii,* extending to the island of Tierra del Fuego. The number of species in South America is small, however, compared to the number in the North Temperate regions. A few species occur in moist tropical forests—for example, two in the Amazon Valley—but bumblebees are always inconspicuous elements in the lowland tropical fauna. They are more numerous at moderate to high altitudes in the mountains and in the subtropical and temperate southern portions of the continent. In the Old World, bumblebees are completely absent from tropical and southern Africa as well as from Australia, although several species have been introduced into New Zealand as pollinators. (A report of a specimen from equatorial Africa is possibly an error, but see Tkalců 1966.) They are also absent from the lowland tropical regions of India and southeast Asia. They do reach North Africa, and at high altitudes in the mountains they extend southeast as far as Java.

There exist several excellent general accounts of bumblebees—books by Free and Butler (1959), Hoffer (1882), Khalifman (1971), May (1959), Plath (1934), and Sladen (1912). More detailed accounts of biologies of particular species are given by Frison (1917, 1918, 1928, 1929, 1930), Hobbs (1964-68), Sakagami and Zucchi (1965), and Stein (1956).

The species of *Bombus* form a morphologically as well as behaviorally cohesive group. Richards (1968) has given a reasonable grouping of species into subgenera. These subgenera seem to me more similar to one another than do subgenera in most other genera of Hymenoptera. There is no justification, so far as I can see, for Milliron's (1971) idea, first published in 1961, that *Bombus* is polyphyletic, some groups derived from xylocopoid ancestors, others from anthophoroid ancestors. A vast array of morphological and behavioral characters, including larval characters and internal characters that could hardly be convergent, demonstrate the common ancestry of all species of *Bombus.* Conversely, all are very different from the Anthophorinae and Xylocopinae in a multitude of characters. The relations suggested in Figure 2.2 seem well justified. It is unfortunate that Milliron's revision of the *Bombus* of the Western Hemisphere, currently being published, will be marred by unfamiliar generic names and the concept of polyphyly of the group.

A. Nesting

1. Nests

Bumblebee nests are totally different from those of the bees described earlier, except for the Euglossini. Instead of consisting of burrows and cells excavated in the substrate, bumblebee nests consist of cells, storage pots for provisions, and other structures made of wax mixed with pollen. These structures are protected from the weather in various ways. Most species of bumblebees commonly select abandoned nests of small rodents, such as field mice, in which to nest. Such a rodent nest consists of a hollow, often in the ground, containing an accumulation of fine pieces of dry grass, moss, leaves, and other fine, dry materials gathered by the rodent. Nests in the ground are reached through the rodent burrow that leads to the nest. The bumblebee nest is made in a hollow in the midst of the fine materials in the rodent's nest. Abandoned birds' nests, especially if made in cavities, may be similarly used. Some species prefer sites approached by underground tunnels, often several feet in length, while others prefer sites closer to the surface of the ground. Figure 28.1 shows a queen at the entrance of an underground site, in this case a hibernating place. Other species usually select sites on the surface of the ground, for example, under masses of vegetation. Some surface species need not find deserted rodent nests, but will use bird nests even if open above, cavities under grass or moss, bundles of hay or straw, or hollows in decaying logs or under the roots of decaying trees. The literature is full of notes on strange places and materials utilized by bumblebees as nest sites, such as the stuffing in old upholstery or pillows.

FIGURE 28.1. Queen of *Bombus nevadensis* at the entrance of her underground hibernaculum. All bumblebees hibernate in the ground, and many also nest in the ground. (Photograph for G. A. Hobbs, Entomologist, Canada Department of Agriculture, by staff photographer E. T. Gushul.)

Bombus transversalis of the Amazon Valley starts its nests in the litter on the forest floor and digs out the soil beneath so that ultimately the nest is in a hollow just below the surface of the ground, roofed over by litter supported by rapidly growing tree roots (Dias, 1958). Bumblebees move material about in order to enlarge or modify the cavity in which they nest, and the surface-nesting species, often called carder bees, such as *B. agrorum* and *ruderatus,* may completely construct a cavity and enlarge it as needed by moving bits of down, grass, and other dry, soft, and insulating materials to cover the cells and other nest structures. In a protected place, such as inside a shed, the cells are sometimes started almost in the open and materials brought to cover them.

A waxy sheet often forms a roof over the cells and cocoons, and may be supported by pillars at various points and by the nesting material with which the sheet of wax is in contact. Such a canopy is more common in underground nests and in nests inhabited by large colonies than in surface nests or nests of small colonies. It is the equivalent of the batumen of meliponine nests, in that it protects the whole nest, but in bumblebees it protects from above only, for it never surrounds the nest beneath.

The construction material used for the cells and for the pots that hold the provisions is, as indicated above, wax mixed with pollen; or the pots may be old cocoons strengthened, waterproofed, and often extended upward with wax. The wax is secreted from between the abdominal segments of the females; at first it is from the queen, later from workers, and it comes from between both sterna and terga (Figure 1.16).

2. Gyne nests

This section concerns nests during the initial establishment stage and during the subsocial phase, before maturation of the workers. When an overwintered gyne has found a suitable nest site, she usually rearranges the nesting material to form a chamber 2 or 3 cm in diameter and lined with the finer fibers among those available. She now feeds regularly, often bringing nectar back to the nest and sometimes regurgitating it on the fibers of the nest floor, for use in bad weather. She also begins to produce wax from between the abdominal segments, and with it constructs her first distendible cell, located on the floor of the nest cavity, and a honey pot, also on the floor and always at or close to the entrance of the cavity. The construction of the cell apparently usually precedes the building of the pot, but it may occur concurrently (Alford, 1970).

There is only one cell in each gyne nest. The cell differs in various ways from cells made later, during the eusocial phase. To begin with, it is on the floor of the nest cavity, not on a cocoon, for there are no cocoons.

The bee may construct the distendible cell as would be expected, that is, a shallow wax cup, and then bring pollen, some of which, in most species, is placed on the floor of the cup (the cell is said to be primed with pollen) while other pollen is mixed with the wax. On the other hand, some species make a pollen lump on the floor of the chamber before wax is added to its surface to form a cell. Alford (1970) noted that in the field *Bombus hortorum* and *agrorum* start the pollen lump before providing wax, while in captivity wax cells were often provided first; thus the differences does not seem to be necessarily a specific character. After the pollen has been provided, either as a temporarily exposed lump on the floor of the nest or in the cell, eggs are laid in it. They are sausage-shaped, white, 2 to 4 mm long and 0.5 to nearly 1 mm in diameter, not at all curved or scarcely so. Their axes are vertical in field nests of all species (Hobbs, 1964-68), unlike the horizontal eggs in subsequent cells. The eggs were more or less horizontal and on top of any pollen present in the first cells of Alford's laboratory nests. The reason for this laboratory abnormality is not obvious to me. In a few species (for example, *B. balteatus* as shown by Hobbs) the eggs of the first cell are in contact with one another in an otherwise empty pollen-wax cell. In most species, however, eggs in the first cell are separated by pollen, unlike those in later cells.

There is confusion in the literature as to what to call a cell in such cases. Some authors consider the pollen-walled space containing a single egg, for example in *Bombus hortorum* and *agrorum*, as a cell. Others use the word cell for the space that is either initially or later enclosed by a pollen-wax wall, which ordinarily surrounds several eggs. I have followed the latter usage, as it corresponds better with the situation in later broods and even in the first brood after pollen has been eaten by young larvae, but prefer the expression distendible cell to show that *Bombus* cells may not be homologous to cells of other bees and wasps. The initial distendible cell, then, in most species, is divided by pollen partitions into subcells, each of which contains a single egg.

Not all the eggs are laid simultaneously in the first cell; in *Bombus hortorum* and *agrorum* a pollen lump for a first cell is often partly enclosed in wax for a time, the part in which eggs have been laid being covered, the rest being open for addition of eggs over the next two or three days. Soon after the last eggs are laid, a wax covering is extended over the pollen and eggs, completing the first cell no matter whether it was initiated with a wax cup or with a pollen lump. The wax penetrates the surface of the pollen, and the result is the usual pollen-wax mixture used by *Bombus* for cell construction.

The number of eggs in the first cell varies with the species, but is less than in later cells; for example, Alford found the usual number laid by *Bombus agrorum* to be 8, with a maximum of 9, while *B. hortorum* reaches a maximum of 16 with a probable usual range of 12 to 16.

The wax honeypot near the entrance of the nest cavity is sometimes over 1 cm wide and nearly 2 cm high, although commonly smaller. Its construction requires a day or two. The pot is then provisioned, and if full is nearly closed; it serves as a food source for the gyne in bad weather.

After hatching, larvae of the first brood eat the pollen around, separating, or under the eggs, or they may feed on the pollen-wax material of the distendible cell, according to the species. Thereafter they are fed honey, from above through openings made in the top of the cell, and pollen, which the gyne pushes under the cell from the sides. This raises the marginal larvae, tending to emphasize the incubation groove (Figure 28.7), the depression on top of the cell or cluster of cocoons in which the gyne nestles to warm the brood. Thus, feeding of pollen to the larvae of the first brood is, even for pollen storers, much like that of pocket makers, except that in most species no recognizable pockets are formed. (For further explanation of these groups of *Bombus,* see the next subsection.) As the larvae in the gyne nest grow, the cell also becomes distended and changes shape, as is explained more fully for later broods in the next subsection.

The subgenus *Bombias* (*B. nevadensis*) is unusual in that the larvae in the first cell, as in subsequent cells, are permanently isolated. The first cell is divided into subcells by pollen or waxy pollen walls, and the larvae maintain the isolation as they grow. As noted below, larvae of subsequent broods also are isolated, each larva in its own cell. *Bombias* is also unusual in that the gyne makes pollen pockets at the sides of the first cell and stuffs pollen into them and thus beneath the larvae of the first brood (Hobbs, 1964-68). This is extraordinary, since these bees are pollen storers which, for subsequent broods, make no pollen pockets and feed each larva by regurgitation through the tops of the cells.

3. The brood during the eusocial phase

This section deals primarily with nests containing colonies of adult bees, that is, after emergence of adult workers. The distendible cells of *Bombus* differ from those of all other bees in that they are somewhat irregular in shape and in almost all species are provided with several eggs, up to 23 or thereabouts for *B. agrorum* (Richards, 1946). At first they are stocked with little food. Although the number of eggs per cell varies with the species, it is typical of *Bombus* that in established colonies the eggs are horizontal and in contact, not vertical and separated by pollen as is common for the first brood. They largely fill the cell which, at this stage, is quite small.

Sakagami and Zucchi (1965) give a detailed description of the initial construction of a cell and of oviposition, based on *Bombus atratus* in Brazil. Cell construction, by the queen, requires about 50 minutes, and is divisible into three phases: (1) scraping and shaping the construction site on top of a rather new cocoon, (2) scraping the pollen-wax mixture from the cocoon,[1] for use in cell construction, and (3) excursions to more distant cocoons for more building material. During (2) and (3) an irregular pile of pollen-wax is formed; gradually this is shaped into a cup. Egg laying then occurs, 7 to 9 eggs being placed horizontally, one after another, into the cup; before each is laid the sting is fully exserted (Figure 28.2) and penetrates the wall of the cup. Laying of the entire group of eggs takes about 10 minutes. Capping of the cell, also done by the queen, begins soon after laying of the last egg, and lasts 30 to 90 seconds.

In some species the eggs for a single cell are deposited at intervals of hours, unlike the situation in *B. atratus*. Meidell (1934) describes how in a colony of *B. agrorum* the queen made the cell (she was working at 3:00 A.M.), using wax brought to her vicinity by workers. During the early morning she laid one or two eggs, another at 7:27, the next at 10:48, the next at 19:52. When laying the queen pushes away any nearby workers, opens the cell, turns and places the apex of the abdomen in the cell. After laying she very quickly turns and closes the cell with her jaws. Meidell noted that sometimes a laying worker will put its eggs in the same cell along with eggs of the queen; at other times such workers make their own cells. Eggs laid by workers are commonly irregularly oriented, neither all horizontal like those of the queen nor all vertical like those of the young gyne (Katayama, 1971).

Although I have repeatedly emphasized the placement of eggs of *Bombus* in clusters, several to a cell, there is an exception in the subgenus *Bombias* (*Bombus nevadensis* and its subspecies *auricomus,* which was long given specific status). In these bees individual eggs and larvae are permanently isolated from one another, each being in a separate brood cell (Figure 28.3). At first thought this suggests other bees, but *Bombias* is a specialized subgenus in several ways and there is no resemblance otherwise between cells of *Bombias* and those of Euglossini or other bees. Species of this subgenus are pollen storers and feed each larva separately, by injections through the tops of the cells.

A layer of rather dry pollen forms the floor of the cell in established colonies of some species. When the eggs hatch, the larvae are able to feed on this material. Alternatively, in established colonies of other species, the cells are not primed with pollen and the young larvae may eat the wax and pollen mixture of which the cell is made.

The manner of subsequent larval feeding depends upon the species of bumblebee as well as the season, as has been explained in part in discussing caste determination (Chapter 10, B). At about the time that the eggs hatch in the group of species called by Sladen the *pocket makers,* one, two, or even three wax pockets are constructed on the sides of the cell. The pockets open into the lower part of the cell in such a way that the firm, pollen-nectar mixture pressed into the pockets by adult bees forms the floor under the larvae and provides a continuing food supply for them (see Figures 28.4 and 28.5). The pockets are thus the source of food for the larvae as well as storage places for pollen that may be used by the adult bees. Only rarely is a little pollen stored in abandoned cocoons (Katayama, 1966) by pocket makers.

After the larvae have finished feeding, the adult bees

1. In some pocket-making species, such as *Bombus atratus* and *diversus,* new cells are made on top of a cell with nearly mature larvae or on new cocoons, while in a pollen storer, *B. ignitus,* new cells are made on top of cocoons one to seven days old (Katayama, 1971).

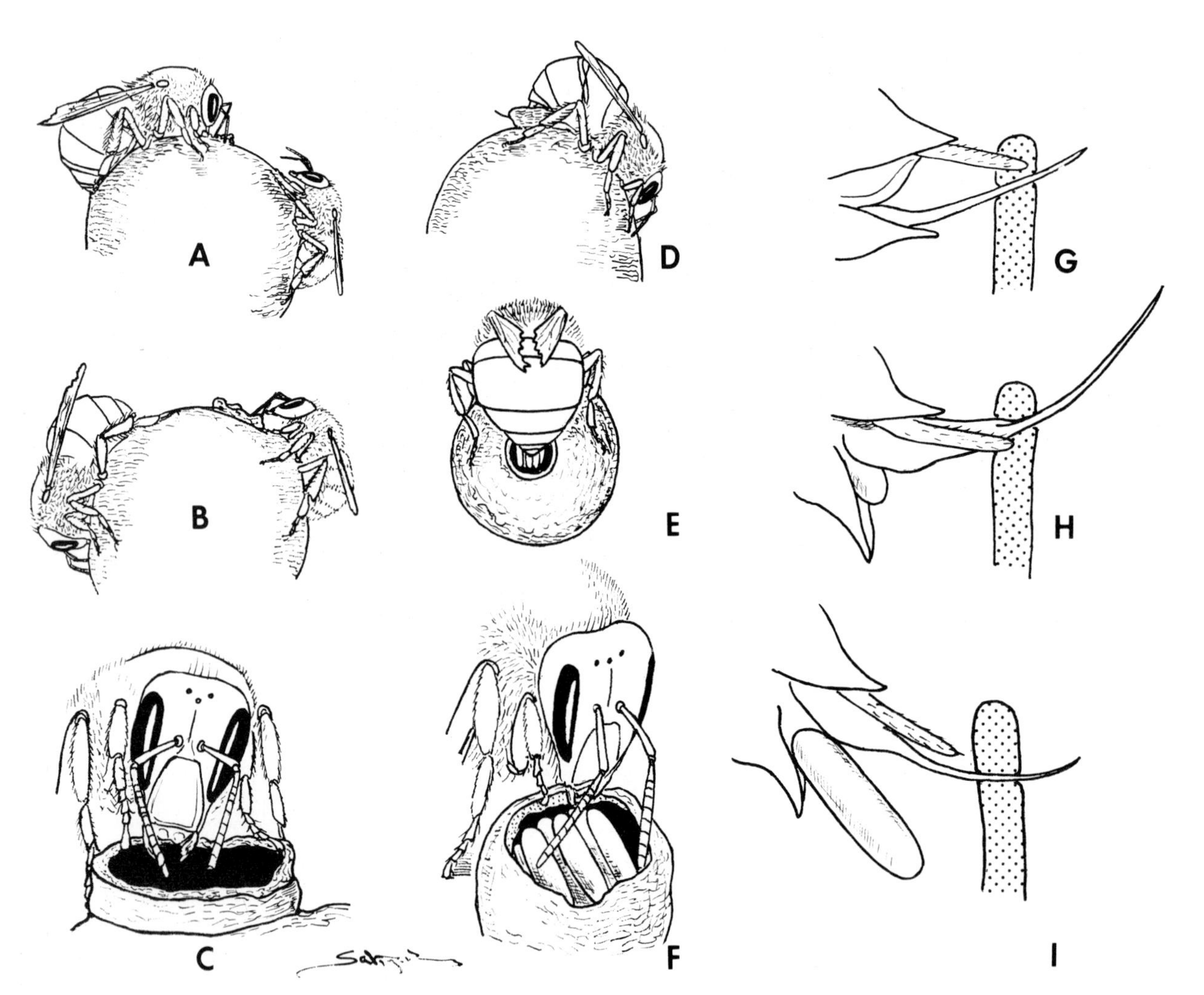

FIGURE 28.2. Cell making and egg laying in an established nest of *Bombus* (*Fervidobombus*) *atratus*. (*A*), (*B*), and (*C*) The queen, sometimes interfered with by a worker, scrapes wax from a large cell containing prepupae or pupae to make the new, small cell on top for her eggs. (*D*), (*E*) Laying. (*F*) The queen closing the new cell. (*G*), (*H*), and (*I*) In laying, the queen exserts her sting and pierces the cell wall (*stippled*); in *H* the egg is appearing and in *I* it is being dropped. (From Sakagami and Zucchi, 1965.)

remove the pockets, utilize the wax elsewhere, and eat the remaining pollen. Occasionally a pollen pocket will be built up, as are old cocoons by pollen storers, to form a cylinder for storage of more pollen. As indicated in Chapter 10 (B), larvae of pocket makers may occasionally be fed by an adult which opens the top of the cell and regurgitates a liquid honey-pollen mixture into it. Indeed, the latter method of feeding is the only one used for feeding larval gynes in most species.

Alternatively, in the other group of the genus *Bombus,* pockets are not constructed and feeding of the larvae is exclusively by regurgitation into the top of the cell, which is often closed up again after each feeding. Such bumblebees were called *pollen storers* by Sladen since, instead of keeping the pollen store in the pockets, it is placed in old cocoons which are often elongated by means of wax until they form cylinders several centimeters high (Figure 28.5). In some species of this group (for example,

FIGURE 28.3. New cells of *Bombus* on top of clumps of cocoons. The cells have been opened to show the eggs. *At left: Bombus* (*Cullumanobombus*) *rufocinctus,* with a clump of eggs in the cell, as is usual in bumblebees. *At right: Bombus* (*Bombias*) *nevadensis,* with one egg in each new cell. (Photographs for G. A. Hobbs, Entomologist, Canada Department of Agriculture, by staff photographer E. T. Gushul.)

B. terrestris, lucorum, and *latreillellus*) the cell tends to become subdivided as the larvae grow, and they may even become rather dispersed so that it is hard to know which mature larvae came from the same egg clump, a tendency which reaches its culmination in the subgenus *Bombias.*

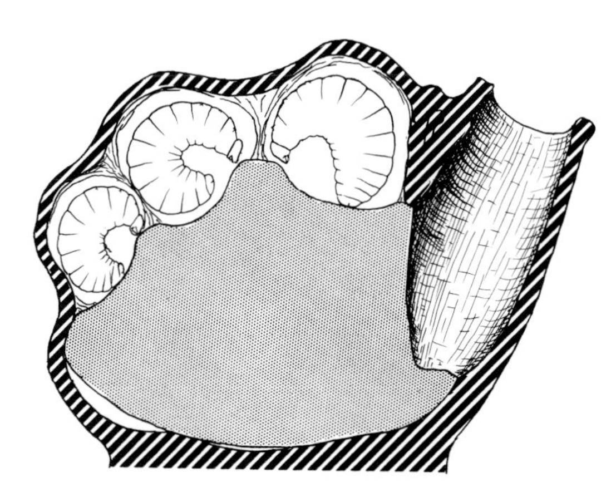

FIGURE 28.4. Sectional view of a pollen pocket (opening at right) and larvae on the pollen, *Bombus diversus. Cross-hatched areas* represent pollen-wax cell and pocket walls cut to expose the interior. *Dotted area* represents pollen. Between and above the larvae is thin silk spun by the larvae. (Modified from Katayama, 1966.)

In all bumblebees and in gyne nests as well as older ones, the initially small cell becomes greatly distended as the larvae grow. This growth can be recognized in Figure 28.5 but shows best in Figure 28.6. To permit growth, additional wax is added to the cell by the adult bees from time to time. In some species larvae are always covered by wax, but in others, such as the three mentioned in the preceding paragraph, holes are left in the tops of cells. Irregular gaping spaces through which the large white larvae are visible commonly occur when larval growth greatly stretches the material of which the cells are built. Large holes made in cells containing eggs or small larvae, however, result in destruction of the young by adult bees. When the larvae reach maturity, each spins a firm silk cocoon around itself. At the time of cocoon construction, the cell becomes less irregular in appearance, since each cocoon, pushing out against the wax wall, results in its own noticeable convexity. After the cocoon is constructed, the accumulated feces are voided by the larva. Thus, as in the other Apidae, feces are found inside the cocoon instead of outside it, as in most cocoon-spinning bees.

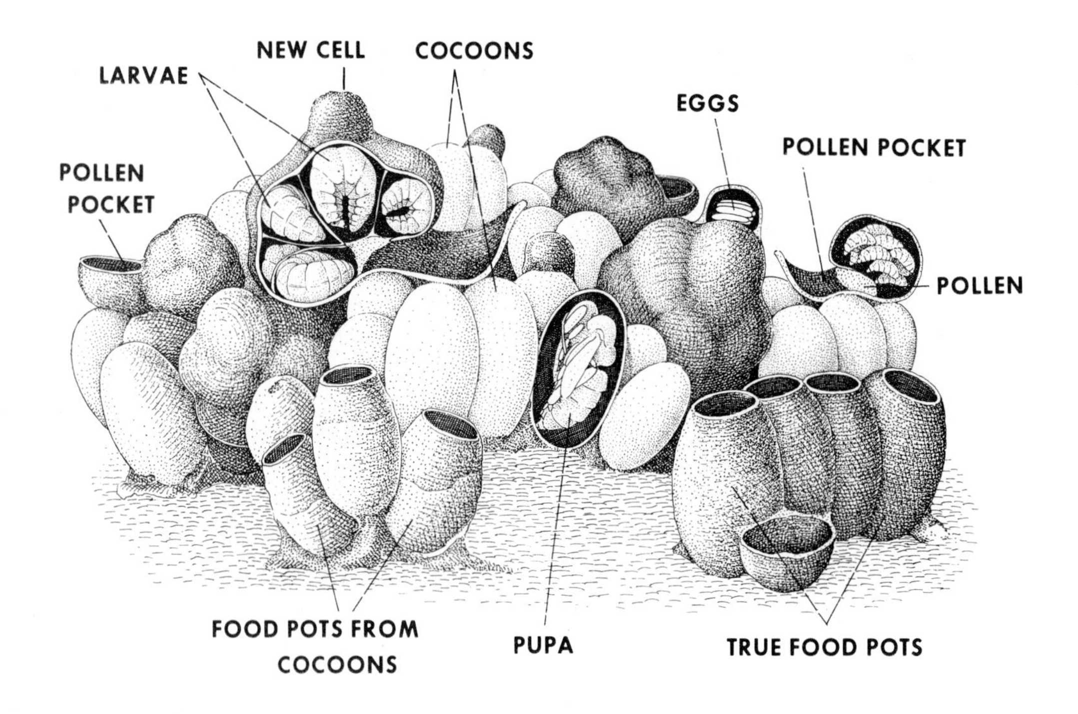

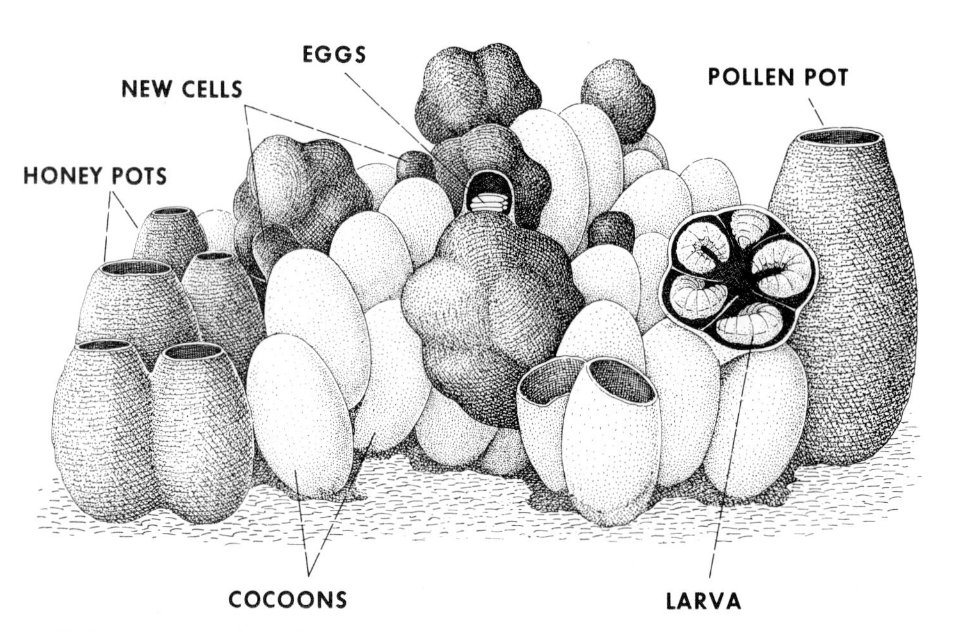

FIGURE 28.5. Views of brood of (*above*) a pocket-making bumblebee, *Bombus* (*Fervidobombus*) *atratus*, and (*below*) a pollen-storing species. In the lower figure the five larvae were exposed in a cell that faced the viewer rather than facing upward. The construction of new cells on large cells is common in *B. atratus* but is not characteristic of all pocket makers, some of which build new cells on cocoons. (Original drawings by J. M. F. de Camargo.)

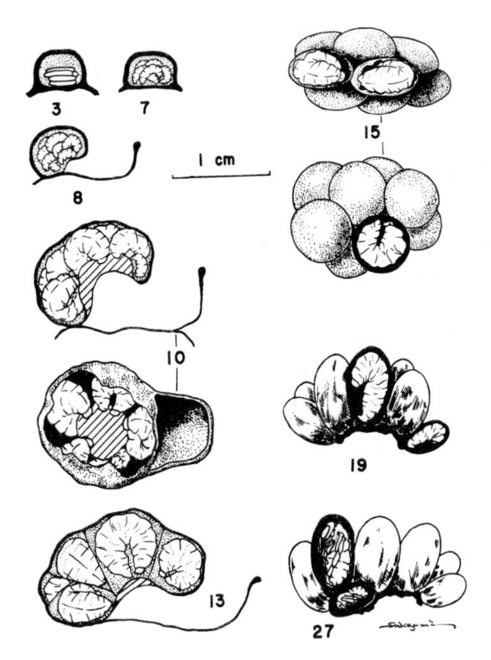

FIGURE 28.6. The growth of a distendible cell of *Bombus atratus* is shown in these diagrams. *Numbers* are the number of days after oviposition. *Upper left:* The initial small cell with eggs and small larvae. Then the pollen pocket is constructed, to the right of the cell, which grows with larval growth. *Slanting lines* represent the pollen. *At lower left,* the larvae are separated by thin, silken walls that they have spun. *At right,* the larvae above are still in wax enclosures, each more or less by itself. *Below,* they have spun the elongate cocoons and most of the wax has been removed by adult bees. All the diagrams represent vertical sections, except the lower figures for 10 and 15 days, which are top views. (From Sakagami, Akahira, and Zucchi, 1967.)

After the cocoons are constructed, the adult bumblebees remove the pollen-wax cell material covering the cocoons, exposing them to view, and use it for new egg cells or other construction in the nest. New cells are generally on top of cocoons, a little to one side, so that emergence of adults does not disturb the cells. The cocoons are not removed on emergence of the adults but remain in position, and are the most conspicuous items in a well-developed *Bombus* nest. They are typically in irregular clusters, each cluster representing the cocoons made by larvae that started life in a single cell. The long axes of the cocoons themselves are more or less vertical, the clusters more or less horizontal and usually at about the same level as other such clusters, so that an appearance of very irregular horizontal layering is often evident. Since the cocoons are not destroyed, the number of cocoons in a nest provides a record of the total productivity of the nest, except that sometimes in old nests some of the oldest cocoons have decomposed. Because of the great size variability in *Bombus,* the cocoons vary notably in size, the lower (oldest) ones averaging smaller than those above; of course the gyne and male cocoons produced near the end of the nesting cycle are markedly larger than those from which workers emerge, although sometimes intermediates exist. Because new cells are made on top of cocoons, the brood usually expands upward, the lowest cocoons representing early stages of colony history and upper layers representing late stages. Lateral expansion often occurs, however, so that especially in old and large nests, there may be much irregularity in the layering.

4. Group characters

Some behavioral characteristics of various groups of *Bombus* with correlates in nest or brood appearance have been listed by Hobbs and are summarized in Table 28.1. This table is based on North American species. Pollen storers and pocket makers are not found in the same subgenus, but there are no obvious morphological correlates for these two major behavioral groups.

B. Life cycle

Bumblebee colonies rarely survive more than a year in temperate or cold climates, where they break up in the autumn. As in most other primitively eusocial bees, overwintering occurs as fertilized young gynes. Most bumblebee gynes disperse from their nests and burrow into the soil far from the nest site, but gynes of a few species burrow near or even beneath the nest. Each gyne makes an oval or spherical chamber 25 to 30 mm in diameter in the soil, either in banks or gentle slopes. Depths in England average 8 cm, some species usually remaining at the litter-soil interface, others going down into the soil. Alford (1969) has made an ecological and physiological study of hibernation, and finds shaded, well-drained sites with a northwest exposure to be most frequented. Hobbs, however, has found that various exposures are utilized if the soil is moist and easy to dig (personal communication); presumably these attributes are common in England on northwest exposures. Alford

TABLE 28.1. Some nest attributes of various groups of *Bombus*.

Attribute	Group
Gynes isolate eggs of first brood in pollen	Most forms
Gynes place eggs of first brood together	Subg. *Alpinobombus*
More than one egg per cell	Most forms
Only one egg per cell after first brood	Subg. *Bombias*
After first brood, larvae largely fed through pockets	Pocket makers
a. All larvae apparently fed through pockets	Subg. *Subterraneobombus*
b. Only worker brood fed through pockets	Most pocket makers
Pockets not used (except for first brood of *Bombias*); food after first brood introduced from above as pollen-honey mixture	Pollen storers
a. Egg cells often primed with pollen	Subg. *Pyrobombus*
b. Egg cells, except for first brood, not primed with pollen	Most pollen storers

SOURCE: Summarized from Hobbs, 1964.

also notes that the honey in the crop and the fat reserves of gynes are used up in autumn, and that winter survival is dependent on glycogen reserves.

Dias (1958) believes that in *Bombus transversalis* of the Amazon, ordinary colonial activity is largely in the dry season, and that gynes survive the wet season alone, doubtless flying some and visiting flowers for food. At least in southern Brazil, in a warm temperate region, *Bombus* gynes hibernate in the soil as in the north; I dug a gyne of *B. bellicosus* from the soil in winter near Curitiba (latitude about 25° S, altitude 900 m).

In the spring in temperate regions bumblebee gynes emerge from hibernation and take food from the spring flowers. Their ovaries, which during and immediately after hibernation are threadlike, begin to enlarge. It is at this season that the gyne bumblebees are most conspicuous, for not only do they visit flowers, but they also search for nest sites. In so doing they spend most of the daylight hours flying over the ground, into dense vegetation, along banks, alighting from time to time to walk into dark cavities of various sorts. Such searching may go on for weeks, and the more conspicuous and suitable sites may be found repeatedly by different females, with resulting fights which lead to considerable mortality among gynes. At this season of the year they are highly intolerant of other individuals in their nests; there is never more than one gyne per new nest.

In the tropics, searching for nest sites by gynes may probably occur at any season of the year, but in any given area, for a particular species, there is a season when such activity is most common.[2] Even in warm temperate regions the typical cycle may be much modified. Thus in Corsica *Bombus terrestris* disappears in the dry, late summer when there are no flowers; gynes fly after the autumn rains, workers appear in December, and males in January, although at least some colony activity continues through the spring (Ferton, 1901). In New Zealand gynes and workers of *B. terrestris* introduced from England can be seen in flight in any month of the year, and some nests are established in midwinter instead of in spring. Another species, *B. ruderatus,* retains a life cycle more comparable to that which it exhibits in England (Cumber, 1954, 1963).

In the Amazon, colonies of *Bombus transversalis* are destroyed by natural enemies, especially stratiomyiid fly larvae, in less than a year—Dias (1958) says five or six months. Colonies of *B. atratus* in the state of São Paulo, however, commonly continue their activity for several years (Moure and Sakagami, 1962; Sakagami and Zucchi, 1965; Zucchi in Kerr, 1969; Saraceni, 1972). When censused such colonies contain from one to over 100 queens, many of them fertilized and with enlarged ovaries. After mating most gynes return to their nests. As explained elsewhere, in autumn in temperate climates young gynes often are active in colony life and work together; moreover, under unusual or experimental conditions, polygynous colonies of mature, laying queens have been made. Establishment of natural polygynous colonies of *B. atratus* perhaps requires decreased aggressiveness by the gynes during the phase equivalent to spring in temperate regions, but the aggressiveness is not entirely eliminated and continues until monogyny is reestablished.

The old nests of *Bombus atratus* are large enough that 10 to 20 young gynes can at first defend their own areas inside the nest. Workers and males are free to move about, and the queens lay and young are reared in each territory. With the growth of the broods there is more

2. Repeated references in the literature to swarming by tropical bumblebees are in error. They are based on the papers by Ihering (1903a, b) in which he gives hearsay reports of swarms in São Paulo, reports which he believed because he noted survival of colonies through the winter and did not observe the establishment of colonies by lone gynes. Ordinary colony establishment has now been observed by various persons in the same and other neotropical areas and there seems to be no probability of swarming.

and more chance of friction between neighboring queens. When a queen approaches another's territory, the owner shakes her wings, buzzes, and usually drives the intruder away. But if the intrusion continues, the bees curl their abdomens under and project large drops of liquid feces at one another. If the intruder is hit, she retreats; it is reported that she never tries to enter any other territory, and may leave the nest. We are not told what happens if the owner is hit, but if neither is hit, the bees also fight with jaws and stings until one or both is killed. Continuation of this process results in gradual reduction of the queen population until there is only one, as in Figure 28.7. The complete polygynic-monogynic cycle requires three to six months. Should the last queen be killed or die, a fertilized worker, of which are commonly several in a colony, begins or continues to lay fertilized eggs and the colony survives.

The coexistence of laying queens in *Bombus atratus* is probably an adaptation to subtropical conditions, and must result in a rapid increase of the worker population during the polygynic phase. Perhaps the reduction in egg production accompanying return to monogyny is the factor that, for this species, leads to an increased worker-larva ratio and a new brood of young gynes. Temperate species, when introduced into mild climates like that of New Zealand, sometimes produce large nests in which numerous young queens may be present, but there is no evidence that they can remain together and reproduce (Cumber, 1949b).

Egg, larval, and pupal durations vary with the weather (in nests in artificial observation containers) and probably with the species. Within a single nest, the growth period varies, so that Brian (1951) found that emergence of some adults produced from a given batch of eggs might antedate the emergence of certain adults from another batch laid earlier. Marginal individuals in the first brood and in later clusters of young often mature later than central ones, which are probably kept warmer. Brian found the egg stage lasting from 4 to 6 days, the larval stage from 10 to 19 days, and the pupal stage from 10 to 18 days in *Bombus agrorum.* For various species, however, Hasselrot (1960) suggests that in undisturbed nests in natural locations with efficient temperature control, duration of immature stages is less variable than has usually been believed; he gives the following estimates of mean values: egg, 3.4 days; larva, 10.8 days; and pupa, 11.3 days.

Some students (for example, Free, 1955a) maintain that the developmental period for queens and workers is the same, while others (like Frison, 1928) find that the larval period is positively related to adult size, for *Bombus bimaculatus* being 11 to 13 days for workers, 13 to 17 days for queens. Perhaps there is variation among species.

The young adult bumblebee chews its way through the top of the cocoon, often with help from nestmates, and crawls out. Newly emerged adults are slow moving and rather wet looking, the wings limp and the hair silvery. Only after a few days do the wings harden and the hair acquire the coloration typical for the species, usually various combinations of black, yellow, white, or red.

Under natural circumstances the first batch of eggs produces only workers, as do the subsequent ones in most species for varying lengths of time, depending on the climate and species. As soon as the first workers reach maturity, they take over all foraging activity, although the queen continues throughout her life to share in the care of brood and to engage in other activities in the nest. Ordinarily only if some catastrophe reduces the size of the worker force will the queen again leave the nest to forage. But in a cool and rainy summer in southern Alberta, when first broods were often small and late, queens were not infrequently seen on flowers at the end of July. While the first batch of young are still large larvae or pupae, the queen constructs one or more new cells on top of their cells or cocoons. The new cells are on the tops of the two ridges that margin the incubation groove (Figure 28.8). As activity continues through the season, new cells are constructed on top of old cells or on cocoons in a progressive way, the nest cavity being enlarged as need be by moving the nest materials. As described in Chapter 7 (A), the number of eggs in a cell is proportional to the number of cocoons in the batch on top of which it was built, so that as the batches become more numerous and contain more and more cocoons, the egg-laying rate increases. This arrangement provides for the relation between the number of larvae requiring feeding and the number of workers that will be present to care for them. Thriving colonies are illustrated in Figure 28.9.

As explained in Chapter 10 (B), a time is reached when there is a shift of production from workers to reproductives. For some species, such as *Bombus pratorum* in Europe and *B. bimaculatus* in North America (Frison, 1928), the reproductives appear in June and July; mated gynes of *B. bimaculatus* go into hibernation as early as July. *B. jonellus* is another such species in Europe and

FIGURE 28.7 Sketch of brood and adults of *Bombus* (*Fervidobombus*) *atratus* of Brazil. (*A*) Queen (all other bees shown are workers). (*B*) Large and small workers working on an egg cell which is on a clump of large larvae, not on cocoons as in most northern species. (*C*) Defecation. (*D*) Pressing pollen into a pollen pocket. (*E*) Inspection of a larval cell. (*F*) Fanning. (*G*) Slight show of aggressiveness. (*H*), (*I*), (*J*) Brooding. (*K*) Resting outside the brood area. (*L*) Assisting emergence of a new worker from a cocoon. (*M*) Grooming herself. (*N*) A forager with orchid pollinia on the thorax. (*O*) Working on honey pots. (*P*) Grooming. (*Q*), (*R*) Defensive positions, usually seen when a nest is opened. (*S*) Feeding a larva with liquid through the top of a cell (enlarged). (From Sakagami and Zucchi, 1965.)

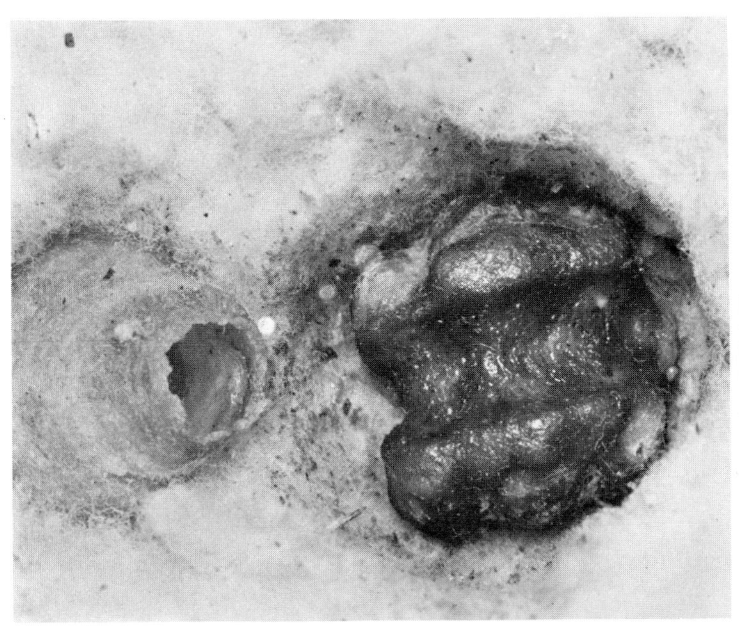

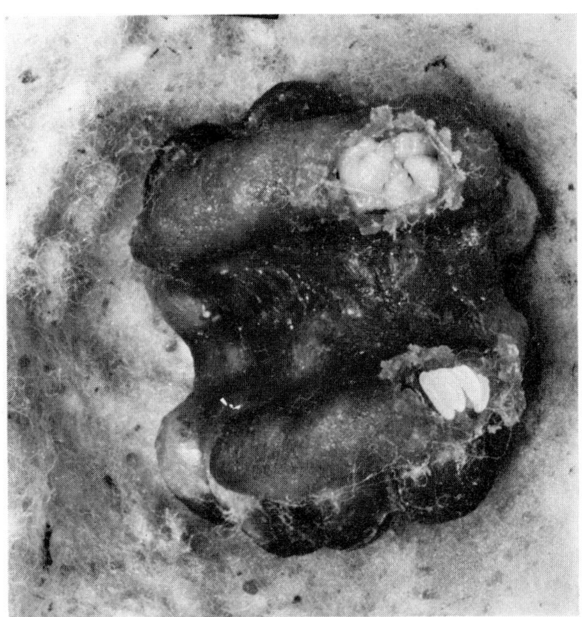

FIGURE 28.8. *Left:* Gyne nest of *Bombus frigidus,* with the honey pot at the left and the cocoons of the first brood at the right, in a clump showing a strong incubation groove. On the summits of the ridges on either side of the groove are new cells, four completely fused on the upper ridge and four recognizably separate on the lower ridge. *Right:* Similar brood of *Bombus rufocinctus,* with new cells constructed along summits of the ridges of cocoons on either side of the incubation groove; these have been opened to show eggs below and small larvae above. The workers of the first brood are nearly ready to emerge from their cocoons. (Both photographs for G. A. Hobbs, Entomologist, Canada Department of Agriculture, by staff photographer E. T. Gushul.)

a recent report indicates that some of its gynes, instead of entering hibernation in July, establish new nests whose first workers appear in August (Meidell, 1968). Young gynes and males of this species are again on the wing in small numbers in September. Thus this species seems to have a partial second generation. In most species, however, reproductives are not produced until late summer or autumn. Males are produced first, from eggs laid when the queen is laying at her maximal rate (Cumber, 1949a). Gynes are produced later, when for various reasons described in Chapter 10 (B), the number of workers present to feed each larva is higher. After production of the sexual brood, workers are not produced again by the queen, at least among temperate species. Laying workers in a mild climate (for instance, where *B. terrestris* is introduced in New Zealand, see Cumber, 1949a) may continue laying and if fertilized may produce more workers after the sexual brood.

As the workers die off and the population of the colony declines, its stored food and temperature control fail, the old queen and her last workers expire, and the young gynes leave to hibernate. The males will have left and for the most part died earlier; they maintain no regular connection with the parental colony after they leave it at the age of only a few days.

Interestingly, before hibernation young gynes may undertake all sorts of colony activities, including feeding the brood, secreting wax, defending the nest, and under certain circumstances even foraging. Lehmensick and Stein (1958) describe the curious case of a colony of *Bombus hypnorum* that developed in the warmth and abundant food of the honeycomb region of an *Apis* hive. By early June (Germany) there were numerous young gynes and males, but only a few tiny workers had been produced. The gynes foraged, cared for the brood, constructed the nest and mated; the workers did nothing. Cumber (1963) found numerous young gynes acting like workers; they had been produced in spring in a large overwintered

colony of *B. terrestris* in the mild climate of New Zealand, where the species has been introduced from England. No males were present and the gynes were therefore unfertilized. Before normal autumnal production of gynes, males are also produced, but this had not occurred in the nest discussed by Cumber.

Mating may occur on the flowers, near the nest (males have been seen to chase a young gyne down the nest burrow, and it is probable that mating may occur in the nest), or well away from the nest, where males of some species select rather striking markers of various sorts (a flower, rock, fencepost, or tree trunk) about which they fly and on which they rest, flying after and if possible pouncing upon any moderate-sized object that flies by, fortunately including young gyne bumblebees (Schremmer, 1972a). Males with the latter type of behavior possess large eyes and ocelli, presumably associated with perception of passing females. Another type of mating behavior is that of certain species such as *Bombus terrestris* and *pratorum* whose males establish definite routes along which they fly repeatedly, hesitating periodically at special places, such as the foot of a tree or a particular leaf or stem. One of Charles Darwin's least-known works, originally published in German and only recently translated into English (Darwin, 1965), concerned this behavior and the observation that male bumblebees would use the same or approximately the same route in successive years. Frank (1941) measured such a circuit and found it to be about 275 m around, with 27 stopping places. Each male has a different route, although routes may overlap. The stopping places, as Haas (1946, 1952) found, are marked by volatile substances secreted by the male's mandibular glands (Stein, 1963). These materials differ among all 13 species of *Bombus* and 6 species of *Psithyrus* studied in Sweden by Kullenberg, Bergström, and Ställberg-Stenhagen (1970). Free (1971) found that young gynes of *B. pratorum* attract males only near the males' scenting places, not between them and that the odor of the gyne is important in inducing the male to grasp and

FIGURE 28.9. Colonies of *Bombus nevadensis* (*top two*) and *B. huntii* (*bottom*). *Top:* Note individual egg cell (lower right) and group blended together (lower center) as well as open storage pots at bottom. *Middle:* Brood mass, covered with new males and gynes. (Photographs for G. A. Hobbs, Entomologist, Canada Department of Agriculture, by staff photographer E. T. Gushul.)

mate with her. Presumably the function of the flight paths is to disperse males and thus improve the chances that females entering the area will be found by males.

C. Economic importance and foraging

Because of their long tongues, bumblebees are important pollinators of various flowers with deep, tubular corollas, and they were imported into New Zealand in order to pollinate red clover, a plant not ordinarily pollinated successfully by honeybees. It was Charles Darwin who found that 100 heads of red clover pollinated by bees produced 2700 seeds, while 100 heads protected from bees produced none, for the flowers are self-sterile.

One of the principal problems in efficiently utilizing bumblebees as pollinators of crops has been successful overwintering of queens under artificial conditions. Milliron (1967) gives references to some of the pertinent papers, but no immediate promise of success on a large scale. Holm (1966), who gives an excellent review of his own work and that of others on utilization and management of bumblebees, is far more optimistic, but finds establishment and successful development of colonies by overwintered queens to be the main problem. Hobbs, Nummi, and Virostek (1962) and Hobbs (1967) felt that the main problem still to be solved in management of bumblebees in Alberta was control of the cuckoo bumblebees, *Psithyrus*.

Flight activity of female bumblebees has been considerably studied because of their importance as pollinators of various leguminous crops, especially in the North Temperate Zone. An interesting and seemingly (for the bee) inefficient bit of *Bombus* behavior that must increase its effectiveness in cross-pollination is the habit of visiting only a few flowers on any one visit to a red clover head. A bumblebee collects from an average of only five or six flowers out of the many in the head before moving on to another head (Free and Butler, 1959). Not all *Bombus* species are equally attracted to any given crop, for example, red clover. The short-tongued species, which cannot easily reach nectar in deep tubular flowers, often bite holes through the corolla tubes and obtain nectar without pollinating the flowers. Hawkins (1956) has shown that red clover seed yield depends principally on the abundance of long-tongued species of *Bombus;* the frequency of short-tongued species, of *Apis,* and of other possible pollinators in the fields (in England) have little influence. As Free and Butler (1959) point out, the importance of *Bombus* as a pollinator of agricultural crops is by no means limited to red clover. Alfalfa, fruit trees, cotton, and many other crops are efficiently pollinated by these insects. Those authors give a good review with appropriate references to the extensive literature on the economic importance of bumblebees, and Free (1970b) cites the more recent papers.

As has been noted by various authors, some *Bombus* workers spend the night away from the nest; Free (1955d) found that about 15 percent do so. The first morning activity observed in Holland at a nest of *B. equestris* in warm weather was the arrival of a few workers returning after sunrise (Den Boer and Vleugel, 1949). About an hour and twenty minutes after sunrise, workers began

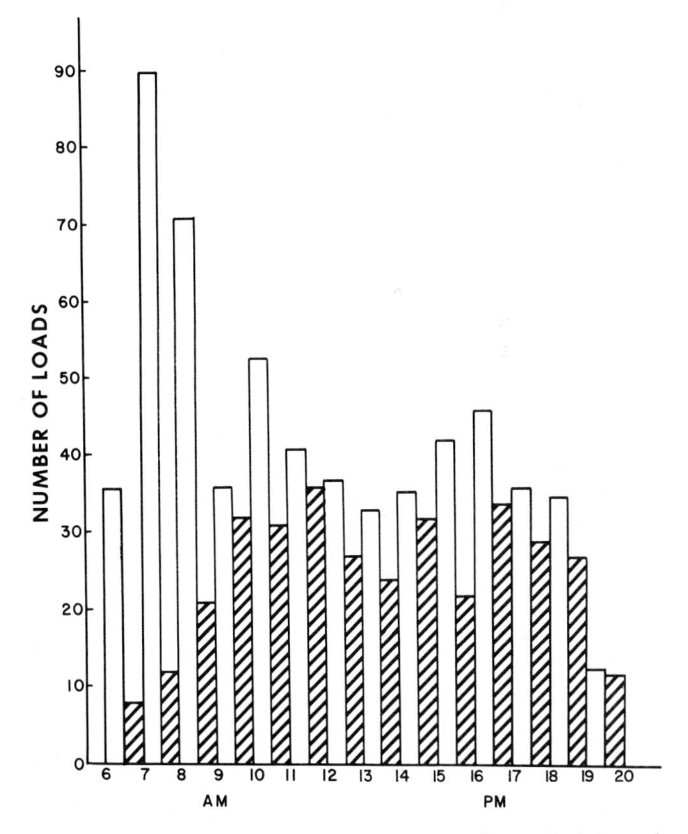

FIGURE 28.10. Number of loads of pollen (*shaded*) and nectar (*white*) collected by workers of a colony of *Bombus lucorum* at different times during three days. (From Free, 1955d.)

TABLE 28.2. Duration (in minutes) of periods which foragers spent inside their nests between foraging trips.

Species	Nectar gatherers		Pollen gatherers	
	No. observations	Mean	No. observations	Mean
Bombus agrorum	226	1.7	91	2.9
B. pratorum	239	1.9	53	2.7
B. sylvarum	145	2.2	308	3.4
B. lucorum	468	3.8	203	5.3

SOURCE: Free, 1955d.

leaving the nest on foraging flights. As with other bees, foraging flights can be divided into two groups, those in which the returning bee carries a load of pollen and those in which she does not. The latter are presumed to carry in the crop either nectar or pollen as food for the forager or nectar for the colony. Den Boer and Vleugel note that most pollen-collecting flights occurred before noon. This corresponds with observations on various other bees, for example, my observations on various halictines; but it is in disagreement with Free's (1955d) data on various species of *Bombus,* an example of which is shown in Figure 28.10. For *B. equestris,* most bees making a pollen-collecting flight did not leave again that day, or if they did so, departed only after two to three hours. The authors conclude that pollen-collecting flights, which lasted about an hour, must be "very tiring." In contrast, nectar-gathering flights were shorter, and the bees often remained in the nest for only about two minutes before departing again on another such flight. For *B. agrorum,* however, Brian (1952) makes no such distinction and indicates that a bee may get rid of her load of pollen as well as nectar and leave the nest within one or two minutes. Her bees evidently found food more easily, for trips from which bees returned with pollen lasted (in 1947) on the average only 18.6 minutes, those on which the bee returned with nectar only 16.3 minutes (381 trips timed). Free (1955d), in an extensive study, noted that mean times between trips were short, but longer for pollen loads than for nectar loads, and varied with the species (Table 28.2). Den Boer and Vleugel as well as other writers noted striking individual differences in the behavior of the bees. Some spent the whole day outside the nest without collecting any pollen or making trips back and forth, while others entered and left the nest many times a day. As indicated in Chapter 12, large workers begin foraging when younger than small ones, and also more commonly carry pollen loads (Brian, 1954); hence they are the principal foragers for a colony.

29 *Stingless Honeybees*

Like the true, stinging honeybees of the genus *Apis*, the stingless honeybees (tribe Meliponini), often called simply stingless bees, are highly eusocial forms living in permanent colonies, often composed of many thousands of individuals. In their own ways, the stingless honeybees and *Apis* have attained approximately equal social levels (Sakagami, 1971). The tribe Meliponini is a much more diversified group than the tribe Apini (*Apis*) and contains many species. One author, Padre J. S. Moure, has divided it into 21 genera, an indication of its diversity. However, I believe that a more useful classification is that which is used here and which corresponds reasonably well with that of most authors. I recognize the following genera: (1) *Melipona,* consisting of a number of tropical American species varying from somewhat smaller than to somewhat larger than *Apis mellifera;* (2) *Meliponula,* consisting of a single African species having the appearance of a small *Melipona;* (3) *Trigona,* a large Pantropical genus of long-winged bees, ranging from almost as large as *A. mellifera* down to the shortest of all bees, 2 mm in length; (4) *Dactylurina,* a genus represented by a single, slender-bodied African species which should perhaps be left in *Trigona;* (5) *Lestrimelitta,* a genus of African and tropical American robber bees, dependent upon the products of nests of *Trigona* or less commonly *Melipona* or even *Apis.* As suggested in Chapter 19 (C), the American and African *Lestrimelitta* quite likely arose from different groups of *Trigona* and should be given generic status, *Lestrimelitta* for the American species and *Cleptotrigona* for the African ones.

The chief difference of opinion among taxonomists concerns the forms here placed in the genus *Trigona.* The situation is very different from the *Bombus* problem; in that genus all species are similar and the breaking of the genus is singularly inappropriate. *Trigona,* however, is a large genus that is very diverse both in morphology and biology. The differences among some of the subgenera of *Trigona* seem nearly as great as the differences between *Trigona* and the related genera listed above. It would be reasonable, therefore, to recognize several genera for the groups placed in *Trigona,* although I prefer not to do so. The following is a list of subgenera of *Trigona,* according to the classification used by Wille and Michener (in press):

Subgenus *Meliplebeia.* This group contains the African relatives of *Plebeia.* It has many primitive characters and contains forms for which the names *Plebeina, Plebeiella,* and *Apotrigona* are sometimes used.

Subgenus *Plebeia.* A large and in many ways primitive group in America, with a few representatives in Australia and New Guinea. The latter are sometimes put under *Austroplebeia,* while some unusual South American species are placed under *Friesella, Mourella,* and *Schwarziana.*

Subgenus *Nogueirapis.* A small American group, perhaps properly included in *Plebeia.* It contains the Oligomiocene fossil from Chiapas discussed in Chapter 2 (B) as well as the modern species.

Subgenus *Axestotrigona.* A small African subgenus.

Subgenus *Hypotrigona.* An Afroasian subgenus of minute bees, probably primitive in many of its features, excepting those associated with small size. Other names have been proposed for certain species of this group, but all apply to species about which no biological data are available. A fossil (Oligocene?) species from the Dominican Republic is the only representative from the Western Hemisphere (Wille and Chandler, 1964).

Subgenus *Trigonisca.* The American equivalent of *Hypotrigona,* although a few species of *Plebeia* are nearly as small.

Trigonisca might have acquired its small size independently from *Hypotrigona,* but the two probably had a minute common ancestor. At least they show similarities other than those that seem to be convergent and related to minute size.

Subgenus *Scaura.* A small American group that includes the species sometimes placed in *Schwarzula.*

Subgenus *Partamona.* A common American subgenus of moderate-sized *Trigona* species.

Subgenus *Paratrigona.* An American group of rather small species usually with conspicuous yellow facial markings. A subgroup called *Aparatrigona* is included.

Subgenus *Scaptotrigona.* A common American subgenus of moderate-sized species.

Subgenus *Nannotrigona.* An American subgenus containing but few species, all of rather small size with coarsely sculptured integument.

Subgenus *Cephalotrigona.* A small American subgenus of large bees with relatively large heads.

Subgenus *Oxytrigona.* A small American subgenus noted for the burning secretion of the mandibular glands.

Subgenus *Tetragona.* An enormous subgenus in America and the Indoaustralian tropics, represented in Africa by the closely related genus *Dactylurina* which could even be included in *Tetragona.* The species mostly have slender abdomens, sometimes markedly elongate. They vary in size from rather small to rather large. Impressive arguments can be made for recognizing the subgroup *Frieseomelitta* and perhaps *Tetragonisca* as subgenera. In fact in Chapter 15, I have treated *Frieseomelitta* as though it were a subgenus to facilitate expression. The following other names have been given to forms here included in *Tetragona: Duckeola, Geotrigona, Heterotrigona, Lepidotrigona, Ptilotrigona, Tetragonilla, Tetragonula.* Other names, applicable to forms about which there is no biological information, are listed by Michener (1965a).

Subgenus *Trigona* proper. This is a large American subgenus, the only one with fully dentate mandibular margins. It contains medium-sized to large species.

There are several general accounts of meliponine biology, as follows: Darchen and Louis (1961); Moure, Nogueira-Neto, and Kerr (1958); Nogueira-Neto (1970a); Sakagami and Zucchi (1966); and Schwarz (1948). The design of nesting boxes in which stingless bee colonies can be kept, either for their honey or for pollination or for experimental purposes, has been described by Nogueira-Neto (1970a), Portugal-Araújo (1955), and Sakagami (1966).

The Meliponini are restricted to the tropical regions of the world. They range considerably farther south of the equator than to the north. Thus on the coasts of Mexico they extend little if at all beyond the Tropic of Cancer, but in South America they extend nearly to the latitude of Buenos Aires, and 14 or 15 species are found on the southern Brazilian Plateau, well south of the Tropic of Capricorn. In South Africa they range southward into the province of Natal; to the north, probably because of aridity, they apparently do not approach the Tropic of Cancer. In the American tropics the Meliponini are extremely numerous in species and individuals; in fact they are the most conspicuous bees. In the Old World tropics they are not so numerous, the number of species as well as individuals being relatively small. Only in Southeast Asia are species and individuals possibly as numerous as in comparable areas in the American tropics.

Although the species of *Trigona* and *Melipona* occurring in the temperate regions of the Southern Hemisphere are able to withstand rather low temperatures (snow falls nearly every winter in some parts of the southern Brazilian Plateau), the bees merely become inactive on cold days, and fly again whenever the temperature rises. There is no evidence of a winter cluster such as occurs in the temperate races of *Apis mellifera.*

Rather closely allied forms are found on distant continents, as shown in the above list of subgenera of *Trigona.* Note the comments under (1) *Plebeia* and its African relative *Meliplebeia,* (2) *Hypotrigona* and *Trigonisca,* and (3) *Tetragona.* Other groups of bees that are limited to the tropics are probably at least equally old, but are restricted to one or another continental area. It seems unlikely that the dispersal of the meliponines was across the Bering Strait in a time of more tropical climates, since other tropical bees could have used the same route; it is more likely that the unique (for bees) pantropical distribution is related to unusual features of meliponine biology. These are their prediliction for nesting in logs and for storing large quantities of food, and their ability to live sealed up in their nest cavities for long periods, as in time of floods. As a rare event a log containing a living colony may be carried across the ocean and might be washed up where the bees could resume activities on a new continent. Admittedly, the relatively late separation of Africa and South America now recognized may explain the resemblance between stingless bees of these continents, and the scarcity of these bees in the Antilles does not support the idea of transoceanic dispersal. But the absence of other pantropical groups of bees suggests something unique about meliponine dispersal.

A. Nest locations

A detailed summary of the nest locations utilized by stingless honeybees in all parts of the world is given by Wille and Michener (in press). Most Meliponini nest in cavities of appropriate size, which they enter through a small aperture. The majority of species nest in hollow tree trunks or branches. If a cavity is too large, it is limited by walls called batumen plates (see section B2 below). For some of the minute forms, whose colonies are frequently small and sometimes contain only a few hundred workers, a suitable cavity may be found in a hollow liana or a cerambycid beetle burrow in a branch only 2 or 3 cm in diameter. Some species, for example *Melipona nigra* and *Trigona fulviventris,* characteristically nest at the foot of a tree in hollow roots or in a hollow among the roots. Other species nest in the soil, taking advantage of the abandoned nests of ants, termites, or subterranean rodents. For example, *T. mombuca* constructs nests in the abandoned nests of leaf-cutter ants (*Atta*). Since nests of these ants are extremely deep, the *Trigona* nests are also deep. Other species, for example, *T. subnuda* and *mirandula,* nest more shallowly in the soil, and it is sometimes not obvious what sort of hollows they take advantage of; but it is most unlikely that the bees excavate their own nesting cavity. No doubt they can enlarge or shape a cavity, but that they ever start digging from the surface of the soil is improbable. Other species live in walled-off parts of occupied nests of ants or termites, and probably gain some protection from the association. *Trigona peltata* invades the leaf and silk nests of the ant *Camponotus senex* in Central America, a bee nest occupying most of the interior of one of the ant nests. The African *Trigona* (*Axestotrigona*) *oyani* was found nesting in cavities made by pangolins in nests of ants of the genus *Crematogaster* (Darchen, 1971), and in Costa Rica *T.* (*Trigona*) *fuscipennis* nests in cavities made by birds in nests of termites of the genus *Nasutitermes* (Wille and Michener, in press). *T. cilipes* is found in nests of both termites and ants (Figure 18.5). Some species nest regularly in termite nests. *T. latitarsis,* in tropical America, seems to establish colonies by constructing typical-looking nest entrance tubes on the surface of an arboreal termite nest. Then, presumably, bees from the parent nest work their way into that of the termites, gradually displacing the termites from a portion of their abode and excavating and walling off a cavity suitable for the bees' occupany (Figure 29.1). Other species take advantage of hollows made in termite nests by birds or other animals.

Although in general each species has a characteristic location in which it nests, some species show considerable variety in nesting sites. For example, Moure, Nogueira-Neto, and Kerr (1958) record nests of *Melipona quadrifasciata* in trunks from 1 to 3 m above the soil surface, in the soil in nests of *Atta,* and in abandoned mud nests of a bird (*Furnarius*) constructed on telephone poles. They also report that *M. marginata* lives in termite nests, in adobe houses, and in trunks of trees from near ground level up to 15 m high.

There are various species, principally *Trigona,* that nest in man-made cavities, in towns and even cities. Thus in the American tropics species like *T. testaceicornis, T. emerina,* and *T. jaty* often nest in cavities in masonry walls or houses. It is in the Asiatic tropics, however, southeast to northern Australia, that species occur which can nest in extensive thin cavities, as between boards less than a centimeter apart. Such forms, for example, *T. clypearis* and *fuscobalteata,* are common about human habitations and often nest in walls of frame buildings.

In addition to those Meliponini which nest in cavities, there are various species which make nests that are, to varying degrees, exposed. The minute African species *Trigona braunsi* and *gribodoi* usually nest in small hollows in trunks or branches, which may or may not be limited by batumen plates, but sometimes a nest is in a deep crevice in a contorted trunk, which is then closed off by batumen. Such nests are, of course, partly exposed (Bassindale, 1955). The neotropical *T. cupira* usually builds partly exposed nests. They may be in a cavity in an earth bank with one side of the nest exposed or only thinly covered by soil, in an open cavity in a wall or cliff, or among the broad bases of living palm leaves where the nest gets excellent protection from rain but nonetheless is largely exposed. Nests of *T. cupira* are also constructed among branches of bushes or trees, but in such cases are protected from severe weathering by bromeliads, orchids, or other epiphytic vegetation. *Dactylurina staudingeri* constructs exposed nests which hang from the undersides of large branches of trees and are therefore protected from the weather. Other species with exposed nests, such as *T. spinipes, corvina,* and *nigerrima,* all in the American subgenus *Trigona* proper, construct large nests that are fully exposed (Figure 29.2), sometimes high in trees where they receive the full force of the wind and the sun. Such

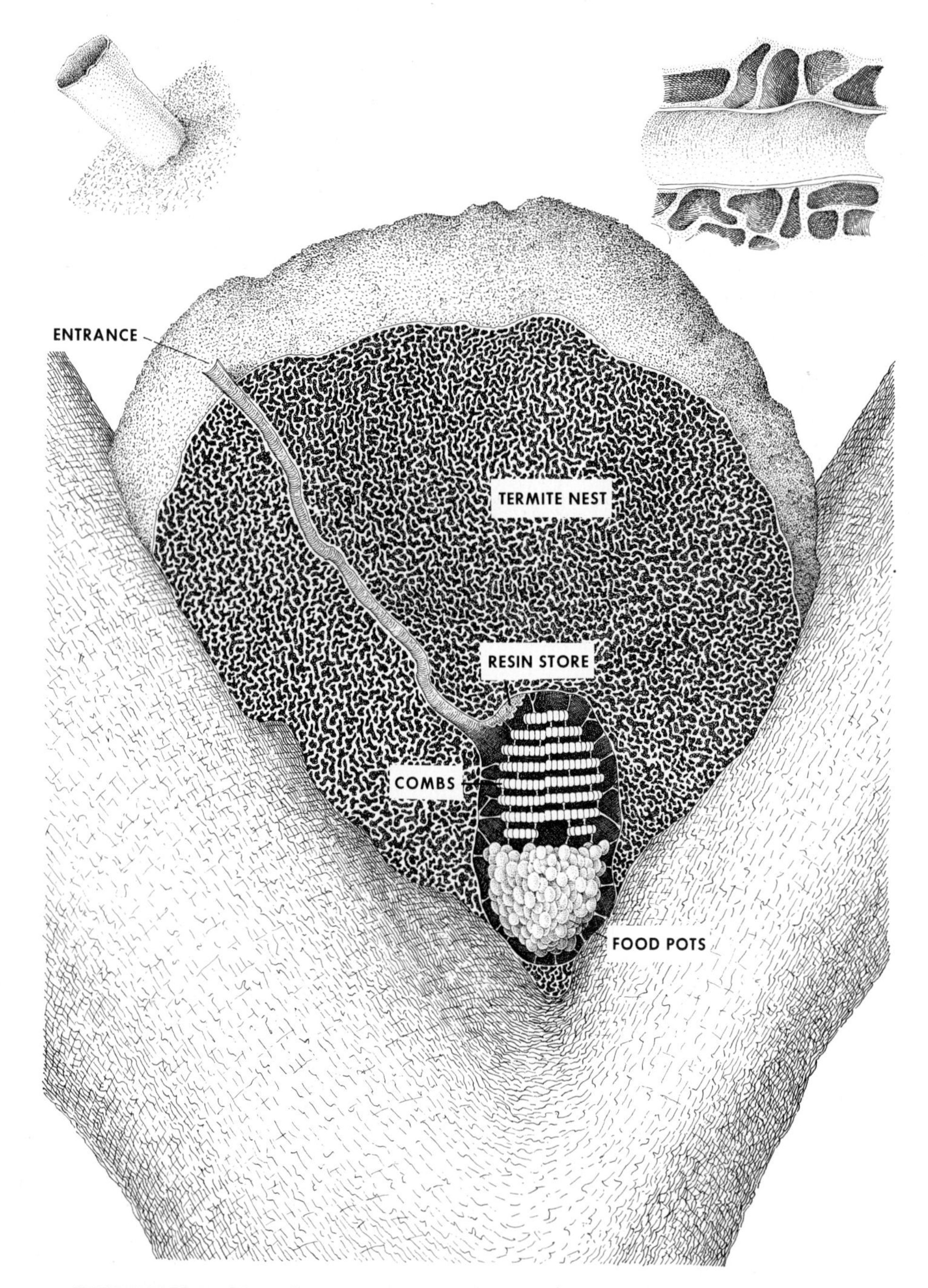

FIGURE 29.1. Nest of *Trigona* (*Scaura*) *latitarsis* in a termite (*Nasutitermes*) nest in the fork of a tree. *Upper left,* an enlargement of the entrance; *upper right,* a section of the entrance tube passing through the termite nest. (From Camargo, 1970; original drawing provided by J. M. F. de Camargo.)

FIGURE 29.2. Nest of *Trigona nigerrima* supported high on small branches of a tree in Costa Rica, opened to show the thick protective batumen layers that must provide considerable insulation. (Photograph by C. D. Michener.)

nests are ovoid, protected by numerous hard layers of nesting material, and reach maximum dimensions up to 85 cm in length and 55 cm in diameter. A nest of *T. corvina* from a tree in Costa Rica weighed 74 kg (162 lb).

One of the attributes of most stingless bee nest sites, or in the case of exposed nests, of the nest itself, is excellent insulation. Nests in large tree trunks or in the soil are particularly well insulated. Many species, particularly those of the moist tropics, are unable to withstand chilling. But a nest of *Trigona emerina* in a poorly insulated wood box in temperate southern Brazil has been recorded withstanding temperatures in the box ranging from −6° C in winter to 45° C (when the summer sun was on the box).

It seems that a major factor limiting the population density of many Meliponini must be the number of suitable nest sites. Moure, Nogueira-Neto, and Kerr (1958) describe an increase from 12 to 42 in the number of nests of one species of *Trigona* on the campus of the Agricultural School of the University of São Paulo at Piracicaba, apparently as a result of the construction of a building, the walls of which contained many holes suitable as nest sites. It seems clear that the populations of meliponine bees in the vicinity of Old Panama City (Michener, 1946) were high because of the numerous holes and cavities suitable for nesting places in the ruins of the abandoned city.

B. Nest architecture and contents

1. Construction materials

Nests of Meliponini are made of wax, which in most parts of the nests is mixed with large amounts of resin (Figure 29.3) gathered by the bees; this mixture is called cerumen. Cerumen is also a name for ear wax, a fact which gives an idea of the commonest color of the bees' cerumen. In some species (*Melipona*) mud is also utilized, by itself or mixed wih wax and resin or vegetable mate-

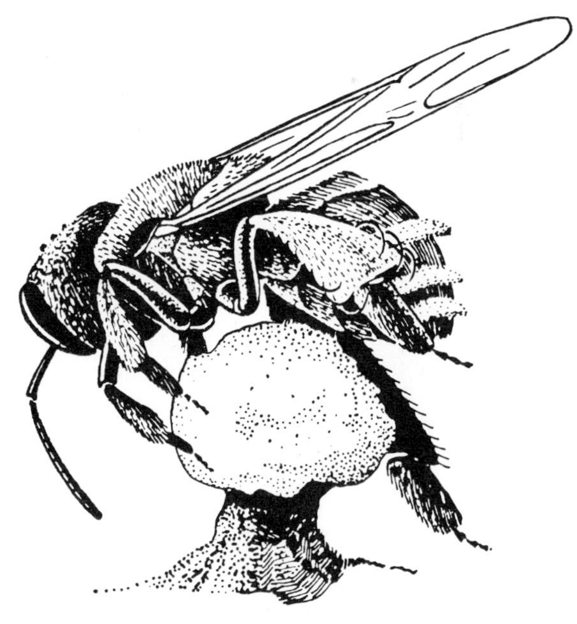

FIGURE 29.3. A worker of *Trigona* (*Scaptotrigona*) *xanthotricha* placing wax scales from the dorsum of the abdomen on top of a wax storage place. (From Hebling, Kerr, and Kerr, 1964.)

rial. The resin or propolis comes from a wide variety of sources, including cut or damaged trees, the surfaces of various plants which secrete resin, and the like. Moreover, some species, principally those that make exposed nests, bite into the soft, growing tissue of plants whose sap or latex becomes viscous on exposure to the air. They then collect the exuding material and carry it back to the nest. Such biting can damage crop plants; for example *Trigona corvina* damages the young growing tips of citrus trees in Central America. Some *Trigona* species also collect oil or grease from about machinery, and one Australian species, *T. hockingsi,* collects fresh paint from newly painted buildings.

Most species collect resins in large quantities, but those which make exposed nests use particularly large amounts. Such species seem to use much of the resin for construction of the protective layers, the supporting pillars, and other parts of the nests which become hard and in some species brittle. The protective layers of exposed nests also often contain chewed leaf material, excrement of animals, and fibers picked out of cattle droppings. Some of the resin is mixed with wax to form cerumen for construction of the softer, more pliable parts of the nest, such as the storage pots and the brood cells, exactly as in species nesting in protected places. It is dificult to say how much resin is added to the wax used in various parts of the nest, but it seems probable that some is added to the material used for nearly all parts of the nest structure.

The terminology for meliponine nest parts used below is that of Michener (1961a; and Wille and Michener, in press). The latter paper summarizes most of what is known of meliponine architecture.

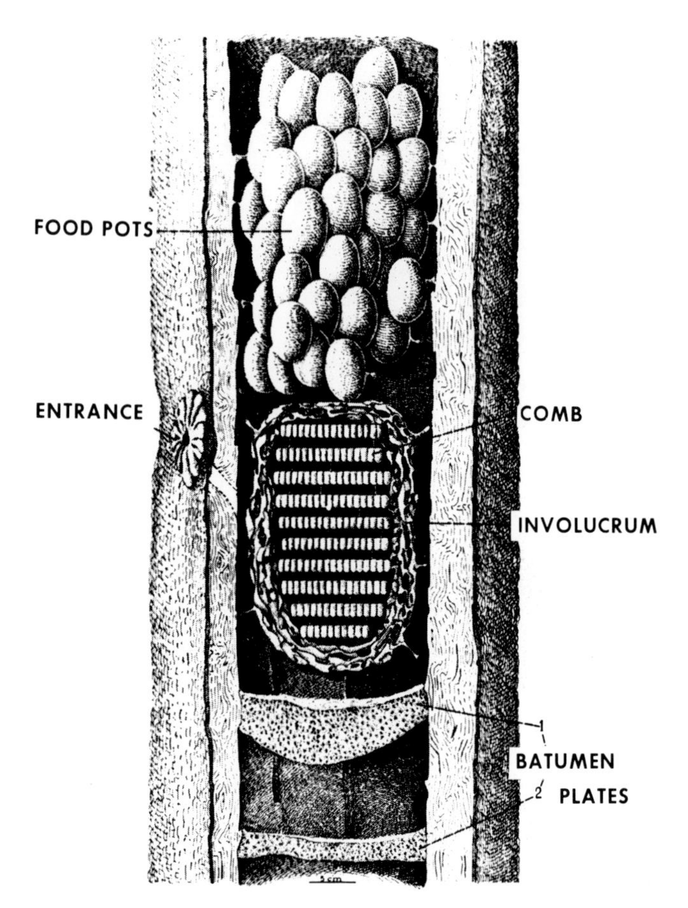

FIGURE 29.4. Nest of the Amazonian *Melipona pseudocentris* in a hollow tree trunk. (From Kerr et al., 1967.)

2. Batumen

Nests of Meliponini are surrounded by a layer of material called batumen. The batumen is made of cerumen (commonly brittle), resin (propolis), or sometimes of vegetable matter or mud mixed with resin or cerumen. The batumen seals the nest cavity, except for the entrance and sometimes minute ventilating perforations. In nests in hollow trees and other enclosed situations the batumen is mainly in the form of thick plates, closing off the hollow above and below to form a cavity of a size suitable for the bees' nest. Such plates are spoken of as batumen plates (Figure 29.4); the word batumen originally referred only to such plates. However, the batumen usually continues around the nesting cavity as a thin lining, often only a single sheet a millimeter or less in thickness, called the lining batumen. In partially exposed nests the batumen is several layers thick, for it forms the laminate wall that separates the nest from the outside. In fully exposed nests the batumen is similarly laminate all the way around the nest. Thus in nests such as those of *Trigona corvina* and *spinipes* there is a thin, brittle outer layer of batumen, the breaking of which allows bees to swarm out upon any intruder, since there are holes and passages through the other layers of batumen. However, it is these other layers which provide protection against heavy blows and against temperature fluctuations. In *T. corvina* they may attain a total thickness of 24 cm at the top and 11 cm at the sides and bottom, and are so strong in old nests, where they are supplemented by earthen material, that one opens them with an axe or machete. Even with such tools it is not always an easy task. In the batumen of the fully exposed nests of *T. spinipes,* insulation on one side is probably provided by a similar mass of earthen material, called the scutellum.

The increase in volume of a meliponine nest with growth of the colony and the quantity of stored food probably is usually rather small. Available space is used more intensively, rather than large amounts of additional space being enclosed. In many natural hollows, such increase is impossible. In others, removal and reconstruction of batumen plates could occur, but I have not seen much evidence of it. Occasionally one sees a batumen plate partially removed, and another farther from the center of the nest, suggesting that the bees have enlarged the space for their nest. Cavities in soil and in termite nests probably are often enlarged by bees. Species making exposed nests probably increase their size, but not enormously, as do vespid wasps which are constantly adding new outer layers and removing inner ones. Young nests of *Trigona corvina* in Costa Rica averaged somewhat smaller than old ones, but nonetheless the principal change was in the strength of the batumen, which was greatly thickened with age. For *Dactylurina,* Darchen (1966) has described increase in size of the nest by formation of blisters of fresh cerumen bulging from aperatures in the outer layer of batumen, but my impression is that this process results in only minor increases in the amount of enclosed space.

3. The entrance

The entrance into the cavity of the nest is often elaborated in various ways. If it is a hole through soil or wood, it is ordinarily lined with resin or cerumen. A short or long entrance tube, varying with the species, often extends into the open air from the nest entrance. Depending on the species, it may be made of brittle cerumen or of soft and pliable cerumen probably containing much wax. Some of the smaller species make entrance tubes which are soft, at least near the opening, and regularly close the nest entrance each evening, only to open it and reform the entrance proper in the morning. In certain species the entrance is ornate, with interesting radiate or other patterns (Kerr, et al., 1967); see Figures 18.5, 29.4, 29.5, 29.6, and 29.7. In some species the entrance tube is continued internally far into the nest cavity, apparently as a device to facilitate the movement of returning foragers directly to the vicinity of the food storage or transfer areas.

4. Pillars and connectives

Inside the nest cavity no structures are free; just as in *Apis* nests, everything is firmly attached to something else with propolis or cerumen. In many species of Meliponini such attachment, as well as support of various nest structures, is achieved by means of conspicuous pillars and connectives of cerumen. They are usually rather soft, but close to the batumen, and especially at the bottom of the nest, pillars and connectives are often thickened and brittle.

5. Food storage

In contrast to *Apis,* in nests of Meliponini honey and pollen are stored in pots quite different from the brood cells (Figures 29.1, 29.4, and others) and made of soft cerumen. These pots are located in clusters outside of the

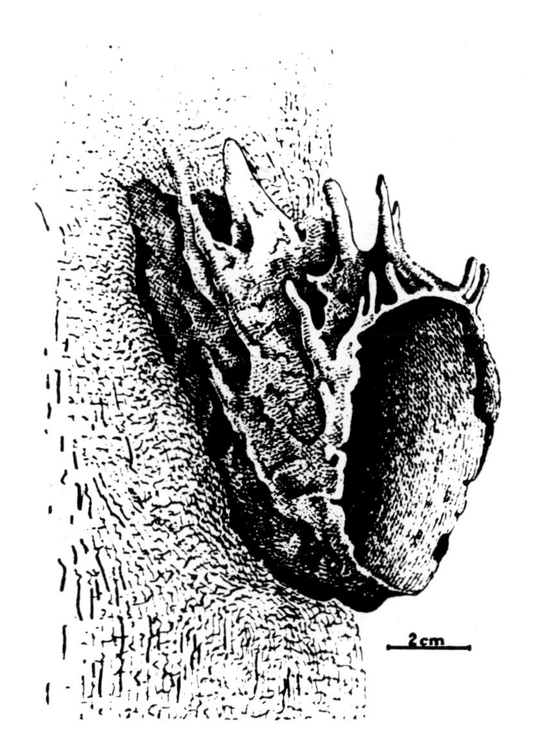

FIGURE 29.5. Nest entrance of *Trigona* (*Tetragona*) *lurida* projecting from the outside of a tree trunk. (From Kerr et al., 1967.)

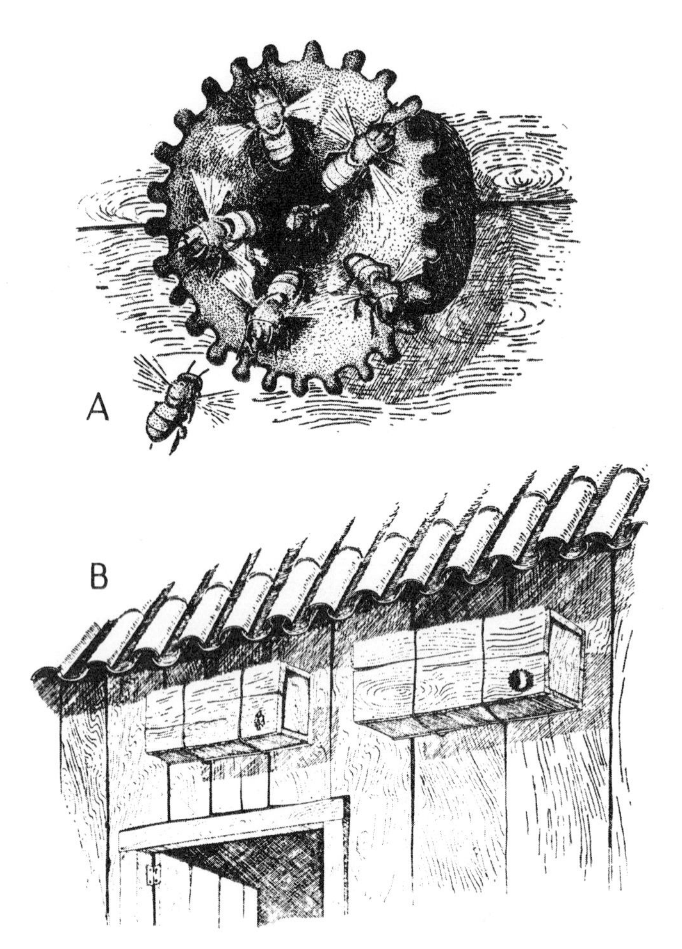

FIGURE 29.6. (*A*) Nest entrance of *Melipona seminigra merrillae.* (*B*) Rustic hives of *M. compressipes manaosensis* and *M. s. merrillae* on the side of a rural house in the state of Amazonas, Brazil. (From Kerr et al., 1967.)

brood chamber. Sometimes honey pots are segregated from pollen pots; in other species they are intermixed. In a few species pots of two distinct forms are made, those for honey being of the usual oval or almost spherical form, those for pollen being elongate, vertical tubes or cones (Figure 29.12). Such tubes or cones are known for only a few species, all American and belonging to a group of *Tetragona* called *Frieseomelitta,* for example, *Trigona silvestrii,* in which the pollen storage tubes are 10 to 20 cm high and resemble miniature organ pipes. Storage pots of the usual form vary from 55 × 48 mm, for a large pollen pot of *Melipona nigra,* or 40 to 50 mm in diameter for pots of *Trigona beccarii,* to subspherical pots only 3.5 mm in diameter for *T. schrottkyi.* In any nest the pots vary considerably in size; it is not uncommon to see small pots only half as large as the larger ones in the same cluster or at least in the same nest.

As explained in Chapter 19 (C), in *Lestrimelitta* there is no separation of pollen and honey pots, all the provisions that are stored consisting of a mixture of the two materials. This is easily understood, since the species of *Lestrimelitta* are robber bees that obtain their provisions from the nests of *Trigona* and *Melipona.* The pollen, as they find it compacted in the nests of their hosts, is not loose and subject to being carried in the usual way in

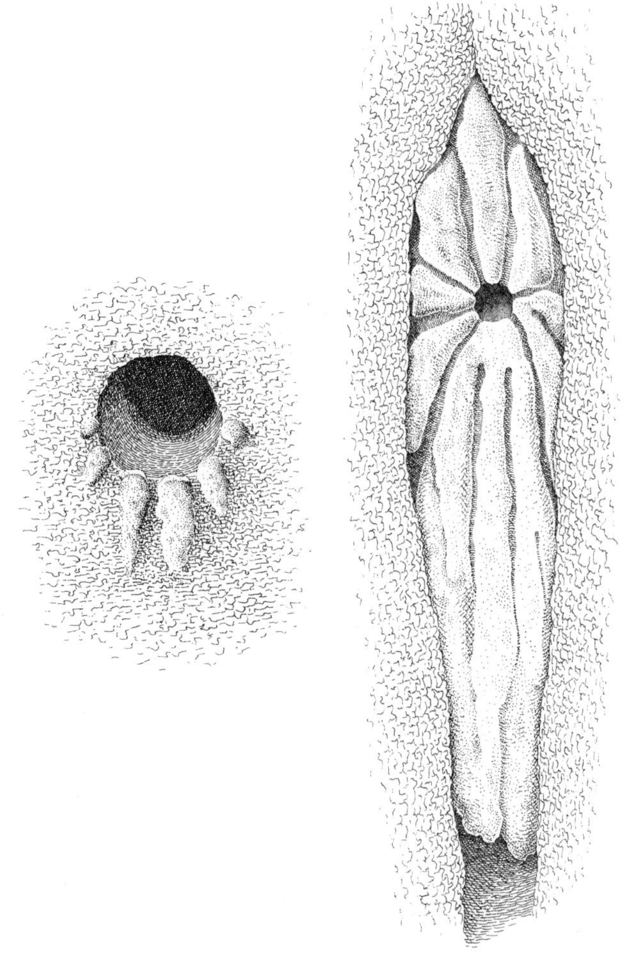

FIGURE 29.7. Nest entrances of *Melipona interrupta grandis* (*left*) and *M. fuscata melanoventer.* (From Camargo, 1970; original drawings provided by J. M. F. de Camargo.)

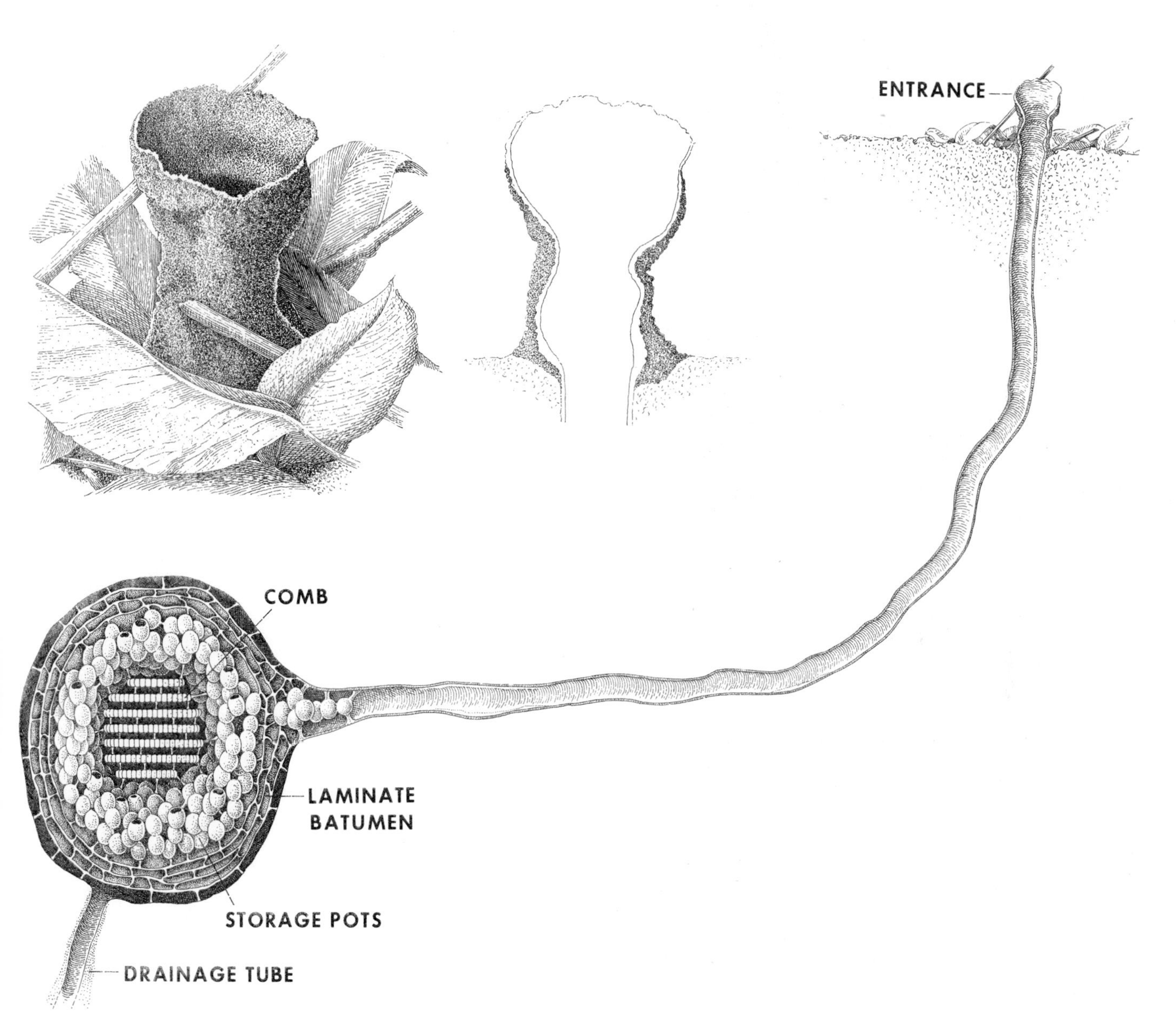

FIGURE 29.8. Nest of *Trigona* (*Trigona*) *recursa* in the soil, with details of the entrance above. Note that there is no involucrum between the brood chamber and the storage pots. (From Camargo, 1970; original drawing provided by J. M. F. de Camargo.)

corbiculae (which in any event are absent). Pollen is therefore carried in the crop, along with honey or brood-cell provisions taken from the nests of the host, and the mixture is placed in the storage cells in the *Lestrimelitta* nests.

6. Brood chamber

The heart of a meliponine nest is the brood chamber, the cavity containing the brood cells. In many species it is surrounded by an involucrum, that is, a sheath of one to several perforated layers of soft cerumen separating the brood chamber from the rest of the nest (Figures 29.4, 29.9, 29.10, 29.11, and others). As Kerr and his associates (1967) and Wille and Michener (in press) show, involucra are occasionally absent in subgenera or species that normally construct them (Figure 29.8).

In most species the cells are arranged in horizontal combs (Figures 29.11, 29.12c, and so on) which may be connected to form a spiral (Figure 29.13). The cells are vertical and open upward, and are made of soft cerumen. As mentioned elsewhere, they are mass provisioned, and each is closed after the egg is laid in it. The egg stands erect on the provisions, which almost always consist of two recognizable layers, an upper, clear layer and a lower, firmer and much thicker layer containing pollen (Figure 29.14). Kerr and Laidlaw (1956) report that provisions consist of glandular secretions of the workers mixed with about 16 percent pollen and 8 percent honey; they are ordinarily quite fluid, so that one wonders how the egg can remain erect. The percentage of pollen is probably much higher, however, in *Meliponula bocandei,* an unusual African species in which the provisions in the brood

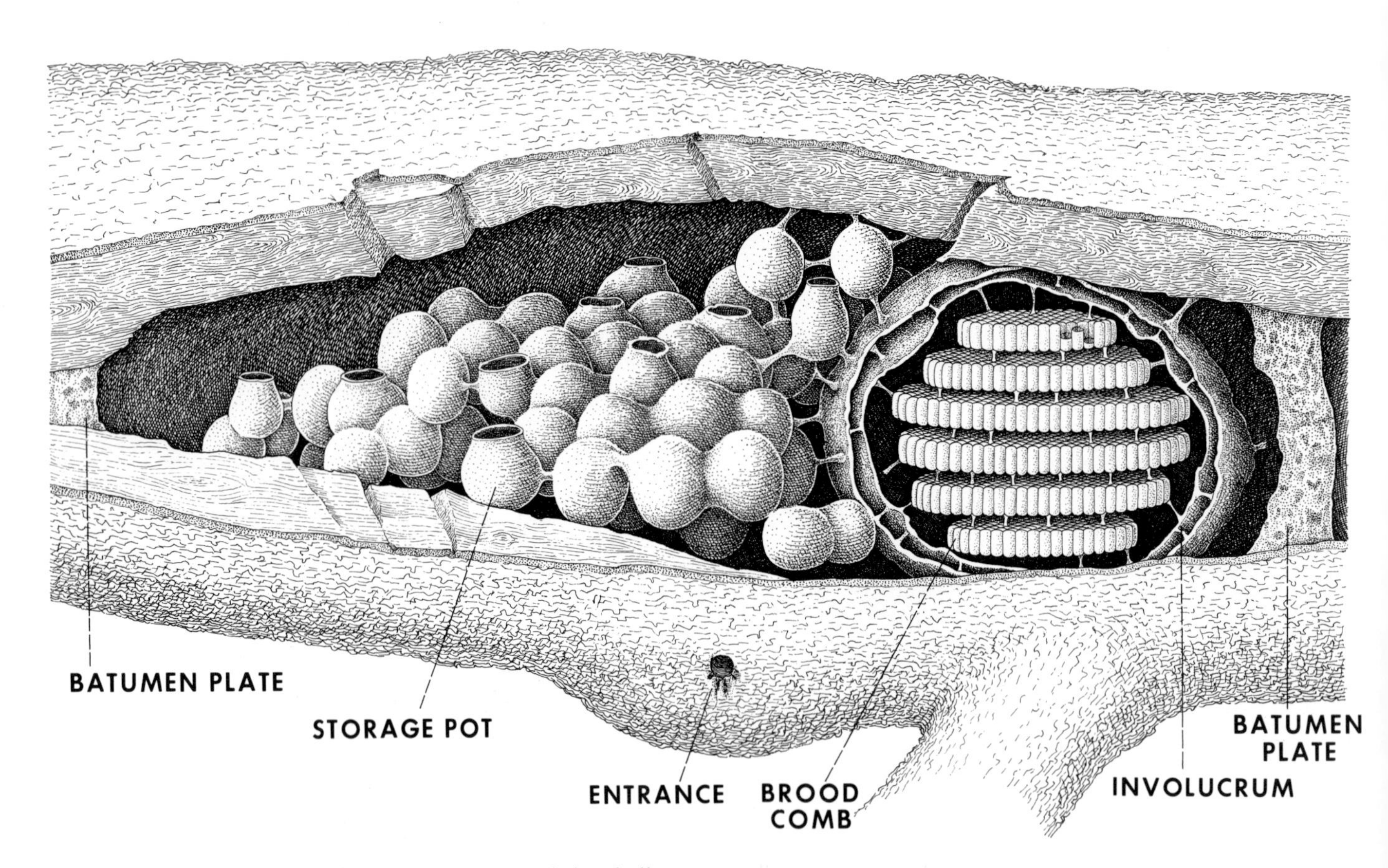

FIGURE 29.9. Nest of *Melipona interrupta grandis* in a hollow branch. (From Camargo, 1970; original drawing provided by J. M. F. de Camargo.)

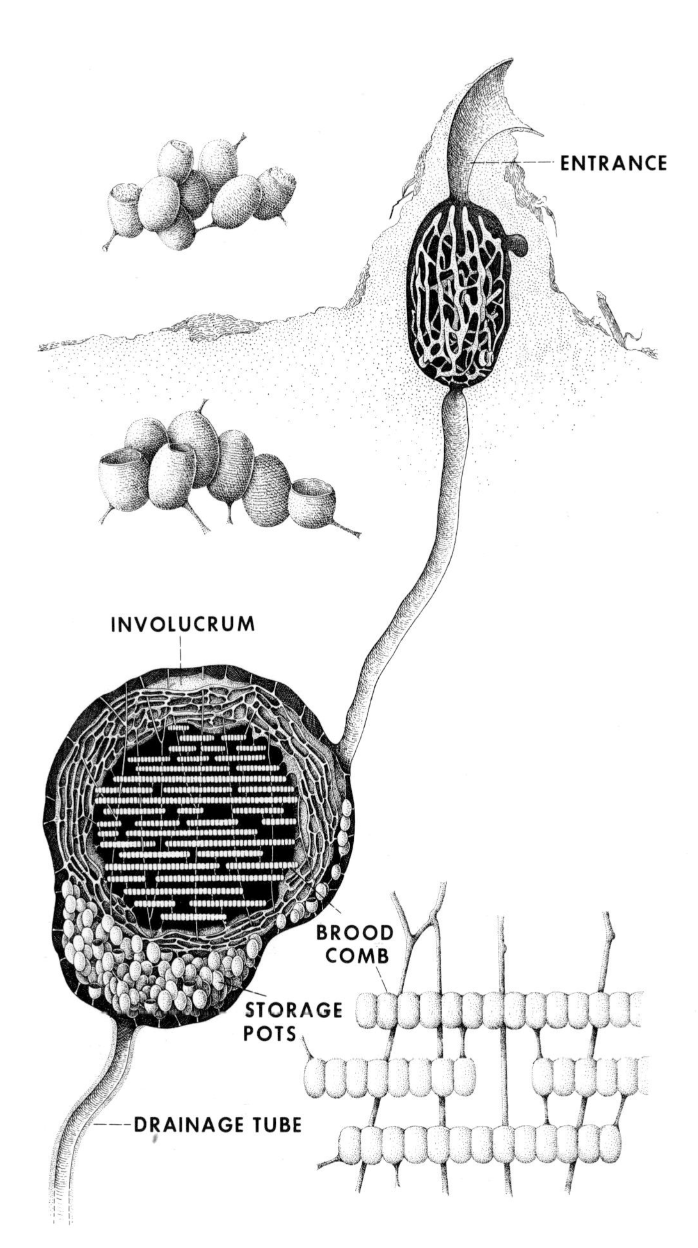

FIGURE 29.10. Nest of *Trigona* (*Partamona*) *testacea* in the soil, with details of storage pots above and of brood comb below. (From Camargo, 1970; original drawing provided by J. M. F. de Camargo.)

cells are firm and not divided into the usual two layers. As noted in Chapter 8 (A3), two or more eggs are sometimes left in the same cell, but in such cases only one bee, probably normally a male, survives.

As soon as they are mature, the larvae spin cocoons inside their cells. Immediately thereafter, as in *Bombus* but unlike other bees, the workers remove the wax from the cocoons wherever they can get at it, leaving the cocoons attached to one another in a layer or comb by the wax which is between them (Figure 16.6). In some species the scattered pillars that support the combs of cells are also removed, so that the combs of cocoons are supported mostly or entirely by their margins.

The cells are constructed at the advancing edges of the newer combs (Figure 29.15), which together form an advancing front of brood cells. A new comb is started near the center of the preceding one or by a spiral continuation. New cells are constructed not only at that point, however, but also at the margins of the one, two, or sometimes three preceding combs, until they reach the diameter permitted by the brood chamber. Since the advancing front of new cells always progresses upward, it soon reaches the top of the brood chamber. By this time, however, bees are emerging from cocoons in the lower part of the chamber; these are then cleared away, and a new advancing front develops there (Figure 29.15,

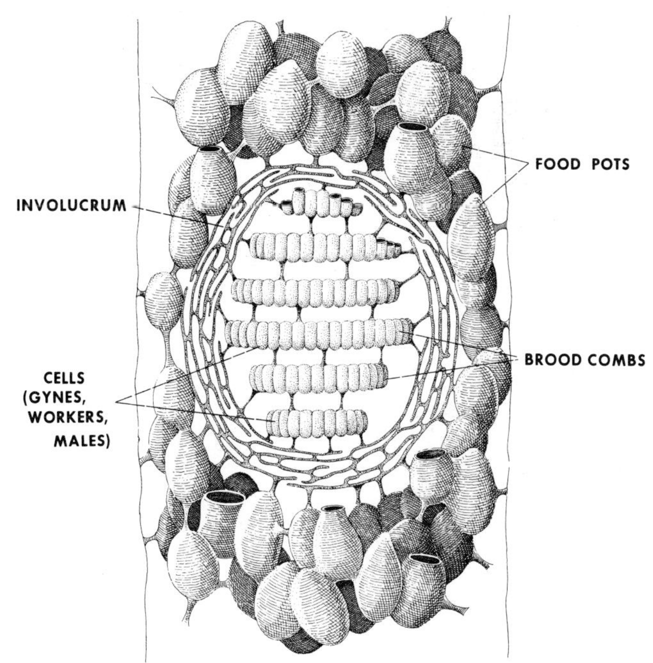

FIGURE 29.11. Diagram of nest of *Melipona*. (Original drawing by J. M. F. de Camargo.)

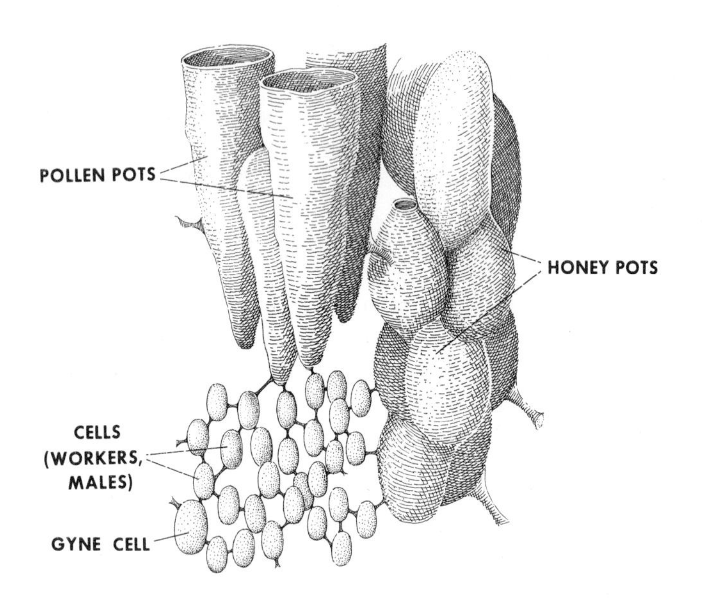

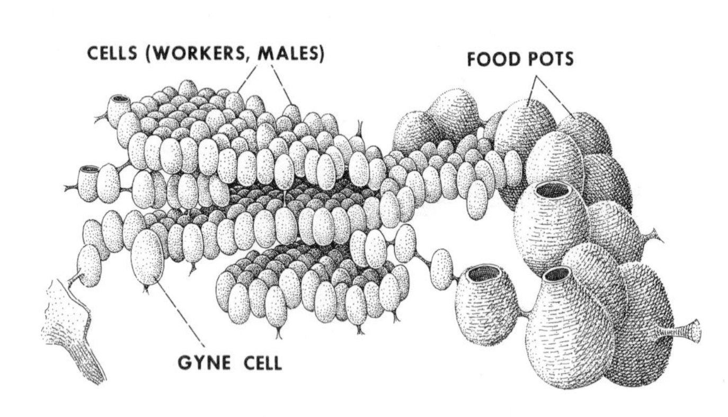

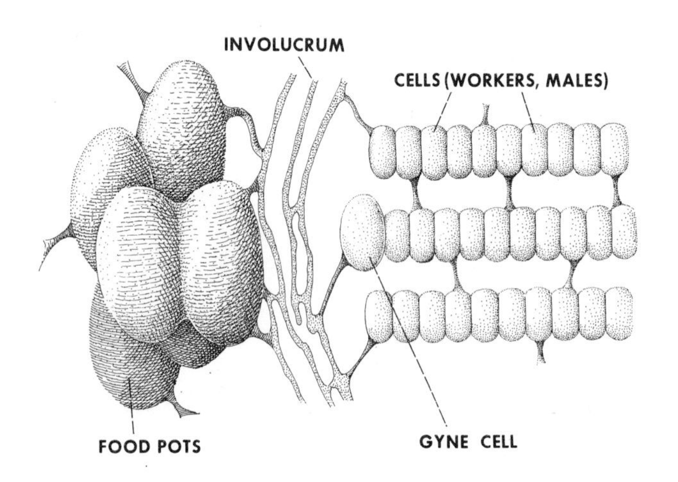

FIGURE 29.13. Brood comb of *Trigona* (*Oxytrigona*) *tataira* removed from the nest, showing the spiral type of comb development found in some species. (Photograph by A. Wille.)

right). In view of the location in which new cells are constructed, it is obvious that adults would appear first in the center of the lowest level of the brood, and then, progressively, outward and upward. The result is that the lower combs become rings, with a cavity resulting from removal of abandoned cocoons inside; then they disappear as the cavity grows outward and upward through the brood chamber. New pillars are constructed as needed to support the cocoons, which are often left temporarily almost free by emergence of bees and removal of adjacent cocoons.

Although the arrangement of cells and combs described above is by far the most common in the Meliponini, it is not universal. One of the most distinctive features of *Dactylurina* is that its combs are vertical, double-layered, with horizontal cells opening on each side (Figure 29.16). This is a remarkable parallelism to the cell arrangement used by *Apis.*

Several apparently unrelated subgenera of *Trigona* and

FIGURE 29.12. Diagrams of brood arrangement and storage pots of *Trigona*. *Above: Trigona* (*Tetragona*) *flavicornis,* with brood cells in a "cluster" arrangement, without involucrum, and with storage pots of two types. *Center: Trigona* (*Tetragona*) *ghilianii,* with cells in irregular combs and without involucrum. *Below: Trigona* (*Trigona*) sp., with cells in regular combs and with an involucrum. (Original drawings by J. M. F. de Camargo.)

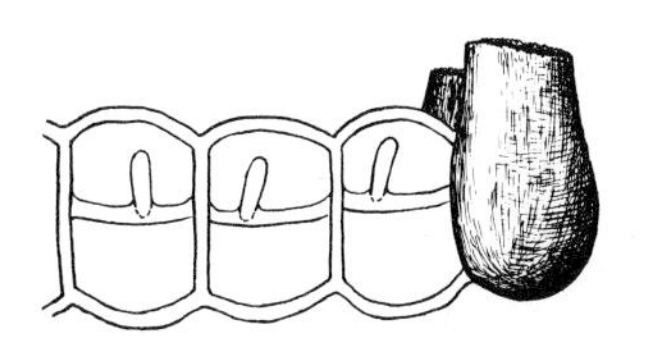

FIGURE 29.14. *Left:* Three brood cells in a comb of *Trigona* (*Tetragona*) *carbonaria* in sectional view to show provisions and eggs; new brood cells, not yet provisioned, extending above level of closed, completed cells. *Right:* Brood cells of *Trigona* (*Plebeia*) *australis,* one opened and one in sectional view to show eggs, the uppermost ready to be provisioned. (Both from Michener, 1961a.)

Lestrimelitta (Table 29.1) consist of species or contain certain species which construct their cells in clusters instead of in horizontal combs. *T.* (*Plebeia*) *australis,* from Australia, has this habit (Figure 29.17). Moreover, its cells are spherical rather than elongated (Figure 29.14). New cells are added to the outside of the cluster. Ultimately the cluster becomes hollow due to emergence of the bees and destruction of the cocoons in the center, at which point new cells are begun in the middle of the cluster (Figure 29.18). Spaces among the cells allow access of the bees to the entire cluster. The closely related *T.* (*P.*) *cincta* of northern Australia and New Guinea has similar cell clusters, except that there is a tendency for the cells to form layers which are irregularly spherical and concentric. These species, belonging to the subgenus *Plebeia,* may show a primitive cell arrangement.

Probably not all cluster arrangements of cells in the Meliponini are primitive, however. Clustered cells are made by various genera and subgenera. If all were primitive, then essentially identical combs must have arisen repeatedly. It is easier to imagine that comb arrangements have repeatedly broken down to form clusters. This view is supported by the observation that, except for Australian *Trigona australis* and *T. cincta,* cells of cluster makers are usually vertically elongated, suggesting those of comb-making species. For the most part, cluster makers nest in irregular or small cavities, like the minute species of the subgenera *Trigonisca* in the Americas, which construct nests in small branches and lianas, and some

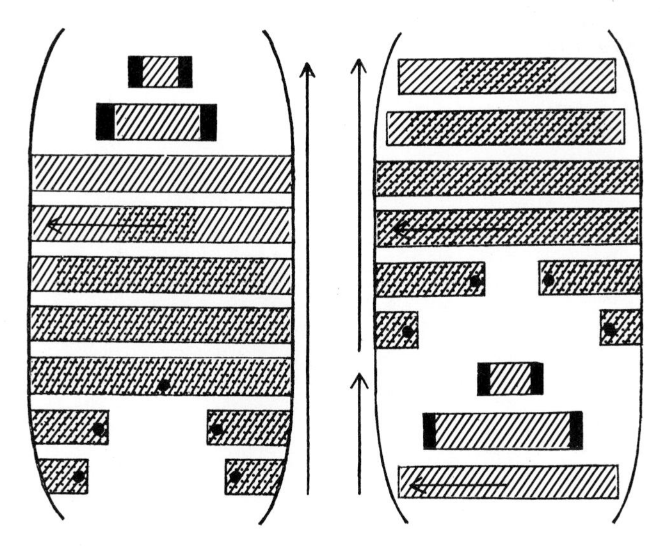

FIGURE 29.15. Diagrammatic vertical sections through the brood of *Trigona* (*Tetragona*) *carbonaria. Black,* the advancing fronts of new cells. *Shaded,* brood; *stippled areas* indicate cocoons are present. *Small black discs* mark the places where young adults are emerging and cocoons (and combs) are being destroyed. This process makes space for the growth of the advancing fronts. *Arrows* indicate progression from older to younger within the brood. (After Michener, 1961a.)

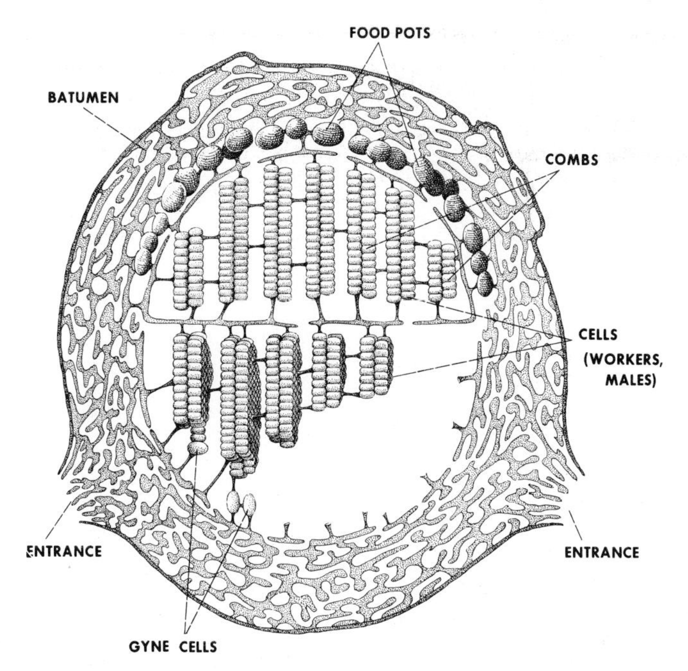

FIGURE 29.16. Vertical section of a nest of *Dactylurina staudingeri,* an African relative of *Trigona* subgenus *Tetragona.* (Original drawing by J. M. F. de Camargo; modified from Darchen and Pain, 1966.)

FIGURE 29.17. *Trigona* (*Plebeia*) *australis* from Australia. *Top row:* Cell cluster (note the spherical cells opening in different directions), storage pots, and old, pitted storage pots. *Bottom row:* Brood chamber largely surrounded with involucrum, with involucrum removed, and cocoons (queen cocoon at right center). (From Michener, 1961a.)

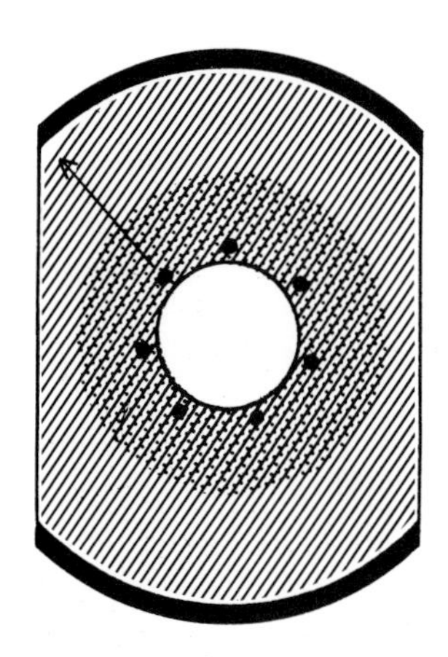

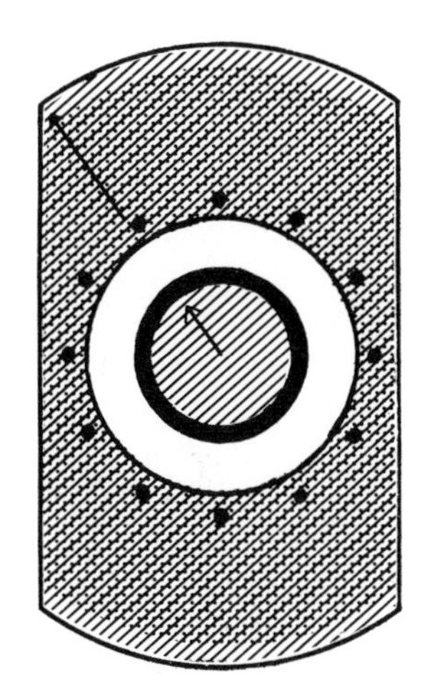

FIGURE 29.18. Diagrammatic sections through the brood of *Trigona* (*Plebeia*) *australis*. The symbols are the same as for Figure 29.15. (After Michener, 1961a.)

species of tropical Asia and the East Indies—like *Trigona* (*Tetragona*) *iridipennis*—which often nest in small irregular cavities. *T.* (*T.*) *clypearis* and *fuscobalteata*, as indicated in section A above, often makes nests spread out to occupy the small space between two boards in wooden buildings (Figure 29.19). A single layer of cells is constructed in such a space, but the nest may extend for a meter or more. Such species have abandoned not only the comb-type arrangement of cells but also the involucrum, the brood cells and the storage pots being in the same space. Presumably these are adaptations to the use of slender or irregular cavities. The African *T.* (*Hypotrigona*) *braunsi* and *gribodoi* typically nest in such irregular cavities in tree limbs or trunks (Pooley and Michener, 1969), and if experimentally given plenty of space, they nonetheless disperse clusters of storage pots and cells, just as in nature (Figure 29.20).

A feature of the comb-making species which such cluster makers retain is advancing fronts of new cells which move upward through the brood chamber. There may be only one such front or, because of irregularity of the nest cavity, there may be several. Rarely, even in a large and regular brood chamber, there is more than one advancing front. For example, four such fronts were noted in a nest of *T.* (*Tetragona*) *hockingsi* in northern Australia. In this species the cells are arranged in a basically cluster pattern but joined to form irregular horizontal combs (Figure 29.21), yet it is a close relative of *T. carbonaria*, which makes very regular combs. The queen must move about the nest from one advancing front to the other in order to provide eggs to all of the new cells. Of the various types of cluster makers, some make widely separated cells, as does *T.* (*T.*) *flavicornis*, shown in Figure 29.12. Others make cells mostly fused in groups; sometimes the groups are comblike patches, as in *T. hockingsi*, or almost complete combs, as in *T.* (*T.*) *ghilianii*. In other species

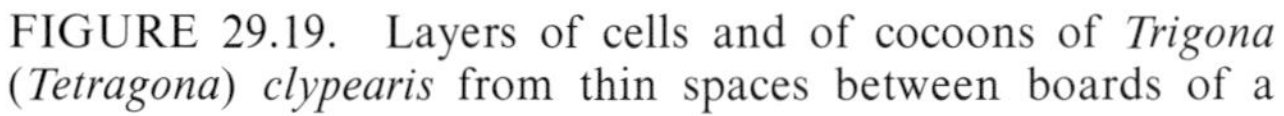

FIGURE 29.19. Layers of cells and of cocoons of *Trigona* (*Tetragona*) *clypearis* from thin spaces between boards of a building in New Guinea. (From Michener, 1961a.)

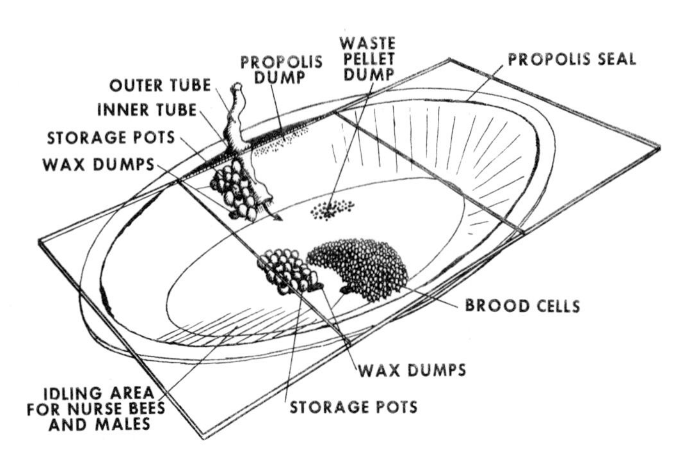

FIGURE 29.20. Observation nest for the African *Trigona* (*Hypotrigona*) *braunsi* made of a basin and glass cover. (From Bassindale, 1955; courtesy of the Zoological Society of London.)

like *T.* (*T.*) *iridipennis* of southern Asia, the cells are mostly fused, but not in flat groups suggesting combs. Thus there exists a variety of kinds of intermediates between the cluster and comb arrangements.

As has been explained in Chapter 10 (C3), in *Melipona* the cells are all of the same size, and young queens are produced in large numbers, most of them being killed by the workers. In all of the other genera, while male and worker cells are of the same size, queen cells are very much larger. In these bees queens are produced infrequently as compared to *Melipona.*

7. *The relation of architecture and classification*

Table 29.1 presents data drawn from Wille and Michener (in press) which show that few of the more striking architectural features make good sense taxonomically. Character 2a (brood cells horizontal, in double vertical combs) characterizes only *Dactylurina,* but it has to be noted that this character is one of the chief reasons that *Dactylurina* is not included in *Tetragona.* Character 4, the size of gyne cells, separates *Melipona* from all other genera. Most of the characteristics, however, seem to ignore the classification. This is presumably because of independent adaptation of the various groups to the conditions in which they live. For example, as indicated in section 6 above, brood cells occur in clusters (character 2b) in presumably primitive forms as well as in various derived forms where the comb arrangement has been broken down to accommodate the nest to slender, flattened, or small sites. The cells are cylindrical (character 1) in species that crowd them into combs, and more or less so in those derived from comb makers. The elaboration of protective layers of involucrum and batumen (characters 6 and 7) probably depends on the ability of a species to withstand temperature fluctuations as well as on the sites that it selects for nests. It is therefore not surprising that these characters vary within subgenera and genera as much as between them. Of course one might claim that errors in classification are responsible for the lack of agreement between architecture and classification. However, the latter is based on many morphological features; no classification based on the architectural features listed would be concordant with one based on the morphology of the adult insects.

C. *Mating*

Mating in both *Trigona* and *Melipona* occurs outside of the nest and presumably in flight. Males often form swarms in the air, frequently near nests or near sites being

FIGURE 29.21. Brood of *Trigona* (*Tetragona*) *hockingsi*—an intermediate between the cluster and the comb arrangement. (From Michener, 1961a.)

TABLE 29.1. Some characteristics of architecture of the various groups of Meliponini.

Architectural characteristic	*Meliplebeia*	*Plebeia*	*Nogueirapis*	*Axestotrigona*	*Hypotrigona*	*Trigonisca*	*Scaura*	*Partamona*	*Paratrigona*	*Scaptotrigona*	*Nannotrigona*	*Cephalotrigona*	*Oxytrigona*	*Tetragona*	*Trigona* s. str.	*Dactylurina*	*Cleptotrigona*	*Lestrimelitta* s. str.	*Meliponula*	*Melipona*
1. Brood cells cylindrical	+	+,±,−^{r}	+	+	±	±	+,−	+	+	+	+	+	+	+,±	+	±	±	+	±?	+
2a. Brood cells in double vertical combs	−	−	−	−	−	−	−	−	−	−	−	−	−	−	−	+	−	−	−	−
2b. Brood cells in clusters	−	+[a]	−	−	+	+	+	−	−	−	−	−	−	+[c]	−	−	+	−	−	−
2c. Brood cells irregularly layered (various ways)	−	+	−	−	+^{r}	−	−	−	−	+^{r}	−	+^{r}	+^{r}	+	+	−	−	−	+	−
2d. Brood cells in horizontal combs (s = joined in spiral)	+,s	+[b],s^{r}	+	+,s	−	−	+	+	+	+,s^{r}	+	+	s	+,s	+,s	−	−	+	−	+,s^{r}
3. Pillars through several combs	−	−	−	−	−	−	−	+	−	−	−	−	−	−	+,−	−	−	−	−	−
4. Gyne cells large	+	+	+	+	+	+	+	+	+	+	+	+	+	+	+	+	+	+	+	−
5. Honey and pollen pots differently shaped	−	−	−	−	−	−	−	−	−	−	−	−	−	−,+	−	−	−	−	−	−
6. Involucrum present	+	+,−	+	−,+	−	−	−	+,−^{r}	+,−^{r}	+	+	+	+	−,+	+,−^{r}	+	−	+	+	+,−^{r}
7. Multiple layers of involucrum and batumen	+	−,+^{r}	+	−,+	−	−	−	+	−,+	+	+,−	+	−,+^{r}	−,+	+,−	+	−	+,−	+,−	+,−
8. Nest exposed, not in cavity	−	−	−	−	−	−	−	−,±	−,+^{r}	−	−	−	−	−	+,−	+	−	−	−	−

Note: The superscript *r* marks rare alternatives when there is variation in a characteristic; + indicates agreement with statement at left; − indicates disagreement; ± indicates an intermediate condition.

[a] Cells occur in clusters in some species of both Australian and American groups of *Plebeia.*

[b] Most species of *Plebeia* have cells in separate horizontal combs.

[c] Certain of the Indoaustralian *Tetragona* have brood cells in clusters, as do all species of the American group called *Frieseomelitta.* Other species in both areas make combs.

developed for new nests. These male swarms, sometimes consisting of hundreds and perhaps even thousands of bees, are the counterpart of the areas in which male *Apis* fly about, but are far more localized and the bees much less fast-flying than in *Apis*. Probably when a virgin queen enters such a swarm, mating occurs quickly. As in *Apis* and unlike all other bees, a queen returning from a mating flight carries the male genitalia or "mating sign" on the end of her abdomen, but unlike *Apis* she mates only once (Kerr, Zucchi, Nakadaira, and Botolo, 1962). Silva, Zucchi, and Kerr (1972) found that queens, at least in *Melipona,* cannot remove the male genitalia until they return to the nest, where they can rub against the comb surfaces.

In contrast to nearly all that has been said in the preceding paragraph is the record by Sakagami and Laroca (1963) of a male of *Lestrimelitta ehrhardti* mating with the gravid, laying queen inside the nest (Figure 29.22). If this represents normal behavior, it is a noteworthy finding, for it implies repeated matings during the egg-laying life of the queen, as in termites. That the behavior was neither an isolated abnormality nor something unique to *Lestrimelitta* is suggested by the fact that a gravid female of *Trigona postica* offered experimentally to males swarming outside of a nest entrance elicited mating attempts. Indeed it seems likely that disturbance of a colony sometimes leads to mating by the gravid queen (Sakagami, personal communication). It is at least evident that gynes, unlike those of *Apis,* need not be flying high in the air to attract the attention of males, and C. A. de Camargo (1972c) has shown that virgin gynes of *Melipona quadrifasciata* will mate in a small plastic box.

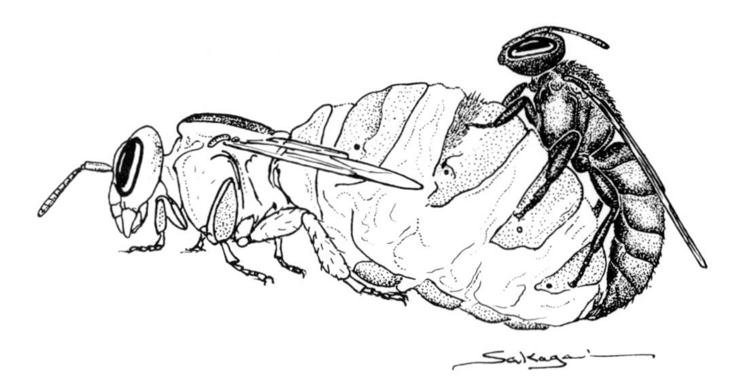

FIGURE 29.22. Gravid queen of *Lestrimelitta ehrhardti* mating with male. (From Sakagami and Laroca, 1963; original drawing provided by S. F. Sakagami.)

D. Importance to man

Being extremely common over enormous areas in tropical America and in southeast Asia, stingless bees must be among the major pollinators in the tropics (Nogueira-Neto, 1970a). Because tropical agriculture has not received the attention that has been given to temperate agriculture, little concrete evidence of the importance of these bees as pollinators is available. However, circumstantial evidence from reductions and failures of certain crops in Central and South America quite clearly indicates the value of these bees. The conservation of limited areas of forest permits survival of many species even in largely agricultural areas, and is important to enhance the value of those areas.

In some Meliponini considerable supplies of honey and pollen are stored. Nests of *Melipona* sometimes contain about 2 kg of honey. Although this is a small quantity compared to the amount found in a thriving *Apis* colony, it is nonetheless a significant reservoir of sweets. Peoples in most tropical American areas rob meliponine nests for the honey as well as for wax, and often for the edible larvae. In the American tropics nests are often kept as hives (Figure 29.6) and, particularly in Yucatan, apiaries of *Melipona beecheii* hives, each in a section of hollow log, are kept by the Indians. The flavor of the honey of Meliponini varies greatly, from acid, watery, and unpleasant to at least as good as that of *Apis,* depending on the species of bee as well as the kinds of flowers used by the bees. Some of the poorest honey is made by those species of the subgenus *Trigona* that make exposed nests, while excellent honey is made by *Melipona* as well as by some *Trigona,* including some common small species such as *T. jaty.*

Conversely, some species of the subgenus *Trigona* are injurious, as indicated in section B1 above, in biting the bark and buds of certain trees to obtain the exudate for nest construction.

30 *True Honeybees*

The true honeybees all belong to the genus *Apis.* Like their relatives, the stingless honeybees, they are all highly social, and thus always live in colonies. They differ from stingless honeybees in some extremely elaborate specialized features, such as the communicative dances and the strangely modified male genitalia, which are unique in the whole order Hymenoptera. At the same time, the genus *Apis* possesses other features that are more primitive than the equivalent features of the stingless bees. Examples are the thoroughly functional sting and the complete venation of the wings. A comparison of the social organization of *Apis* with that of stingless bees is given by Sakagami (1971).

More has been written about the common honeybees, *Apis melifera,* than any other insect. Tens of thousands of titles on the subject are known from the earliest writings down to the present time. A wealth of books exists on beekeeping—how to do it in each of many climatic and vegetational areas, how to make money from honey, wax, royal jelly, and even bee venom, and how to serve agriculture and also make money by renting hives of bees as pollinating agents. A few major books on beekeeping in various areas are those of Grout (1963), Ordetx and Espina (1966), Singh (1962), and Smith (1960). There are also many beekeepers' journals, the key one in English being *Bee World,* the key abstracting journal being *Apicultural Abstracts,* both published by the Bee Research Association in Britain.

Many important books on *Apis mellifera* emphasizing its biology rather than practical value have also been published, of which the following are some that will prove useful: Butler (1949, 1954a), Buttel-Reepen (1903, 1915), Chauvin (1968), Frisch (1950, 1953, 1955, 1965, 1967a), Lindauer (1961), Ribbands (1953), and Zander (1964). Anatomical and physiological accounts of *Apis* include Chauvin (1968), Nelson (1915), Snodgrass (1925, 1942, 1956), and Zander (1951). Relatively little has been published on the other species, but bibliographies of *A. florea* and *dorsata* have been compiled by Morse (1970).

A. The species and races of honeybees

The genus *Apis* is native to Eurasia and Africa. Before European settlement it was not found anywhere in the Western Hemisphere or in Australia or the Pacific except for continental islands such as Japan, Formosa, the Philippines, and Indonesia. There are probably only four, or perhaps five or six species of *Apis,* although Maa in a revisional study recognized twenty-four. In any one species group, however, most of his "species" replace one another geographically and are therefore probably only geographical races; his groups (that is, his genera and subgenera) probably more nearly correspond to species in the usual sense. The species and races of the genus were most recently reviewed by Ruttner (in Chauvin, 1968). The best-known species is the common honeybee, *A. mellifera* (Figure 30.1), which has been introduced to most parts of the world as a honey and wax producer and pollinator. It occurs as a native in Europe, western Asia, and throughout Africa except for the great desert areas. Several of the local variants or subspecies have been domesticated and, along with hybrids of various sorts, constitute the strains or races used in apiculture. It is these strains of European origin that have been spread over the world by European man.

FIGURE 30.1. A honeybee of the South African race *Apis mellifera capensis* collecting pollen from a flower of the family Compositae. (Photograph by E. S. Ross.)

The closest relative of *Apis mellifera* is *A. cerana,* the southern form of which is often known as *A. indica. A. cerana* occurs in southern and eastern Asia, from Ceylon and India to China and Japan, and southeast to the Moluccas. The other two types of honeybees, each strikingly different from *A. mellifera* and *A. cerana,* are the small *A. florea* and the large *A. dorsata.* Each is restricted to southern Asia, including adjacent islands such as Ceylon, the western parts of the East Indian Archipelago, and the Philippines; *A. dorsata* reaches Timor, but no species of *Apis* exists as a native as far east as New Guinea or Australia. *A. florea* ranges west to Iran. A species named by Rayment from southeastern Australia appears to have been either imaginary or a result of misunderstanding of local oral accounts. *A. florea* and possibly also *A. dorsata* and *A. cerana* may be divisible into two or possibly more distinct species, but a careful new revisional study will be necessary to establish this. I have seen male specimens of the *A. florea* group that seemed to belong to two different species (*A. florea, A. andrenoides*). The problem is that males are not available from enough populations to make a proper study worthwhile at this time; the most distinctive of the supposed differentiating characters are those of males.

There are good reasons to believe that several characteristics of *A. florea* are ancestral. For example, its communicative dances occur on a horizontal surface, are oriented by the sun, and the straight runs are toward the food source (Chapter 15, D3). Moreover, the relation between its orientation to light and its type of orientation to gravity is as in many other bees but differs from the other three species of *Apis*. Thus it seems probable that an ancestral *Apis* species gave rise to two phyletic lines, one leading to *A. florea*, the other to the remaining species of the genus. The latter line evidently divided again, giving rise to (1) an ancestor of *dorsata* that retained the single, exposed nest comb and danced in the light, as does *dorsata*, and (2) an ancestor of *cerana* and *mellifera* that constructed multiple combs in dark cavities and danced in the dark.

Each of the species varies strikingly over its range, suggesting the adaptation of each local population to the conditions under which it lives. The variations involve not only color and details of structure, but also social behavioral attributes, some of which, in the case of *Apis mellifera*, turn out to be of great practical importance to beekeepers. An example of a structural difference of some importance in that species is tongue length, which varies in a geographical gradient across Russia. Tongue length influences types of flowers which bees use, some tubular corollas being too deep for bees with short tongues.

Behavioral differences tend to be more interesting, however. Some races of *Apis mellifera*, for example the African and Caucasian subspecies (*A. m. adansonii* and *A. m. caucasica*), tend to use large quantities of propolis. This resin so gums the frames to the hive as to make rapid manipulation on a commercial scale difficult and expensive. For this reason, although it is in many ways an excellent bee, good for hobby beekeeping because of its gentleness, the Caucasian race is little used commercially in the United States. Aggressiveness, that is, readiness to attack and sting, varies widely among races and also among individual colonies. Some strains of the African race are excessively aggressive, German and English bees (*A. m. mellifera*) are also quite ready to sting, while Italian bees (*A. m. ligustica*), the chief commercial race in most areas, are more gentle although not so calm as the Caucasians. Races also vary in such strange behavioral features as the extent to which they run about on the combs when a hive is being manipulated. Bees that run a lot are inferior from the viewpoint of beekeepers because the bees may fall in groups from the bottom edge of a comb or form festoons there. One result is that bees get crushed when the comb is returned to the hive. The extreme of this sort of behavior is found in some colonies of the African race, in which bees not only run on combs but take flight.

Races of *Apis mellifera* also differ in their susceptibility to certain bee diseases (Chapter 18, B), and in quantitative aspects of their communicative dances (Chapter 15, D3). The Cape race from South Africa (*A. m. capensis*) differs from others in the frequency with which workers lay diploid, female-producing eggs (Chapter 8, A1). Thus the races of the honeybee, which are morphologically much alike, differ in a striking variety of behavioral and physiological attributes.

A race that is currently causing concern in the Western Hemisphere is the African *Apis mellifera adansonii*. In 1956 African strains were carried to Brazil in an effort to improve the adaptation of the honeybees in Brazil to tropical conditions. African bees appear to produce more honey in the tropics than do European strains. The African bees are said to work earlier and later each day and in poorer weather than do European strains. They have, however, three major drawbacks: they are highly aggressive, stinging readily; they abscond freely; and they form enormous feral populations. The effort to improve the honeybees of South America was a reasonable approach to the problem of developing a good strain of *A. mellifera* well adjusted to the Brazilian tropics. African bees, however, unexpectedly escaped in 1957; we will never know how different the results would have been had selected hybrids been distributed to beekeepers instead.

The twenty-six queens from Africa that absconded with their workers, plus the sperm cells carried in their spermathecae, spawned a revolution in beekeeping in South America which may ultimately involve the whole Western Hemisphere. Because of frequent and severe stinging of animals and people, animal husbandry and public health are also involved. Extremely aggressive *Apis* are spreading widely in South America (Kerr, 1969). They are by now probably all hybrids with the European strains kept earlier in South America, and these hybrids, called the Brazilian honeybee, have replaced the European bees entirely in most parts of a region about as large as the United States (minus Alaska): most of Brazil, Paraguay, Uruguay, and parts of Argentina, Bolivia, and Peru. The aggressive bees make apiculture difficult. There are records of deaths of domestic animals and of man attributable to honeybees, and it would appear that the frequency of such occurrences has increased greatly as a result of the introduction of African genes into South America. Many beekeepers have been forced out of business because of the difficulties of handling the bees and the need to place hives far from people and domestic animals.

Kerr (1967) maintained that through distribution to beekeepers of pure Italian queens, aggressiveness of the population in his area was being reduced while the high honey production of the African race was not much reduced. However, many of the Brazilian beekeepers who have learned to handle their Africanized bees do not favor introduction of more European genetic material because it reduces honey production.

More or less massive stingings (as distinguished from one or a very few stings, serious only to allergic persons) probably result from two quite distinct phenomena. First, bees may be aggressive near their nests and may attack nearby people and animals. Manipulation of hives or inadvertent disturbance of a colony may be preludes to such stinging. Second, swarms or absconding colonies may start to land on a person. If he brushes them off and crushes a few, alarm pheromones will be liberated, and stings may result. Each sting liberates more alarm pheromone, so that a chain reaction and mass stinging can occur. This is an interpretation based on accounts of stingings in Brazil, not on scientific studies.

One reason that such occurrences are more common with the Brazilian honeybee than with European races is the high population of wild colonies attained. In parts of Africa one can put out a box or pieces of hollow tree with the expectation that they will be occupied by swarms of *Apis* within a few days. In South America, whereas the feral colonies of *Apis* were relatively scarce when the honeybees were of the European races, colonies of Brazilian bees are now very abundant, not only in hollow trees but in holes in the ground, in termite nests hollowed by armadillos, and in any such cavities. Kerr (1971) recorded an average of 107.5 nests of Brazilian bees per square kilometer in a savanna area in the state of Mato Grosso. Much of the increase in swarms and absconding colonies seen in Brazil is due to the small cavities often used by this bee. There is evidence from beekeepers' statements that, once established in adequate hives and if provided with adequate food and water, Brazilian bees neither swarm nor abscond more often than European bees.

Brazilian bees have crossed the Amazon and appear to be moving northward at a rate of about 200 miles per year. There is no evident climatic or topographic barrier to prevent their spread into North America. To judge by the distribution of *Apis mellifera adansonii* in Africa and the continuing but slow spread of Brazilian bees in temperate South America, such bees could occupy the southern United States. Because of hybridization and selection as they move, no one can predict the attributes of such bees when or if they reach North America. Even if some of their undesirable features are naturally or purposefully diluted, many beekeepers' practices will need to be altered because of differences in behavior among races of

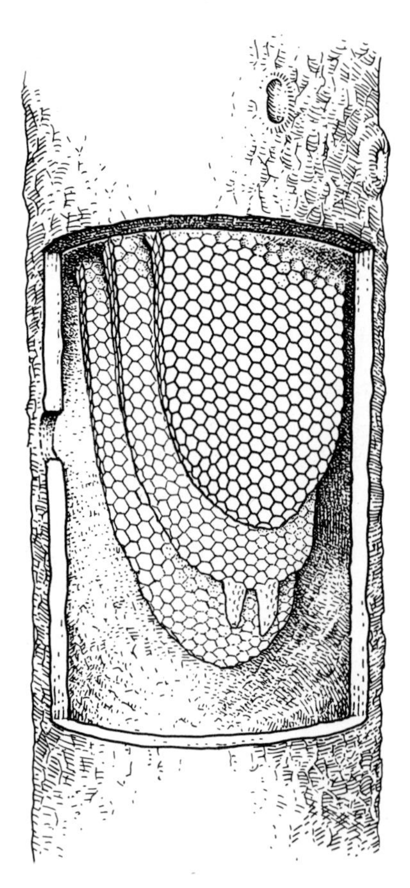

FIGURE 30.2. Nest of the common honeybee, *Apis mellifera,* in a hollow tree. The nest consists of double-layered vertical combs of subhorizontal wax cells (see Figure 30.3), usually supported from above, sometimes also from lateral margins. Two gyne cells hang from the edge of the middle comb. (From Wilson, 1971.)

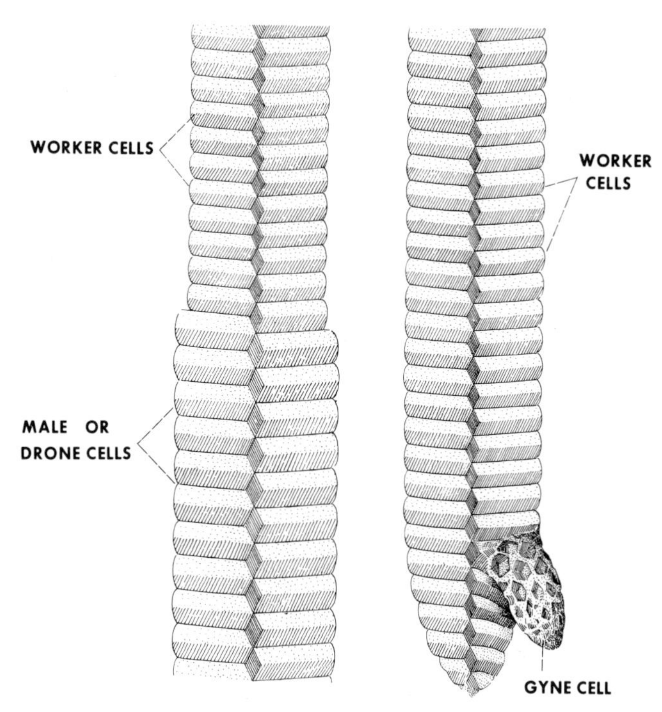

FIGURE 30.3. Sectional views of two combs of *Apis mellifera.* (Original drawing by J. M. F. de Camargo.)

A. mellifera. An account of current practices in Brazil is given by Wiese (1972).

If the Brazilian honeybees arrive in North America still as aggressive as those now spreading in northern Brazil, there will not only be severe problems for the beekeeping industry and problems of stingings, but commercial pollination will also be hampered. Very aggressive bees cannot be placed as pollinators in orchards and fields where people and machines are working (see section J below). The influence on agriculture in North America could be significant, not to mention the effect on people's enjoyment of the out-of-doors.

B. Architecture

As is well known, the combs in the genus *Apis* are vertical and each is made of two layers of wax cells, back to back and opening on opposite sides of the comb (Figures 30.2 to 30.4). Such cells are both for storage of food and for rearing of workers, and while nearly horizontal, they slope upward slightly toward their openings. The cells themselves are elegantly hexagonal and form a beautiful pattern of great uniformity (Figures 30.6 and 30.9) except around the edges or where larger cells are being constructed for queen or male production (Figure 7.2). Unlike the Meliponini, the cells (hence combs) are

FIGURE 30.4. Sectional view of a part of a comb of *Apis mellifera,* showing pupae, a prepupa (*near upper left*), and a large larva (*lower right*), all in capped cells. (Photograph by E. S. Ross.)

FIGURE 30.5. A nest of *Apis dorsata* in the Philippines, showing the normal blanket of bees, somewhat disturbed by the observer. As soon as the blanket is lifted with the paddle, the bees take flight. (Photograph by R. A. Morse.)

used repeatedly, instead of being destroyed and rebuilt with each usage.

The principal material used in construction is wax, which is secreted from between the ventral abdominal segments of the young adult workers. Foreign resins and gums placed in nests of *Apis mellifera* may be incorporated with wax in comb construction, and old wax used in capping cells often contains pollen, propolis, and bits of cocoons, and hence is darker than wax cappings on new comb. *Apis* can thus produce mixtures like the cerumen and pollen-wax mixture of stingless bees and bumblebees, respectively. Presumably the common use of nearly pure wax by *Apis* is a specialization associated with more adequate wax production.

Resins from various plant sources are collected by honeybees and are used by *Apis mellifera* and *cerana* for closing any cracks in the nesting cavity, sticking down any loose objects in the nests, and the like. Resins are also mixed with wax for strengthening around the edges of honeybee combs, and this mixture can be called cerumen, like that used by stingless bees.

In *Apis florea* and *dorsata* each colony constructs a single comb in a situation exposed to the light. *A. dorsata,* the giant honeybee, hangs its combs from overhanging rocks or cliffs, from large, more or less horizontal branches of trees (Figure 30.5 and 30.6), and even from the eves of large buildings. Sometimes numerous colonies nest in the same vicinity, for example, in one

FIGURE 30.6. A small comb of *Apis dorsata,* from which the adult bees have been removed. The lower part of the comb consists of capped brood cells. (Photograph by R. A. Morse.)

tree. A single comb may be two meters long and a meter across. *A. florea,* the little honeybee, also constructs its combs in trees or sometimes hanging from other supports (Figures 30.7 and 30.8), but they are small, the size of a man's palm, and supported by small branches usually in dense vegetation. Such a comb, unlike that of other *Apis,* is so built that its upper part surrounds the support and provides a flat space on top (Figure 30.8), where the dances concerned with foraging can take place (see Chapter 15, D3). Although the nest sites of these species are exposed, the comb itself is normally hidden by a blanket of bees (see Figures 17.3, 30.5, and 30.7) and thus protected from rain, sun, and wind.

Apis mellifera and *A. cerana,* on the other hand, build nests consisting of multiple parallel combs, and while an occasional series of such combs may be constructed by a colony in the open, the typical situation is a hollow tree, a cavity in the ground or in a cliff or bank, or in a box (hive). The combs are constructed in the dark and the nest entrance is a relatively small hole through which the bees pass on their way in and out and which can be readily guarded by the bees. It is a strikingly different situation than the exposed sites of *A. florea* and *A. dorsata.*

As is well known, honeybees construct three kinds of cells. Most of them are the standard hexagonal cells of the comb for food storage or for rearing workers. In *Apis*

FIGURE 30.7. A nest of *Apis florea* on a hut in India. The comb is solidly covered with a blanket of bees. (Photograph by E. S. Ross.)

FIGURE 30.8. A comb of *Apis florea* from which bees have been removed, showing the expanded upper part, on the horizontal surface of which dances occur. (Photograph by R. A. Morse.)

dorsata, those used for storage are sometimes extraordinarily deep, the upper or storage part of the comb being very thick. Similar hexagonal cells are used for producing males. In *A. dorsata* males come from cells scattered among other brood cells, and therefore of the same diameter as the others but longer, so that when capped they project beyond the worker cells. In other *Apis* species the drone cells form separate portions of the comb and are larger in diameter than worker cells (Figure 7.2). Gynes in all species are reared in much larger, irregularly shaped cells, never forming any sort of comb, and typically hanging from the edge of the comb or from the upper margins of any holes in the comb. As explained in Chapter 11 (A), however, gyne cells may be constructed elsewhere in the comb to produce emergency replacement queens. The statement that has often appeared in the literature to the effect that *A. dorsata* does not construct large gyne cells is incorrect. Viswanathan (1950b) and Thakar and Tonapi (1961) show that this species has gyne cells much as in other *Apis.*

In the giant honeybee, *Apis dorsata,* the worker cells are scarcely larger in diameter than those of *A. mellifera.* They are, however, much deeper..Not surprisingly, the workers of *A. dorsata* are scarcely larger in diameter than those of *A. mellifera* but are longer and more slender. The proportions of the cells in *A. florea* are more like those of *A. mellifera.*

The number of cells in a honeybee nest varies greatly, but there are commonly many thousands, perhaps 100,000 in a strong colony of *Apis mellifera.* According to Ribbands' (1953) review, one pound of wax made from about 450,000 wax scales can be built into some 35,000 cells, in which an equal number of young workers could be developing at any one time, or in which 10 kg (22 lb) of honey could be stored.

C. Brood

The elongate, slightly curved honeybee eggs are ordinarily laid one per cell, attached to the bottom of the cell by one end, and projecting outward toward the cell entrance. The cells are left open during larval growth and larvae are fed progressively. Bee milk,[1] a hypopharyngeal and mandibular gland secretion of young adult worker bees, is placed in superabundance in the cells containing young larvae (Figure 30.9), which are therefore able to feed continuously. The gyne cell is large enough that it can always accommodate a similar abundance of food, and gyne larvae are continuously fed in the same way; there is royal jelly[1] in the cell with the larva at all times until it reaches maturity. In contrast, larvae of workers and drones, as they grow, fit their cells snugly, and there is no room for food around them. After the third post-hatching day, the larvae of workers and males are fed directly with small quantities of mixed worker food (pollen, honey, mixed with bee milk) so that there are never large accumulations in the cells. Dried food still remains behind in gyne cells after emergence of adults, whereas little or no such material remains in the cells of the less generously fed larvae of workers and males, which have nothing upon which to feed after the cells are closed by the worker bees. Worker and presumably male larvae lose weight between that time and pupation. Jay (1963b) has given a review of behavior during the whole developmental process. Table 30.1 indicates the approximate duration of various stages for each caste and sex. Comparable data for *Apis dorsata* are provided by Qayyum and Ahmad (1967).

Two or three days before pupation the worker bees cap the cells with wax. After this the larva within the cell spins its cocoon and the cell is not opened again until the bee reared in it reaches maturity and emerges.

D. The unfavorable season

In the tropics honeybees are active throughout the year. In all species there is considerable food storage in the upper parts of the combs (Figure 30.10) and therefore the colonies are able to exist for considerable periods, for example during the dry seasons in some tropical areas, when little or no food is available from flowers. As explained in Chapter 17 (G), however, tropical forms of *Apis* are quite capable of migrating, and at certain seasons abscond from their nesting sites and establish themselves elsewhere. Thus the African race *A. mellifera adansonii* absconds when there is a shortage of either food or water, and other tropical forms of *Apis* do likewise. Temperate climate races of *A. mellifera,* however, less commonly respond to unfavorable conditions in the same way, and colonies often die off, if conditions are bad enough, without abandoning their nests. No doubt apiculture has

1. For an explanation of this terminology see Chapter 10 (C2).

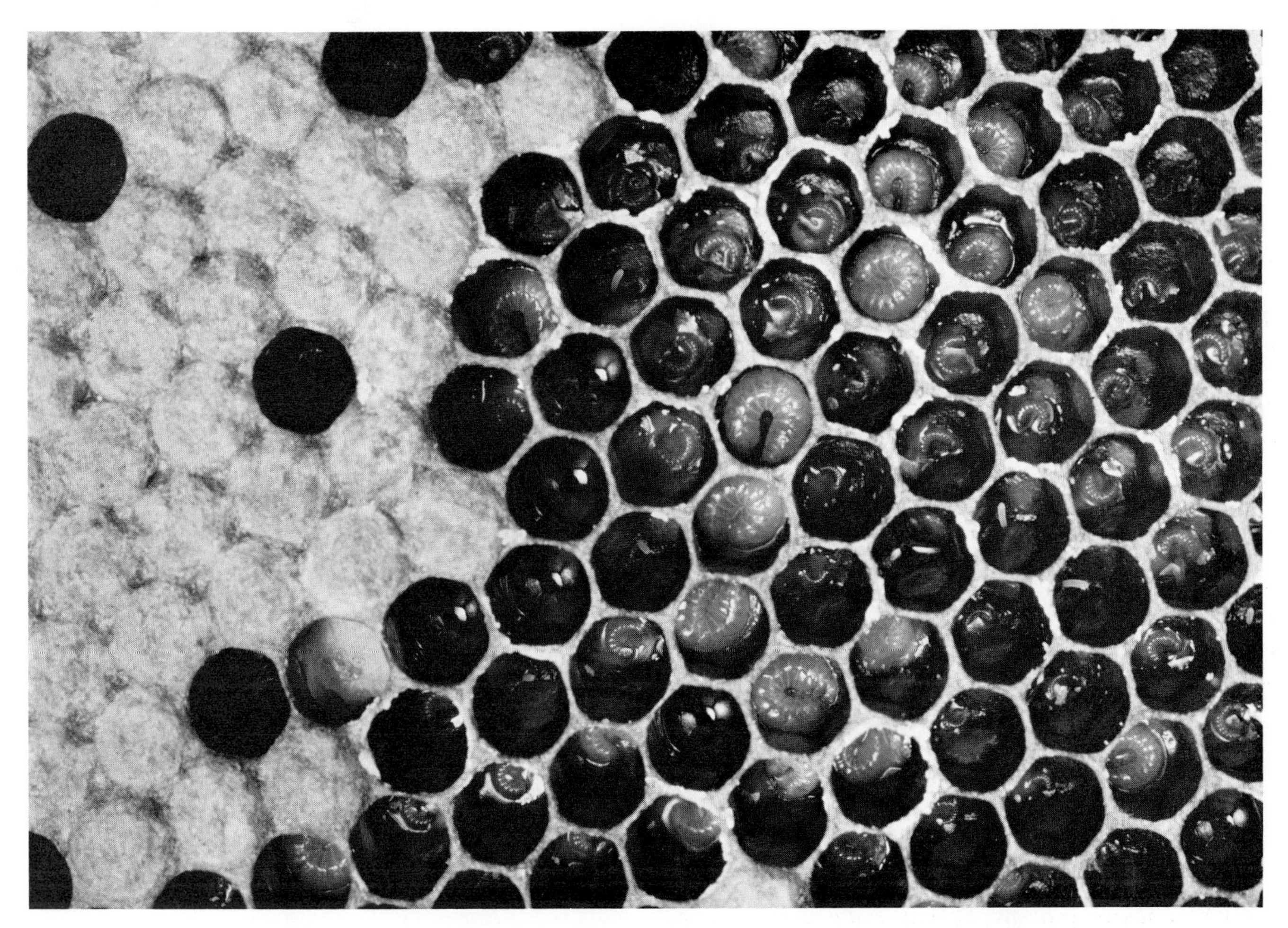

FIGURE 30.9. A brood comb of *Apis mellifera* from which adult bees have been removed. Note the elongate eggs (*center right*), small larvae in superabundance of food, large larvae occupying much of their cells, and *at left,* capped cells containing prepupae or pupae. (Photograph by E. S. Ross.)

TABLE 30.1. Durations of developmental stages of *Apis mellifera,* shown as number of days after egg laying.

	Number of days for —		
	worker	gyne	male
Laying	0	0	0
Hatching	3	3	3–5
Larval feeding finished	7–9	7–9	9–10
Cocoon spinning finished	9–11	8–10	11–12
Pupation	11–14	10–12	14–17
Emergence of adult from cell	19–22	15–17	24–25

SOURCE: Jay, 1963b.

tended to select strains, both consciously and inadvertently, that do not often abscond. Bees that readily do so got away in the early stages of domestication. Absconding is sociophysiologically very like swarming, except that all or nearly all the bees leave instead of some staying behind with a young queen (Martin, 1963).

The unfavorable condition to which northern races of *Apis mellifera* are best adapted is cold. During the winter, absconding could have no favorable effect, providing the bees are in a reasonably well-insulated situation, and probably would lead to death of the bees. It is easy to see that absconding behavior would be selected against in cool weather. The northern races have a different and unique way of surviving through the cold of winter,

FIGURE 30.10. A honey storage area of a comb of the common honeybee *Apis mellifera.* The freshly stored honey can be seen (by reflections of the light) in the upper part of the photograph; some cells are being capped (white) for storage. (Photograph by E. S. Ross.)

namely warming themselves in the winter cluster (Chapter 17, D3).

Longevity and physiology of workers at various seasons become interesting in connection with the winter cluster. In summer, if workers avoid spiders, birds, insecticides, and other hazards, most live about six weeks as adults, although there are records of 55 days. Shorter average life spans (about 33 days) are presumably due at least in part to losses in the field (Ribbands, 1953; Sekiguchi and Sakagami, 1966). In cool temperate climates winter longevity has to be much longer, for example, September to May. Furthermore, there are records of workers emerging in autumn that attained maximum longevities of 300 to 400 days. That is, they lived not only through the winter but also through much or all of the following summer (Maurizio, 1959). All this may seem surprising in view of the winter activity necessary to keep warm.

Bees emerge in autumn into a very different world than those emerging at other seasons. There is little or no brood to feed and there are plenty of provisions. The young bee feeds on pollen for several weeks. Probably for this reason, unlike ordinary summer workers, winter ones have enlarged fat bodies with many globules of protein and enlarged hypopharyngeal glands, like nurse bees in summer (see Chapter 12, B2). Along with enlargement of these structures with autumnal consumption of pollen goes an increase in life expectancy to six or eight months, with the even longer extremes as indicated above.

Even in summer, artificial prevention of brood rearing can cause similar gland and fat development, and in preswarming and swarming colonies (in which there is little or no brood rearing) the same thing occurs to a lesser extent. In such cases the life expectancy is also increased. Maurizio (1959) has summarized studies of various other organs that differ between summer and winter bees. She finds that among the organs studied, the best visible correlate of longevity is the fat body. Winter bees have strikingly reduced metabolism (half the CO_2 production) compared to summer bees at the same temperature, and the same is likely for broodless and swarming summer bees.

In tropical climates with a strong dry season, the long-lived brood of workers occurs at the time of the dry season instead of in winter (Maurizio, 1961). It is a response to the condition in the nest suggested above—plenty of stored pollen but little brood rearing. Thus with relatively little behavioral adaptation (cutting down on brood rearing at the proper season), populations of *Apis* can produce a long-lived worker brood to survive almost any sort of recurring unfavorable period, irrespective of the season when it occurs.

As noted above, the glandular and fat development is associated with increased longevity, obviously biologically important both in winter and at swarming time when the swarm does no foraging. In winter, work is necessarily reduced by external conditions, but the same thing happens in summer in artificially broodless and in swarming colonies. It is as though the pharyngeal gland and fat-body enlargement were associated with an actual antidote to aging, for overwintered workers may live far longer during the active season of the following year than workers that emerge during that season (Maurizio, 1959; Sekiguchi and Sakagami, 1966).

Of course average age is quite different from the unusually long lives of the bees that live through a winter and the following summer. Figure 30.11 shows average longevities for bees in Switzerland and Egypt. Data for northern Japan are provided by Fukuda and Sekiguchi (1966).

It is probable that the increased protein consumption and resultant increased longevity would lead to a greater total work output per bee. *Apis* behavioral evolution, however, has led to shorter lives and less protein consumption. Perhaps the larger population attainable by

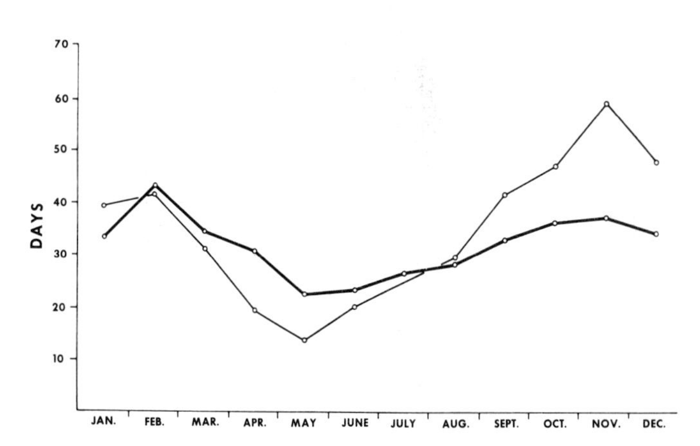

FIGURE 30.11. Average longevity (in days) of workers of *Apis mellifera* at various times of year in Switzerland (*heavy line*) and Egypt (*light line*). (From A. Maurizio, in Chauvin, 1968, courtesy of Masson et Cie, Paris.)

this arrangement has advantages in terms of defense, environmental control, and storage of adequate supplies for overwintering. After midwinter in cool temperate climates, brood rearing begins in the center of the winter cluster and temperatures are then maintained at higher levels, 32 to 35° C. The ability of the northern subspecies of *A. mellifera* to survive through the winter in this way must be the major factor that has made possible the range extension of this basically tropical insect into areas of cold winters. In Japan and China *A. cerana* shows a similar ability, also forming winter clusters.

E. Spring growth in temperate regions

Egg laying by *Apis mellifera* queens in the late winter and early spring is limited to the centers of some of the combs, that is, the center of the winter cluster area, where the temperature can be maintained at a constant high level in spite of variable cold weather. As the season advances, the brood area increases as the ability of the bees to maintain uniform temperatures improves with improving weather conditions. In late winter and early spring food for the larvae is prepared from the overwintering stores of the colony; later, with spring weather, the food is prepared from provisions obtained from spring flowers.

The late winter and early spring brood rearing results in a buildup of young workers, so that the colony population is moderately high when the spring flowers appear and exploitable food becomes abundant outside. The rearing of workers continues throughout the spring and summer at rates that vary with weather and food supply but which may be very high. A queen has been timed laying as many as six eggs per minute although this rate is not maintained for long (Chapter 6, F), and in her usual lifetime of two or three years she may lay 600,000 eggs.

Although in normal hives about half the queens are superseded when two or three years old, there are many records of queens living for four to six years and even much longer, perhaps ten years, in which case her total production must of course be far larger. Bozina (1961) reports that in normal colonies in hives, 30 to 35 percent of the queens lived four to six years, that two lived eight years, and one was still living and nine years old.

F. Gyne production[2]

In late spring or early summer bees commonly build new gyne cells. Their appearance is the most obvious sign of the beginning of a series of events that will lead to fission of the colony. As she moves around within the brood area of her nest, that is, on those parts of any combs which are maintained at the brood temperatures of about 35° C, the queen lays eggs in whatever clean empty cells she encounters, be they worker, gyne, or drone cells. There are reports of workers almost chasing the queen to gyne cell cups, which are out of the area where she would most often wander. Thus the queen lays the egg in the cell that will produce her own replacement.

The young gyne emerges from her cell by cutting about three-quarters of the way around it near the tip and pushing the resulting cap open, unless the cap has already been removed by workers; she climbs out among the worker bees. They pay little attention to her while she is virgin, although in an otherwise queenless colony her presence prevents the signs of queenlessness. The new gyne helps herself to honey within the hive and sometimes solicits food from workers, but she is not known to eat pollen. She wanders around the brood area, rarely leaving that part of the nest.

As is well known, young gynes are deadly enemies to one another, grappling and stinging until one is killed (Chapter 11, B). They may also be killed by workers. Nevertheless it sometimes happens that several young gynes or queens will temporarily be present in a colony, and swarms containing several queens are well known, especially in the Brazilian honeybee (Cosenza, 1972). Swarms with multiple queens are likely to be afterswarms. An afterswarm leaves a colony after the first swarm in spring or early summer. The first swarm is headed by the old, overwintered queen. Afterswarms are headed by relatively young daughter queens, or some-

2. In nearly all literature on honeybees, the word "queen" is used instead of "gyne." In highly eusocial forms this is perfectly practical, since one can distinguish the potential gynes from the potential workers even before maturation. In primitively eusocial forms, however, one commonly cannot distinguish a potential queen (gyne) from workers until it actually becomes a queen. An effort has therefore been made to develop a widely applicable terminology and to apply it consistently. According to this terminology, a reproductive female honeybee is called a gyne, and becomes a queen only when she heads a colony.

times by several. They commonly fight it out in a new site, but there are reports from Brazil of such a swarm dividing into small monogynous swarms, each of which presumably seeks a nest site.

When a few days old a virgin gyne usually begins making short flights from her nest or hive. At first such flights last only a minute or two, but gradually they are extended to 10 or 15 minutes or more. Presumably they are orientation flights (like those of workers) that enable the gyne to learn the location of her nest and the topography of the countryside, essential information if she is to return successfully after mating flights.

G. Mating

When 3 to 16 days old, gynes fly from their hives to mate. These flights last from 5 to 50 minutes, and are mostly in the afternoon. In 1792 Huber reported that gynes mated when over 21 days old became male-producers (Ribbands, 1953). We now know that this means that mating was not successful. Some 170 years later Zmarlicki and Morse (1963) verified the observation that older queens do not often mate; in their case 11 out of 13 gynes prevented from flying until a month old became male-producers. Although at one time it was believed that each gyne mated but once in her life, it is now clear that they regularly mate repeatedly, perhaps an average of seven to ten times (Taber and Wendel, 1958). All of these matings may occur on a single flight, or if bad weather or lack of males cause a gyne to return to her colony before the spermatheca is full, she may make a second flight a day or two later. Apparently insufficient sperm or seminal fluid stimulates subsequent flights (Woyke, 1964). All of the matings occur within a brief period in the gyne's life, after which she has an adequate supply of sperm cells to permit her to lay female-producing eggs for a period of years, the sperm cells being kept alive in the spermatheca for this whole period.

Males pay no attention to gynes in the nest. The males usually move about but little and often solicit food which is thought to be mostly bee milk (Ribbands, 1953), but they do not feed other members of the colony (Hoffmann, 1966).

Mating may involve no evident special concentrating areas, but in many regions such areas occur and are used year after year. They are places where the males gather on sunny afternoons, flying about at high speeds usually far above the level of the ground so that they are not readily seen by human observers. With balloons and tethered gynes Ruttner (1966) found maximum male activity at 15 to 25 m above the ground, with some gynes followed above 40 m but none to 60 m. Rarely males follow gynes close to the ground, especially on windy days. Males are powerful flyers; Ruttner and Ruttner (1966, 1968) say that they sometimes come to an aggregation site from distances of over 6 km. Difficulties of getting pure matings of specific strains of *Apis* further demonstrate that males must range for great distances. Few places lack feral *Apis* capable of mating with the gynes for which one may be trying to make particular matings. Partly for this reason, to facilitate breeding for desired traits, the technique of artificial insemination of gynes has been developed.

Males usually make their first flights when 9 to 18 days old (Drescher, 1969) or 6 to 12 days old under Witherell's (1971) conditions. The flights, for unsuccessful males, are mostly about half an hour in duration (ranging up to 207 minutes). They average longer the older the drone (up to 31–40 days old) and terminate with his return to the hive (or occasionally to a different hive). A male may make several flights in one afternoon if the weather is good. For a successful male the flight terminates in dismemberment and death. So far as known the males are continually in flight when away from the hive, except that occasionally one comes down apparently exhausted, like the one shown in Figure 30.12; they are not known to take food from flowers.

A pheromone from the gyne's mandibular gland is the attractant. *Trans*-9-keto-2-decenoic acid, or "queen substance," is the principal material involved, and is considered by Butler not only as an attractant but as an aphrodisiac, for it stimulates the male to mate once he reaches a queen with an open sting chamber. *Trans*-9-hydroxy-2-decenoic acid, the swarm-clustering pheromone, is a minor attractant according to Butler (1967b, 1969) and Butler and Simpson (1967). In a later account of mating behavior, however, Butler (1971) ignores the latter substance.

Normal mating of honeybees has rarely been observed. However, as a result of occasional and almost accidental observations, and more recently observations with virgin

FIGURE 30.12. A male (drone) honeybee, *Apis mellifera.* Occasionally drones, possibly lost or exhausted, descend from mating flights and alight on vegetation or on the ground. There is no evidence that they ever feed from flowers; this one was not doing so. (Photograph by E. S. Ross.)

gynes tethered on wires or on balloons high above the ground (Gary, 1963; Pain and Ruttner, 1963; Butler, 1971), it is possible to give some idea of the happenings during the mating. The details of mating are also described by Woyke (1958), Woyke and Ruttner (1958), Ruttner (1956; in Chauvin, 1968), and Gary and Marston (1971). The gyne must fly into an area in which numerous males are coursing about in the air. When this occurs, males are attracted, perhaps from as far as 60 m, by her sex pheromone, see her with their enormous eyes, and follow her, quickly forming a "comet" behind her. Such a comet may consist of dozens or hundreds of males; one of them reaches the gyne and, if she possesses certain unidentified contact pheromones, he mates with her in the air, a process that is completed very quickly. The male genitalia are enormous and very different from those of other Hymenoptera. As part of the mating, the distal portion of the male genitalia virtually explodes with a loud pop into the sting chamber of the female and breaks off there. This part of the penis remains in the sting chamber of the female. This "mating sign" can be worked free and does not prevent additional matings on the same flight, but the gyne characteristically returns to the hive carrying a mating sign if she has mated on a flight.

Sometimes a group of workers, a mating swarm, leaves the hive with the gyne, settles somewhere nearby, and the gyne after mating returns to it. The swarm then returns to the hive. Such a gyne might already be called a queen. Usually, however, the gyne leaves the colony and returns directly to it.

H. Swarming

Swarming is a characteristic of all species of honeybees, since it is the method by which the colonies reproduce. The various strains of *Apis mellifera* differ in the frequency with which they swarm, a matter of practical importance to beekeepers since large colonies are more efficient honey producers than small ones, and repeated swarming results in numerous small colonies.

The "swarming impulse" of a colony normally leads to production of gynes (Chapter 13, B). As they grow in their cells, the old mated queen produces fewer and fewer eggs each day, and may stop altogether. Her abdomen shrinks, since it no longer contains fully formed eggs. Usually, soon after the cells in which young queens are being produced have been sealed, the old queen leaves the hive, accompanied by a swarm of workers. The departure may be delayed by bad weather until after the emergence of the young queens, but ordinarily it precedes their emergence. Sometimes a few drones accompany the old queen and the numerous workers. After flying about for a short time the swarm settles, frequently on the limb of a tree or some other support (Figure 13.1). After exploratory flights by individual bees, and communication as to locations of suitable sites, the whole swarm flies again to occupy an appropriate cavity, in which they will establish a new nest. In the new situation the workers must quickly secrete enough wax to construct combs for storage and brood, the ovaries of the old queen enlarge again, and a functional colony is established. Meanwhile, at the old location, the surviving young gyne mates, her ovaries enlarge, and she becomes the queen for the colony at the original site. The two colonies, if successful, continue through the summer and fall, each storing enough for its winter suvival.

The proportion of bees leaving to those that remain at the old site varies, but usually more leave than stay. If a colony of under 20,000 bees swarms, only a small percentage remains behind, usually too few for survival of the colony (Martin, 1963). As indicated in section F above, afterswarms often also issue from a colony. This results in further subdivision of the group of bees left at the original site.

It is obvious from the above that if a queen bee lives for several years, it is her fate to move as frequently as a normal swarm issues from her hive. Typically this would be at least annually. In each site she would leave a colony, headed by one of her daughter queens.

I. Supersedure and emergency gyne production

As a queen becomes old she may use up her entire supply of sperm cells, after which she is able to lay only male-producing eggs. Usually before this happens, she becomes less effective than previously, a condition that is probably sensed by the worker bees due to the decrease in her production of queen substance. In any case, old queens are usually replaced by supersedure. Supersedure is much like swarming, except that the colony does not divide. The old queen herself lays the eggs in the queen cell cups, which are normal in appearance and location. The young gynes, on maturation, fight it out, and the survivor mates as described above. On returning to the colony, the survivor may live for a period of months with her mother, who will not have left with a swarm; the two queens may lay eggs in the same brood area until the old queen dies or is killed.

As has been explained in Chapter 11 (A), in the event of death or removal of the queen emergency gyne cells may be constructed around female eggs or young larvae. Such cells therefore differ in location from normal and supersedure gyne cells, but the production of the gynes and their subsequent behavior are the same. Although normal gyne production is mostly in late spring and early summer, supersedure and emergency gyne production may occur at any time during the warm season of the year; the continued presence of males in honeybee colonies throughout the warm season provides for fertilization of such gynes as are produced outside the period of normal gyne production.

J. Importance of *Apis* to man

Man has no doubt been robbing honeybee nests for honey and edible larvae and pupae of bees since early in human history, probably a continuation of his ancestors' activities before they were men at all. For at least four thousand years he has been keeping bees in hives of one sort or another. Records of approximately this antiquity for hives of bees come from both China and Egypt. *Apis mellifera* and *cerana* have both been involved in this history, although the former is by all means the more important and the better bee for beekeepers. In temperate areas where the two species meet, as in Japan where the native bee is *A. cerana* but where *A. mellifera* has been imported, *A. mellifera* replaces *A. cerana* not only as the domestic bee, but also in the more or less wild situations. Thus, temperate climate races of *A. cerana* are perhaps on the way to extinction.

Although *Apis mellifera* may be spoken of as domesticated where it is under the control of beekeepers, its behavior is not modified like that of domestic vertebrate animals. The latter, as individual animals, learn how to get along with man, and frequently are taught to do things which the animal would never do under natural circumstances. The honeybees learn no new behavior from man; beekeepers simply take advantage of their knowledge of bee behavior to get desired products (honey, wax, sometimes royal jelly, sometimes venom) and services (pollination) from the bees. Man can modify bees by selection for improved strains but cannot add to the normal behavioral repertoire of individual bees. The beekeeper does promote desired bee activities by the opportunities which he provides or withholds. For example, he locates the hives in places where he believes there will be a good crop of nectar or in places where pollination service is needed by a farmer. Also, he can control swarming and thus maintain large colonies by removing queen cells from the combs as they are produced, so that the colony is unable to rear young queens.

The beekeeper also arranges matters to facilitate his use of the bees. A natural honeybee comb hangs from the roof of the cavity in which it is constructed and consists, in its upper portion, of a large number of cells in which honey is stored. This part of the comb supports the lower parts, in which pollen may be stored and in which brood is reared. Robbing such a nest of its honey therefore results in removing the supporting parts of the combs; the lower parts containing the brood, if not destroyed, at best are left lying about on the bottom of the nest cavity or hive. This problem exists whether the nest is in a hollow tree or in an old-fashioned straw hive such as was used for many centuries and is used even now in certain areas of the world. Modern bee hives, however, contain combs in frames which can be lifted out of the boxes which constitute the hive. The lower box, or hive body, contains those combs in which brood is kept and need not be disturbed when the honey-containing combs in the upper box or boxes (called supers) are removed. By providing independent support for what in nature would be the lower parts of large combs, it is possible to get the honey without slowing down the bee-rearing processes in the lower part of the nest.

Another way in which a beekeeper controls bee activity is in connection with comb maintenance and construction. He can insert into the frames of the hive sheets of wax stamped with the hexagonal pattern of cells. The bees can then "draw out" the ridges which form the pattern to make complete combs of cells. In this way the need for wax production in the colony is reduced, and the bees are encouraged to make regular, parallel combs with no interconnecting spurs. This greatly facilitates the taking of honey and other manipulations. Moreover, by removing the cell caps with a hot knife and spinning the honey out of the cells by centrifugal force, the beekeeper can obtain and return to the hive nearly complete combs which, with very little additional wax provided by the colony, can be re-used for storage of another batch of honey.

Although individual bees are not taught useful behavior by beekeepers, selection for strains with useful attributes has been carried on in recent years. It is perhaps surprising that little such selection occurred, so far as can be judged by comparing wild and domestic bees, over the thousands of years that honeybees have been managed by people. The reason presumably is that each gyne mates high in the air with several males from unknown and often different sources, often from wild colonies, and in any case usually not from the colony in which the gyne originated. It has therefore been difficult or impossible to cross individuals from colonies with known desired characteristics. In the last twenty years, however, methods of artificial insemination of queen bees have been devised so that semen from drones of known backgrounds can be used to fertilize gynes of known origins. This has made possible crossing and subsequent selection for desired traits, and hence the development of strains of bees having desired honey production, disease resistance, and other characteristics. A strain has even been developed with improved abilities to pollinate alfalfa, a plant which ordinarily is visited by honeybees principally for nectar but not for pollen and for which they are therefore, under most circumstances, ineffective pollinators (Nye and Mackensen, 1968–70). (This strain, however, is much less effective than certain kinds of solitary bees which are efficient and regular visitors of small legumes, including alfalfa.)

Beekeeping has become a business of substantial importance. Some beekeepers own thousands of hives and travel with the season from the south, for example southern Texas, far to the north, for example into southern Canada, remaining as long as possible in the moving zone of spring flowers which provides excellent nectar and pollen sources. In the autumn they return south to the brief winter of moderate climates. Such an operation involves large quantities of machinery for transport and for removing and processing the honey.

To give an idea of the size of the apiculture industry, there are about 4,700,000 hives in the United States, tended and exploited by beekeepers, in addition to uncounted feral colonies. The hives produce some 230 million pounds of honey (based on Anonymous, 1971) valued at 45 million dollars annually. Beeswax production amounts to some 2 million dollars annually in the United States. World honey production is about 900 million pounds (based on Hannawald, 1967).

In middle latitudes of the United States it is usually necessary to leave about 40 pounds of honey in a hive as it goes into the winter in order to be certain that the colony will have enough food to maintain its warmth in the winter cluster and to start rearing brood in late winter so as to have a good supply of young workers by the time the spring honey flow begins. The amount of honey that can be removed from the hive, nonetheless, is surprising. Under excellent conditions many hives can be robbed of 100 pounds of honey per year and some produce much more, although usually only under careful management. The average yield per hive is about 50 pounds in the United States. When one considers the minute quantities of nectar available in each flower, one recognizes the enormous labor involved in assembling 100 pounds of honey in one place, especially when, through the year, the bees have been eating as well as accumulating and storing.

Because of the importance of bees in pollination, and because in the United States many of the native pollinators have been depleted by destruction of their habitats through cultivation and poisoning by insecticides, honeybees are frequently in demand for pollination purposes. Farmers who need them to pollinate their crops have to pay for the service, since excellent pollination is attainable only by crowding the colonies to a point where they produce less honey than if they were dispersed. If the farmer obtains the cooperation of a beekeeper and puts enough honeybee hives into his apple orchard, for example, to obtain optimum pollination, the beekeeper will get less honey production than if his bees were scattered over a larger area. The farmer has to pay the beekeeper enough to compensate for this decrease in honey production.

The monetary importance of the honeybee as a honey and wax producer, as indicated above, is small compared to the importance of this and other species as pollinators. Levin (1967) gives the 1963 farm value of crops requiring bee pollination as 1 billion dollars in the United States and the farm value of additional crops benefited by but not requiring bee pollination as 6 billion dollars.

These figures do not consider natural vegetation, which is enormously important. Many plants of prairies, forest, meadows, deserts, and swamps are bee pollinated. Such plants are among those that produce food for stock and for wildlife, maintain watersheds, prevent erosion, and constitute part of the natural environments to be studied, enjoyed, and used by people. Floras evolved with and (aside from wind-pollinated species) were pollinated by the native pollinators, especially solitary and social bees. Destruction of nesting sites and the effects of insecticides have reduced the native bee populations even in many nonagricultural regions. It is very likely that for some plants in many areas introduced honeybees have replaced, at least in part, the native species as pollinators. At least we are as yet largely ignorant of any failure of reproduction of native vegetation because of lack of pollinators. Where they are introduced, honeybees no doubt also compete with native bees for food resources, so that honeybees, too, may reduce populations of native species. Not a single detailed study of such matters has yet been made in any part of the world. It seems reasonable to believe, however, that bees, both native and introduced, are important in the maintenance of natural vegetation in most parts of the world.

Appendix, Glossary, Literature Cited, and Index

Appendix: Scientific Names

Those who wish to look into the published accounts for information not included in this book may encounter problems with the names of bees. Because the nomenclature used in this book has been brought up to date, there are discrepancies between names used here and those used for the same bees in some other publications.

The names of organisms are given so that users of the names can know what kinds of creatures are being discussed. If the scientific names amounted only to names, they could be relatively stable. The most common change would be when two names were found to represent one kind of creature—one of the names would then be abandoned as a synonym—or the reverse, when organisms thought to be of one kind were recognized as two, and a new name given to one of them.

Very unfortunately, scientific names are more than just names, for the genus-species combination states an opinion about the classification of the organism.[1] As knowledge of a group of organisms improves, its classification and hence some of the names involved may change. The result is considerable instability, especially in generic names. In groups like bees that are not yet very well studied, such changes are unavoidable unless we either abandon the Linnaean system of nomenclature or stop learning more about comparative aspects of biology.

The following notes will assist in relating the contents of this book to earlier publications.

Halictidae

Sakagami and Michener (1962) tabulated the names used in earlier biological studies of halictines, along with the names that seemed correct in 1962. There are two errors: the species *eurygnathus* and *maculatus* should both be in *Halictus* rather than *Lasioglossum*. Also, the species *mutabilis* is now placed in the genus *Ruizantheda.*

Usage of the generic name *Halictus* has varied enormously. One author (Vachal) applied it to virtually the entire subfamily Halictinae. Some authors still use it in a broad sense to include the species here placed in the genera *Halictus* and *Lasioglossum;* thus *Lasioglossum* is included in *Halictus* by these authors. For this reason bees sometimes called *Halictus calceatus, H. malachurus* (workers used to be called *H. longulus*), *H. marginatus,* and *H. duplex* are here called *Lasioglossum calceatum, L. malachurum, L. marginatum,* and *L. duplex.* In short, some of the species put in *Halictus* by some authors are here put in *Lasioglossum.*

On the other hand, the genus *Lasioglossum* as understood for this book is broken up by other authors into several genera. This means that species placed under the generic names *Evylaeus* and *Dialictus* by some authors (Mitchell, Knerer) are here included in the genus *Lasioglossum,* with *Evylaeus* and *Dialictus* as subgenera. For further comment, see Chapter 25.

Older literature included both *Augochlorella* and *Augochloropsis* in *Augochlora.*

The following synonymy in the Halictinae may help to explain some nomenclatural problems (italicized names are valid):

Dialictus = Chloralictus.

Lasioglossum (*Dialictus*) *imitatum* = L. inconspicuum = L. stultum.

Xylocopinae

Nearly all the allodapine bees except *Exoneura* have in the past been included in *Allodape.* More recently, after separation of *Allodapula* from *Allodape,* the former re-

1. A subgeneric name placed in parentheses between the generic and specific names gives more classificatory information. It can always be omitted, as is done in this book except where it has some interest.

mained a composite group and *Braunsapis* and *Halterapis* were therefore separated from *Allodapula.* All Asiatic or Australian references to *Allodape* or *Allodapula* refer to species now placed in *Braunsapis.* Biological observations on *Allodape marginata* in Formosa are properly attributed to *B. sauteriella.* For African species the important nomenclatural points are indicated by Michener (1971). Accounts of *Allodapula foveata* and *grandiceps* should be attributed to *Braunsapis foveata* and *leptozonia.*

Synonymies important in interpreting earlier papers on behavior of Xylocopinae are as follows (valid names in italics):

Allodape mucronata = A. angulata.
Allodapula variegata = Allodape (or Allodapula) pringlei.
Allodapula acutigera = Allodape halictoides.
Braunsapis leptozonia = Allodapula grandiceps.
Compsomelissa = Exoneurula.
Xylocopa tranquebarorum = X. orichalcea.

Apidae

All of the nonparasitic Euglossini were formerly placed in the genus *Euglossa.* Therefore there are old biological reports on species of *Eulaema* and *Euplusia* published as *Euglossa.*

Bremus is a synonym of *Bombus.* Various combinations with *bombus,* such as *Megabombus,* represent groups of species of *Bombus* that should not be recognized as genera, although some are useful as subgenera. See Chapter 28.

Among the tropical stingless honeybees (Meliponini) there is great difference of opinion as to the amount of generic splitting that should be recognized among the forms here placed in the genus *Trigona.* Older authors even placed all the species of *Trigona* under *Melipona. Plebeia* and names formed in combination with it, and combinations with *Trigona,* are all considered by me as subgeneric names under the genus *Trigona* or as synonyms. So also are *Duckeola, Friesella, Frieseomelitta, Mourella, Nogueirapis, Partamona, Scaura, Schwarziana,* etc. See Chapter 29. However, in contrast to *Bombus, Trigona* is morphologically and behaviorally very diverse, and splitting it into several units can easily be justified. I happen to think that larger, less split genera are more useful to all concerned, but some authors strongly defend the opposite view.

Well-known specific names now considered obsolete are listed by Wille and Michener, in press. Synonymies important for species referred to in this book are as follows (valid names in italics):

Trigona australis = T. cassiae
Trigona cilipes = T. compressa
Trigona clypearis = T. wybenica
Trigona spinipes = T. ruficrus
Melipona nigra = M. schencki.

Two African forms, perhaps not specifically distinct, have often been confused. East African data relate to *Trigona gribodoi,* west African data concern *T. braunsi.*

Apis mellifica is a synonym of *A. mellifera.* The names *Megapis* and *Micrapis* do not seem to me justified at the generic level. They apply to the species or groups of *Apis dorsalis* and *A. florea,* respectively.

Glossary

Among the terms explained below are not only specialized ones used in insect sociobiology but also some entomological terms which may be helpful to persons with no more than an elementary background in biology. Certain other entomological terms are explained in Chapter 1. The objective is to eliminate the need for nonbiologists or nonentomologists to consult general entomological textbooks. Some of the terms, such as swarm and dance, are defined as they are used for bees only, although they have broader entomological, biological, or general meanings.

ABSCONDING. Departure of a whole colony of highly eusocial bees for a new nest site.

ACULEATE HYMENOPTERA. Hymenoptera with stings: the bees, wasps, and ants. The stings may be reduced; "stingless" bees are included in the aculeate Hymenoptera.

AGGREGATION. A group of nests. While the commonest aggregations are of nests of solitary bees, colonies of any social level may also form aggregations.

ALARM PHEROMONE. A substance exchanged among individuals of a species that causes alertness and readiness to attack a moving object or potential enemy.

ALTRUISM. Dangerous behavior or behavior disadvantageous to the individual that benefits other individuals of the species. Often used with respect to reproduction, in which case the altruistic individual reduces its reproductivity while enhancing that of another.

ANTENNATE. To touch, usually repeatedly, with the antennae.

APIS. Honeybees, and for this book, the common domesticated honeybee, *Apis mellifera,* unless another species or the generic concept is indicated.

ARRHENOTOKY. Production of males from unfertilized eggs. In bees and most other Hymenoptera this is production of haploid males, the diploid females being from fertilized eggs.

AUXILIARIES. Gynes which, in association with a queen, become workers.

BASITARSUS. The first or basal tarsal segment.

BASITIBIAL PLATE. A small plate or scalelike projection at the base of the hind tibia.

BATUMEN. The layer of material (cerumen, mud) surrounding the nest cavity of stingless honeybees.

BATUMEN PLATES. Thick batumen isolating portions of a larger cavity from the part that is used as a nest cavity by stingless honeybees.

BEE MILK. Brood food of *Apis,* secreted by nurse bees, probably mixed with some crop contents. (Royal jelly has sometimes been used in this sense).

BOMBUS. The genus to which the nonparasitic bumblebees belong.

BROOD. (1) A collective term for the immature stages (eggs, larvae, pupae) in any nest. (2) Temporally separated batches of young produced by a colony. Thus one may speak of the first brood of young, the second, and so forth.

BROOD FOOD. Any food of larval bees.

BUMBLEBEE. A member of the genus *Bombus* or *Psithyrus.*

CALLOW. Teneral. A young adult that has not yet attained its mature coloration. Sometimes the wings are milky and sometimes the body coloration is paler than in the mature adults.

CASTES. Functionally different groups among the females of a colony. The differences may be only behavioral and physiological, or may also involve structure. The differences are permanent, not merely due to age. The two castes are worker and queen.

CELL. A prepared space in which a single immature bee is reared. Less commonly occupied by more than one developing bee, especially in the case of the distendible cells of *Bombus*, which may or may not be homologous to ordinary cells.

CERUMEN. A mixture of resin and wax, the principal material used for nest construction by stingless honeybees. Cerumen is used to a minor extent by *Apis.*

CHEMORECEPTOR. A sensory end organ or sensillum that distinguishes substances chemically, that is, by taste or odor.

CHORION. The egg covering. In bees it is membranous rather than shell-like.

COCOON. The protective secreted covering, made largely of silk, spun by mature larvae of some bees.

COLONY. The mature female bees working in a nest (plus immature stages if being actively cared for, usually progressively fed).

COMB. A layer of more or less regularly arranged, contiguous cells or cocoons.

COMMUNAL. Living as a colony consisting of females of a single generation, each making, provisioning, and laying eggs in her own cells.

COMMUNICATION. Sending of signals that influence the behavior or development of others, usually in the same colony.

COMPOSITE NEST. A nest inhabited by a communal colony.

CONSTANCY, FLOWER, *see* FLOWER CONSTANCY.

CORBICULA. The pollen basket on the hind tibia of Apidae. It consists of a smooth surface surrounded by a row of long scopal hairs, and serves for carrying pollen and other materials to the nest.

COURT. A group of workers that forms a circle around a queen of a highly eusocial colony, ordinarily antennating and licking her and sometimes feeding her.

CROP. The posterior part of the foregut, an enlarged, thin-walled sac for transportation and holding of liquids.

CUTICLE. The secreted exosketeton of an insect, including the flexible portions between sclerites.

DANCE. Movements of communicative importance usually performed on the combs of honeybees.

DIVISION OF LABOR. Differing activities of the members of a colony. The expression may be applied either to differences between castes or to differing activities of individuals of the same caste. The differences may be either permanent or temporal.

DRONE. A male honeybee.

EUSOCIAL. Living as a colony consisting of adults of two generations, mother(s) and daughters—the former the queen(s), the latter the workers.

EXUVIA. Cuticle shed at the molt.

FEMALES. Both queens and workers.

FIELD BEE. A bee that forages for either nectar or pollen.

FLOWER CONSTANCY. The tendency of a foraging bee to use the same species of flower as a food source, especially a pollen source, on any one foraging trip or during a longer period of time.

GALEA. The major bladelike distal part of the maxilla.

GLOSSA. The median, flexible, distal part of the labium.

GYNE. A potential or actual queen. The word is largely used for potential queens, females that will or that may become queens. Thus honeybee females that will become queens are called gynes. All females of some halictid bees are at first gynes, for all are potential queens. Later, they become either workers or queens.

GYNE CELL. The cell in which a new young gyne is reared. An emergency gyne cell is one constructed in a nest that has lost its queen and in which the colony is therefore queenless.

HALICTINE BEE. A sweat bee, subfamily Halictinae.

HAPLODIPLOID. A genetic system in which males have the haploid number of chromosomes and are produced from unfertilized eggs, while females have the diploid number of chromosomes and are produced from fertilized eggs.

HEMITERGITES. The right and left portions of terga that are divided longitudinally.

HIGHLY EUSOCIAL. Living as a eusocial colony in which the castes are morphologically dissimilar and food exchange among adults is extensive.

HIVE. A man-made container in which a colony of bees lives.

HONEY. Nectar that has been collected by bees and partly digested, that is, the sweets broken down into simple sugars, and from which part of the water has been evaporated.

HONEYBEE. A bee of the genus *Apis*. *See also* STINGLESS HONEYBEE.

HONEY STOMACH. Crop.

HOUSE BEE. A bee that spends most or all of its time in the nest.

HYMENOPTERA. The order of insects to which bees belong, including also wasps, ants, ichneumonids, sawflies, and their relatives.

HYPOPHARYNGEAL GLANDS. Paired glands in the facial part of the head, the ducts of which open through the lateral parts of the hypopharynx into the food channel just behind the mouth. Called pharyngeal glands in much of the older literature and food glands in some literature on *Apis*.

INVOLUCRUM. Sheet or sheets of cerumen surrounding the brood chamber in nests of most stingless honeybees.

LABIAL PALPUS. A sensory appendage of the labium.

LABIUM. The posteriormost mouthparts, which, together with the maxillae, form the proboscis in adult bees.

LABRUM. An anterior flap of the head over the area of the mouth opening—an upper lip.

LINING BATUMEN. Thin batumen that lines the walls of a cavity inhabited by stingless honeybees.

MALPHIGIAN TUBULES. Organs for the concentration and excretion of nitrogenous wastes into the gut. They may also secrete part of the material that becomes the cocoon.

MANDIBULAR GLANDS. Glands in the head which open through the membrane at the mandibular base and which commonly secrete food, defense substances, or pheromones important in mating behavior or social organization of bees.

MASS PROVISIONING. The method of feeding larvae in which

food for the entire larval growth is enclosed in a cell which is ordinarily not opened until emergence of the young adult. Used in contrast to progressive feeding.

MATRIFILIAL. Having colonies consisting of mother(s) and daughters.

MAXILLAE. Paired mouth appendages which, with the labium, form the proboscis of adult bees.

MAXILLARY PALPUS. A sensory appendage of the maxilla.

MECHANORECEPTOR. A sensory end organ or sensillum that responds to mechanical stimulation such as touch and vibrations.

MELIPONINE BEES. Stingless honeybees.

METAMORPHOSIS. The change from larva to adult, including that during the pupal stage and the molts before and after it.

METASOMA. The last body tagma, consisting of the second and following abdominal segments.

MIXED FOOD OF WORKERS. Food placed in the cells of older worker larvae of *Apis*. It includes honey and pollen as well as bee milk.

MOLT. Ecdysis, or the shedding of the cuticle that occurs with change of size or form in insect growth.

MONOGYNOUS. Referring to a colony that contains only a single gyne, that is, a single queen (Haplometrotic).

NECTAR. The sweet fluid secreted by nectaries of plants, commonly in flowers, that helps attract bees and that is the raw material from which honey is made.

NECTAR GUIDES. Lines or marks on the throats of flowers which serve to guide bees to the nectaries.

NEST. A construct made by the bees in which young are reared and in which adult females usually live. It may be a burrow or hollow in some substrate or may be built of materials brought to the nesting site.

NURSE BEE. A worker bee, principally in *Apis*, that feeds and cares for larvae.

OCELLI. Three simple eyes on the top of the head.

OLIGOLECTIC. Taking pollen from flowers of only a few species or genera of plants. This word is applied to species, not to individuals. An individual of a polylectic species that takes pollen from only a few kinds of plants is showing flower constancy rather than oligolecty.

OMMATIDIUM. One of the many optical units which make up a compound eye.

ONTOGENY. The development of an individual organism or of an individual colony from inception to maturity and old age.

ORPHAN COLONY. A colony that has lost its queen.

OVARIOLE. One of several ovarian tubules in which eggs are formed.

OVIPOSITOR. The egg-laying appendages, modified in bees into a sting.

PARASOCIAL. Living as a colony in which adult females belong to a single generation. Communal, quasisocial, and semisocial categories are included under parasocial.

PHEROMONE. A substance, small quantities of which serve for chemical communication among individuals of a species.

PHYSOGASTRIC. Much swollen with eggs, like queens of various social insects, including meliponine bees.

POLLEN BASKET. Corbicula.

POLLEN POCKET. A reservoir for pollen beside a cell in certain species of bumblebees. Both larvae and adults have access to the contents of the pocket.

POLLEN POT. A pollen storage reservoir in nests of bumblebees or stingless honeybees. Larvae do not have direct access to the pollen.

POLYGYNOUS. Referring to a colony that contains two or more gynes (Pleometrotic). Commonly this means that the colony contains two or more queens, but they may be only potential queens.

POLYLECTIC. Taking pollen from a wide variety of kinds of flowers. The term is applied to species; polylectic species have a broad range of pollen sources.

PREPUPA. The last part of the last larval stadium, when the larva has defecated, ceased feeding, and straightened its body. For many bees it is a resistant stage of long duration, for example, for hibernation or estivation.

PRIMITIVELY EUSOCIAL. Living as a eusocial colony in which the castes are morphologically similar and food exchange among adults is absent or minimal.

PROBOSCIS. A compound structure (made up of maxillae and labium) which serves for taking up liquids such as nectar.

PROGRESSIVE FEEDING. Progressive provisioning. Feeding of larvae at intervals during their growth. Used in contrast to mass provisioning.

PRONOTUM. The dorsum of the first thoracic segment.

PROPODEUM. The first abdominal segment, which in most Hymenoptera has become an integral part of the second body tagma along with the thorax proper.

PROPOLIS. Resins ordinarily collected from plants, used in the nest by honeybees.

PROPRIOCEPTOR. A sensory end organ or sensillum that responds to positions of the body, making the insect aware of the relative positions of its parts.

PROTERANDRY. The appearance or emergence of males before females.

QUASISOCIAL. Living as a colony in which a group of females of the same generation cooperatively construct and provision cells; all the females have enlarged ovaries and presumably lay.

QUEEN. The member (or members) of a colony that is primarily active in egg laying and relatively or totally inactive in foraging.

Queen substance. The inhibitory pheromones secreted by queen honeybees that prevent construction of queen cells by workers and enlargement of the ovaries of workers.

Replacement. Production of a new queen in a highly eusocial colony to replace a queen that has died or been lost. Often spoken of as emergency replacement. *See also* Supersedure.

Reproductives. Gynes (including queens) and males.

Retinue. Court.

Royal jelly. Bee milk deposited in gyne cells of *Apis*.

Sclerite. A hardened plate or portion of the exoskeleton.

Sclerotized. Hardened to form a sclerite.

Scopa. The brush of hairs with which pollen is carried to the nest from flowers. In Apidae the scopa is reduced to the fringe that rims the corbicula.

Semisocial. Living as a colony consisting of a group of females of the same generation showing castes and division of labor, one or more being queens, the others workers (auxiliaries).

Sensillum. A sensory end organ.

Silk. Secretion of larvae with which cocoons are constructed.

Sleeping cluster. A group of usually male bees that assemble for the night away from the nest.

Social. There are various legitimate usages of this term for bees; see Chapter 5. For this book any members of a colony constitute a social group.

Social interaction. Interrelations among individuals of a colony such that behavior of one or more influences behavior or development of others (see Chapter 4).

Solitary. Living alone in a nest. In solitary species each female makes her own nest or nests, provisions and lays in the cells, and has little interaction with other females either of her own or the daughter generation.

Spermatheca. The reservoir in the female in which sperm cells are stored after mating and from which they are released to fertilize eggs.

Sphecoidea. A superfamily of wasps from which the bees are believed to have originated.

Stadium. The stage or interval between molts during the growth of an insect.

Sternum. The ventral portion of a body segment.

Stigmergy. The influence of constructs already made on subsequent construction activity. Instead of direct communication among nest mates, stigmergy involves social interaction mediated by previously accomplished work.

Stingless bee. Stingless honeybee.

Stingless honeybee. A bee of the tribe Meliponini.

Subsocial. Living as a colony consisting of an adult female and her immature offspring, which are progressively fed.

Supersedure. The production of a new queen to replace an aging one, still present in a colony of highly eusocial bees. *See also* Replacement.

Swarm. A queen and workers that establish a new colony of highly eusocial bees.

Tarsus. The last portion of the leg, beyond the tibia. It is divided into five subsegments.

Teneral. Callow.

Tergum. The dorsal portion of a body segment.

Thelytoky. Production of females from unfertilized eggs.

Tibia. The second long segment of the leg, between the femur and the tarsus.

Trophallaxis. Interchange of material among the members of a colony.

Trophic eggs. Eggs produced for food rather than for reproduction.

Tumulus. The pile of loose earth around the entrance of a nest burrow, pushed out by bees excavating beneath the surface of the soil.

Wax. The secretion from between the abdominal terga or sterna of most Apidae. It is used in nest building.

Worker. A colony member that is active in foraging or nesting and that lays few eggs relative to the queen.

Young worker food. Bee milk supplied to young worker larvae of *Apis*.

Literature Cited

Akahira, Y., S. F. Sakagami, and R. Zucchi. 1970. Die Nähreier von Arbeiterinnen einer stachellosen Biene, *Trigona* (*Scaptotrigona*) *postica,* die von der Königin kurz vor der eigenen Eiablage gefressen werden. *Zool. Anzeiger* 185:85–93.

Alford, D. V. 1969. A study of the hibernation of bumblebees in southern England. *J. Anim. Ecol.* 38:149–170.

———. 1970. The incipient stages of development of bumblebee colonies. *Insectes Sociaux* 17:1–10.

Allen, M. D. 1960. The honeybee queen and her attendants. *Anim. Behav.* 8:201–208.

———. 1965a. The role of the queen and males in the social organization of insect communities. *Symp. Zool. Soc. London* no. 14:133–157.

———. 1965b. The ages of worker honeybees (*Apis mellifera* L.) engaged in "fanning" and in "shiver dances." *Anim. Behav.* 13:347.

Altenkirch, G. 1962. Untersuchungen über die Morphologie der abdominalen Hautdrüsen einheimischer Apiden. *Zool. Beitr.* (n.f.)7:161–238.

Anderson, R. H. 1963. The laying worker in the Cape honeybee, *Apis mellifera capensis. J. Apicult. Res.* 2:85–92.

Anonymous. 1971. World honey crop reports. *Bee World* 52:109.

Aptel, E. 1931. Étude sur les nidifications de l'*Halictus malachurus* K. (forma *longulus*) génération d'été. *Bull. Soc. Entomol. France* 1931:219–222.

Atwal, A. S., and O. P. Sharma. 1968. The introduction of *Apis mellifera* queens into *Apis indica* colonies and the associated behavior of the two species. *Indian Bee J.* 30:41–56.

Autrum, H. J., and M. Stoeker. 1950. Die Verschmelzungsfrequenz des Bienenauge. *Zeitschr. Naturforsch.* 5b:38–43.

Avdeeva, O. I. 1965. Influence of royal jelly on the development of bees in the embryonic period [in Russian]. *Pchelovodstvo* 85(7):10–11.

Baker, H. G., and P. D. Hurd. 1968. Intrafloral ecology. *Annu. Rev. Entomol.* 13:385–414.

Barbier, M., and E. Lederer. 1960. Structure chemique de la "substance royale" de la reine d'abeille (*Apis mellifica*). *Compt. Rend. Acad. Sci.* [Paris] 250:4467-4469.

———, E. Lederer, and T. Nomura. 1960. Synthèse de l'acide céto-9-décène-2-trans-oique (substance royale) et de l'acide céto-8-nonène-2-trans-oique. *Compt. Rend. Acad. Sci.* [Paris] 251:1133–1135.

———, and J. Pain. 1960. Étude de la secretion des glandes mandibulaires des reines et des ouvrieres d'abeilles (*Apis mellifica* L.) par chromatographie en phase gazeuse. *Compt. Rend. Acad. Sci.* [Paris] 250:3740.

Bassindale, R. 1955. The biology of the stingless bee *Trigona* (*Hypotrigona*) *gribodoi* Magretti. *Proc. Zool. Soc. London* 125:49–62.

Batra, L. R., S. W. T. Batra, and G. E. Bohart. 1973. The mycoflora of domesticated and wild bees (Apoidea). *Mycopathologia et Mycologia applicata* 49:13–44.

Batra, S. W. T. 1964. Behavior of the social bee, *Lasioglossum zephyrum,* within the nest. *Insectes Sociaux* 11:159–186.

———. 1965. Organisms associated with *Lasioglossum zephyrum. J. Kansas Entomol. Soc.* 38:367–389.

———. 1966a. Social behavior and nests of some nomiine bees in India. *Insectes Sociaux* 13:145–154.

———. 1966b. Life cycle and behavior of the primitively social bee, *Lasioglossum zephyrum. Univ. Kansas Sci. Bull.* 46:359–423.

———. 1966c. Nesting behavior of *Halictus scabiosae* in Switzerland (Hymenoptera, Halictidae). *Insectes Sociaux* 13:87–92.

———. 1968. Behavior of some social and solitary halictine bees within their nests; a comparative study. *J. Kansas Entomol. Soc.* 41:120–133.

———, and G. E. Bohart. 1969. Alkali bees: response of adults to pathogenic fungi in brood cells. *Science* 165:607-608.

Beier, W., and M. Lindauer. 1970. Der Sonnenstand als Zeitgeber für die Biene. *Apidologie* 1:5–28.

———, I. Medugorac, and M. Lindauer. 1968. Synchronisation

et dissociation de "l'horlage interne" des abeilles par des facteurs externes. *Ann. Epiphyties* 19:133–144.

Beig, D. 1971. Desenvolvimento embrionario de abelhas operárias de *Trigona* (*Scaptotrigona*) *postica* Latreille. *Arq. Zool.* [São Paulo] 21:179–234.

———. 1972. The production of males in queenright colonies of *Trigona* (*Scaptotrigona*) *postica*. *J. Apicult. Res.* 11:33–39.

———, and S. F. Sakagami. 1964. Behavior studies of the stingless bees, with special reference to the oviposition process II *Melipona seminigra merrillae* Cookerell (sic). *Annot. Zool. Jap.* 37:112–119.

Beling, I. 1929. Ueber das Zeitgedächtnis der Biene. *Zeitschr. vergl. Physiol.* 9:259–338.

Bennett, F. D. 1965. Notes on a nest of *Eulaema terminata* Smith with a suggestion of the occurrence of a primitive social system. *Insectes Sociaux* 12:81–91.

Bienyenu, R. J., F. W. Atchison, and E. A. Cross. 1968. Microbial inhibition by the prepupae of the alkali bee, *Nomia melanderi*. *J. Invert. Pathol.* 12:278–282.

Bischoff, H. 1927. *Biologie der Hymenopteren.* Berlin: J. Springer.

Blackith, R. E. 1957. Social facilitation at the nest entrance of some Hymenoptera. *Physiol. Comp. Oecologia* 4:388–402.

Blagoveschenskaya, N. N. 1963. Giant colony of the solitary bee *Dasypoda plumipes* Pz. [in Russian]. *Entomol. Obozrenie* 42:115–117.

Blum, M. S. 1966. Chemical releasers of social behavior. VIII. Citral in the mandibular gland secretion of *Lestrimelitta limao*. *Ann. Entomol. Soc. Amer.* 59:962–964.

———. 1970. The chemical basis of insect sociality. In *Chemicals controlling insect behavior,* ed. M. Beroza, pp. 61–94. New York: Academic Press.

———, R. M. Crewe, W. E. Kerr, L. H. Keith, A. W. Garrison, and M. M. Walker. 1970. Citral in stingless bees: isolation and functions in trail-laying and robbing. *J. Insect Physiol.* 16:1637–1648.

Boch, R., and D. A. Shearer. 1963. Production of geraniol by honeybees of various ages. *J. Insect Physiol.* 9:431–434.

———, and D. A. Shearer. 1966. Iso-pentyl acetate in stings of honeybees of different ages. *J. Apicult. Res.* 5:65–70.

———, and D. A. Shearer. 1967. 2-Heptanone and 10-hydroxy-*trans*-dec-2-enoic acid in the mandibular glands of worker honeybees of different ages. *Zeitschr. vergl. Physiol.* 54:1–11.

———, D. A. Shearer, and A. Petrasovits. 1970. Efficacies of two alarm substances of the honeybee. *J. Insect Physiol.* 16:17–24.

Bohart, G. E. 1955. Gradual nest supersedure within the genus *Osmia. Proc. Entomol. Soc. Washington* 57:203–204.

———, W. P. Stephen, and R. K. Eppley. 1960. The biology of *Heterostylum robustum,* a parasite of the alkali bee. *Ann. Entomol. Soc. Amer.* 53:425–435.

Bonelli, B. 1948. Osservazioni biologiche sull' "*Halictus malachurus*" K. *Boll. Ist. Entomol. Univ. Bologna* 17:22–44.

———. 1964. Osservazioni biologiche sugli imenotteri melliferi e predatori della Val di Fiemme, VI. *Boll. Ist. Entomol. Univ. Bologna* 27:33–48.

———. 1965a. Osservazioni biologiche sugli imenotteri melliferi e predatori della Val di Fiemme, VIII contributo. *Studi Trentini Sci. Natur.* (B)42:97–122.

———. 1965b. Osservazioni biologiche sugli imenotteri melliferi e predatori della Val di Fiemme, VII contributo, *Halictus calceatus* Scop. (sin. *Lasioglossum cylindricum*). *Studi Trentini Sci. Natur.* (B)42:5–54.

———. 1968. Osservazioni biologiche sugli imenotteri melliferi e predatori della Val di Fiemme, XXVIII contributo, *Halictus calceatus* Scop. *Studi Trentini Sci. Natur.* (B)45:42–47.

Bott, R. 1937. Der Nestbau der Furchenbiene (*Halictus malachurus* Kirby). *Natur und Volk* 67:73–83.

Bozina, K. D. 1961. How long does the queen live? *Pchelovodstvo* 38(6):13.

Brian, A. D. 1951. Brood development in *Bombus agrorum. Entomol. Monthly Mag.* 87:207–212.

———. 1952. Division of labor and foraging in *Bombus agrorum* Fabricius. *J. Anim. Ecol.* 21:223–240.

———. 1954. The foraging of bumble bees. *Bee World* 35:61–67, 81–91.

Brian, M. V. 1965a. Caste differentiation in social insects. *Symp. Zool. Soc. London* no. 14:13–38.

———. 1965b. *Social insect populations.* London: Academic Press.

———, and A. D. Brian. 1948. Regulation of oviposition in social Hymenoptera. *Nature* 161:854–856.

Buschinelli, A., and A. C. Stort. 1965. Estudo do comportamento de "*Melipona pseudocentris pseudocentris*" (Cockerell, 1912). *Rev. Brasileira Biol.* 25:67–80.

Butler, C. G. 1949. *The honeybee.* Oxford: Clarendon Press.

———. 1954a. *The world of the honeybee.* London: Collins.

———. 1954b. The method and importance of the recognition by a colony of honeybees (*A. mellifera*) of the presence of its queen. *Trans. Roy. Entomol. Soc. London* 105:11–29.

———. 1957. The process of queen supersedure in colonies of honeybees (*Apis mellifera* Linn.). *Insectes Sociaux* 4:211–223.

———. 1959. Queen substance. *Bee World* 40:269–275.

———. 1960. The significance of queen substance in swarming and supersedure in honeybee (*Apis mellifera* L.) colonies *Proc. Roy. Entomol. Soc. London* (A)35:129–132.

———. 1961. The scent of queen honeybees (*A. mellifera* L.) that causes partial inhibition of queen rearing. *J. Insect Physiol.* 7:258–264.
———. 1965. Sex attraction in *Andrena flavipes* Panzer, with some observations on nest-site restriction. *Proc. Roy. Entomol. Soc. London* (A)40:77–80.
———. 1967a. Insect pheromones. *Biol. Rev.* 42:42–87.
———. 1967b. A sex attractant acting as an aphrodisiac in the honeybee (*Apis mellifica* L.). *Proc. Roy. Entomol. Soc. London* (A)42:71–76.
———. 1969. Some pheromones controlling honeybee behavior. *Proc. VI Congr. Internat. Union Study Soc. Ins.* p. 19–32, Bern.
———. 1971. The mating behavior of the honeybee (*Apis mellifera* L.). *J. Entomol.* (A)46:1–11.
———, and D. H. Calam. 1969. Pheromones of the honeybee—the secretion of the Nassanoff gland of the worker. *J. Insect Physiol.* 15:237–244.
———, and R. K. Callow. 1968. Pheromones of the honeybee (*Apis mellifera* L.): the "inhibitory scent" of the queen. *Proc. Roy. Entomol. Soc. London* (A)43:62–65.
———, R. K. Callow, and N. C. Johnston. 1961. The isolation and synthesis of queen substance, 9-oxodec-*trans*-2-enoic acid, a honeybee pheromone. *Proc. Roy. Soc.* [London] (B)155:417–432.
———, and E. M. Fairey. 1963. The role of the queen in preventing oogenesis in worker honeybees. *J. Apicult. Res.* 2:14–18.
———, D. J. C. Fletcher, and D. Watler. 1969. Nest-entrance marking with pheromones by the honeybee *Apis mellifera* L., and a wasp, *Vespula vulgaris* L. *Anim. Behav.* 17:142–147.
———, D. J. C. Fletcher, and D. Watler. 1970. Hive entrance finding by honeybee (*Apis mellifera*) foragers. *Anim. Behav.* 18:78–91.
———, and J. Simpson. 1958. The source of the queen substance of the honeybee (*Apis mellifera* L.). *Proc. Roy. Entomol. Soc. London* (A)33:120–122.
———, and J. Simpson. 1965. Pheromones of the honeybee (*Apis mellifera* L.). An olfactory pheromone from the Koschewnikow gland of the queen. *Vědecké práce Výzkumného ústavu včelařského v Dole,* 1965, pp. 33–36; also cited by Butler, 1967a, as *Scientific Studies,* Univ. Libeice, Czechoslovakia, 4:33–36.
———, and J. Simpson. 1967. Pheromones of the queen honeybee (*Apis mellifera* L.) which enable her workers to follow her when swarming. *Proc. Roy. Entomol. Soc. London* (A)42:149–154.
Buttel-Reepen, H. von. 1903. *Die stammesgeschichtliche Entstehung des Bienenstaates.* Leipzig: G. Thieme.
———. 1915. *Leben und Wesen der Bienen.* Braunschweig: Vieweg und Sohn.
Camargo, C. A. de. 1972a. Determinação de castas em *Scaptotrigona postica* Latreille. *Rev. Brasileira Biol.* 32:133–138.
———. 1972b. Produção "in vitro" de intercastas em *Scaptotrigona postica* Latreille. In *Homenagem à Warwick E. Kerr,* pp. 37–54. Rio Claro, Brazil.
———. 1972c. Mating of the social bee *Melipona quadrifasciata* under controlled conditions. *J. Kansas Entomol. Soc.* 45:520–523.
Camargo, J. M. F. de. 1970. Ninhos e biologia de algumas espécies de Meliponideos da região de Pôrto Velho, Território de Rondônia, Brasil. *Rev. Biol. Trop.* [San José, Costa Rica] 16:207–239.
———, W. E. Kerr, and C. R. Lopes. 1967. Morfologia externa de *Melipona* (*Melipona*) *marginata* Lepeletier. *Papéis Avulsos Dept. Zool.* [São Paulo] 20:229–258 + 1 plate.
Camillo, C. 1972. Alguns aspectos do comportamento de *Melipona rufiventris rufiventris* Lepeletier. In *Homenagem à Warwick E. Kerr,* pp. 57–62. Rio Claro, Brazil.
Canetti, S. J., R. W. Shuel, and S. E. Dixon. 1964. Studies on the mode of action of royal jelly in honeybee development IV. *Canadian J. Zool.* 42:229–233.
Cardale, J. 1968. Nests and nesting behaviour of *Amegilla* (*Amegilla*) *pulchra* (Smith). *Australian J. Zool.* 16:689–707.
Carrière, J., and O. Bürger. 1897. Die Entwickelungsgeschichte der Mauerbiene (*Chalicodoma muraria* Fabr.) im Ei, *Abhandlungen Kaiser Leopold.-Carolin. Deut. Akad. Naturforsch.* 69:253–420, plates xiii–xxv.
Chauvin, R. (ed.). 1968. *Traité de biologie de l'abeille,* Vols. I–V. Paris: Masson et Cie.
Claude-Joseph, F. 1926. Recherches biologique sur les Hyménoptères du Chili. *Ann. Sci. Natur., Zool.* (10)9:113–268. [Spanish version, 1960, *Publ. del Centro de Estudios Entomol., Univ. Chile* no. 1, pp. 1–60, 18 plates.]
Comstock, J. H. 1924. *An introduction to entomology.* Ithaca, New York: Comstock Publishing Co.
Cosenza, G. W. 1972. Estudo dos enxames de migração de abelhas africanas. *Proc. 1st Congr. Brasileiro de Apicult.* [Florianopolis] pp. 128–129.
Crozier, R. H. 1970. On the potential for genetic variability in haplo-diploidy. *Genetica* 41:551–556.
Cruz Landim, C. da. 1963. Evolution of the wax and scent glands of the Apinae. *J. New York Entomol. Soc.* 71:2–13.
———. 1967. Estudo comparativo de algumas glândulas das abelhas (Hymenoptera, Apoidea) e respectivas implicações evolutivas. *Arq. Zool.* [São Paulo] 15:177–290.
———, M. A. da Costa Cruz, A. C. Stort, and E. W. Kitajima. 1965. Orgão tibial dos machos de Euglossini, estudo ao

microscópio óptico e electrônico. *Rev. Brasileira Biol.* 25:323–342.

———, and A. Ferreira. 1968. Mandibular gland development and communication in field bees of *Trigona* (*Scaptotrigona*) *postica*. *J. Kansas Entomol. Soc.* 41:474–481.

Cumber, R. A. 1949a. The biology of humble-bees, with special reference to the production of the worker caste. *Trans. Roy. Entomol. Soc. London* 100:1–45.

———. 1949b. An overwintering nest of the humble-bee *Bombus terrestris* (L.). *New Zeal. Sci. Rev.* 7:96–97.

———. 1954. The life cycle of humble-bees in New Zealand. *New Zeal. J. Sci. Technol.* (B)36:95–107.

———. 1963. Studies of an unusually large nest of *Bombus terrestris* (Hymenoptera, Apidae) transferred to an observation box. *New Zeal. J. Sci.* 6:66–74.

Custer, C. P. 1928a. The bee that works in stone; *Perdita opuntiae* Cockerell. *Psyche* 35:67–84.

———. 1928b. On the nesting habits of *Melissodes* Latr., *Can. Entomol.* 60:28–31.

Dade, H. A. 1962. *Anatomy and dissection of the honeybee.* London: Bee Research Association.

Darchen, R. 1957. La reine d'*Apis mellifica,* les ouvrières pondeuses et les constructions cirières. *Insectes Sociaux* 4:321–325.

———. 1962. Observation directe du développement d'un rayon de cire. Le role des chaines d'abeilles. *Insectes Sociaux* 9:103–120.

———. 1966. Sur l'éthologie de *Trigona* (*Dactylurina*) *staudingeri* Gribodoi (*sic*). *Biol. Gabonica* 2:37–45.

———. 1969. Le comportement de ponte des reines des deux espèces d'abeilles sans dard (Trigones). *Gaz. Apicult.* No. 743:103–104.

———. 1970. La division du travail chez quelques apides sociaux—abeilles domestiques et méliponides. *Gaz. Apicult.* no. 754:48–51.

———. 1971. *Trigona* (*Axestotrigona*) *oyani* Darchen, une nouvelle espece d'abeille africaine—description du nid inclus dan une fourmilière. *Biol. Gabonica* 7:407–421.

———. 1973. Essai d'interprétation du déterminisme des castes chez les Trigones et les Mélipones. *Compt. Rend. Acad. Sci.* [Paris] 276:607–609.

———, and B. Delage. 1970. Facteur déterminant les castes chez les Trigones. *Compt. Rend. Acad. Sci.* [Paris] 270:1372–1373.

———, and B. Delage-Darchen. 1971. Le déterminisme des castes chez les Trigones (Hyménoptères Apidés). *Insectes Sociaux* 18:121–134.

———, and J. Lensky. 1963. Quelques problèmes soulevés par la création de sociétés polygynes d'abeilles. *Insectes Sociaux* 10:337–357.

———, and J. Louis. 1961. Les mélipones et leur élevage. *Melipona-Trigona-Lestremelitta* (sic). *Ann. Abeille* 4:5–39.

———, and J. Pain. 1966. Le nid de *Trigona* (*Dactylurina*) *staudingeri* Gribodoi (*sic*). *Biol. Gabonica* 2:25–35.

Darwin, C. 1876. *The effects of cross and self fertilisation in the vegetable kingdom.* London: Murray.

———. 1965. On the flight paths of male humble bees. In *The works of Charles Darwin,* ed, R. B. Freeman, pp. 70-73. London: Dawsons. [See also 1968, *Bull. British Mus.* (Natur. Hist.), Hist. Ser. 3:179–189.]

Daumer, K. 1956. Reizmetrische Untersuchung des Farbensehens der Biene. *Zeitschr. vergl. Physiol.* 38:413–478.

———. 1958. Blumenfarben, wie sie die Bienen sehen. *Zeitschr. vergl. Physiol.* 41:49–110.

———. 1963. Kontrastempfindlichkeit der Bienen für "Weiss" verschiedenen UV-Gehalts. *Zeitschr. vergl. Physiol.* 46:336–350.

Deegener, P. 1918. Die Formen der Vergesellschaftung im Tierreich. Leipzig: Englemann.

Deleurance, E. P. 1949. Phénomène social chez *Osmia emarginata* Lep. *Bull. Soc. Entomol. France* 54:9–10.

Delvert-Salleron, F. 1963. Étude, au moyen de radio-isotopes, des échanges de nourritures entre reines, mâles et ouvrières d'*Apis mellifica* L. *Ann. Abeille* 6:201–227.

Den Boer, P. J., and D. A. Vleugel. 1949. Ethologische vaarnemingen aan een nest van *Bombus e. equestris* (F.). *Tijdschr. voor Entomol.* 91:121–134.

Dias, D. 1958. Contribuição para o conhecimento da bionomia de *Bombus incarum* Franklin da Amazônia. *Rev. Brasileira Entomol.* 8:1–20.

Dias, L. B. L., and D. Simões. 1972. Releção entre estrutura etária (normal e anormal) de colonias de *Scaptotrigona postica* Latreille e desenvolvimento glandular. In *Homenagem à Warwick E. Kerr,* pp. 135–142 Rio Claro, Brazil.

Diwan, V. V., and S. R. Salvi. 1965. Some interesting behavioural features of *Apis dorsata* Fabr. *Indian Bee J.* 27–52.

Dixon, S. E., and R. W. Shuel. 1963. Studies in the mode of action of royal jelly in honeybee development, III. *Can. J. Zool.* 41:733–739.

———, and R. W. Shuel. 1965. Major events in the first three days in the larval life of the honeybee. *Proc. XII Internat. Congr. Entomol.* [London] 12:166.

Dodson, C. H. 1966. Ethology of some bees of the tribe Euglossini. *J. Kansas Entomol. Soc.* 39:607–629.

———, R. L. Dressler, H. G. Hills, R. M. Adams, and N. H. Williams. 1969. Biologically active compounds in orchid fragrances. *Science* 164:1243-1249.

Dolphin, R. E. 1966. The ecological life history of *Halictus* (*Seladonia*) *confusus* Smith. *Diss. Abstr.* 27(B):1968.

Dreischer, H. 1956. Untersuchungen über die Arbeitstätigkeit und Drüsenentwicklung altersbestimmter Bienen im weisellosen Volk. *Zool. Jahrbucher, Physiol.* 66:429–472.

Drescher, W. 1969. Die Flugaktivität von Drohnen der Rasse *Apis mellifica carnica* L. und. *A. mell. ligustica* L. in Abhängigkeit von Lebensalter und Witterung. *Zeitschr. Bienenforschung* 9:390–409.

DuPraw, E. J. 1961. A unique hatching process in the honeybee. *Trans. Amer. Microscop. Soc.* 80:185–191.

———. 1967. The honeybee embryo. In *Methods in developmental biology,* ed. F. Wilt and N. Wessels, pp. 183–217. New York: Crowell Co.

Eberhard, M. J. W. 1969. The social biology of polistine wasps. *Misc. Pub. Mus. Zool., Univ. Michigan* no. 140:1–101.

Eickwort, G. C. 1969a. A comparative morphological study and generic revision of the augochlorine bees. *Univ. Kansas Sci. Bull.* 48:325–524.

———. 1969b. Tribal positions of Western Hemisphere green sweat bees, with comments on their nest architecture. *Ann. Entomol. Soc. Amer.* 62:652–660.

———, and K. R. Eickwort. 1971. Aspects of the biology of Costa Rican halictine bees, II, *Dialictus umbripennis* and adaptations of its caste structure to different climates. *J. Kansas Entomol. Soc.* 44:343–373.

———, and K. R. Eickwort. 1972a. Aspects of the biology of Costa Rican halictine bees, IV, *Augochlora* (*Oxystoglosella*). *J. Kansas Entomol. Soc.* 45:18–45.

———, and K. R. Eickwort. 1972b. Aspects of the biology of Costa Rican halictine bees, III. *Sphecodes kathleenae,* a social cleptoparasite of *Dialictus umbripennis. J. Kansas Entomol. Soc.* 45:529–541.

———, and K. R. Eickwort. 1973a. Aspects of the biology of Costa Rican halictine bees, V. *Augochlorella edentata* (Hymenoptera: Halictidae). *J. Kansas Entomol. Soc.* 46:3–16.

———, and K. R. Eickwort. 1973b. Notes on the nests of three wood-dwelling species of *Augochlora* from Costa Rica (Hymenoptera: Halictidae). *J. Kansas Entomol. Soc.* 46:17–22.

Emerson, A. E. 1950. The supraorganismic aspects of the society. In *Colloques Internationaux du Centre National de la Recherche Scientifique,* vol. 34, *Structure et Physiologie des Sociétés Animaux,* pp. 333–353, plate XVII. Paris.

Erickson, R., and T. Rayment. 1951. Simple social bees of Western Australia. *Western Australian Natur.* 3:45–59.

Erp, A. van. 1960. Mode of action of the inhibitory substance of the honeybee queen. *Insectes Sociaux* 7:207–211.

Esch, H. 1961. Über die Schallerzeugung beim Werbetanz der Honigbiene. *Zeitschr. vergl. Physiol.* 45:1–11.

———. 1963. Über die Auswirkung der Futterplatzqualität auf die Schallerzeugung im Werbetanz der Honigbiene. *Verhandl. Deut. Zool. Ges. Wien* 1962:302–309.

———. 1964a. Beiträge zum Problem der Entfernungsweisung in den Schwänzeltänzen der Honigbiene. *Zeitschr. vergl. Physiol.* 48:534–546.

———. 1964b. Über den Zusammenhang zwischen Temperatur, Aktionspotentiaten und Thoraxbewegungen bei der Honigbiene (*Apis mellifica*). *Zeitschr, vergl. Physiol.* 48:547–551.

———. 1967a. The sound produced by swarming honeybees. *Zeitschr. vergl. Physiol.* 56:408–411.

———. 1967b. Die Bedeutung der Lauterzeugung für die Verständigung der stachellosen Bienen. *Zeitschr. vergl. Physiol.* 56:199–220.

———, and J. A. Bastian. 1970. How do newly recruited honeybees approach a food site. *Zeitschr. vergl. Physiol.* 68:175–181.

———, I. Esch, and W. E. Kerr. 1965. An element common to communication of stingless bees and to dances of honeybees. *Science* 149:320–321.

Eskov, E. K. 1968. The sound component in the dances of various strains. *Pchelovodstvo* 88:25.

———. 1969. Evidence of the informative role of the sound components in honeybee mobilization dances. *Zh. Obshch. Biol.* 30:317–323.

Fabre, J. H. 1914. *The mason-bees.* New York: Dodd, Mead, and Co.

———. 1915. *Bramble bees and others.* New York: Dodd, Mead, and Co.

Ferton, C. 1901. Sur l'epoque du réveil des bourdons et des psithyres à Bonifacio. *Ann. Soc. Entomol. France* 70:84–85.

Frank, A. 1941. Eigenartige Flugbahnen bei Hummelmannchen. *Zeitschr. vergl. Physiol.* 28:467–484.

Free, J. B. 1955a. Queen production in colonies of bumblebees. *Proc. Roy. Entomol. Soc. London* (A)30:19–25.

———. 1955b. The division of labor within bumblebee colonies. *Insectes Sociaux* 2:195–212.

———. 1955c. The behavior of egg-laying workers of bumblebee colonies. *Brit. J. Anim. Behav.* 3:147–153.

———. 1955d. The collection of food by bumblebees. *Insectes Sociaux* 2:303–311.

———. 1957. The effect of social facilitation on the ovary development of bumblebee workers. *Proc. Roy. Entomol. London* 32:182–184.

———. 1958. The defense of bumblebee colonies. *Behaviour* 12:233–242.

———. 1961. Hypopharyngeal gland development and division of labour in honeybee (*Apis mellifera* L.) colonies. *Proc. Roy. Entomol. Soc. London* (A)36:5–8.

———. 1965. The allocation of duties among worker honeybees. *Symp. Zool. Soc. London* no. 14:39–59.

———. 1966. The foraging areas of honeybees in an orchard of standard apple trees. *J. Appl. Ecol.* 3:261–268.

———. 1967a. The production of drone comb by honeybee colonies. *J. Apicult. Res.* 6:29–36.

———. 1967b. Factors determining the collection of pollen by honeybee foragers. *Anim. Behav.* 15:134–144.
———. 1968. The conditions under which foraging honeybees expose their Nasonov gland. *J. Apicult. Res.* 7:139–145.
———. 1969. Influence of the odour of a honeybee colony's stores on the behaviour of its foragers. *Nature* 222:778.
———. 1970a. The flower constancy of bumblebees. *J. Anim. Ecol.* 39:395–402.
———. 1970b. *Insect pollination of crops.* London and New York: Academic Press.
———. 1970c. Effect of flower shape and nectar guides on the behaviour of foraging honeybees. *Behaviour* 37:269–285.
———. 1971. Stimuli eliciting mating behavior of bumblebee (*Bombus pratorum*) males. *Behaviour* 40:55–61.
———, and C. G. Butler. 1959. *Bumblebees.* London: Collins.
———, and J. Simpson. 1968. The alerting pheromones of the honeybee. *Zeitschr. vergl. Physiol.* 61:361–365.
———, I. Weinberg, and A. Whiten. 1969. The egg-eating behaviour of *Bombus lapidarius* L. *Behaviour* 35:313–317.
———, and I. H. Williams. 1970. Exposure of the Nasonov gland by honeybees (*Apis mellifera*) collecting water. *Behaviour* 37:286–290.
———, and I. H. Williams. 1972. The role of the Nasonov gland pheromone in crop communication by honeybees. *Behaviour* 41:316–318.
Friese, H. 1891. Beiträge zur Biologie der solitären Blumenwespen. *Zool. Jahrbucher* (*Syst.*) 5:751–860, plate xlviii.
———. 1923. *Die europaeischen Bienen.* Berlin and Leipzig: Walter de Gruyter & Co.
Frisch, K. von. 1950. *Bees, their vision, chemical senses, and language.* Ithaca, New York: Cornell Univ. Press.
———. 1953. *Aus dem Leben der Bienen.* Revised ed. Berlin: Springer-Verlag.
———. 1955. *The dancing bees.* New York: Harcourt, Brace & Co.
———. 1965. *Tanzsprache und Orientierung der Bienen.* Berlin: Springer-Verlag.
———. 1967a. *The dance language and orientation of bees.* Cambridge, Massachusetts: Harvard Univ. Press.
———. 1967b. Honeybees: do they use direction and distance information provided by their dancers. *Science* 158:1072–1076.
———. 1968. The role of dances in recruiting bees to familiar sites. *Anim. Behav.* 16:531–533.
———, and R. Jander. 1957. Über den Schwänzeltanz der Bienen. *Zeitschr. vergl. Physiol.* 40:239–263.
Frison, T. H. 1917. Notes on Bombidae, and on the life history of *Bombus auricomus* Robt. *Ann. Entomol. Soc. Amer.* 10:277–286, plates xxiii–xxiv.
———. 1918. Additional notes on the life history of *Bombus auricomus* Robt. *Ann. Entomol. Soc. Amer.* 11:43–48, plate iii.
———. 1928. A contribution to the knowledge of the life history of *Bremus bimaculatus* (Cresson). *Entomol. Amer.* (n.s.) 8:159–223.
———. 1929. A contribution to the knowledge of the bionomics of *Bremus impatiens* (Cress.). *Bull. Brooklyn Entomol. Soc.* 24:261–285.
———. 1930. A contribution to the knowledge of the bionomics of *Bremus americanorum* (Fabr.). *Ann. Entomol. Soc. Amer.* 23:644–665.
Fukuda, H., and S. F. Sakagami. 1968. Worker brood survival in honeybees. *Res. Population Ecol.* 10:31–39.
———, and K. Sekiguchi. 1966. Seasonal change of the honeybee worker longevity in Sapparo, north Japan, with notes on some factors affecting the life-span. *Jap. J. Ecol.* 16:206–212.
Fye, R. E., and J. T. Medler. 1954. Temperature studies in bumblee (*sic*) domiciles. *J. Econ. Entomol.* 47:847–852.
Gary, N. E. 1961. Queen honeybee attractiveness as related to mandibular gland secretion. *Science* 133:1479–1480.
———. 1962. Chemical mating attractants in the queen honeybee. *Science* 136:773.
———. 1963. Observations of mating behaviour in the honeybee. *J. Apicult. Res.* 2:3–13.
———. 1967. Diurnal variations in the intensity of flight activity from honeybee colonies. *J. Apicult. Res.* 6:65–68.
———. 1970. Pheromones of the honeybee, *Apis mellifera* L. In *Control of insect behavior by natural products,* pp. 29–53. New York: Academic Press.
———, and J. Marston. 1971. Mating behavior of drone honeybees with queen models. *Anim. Behav.* 19:299–304.
———, and R. A. Morse. 1962a. Queen cell construction in honeybee (*Apis mellifera* L.) colonies headed by queens without mandibular glands. *Proc. Roy. Entomol. Soc. London* (A)37:76–78.
———, and R. A. Morse. 1962b. The events following queen cell construction in honeybee colonies. *J. Apicult. Res.* 1:3–5.
Gasbichler, J. 1968. Die optische Orientierung der Honigbiene beim Rückflug zum Stock. *Zeitschr. Bienenforschung* 9:344–355.
Gast, R. 1967. Untersuchungen über den Einfluss der Koniginnensubstanz auf die Entwicklung der endokrinen Drüsen bei der Arbeiterin der Honigbiene (*Apis mellifica*). *Insectes Sociaux* 14:1–12.
Gerber, H. S., and E. C. Klostermeyer. 1970. Sex control by bees: a voluntary act of egg fertilization during oviposition. *Science* 167:82–84.
Girard, M. 1874. Note sur les moeurs des melipones et des

trigones du Brésil. *Ann. Soc. Entomol. France* 43:567–573.
Glushkov, N. M. 1958. Problems in beekeeping in the U.S.S.R. in relation to pollination. *Bee World* 39:81–92.
———. 1964. Increase in the productivity of bees by rearing them in combs with enlarged cells [in Russian]. *Trud. Nauch.—issled. Inst. Pchelovodstva* pp. 34–57.
Goetsch, W. 1940. *Vergleichende Biologie der Insekten-Staaten.* Leipzig: Becker and Erler.
Gonçalves, L. S. 1969. A study of orientation information given by one trained bee by dancing. *J. Apicult. Res.* 8:113–132.
Gonnet, M., P. Lavie, and P. Nogueira-Neto. 1964. Étude de quelques caractéristiques des miels récoltés par certains méliponines brésiliens. *Compt. Rend. Acad. Sci.* [Paris] 258:3107–3109.
Gould, J. L., M. Henerey, and M. C. Mac Leod. 1970. Communication of direction by the honeybee. *Science* 169:544–554.
Grandi, G. 1964. Studio sull'*Osmia emarginata* Lepel. *Boll. Istituto entomol. Univ. Bologna* 27:127–144.
Grassé, P.-P. 1959. La reconstruction du nid et les coordinations individuelles chez *Bellicositermes natalensis* et *Cubitermes* sp., La théorie de la stigmergie, etc. *Insectes Sociaux* 6:41–80.
Gribakin, F. G. 1969. Cellular basis of color vision in the honeybee. *Nature* 223:639–640.
Grout, R. A. 1963. *The hive and the honeybee.* Revised. Hamilton, Illinois: Dadant and Sons.
Grozdanić, S. 1966. Nova posmatranja na nekim vrstama roda *Halictus* (Hymenoptera) prilog poznavanju filogenetskog postanka pčelinje porodice. [New studies concerning certain species of the genus *Halictus.*] *Inst. Biol. Istrazivanja [Beograd], Zbornik radova* 10(15):1–29.
———. 1971. Biologische untersuchungen an den Bienen *Halictus sajoi* (Blüthgen). *Bull. Mus. Hist. Natur.* [Beograd] (B)26:151–157.
———, and Z. Mučalica. 1968. Biologische Untersuchungen an den Bienen—*Halictus interruptus* Panz. *Arh. Biol. Nauka* [Beograd] 20:9p–10p.
———, and Z. Vasić. 1970. Biološka ispitivanja na Pčelama—*Halictus malachurus* [German summary]. *Bull. Mus. Hist. Natur.* [Beograd] (B)25:271–303.
Gutbier, A. 1915. Essai sur la classification et sur le développement des nids des guêpes et des abeilles [in Russian, French summary]. *Horae Soc. Entomol. Rossicae* 41:1–57, plates I–II.
Haas, A. 1946. Neue Beobachtungen zum Problem der Flugbahnen bei Hummelmannchen. *Zeitschr. Naturforsch.* 1:596–600.
———. 1952. Die Mandibeldrüse als Duftorgan bei einigen Hymenopteren. *Naturwissenschaften* 39:484.
Hackwell, G. A., and W. P. Stephen. 1966. Eclosion and duration of larval development in the alkali bee *Nomia melanderi* Cockerell. *Pan-Pacific Entomol.* 42: 196–200.
Hamilton, W. D. 1964. The genetical evolution of social behavior, I and II. *J. Theoretical Biol.* 7:1–52.
———. 1973. Altruism and related phenomena, mainly in social insects. *Annu. Rev. Ecol. Syst.* 3:193–232.
Hannawald, E. B. 1967. Statistics on bees and honey. In *Beekeeping in the United States.* U.S. Dept. Agr. Handbook, no. 335, pp. 139–142.
Hartl, D. L., and S. W. Brown. 1970. The origin of male haploid genetic systems and their expected sex ratio. *Theoretical Population Biology* [Academic Press] 1(2):165–190.
Hasselrot, T. B. 1960. Studies on Swedish bumblebees (genus *Bombus* Latr.) their domestication and biology. *Opuscula Entomol.* Suppl. 17:1–203.
Hawkins, R. P. 1956. A preliminary survey of red clover seed production. *Ann. Appl. Biol.* 44:657–664.
Hebling, N. J., W. E. Kerr, and F. S. Kerr. 1964. Divisão de trabalho entre operárias de *Trigona* (*Scaptotrigona*) *xanthotricha* Moure. *Papéis Avulsos Dept. Zool.* [São Paulo] 16:115–127.
Heinrich, B. 1972a. Patterns of endothermy in bumblebee queens, drones and workers. *J. Comp. Physiol.* 77: 65–79.
———. 1972b. Physiology of brood incubation in the bumblebee queen, *Bombus vosnesenskii. Nature* 239:223–225.
———. 1972c. Temperature regulation in the bumblebee *Bombus vagans:* a field study. *Science* 175:185–187.
Hirashima, Y. 1958. Comparative studies of the cocoon-spinning habits of *Osmia excavata* Alfken and *Osmia pedicornis* Cockerell. *Sci. Bull. Fac. Agr., Kyushu Univ.* 16:481–497.
———. 1959. Biological studies on the parthenogenesis in *Osmia excavata* Alfken and *Osmia cornuta* (Radoszkowski). *Sci. Bull. Fac. Agr., Kyushu Univ.* 17:55–68.
———. 1961. Monographic study of the subfamily Nomiinae of Japan (Halictidae). *Acta Hymenopterologica* 1:241–303.
Hobbs, G. A. 1964. Phylogeny of bumblebees based on brood rearing behaviour. *Can. Entomol.* 96:115–116.
———. 1964–68. Ecology of species of *Bombus* (Hymenoptera: Apidae) in southern Alberta. *Can. Entomol.* 96:1465–1470; 97:120–128, 1293–1302; 98:33–39, 288–294; 99:1271–1292; 100:156–164.
———. 1967. Obtaining and protecting red-clover pollinating species of *Bombus. Can. Entomol.* 99:943–951.
———, W. O. Nummi, and J. F. Virostek. 1962. Managing

colonies of bumblebees (Hymenoptera, Apidae) for pollination purposes. *Can. Entomol.* 94:1121–1132.

Hodges, D. 1952. *The pollen loads of the honeybee.* London: Bee Research Association Ltd.

Hoffer, E. 1882. Die Hummeln Steiermarks, I. *Jahresberichte der steiermarkischen Landes-Oberrealschule in Graz über das Studienjahr, 1881/82* pp. 1–92, plates A and 1–2.

Hoffmann, I. 1961. Über die Arbeitsteilung in weiselrichtigen und weisellosen Kleinvolkern der Honigbiene. *Zeitschr. Bienenforschung* 5:267–279.

———. 1966. Gibt es bei Drohnen von *Apis mellifica* L. ein echtes Füttern oder nur eine Futterabgabe? *Zeitschr. Bienenforschung* 8:249–255.

Holm, S. N. 1966. The utilization and management of bumblebees for red clover and alfalfa seed production. *Annu. Rev. Entomol.* 11:155–182.

Houston, T. F. 1970. Discovery of an apparent male soldier caste in a nest of a halictine bee with notes on the nest. *Australian J. Zool.* 18:345–351.

Ihering, R. von. 1903a. Zur Frage nach dem Ursprung der Staatenbildung bei den sozialen Hymenopteren. *Zool. Anzeiger* 27:113–118.

———. 1903b. Biologische Beobachtungen an brasilianischen *Bombus*-Nestern. *Allg. Zeitschr. Entomol.* 8:447–453.

Iwata, K. 1938. Habits of some bees in Formosa (II) and (IV). *Trans. Natur. Hist. Soc. Formosa* 28:205–215, 373–379.

———. 1964. Egg giantism in subsocial Hymenoptera, with ethological discussion on tropical bamboo carpenter bees. *Nature and Life in Southeast Asia* [Kyoto] 3:399–434, plate I.

Jacobs-Jessen, U. F. 1959. Zur Orientierung der Hummeln und einiger anderer Hymenopteren. *Zeitschr. vergl. Physiol.* 41:597–641.

Jander, R., and U. Jander. 1970. Über die Phylogenie der Geotaxis innerhalb der Bienen. *Zeitschr. vergl. Physiol.* 66:355–368.

Janzen, D. H. 1971. Englossine bees as long-distance pollinators of tropical plants. *Science* 171:203–205.

Jay, S. C. 1963a. The longitudinal orientation of the larval honeybees (*Apis mellifera*) in their cells. *Can. J. Zool.* 41:717–723.

———. 1963b. The development of honeybees in their cells. *J. Apicult. Res.* 2:117–134.

———. 1964. The cocoon of the honeybee, *Apis mellifera* L. *Can. Entomol.* 96:784–792.

Jaycox, E. R. 1970. Honeybee foraging behavior: Responses to queens, larvae, and extracts of larvae. *Ann. Entomol. Soc. Amer.* 63:1689–1694.

Johnson, D. L. 1967. Honeybees: Do they use the direction information contained in their dance maneuver. *Science* 155:844–847.

———, and A. M. Wenner. 1970. Recruitment efficiency in honeybees: studies of the role of olfaction. *J. Apicult. Res.* 9:13–18.

Johnston, N. C., J. H. Law, and N. Weaver. 1965. Metabolism of 9-ketodec-2-enoic acid by worker honeybees (*Apis mellifera* L.). *Biochemistry* 4:1615–1621.

Jones, R. L., and W. C. Rothenbuhler. 1964. Behaviour genetics of nest cleaning in honeybees II. *Anim. Behav.* 12:584–588.

Juliani, L. 1962. O aprisionamento de rainhas virgens em colônias de Trigonini. *Bol. Univ. Paraná* (Zool.) no. 20:1–11.

———. 1967. A descrição do ninho e alguns dados biológicos sôbre a abelha *Plebeia julianii* Moure, 1962 (Hymenoptera, Apoidea). *Rev. Brasileira Entomol.* [São Paulo] 12:31–58, 2 plates.

Jung-Hoffmann, I. 1966. Die Determination von Königin und Arbeiterin der Honigbiene. *Zeitschr. Bienenforschung* 8:296–322.

Kaissling, K.-E., and M. Renner. 1968. Antennale Rezeptoren für Queen Substance and Sterzelduft bei der Honigbiene. *Zeitschr. vergl. Physiol.* 59:357–361.

Kalmus, H. 1954. The clustering of honeybees at a food source. *Brit. J. Anim. Behav.* 2:63–71.

Kapil, R. P., and S. Kumar. 1969. Biology of *Ceratina binghami* Ckll. *J. Res.* 6:359–371. Ludhiana: Punjab Agric. Univ.

Katayama, E. 1966. Studies on the development of broods of *Bombus diversus* Smith, II. Brood development and feeding habits. *Kontyû* 34:8–17

———. 1971. Observations on the brood development in *Bombus ignitus,* I. Egg-laying habits of queens and workers. *Kontyû* 39:189–204.

Kelner-Pillault, S. 1969. Abeilles fossiles ancetres des apides sociaux. *Proc. VI Congr. Internat. Union Study Social Ins.* [Bern] pp. 85–93.

Kempff Mercado, N. 1962. Mutualism between *Trigona compressa* Latr. and *Crematogaster stolli* Forel. *J. New York Entomol. Soc.* 70:215–217.

Kerr, W. E. 1946. Formação das castas no gênero Melípona. *An. Escola Superior Agr. "Luiz de Queiroz"* 3:299–312.

———. 1951. Bases para o estudo da genética de populações dos Hymenoptera em geral e dos Apinae sociais em particular. *An. Escola Superior Agr. "Luiz de Queiroz"* 8:219–354, figs. 3–15.

———. 1959. Bionomy of meliponids—VI—Aspects of food gathering and processing in some stingless bees. Food gathering in Hymenoptera, *Symp. Entomol. Soc. Amer.* Detroit meeting, sponsored by Apiculture Subsection, mimeographed, pp. 24-31.

———. 1967. The history of the introduction of African bees to Brazil. *South African Bee J.* 39:(2):3–5 and *Apiculture in Western Australia* 2:53–55.

———. 1969. Some aspects of the evolution of social bees.

Evolut. Biol. 3:119–175. New York: Appleton-Century-Crofts.

———. 1971. Contribuição à ecogenetica da algumas espécies de abelhas. *Ciencia e Cult.* [São Paulo] 23(suppl.): 89–90.

———, and C. da Costa Cruz. 1961. Funções diferentes tomadas pela glândula mandibular na evolução das abelhas em geral e em "*Trigona* (*Oxytrigona*) *tataira*" em especial. *Rev. Brasileira Biol.* 21:1–16.

———, and H. Esch. 1965. Comunicação entre as abelhas sociais brasileiras e sua contribuição para o entendimento da sua evolução. *Ciencia e Cult.* [São Paulo] 17:529–538.

———, A. Ferreira, and N. S. de Mattos. 1963. Communication among stingless bees—additional data. *J. New York Entomol. Soc.* 71:80–90.

———, and N. J. Hebling. 1964. Influence of weight of worker bees on division of labor. *Evolution* 18:267–270.

———, and H. H. Laidlaw. 1956. General genetics of bees. *Advances in genetics* 8:109–153. New York: Academic Press.

———, and E. de Lello. 1962. Sting glands in stingless bees—a vestigial character. *J. New York Entomol. Soc.* 70:190–214.

———, and R. A. Nielsen. 1966. Evidences that genetically determined *Melipona* queens can become workers. *Genetics* 54:859–866.

———, and R. A. Nielsen. 1967. Sex determination in bees (Apinae). *J. Apicult. Res.* 6:1–9.

———, S. F. Sakagami, R. Zucchi, V. de Portugal-Araújo, and J. M. F. de Camargo. 1967. Observacçẽs sôbre a arquitetura dos ninhos e comportamento de algumas espécies de abelhas sem ferrão dos vizenhanças de Manaus, Amazonas. *Atas do Simposio sôbre a Biota Amazônica* 5(Zool.):255–309.

———, and G. R. dos Santos Neto. 1956. Contribuição para o conhecimento da bionomia dos Meliponini. 5. Divisão de trabalho entre operarias de *Melipona quadrifasciata quadrifasciata* Lep. *Insectes Sociaux* 3:423–430.

———, A. C. Stort, and M. J. Montenegro. 1966. Importância de alguns fatôres ambientais na determinação das castas do gênero *Melipona. An. Acad. Brasileira Cien.* 38:149–168.

———, R. Zucchi, J. T. Nakadaira, and J. E. Botolo. 1962. Reproduction in the social bees. *J. New York Entomol. Soc.* 70:265–276.

Khalifman, I. 1971. *Trubachi Igrayut Sbor.* Moscow.

Kiechle, H. 1961. Die soziale Regulation der Wassersammeltätigkeit im Bienenstaat und deren physiologische Grundlage. *Zeitschr. vergl. Physiol.* 45:154–192.

Knaffl, H. 1953. Über die Flugweite und Entfernungsmeldung der Bienen. *Zeitschr. Bienenforschung* 2:131–140.

Knee, W. J., and J. T. Medler. 1965. The seasonal size increase of bumblebee workers. *Can. Entomol.* 97:1149–1155.

Knerer, G. 1969a. Stones, cement and guards in halictine nest architecture and defense. *Entomol News* 80:141–147.

———. 1969b. Brood care in halictine bees. *Science* 164:429–430.

———. 1969c. Sozialstruktur und ihre Rolle in der Populationsdynamik von Furchenbienen. *Proc. VI Congr. Internat. Union Study Social Ins.* [Bern] pp. 101–107.

———, and C. E. Atwood. 1966. Polymorphism in some nearctic halictine bees. *Science* 152:1262–1263.

———, and C. E. Atwood. 1967. Parasitization of social halictine bees in southern Ontario. *Proc. Entomol. Soc. Ontario* 97:103–110.

———, and C. Plateaux-Quénu. 1966a. Sur la polygynie chez les Halictinae. *Compt. Rend. Acad. Sci.* [Paris] 263:2014–2017.

———, and C. Plateaux-Quénu. 1966b. Sur le polymorphisme des femelles chez quelques Halictinae. *Compt. Rend. Acad. Sci.* [Paris] 263:1759–1761.

———, and C. Plateaux-Quénu. 1966c. Sur l'importance de l'ouverture des cellules á couvain dans l'évolution des Halictinae. *Compt Rend. Acad. Sci.* [Paris] 263:1622–1625.

———, and C. Plateaux-Quénu. 1967a. Usurpation de nids étrangers et parasitisme facultatif chez *Halictus scabiosae* (Rossi). *Insectes Sociaux* 14:47–50.

———, and C. Plateaux-Quénu. 1967b. Comparaison de la construction et de l'architecture de quelques nids d'*Evylaeus* à rayon de cellules. *Compt. Rend. Acad. Sci.* [Paris] 265:455–458.

———, and C. Plateaux-Quénu. 1967c. Sur la production continue ou périodique de couvain chez les Halictinae. *Compt. Rend. Acad. Sci.* [Paris] 264:651–653.

———, and C. Plateaux-Quénu. 1967d. Sur la production de males chez les Halictinae sociaux. *Compt. Rend. Acad. Sci.* [Paris] 264:1096–1099.

———, and C. Plateaux-Quénu. 1970. The life cycle and social level of *Evylaeus nigripes,* a Mediterranean halictine bee. *Can. Entomol.* 102:185–196.

Koeniger, N. 1969. Experiments concerning the ability of the queen (*Apis mellifica* L.) to distinguish between drone and worker cells. *XXII Internat. Beekeeping Congr. Sum.* p. 138.

———. 1970. Factors determining the laying of drone and worker eggs by the queen honeybee. *Bee World* 51:166–169, and Über die Fähigkeit der Bienenkönigin (*Apis mellifica*) zwischen Arbeiterinnen und Drohnenzellen zu unterscheiden. *Apidologie* 1:115–142.

Koltermann, R. 1969. Lern- und Vergessensprozesse bei der Honigbiene—aufgezeigt anhand von Duftdressuren. *Zeitschr. vergl. Physiol.* 63:310–334.

Krombein, K. V. 1967. *Trap-nesting wasps and bees: Life histories, nests, and associates.* Washington, D. C.: Smithsonian Press.

Kropáčová, S., and H. Hasslbachová. 1971. The influence of queenlessness and of unsealed brood on the development of ovaries in worker honeybees. *J. Apicult. Res.* 10:57–61.

Krunić, M. 1959. Prelazne forme ismedu solitarnih i socijalnik pčela. [Transition forms between solitary and social bees.] *Zbornik Matice Srpske* (Novi Sad) 17:102–111.

Kugler, H. 1943. Hummeln als Blütenbesucher. *Ergebnisse Biol.* 19:143–323.

Kullenberg, B., G. Bergström, and S. Ställberg-Stenhagen. 1970. Volatile components of the cephalic marking secretion of male bumblebees. *Acta Chem. Scand.* 24:1481–1482.

Lacher, V. 1967. Verhaltensreaktionen der Bienenarbeiterin bei Dressur auf Kohlendioxid. *Zeitschr. vergl. Physiol.* 54:75–84.

Laidlaw, H. H., F. P. Gomes, and W. E. Kerr. 1956. Estimation of the number of lethal alleles in a panmitic population of *Apis mellifera* L. *Genetics* 41:179–188.

Lau, D. 1959. Beobachtungen und Experimente über die Entstehung der Bienenwabe (*Apis mellifica* L.). *Zool. Beitr.* (n.f.)4:233–306.

Lauer, J., and M. Lindauer. 1971. Genetisch fixierte Lerndispositionen bei der Honigbiene. In *Informationsaufnahme und Informations-verarbeitung im lebenden Organisms, Akad. Wissensch. Lit., Math.-Naturwissensch. Klasse* pp. 75–102. [Mainz.]

Lavie, P. 1960. Les substances antibactériennes dans la colonie d'abeilles (*Apis mellifica* L.). *Ann. Abeille* 3:103–183, 201–305.

Lecomte, J. 1963. Étude des échanges de nourriture de la colonie de bourdons au moyen de radioisotopes. *Compt. Rend. Acad. Sci.* [Paris] 257:3664–3665.

Legewie, H. 1925a. Zur Theorie der Staatenbildung I. Teil: Die biologie der Furchenbiene *Halictus malachurus* K. *Zeitschr. Morphol. Ökologie Tiere* 3:619–683.

———. 1925b. Zur Theorie der Staatenbildung II. Teil. *Zeitschr. Morphol. Ökologie Tiere* 4:246–300.

———. 1925c. Zum Problem des tierischen Parasitismus I. Teil: Die Lebensweise der Schmarotzerbiene *Sphecodes monilicornis* K. (= *subquadratus* Sm.). *Zeitschr. Morphol. Ökologie Tiere* 4:430–464.

Lehmensick, R., and G. Stein. 1958. Biologische Beobachtungen an einen polygynen Nest von *Bombus hypnorum* L. *Zool. Jahrbucher* (*Syst.*) 86:71–84.

Le Masne, G. 1952. Classification et caractéristiques des principaux types de groupements sociaux réalisés chez les invertébrés. *Colloque International sur la Structure et la Physiologie des Sociétés Animales* [Paris] pp. 19–70, plates I–II.

Levchenko, I. A., I. I. Shalimov, I. G. Bagrii, V. N. Olifir, and V. A. Gubin. 1969. Comparative study of signalling in some races of honeybees. *XXII Internat. Beekeeping Congr. Sum.* p. 146.

Levin, M. D. 1966. Orientation of honeybees in alfalfa with respect to landmarks. *J. Apicult. Res.* 5:121–125.

———. 1967. Pollination. In *Beekeeping in the United States.* U. S. Dept. Agr. Handbook 335:77–85.

———, and S. Glowska-Konopacka. 1963. Responses of foraging honeybees in alfalfa to increasing competition from other colonies. *J. Apicult. Res.* 2:33–42.

Lin, N. 1964. Increased parasitic pressure as a major factor in the evolution of social behavior in halictine bees. *Insectes Sociaux* 11:187–192.

———, and C. D. Michener. 1972. Evolution and selection in social insects. *Quart. Rev. Biol.* 47:131–159.

Lindauer, M. 1949. Ueber die Einwirkung von Duft- und Geschmacksstoffen sowie anderer Faktoren auf die Tänze der Bienen. *Zeitschr. vergl. Physiol.* 31:348–412.

———. 1952. Ein Beitrag zur Frage der Arbeitsteilung im Bienenstaat. *Zeitschr. vergl. Physiol.* 34:299–345. Translation, *Bee World* 34:63–73, 85–90.

———. 1955. Schwarmbienen auf Wohnungssuche. *Zeitschr. vergl. Physiol.* 37:263–324.

———. 1957a. Communication among the honeybees and stingless bees of India. *Bee World* 38:3–14, 34–39.

———. 1957b. Sonnenorientierung der Bienen unter der Aquatorsonne und zur Nachtzeit. *Naturwissenschaften* 44:1–6.

———. 1961. *Communication among social bees.* Cambridge, Massachusetts: Harvard Univ. Press.

———. 1967. Recent advances in bee communication and orientation. *Annu. Rev. Entomol.* 12:439–470.

———. 1969. Lernen und Vergessen bei der Honigbiene. *Proc. VI Congr. Internat. Union Study Social Ins.* pp. 153–158.

———. 1971. The functional significance of the honeybee waggle dance. *Amer. Natur.* 105:89–96.

———, and W. E. Kerr. 1958. Die gegenseitige Verständigung bei den stachellosen Bienen. *Zeitschr. vergl. Physiol.* 41:405–434.

———, and W. E. Kerr. 1960. Communication between the workers of stingless bees. *Bee World* 41:29–41, 65–71.

———, and H. Martin. 1968. Die Schwereorientierung der Bienen unter dem Einfluss des Erdmagnetfeldes. *Zeitschr. vergl. Physiol.* 60:219–243.

———, and J. O. Nedel. 1959. Ein Schweresinnesorgan der Honigbiene. *Zeitschr. vergl. Physiol.* 42:334–364.

Lopatina, N. G., and E. G. Chesnokova. 1969. The role of training in receiving information on space location of a food source by honeybees [in Russian, English summary]. *XXII Internat. Beekeeping Congr. Sum.* p. 147.

Lotmar, R. 1945. Die Metamorphose des Bienendarmes. *Beih. Schweiz. Bienen-Zeit.* 1:443–506.

Lukoschus, F. 1956. Untersuchungen zur Entwicklung der Kasten-Merkmale bei der Honigbiene (*Apis mellifica* L.). *Zeitschr. Morphol. Ökologie Tiere* 45:157–197.

Lüscher, M., and I. Walker. 1963. Zur Frage der Wirkungsweise der Koniginnenpheromone bei der Honigbiene. *Rev. Suisse Zool.* 70:304–311.

MacKay, D. M. 1972. Formal analysis of communicative processes. In *Non-verbal communication,* ed. R. A. Hinde, pp. 3–25. Cambridge, England: Cambridge Univ. Press.

Maidl, F. 1934. *Die Lebensgewohnheiten und Instinkte der staatenbildenden Insekten.* Wien: Fritz Wagner.

Malyshev, S. I. 1924. The nesting habits of *Panurginus* Nyl. [in Russian, English summary]. *Izv. Nauch. Inst. Imeny P. F. Lesshafta* 9:196–200.

———, 1935. The nesting habits of solitary bees, a comparative study. *Eos* 11:201–309, plates III-XV.

———. 1968. *Genesis of the Hymenoptera.* London: Methuen and Co.

Markl, H. 1971. Proprioceptive gravity perception in Hymenoptera, in *Gravity and the organism,* ed S. A. Gordon and M. J. Cohen, pp. 185–194. Chicago: Univ. Chicago Press.

Martin, H., and M. Lindauer. 1966. Sinnesphysiologische Leistungen beim Wabenbau der Honigbiene. *Zeitschr. vergl. Physiol.* 53:372–404.

Martin, P. 1963. Die Steuerung der Volksteilung beim Schwärmen der Bienen. Zugleich ein Beitrag zum Problem der Wanderschwärme. *Insectes Sociaux* 10:13–42.

Maschwitz, U. W. 1964a. Alarm substances and alarm behavior in social Hymenoptera. *Nature* 204:324–327.

———. 1964b. Gefahrenalarmstoffe und Gefahrenalarmierung bei sozialen Hymenopteren. *Zeitschr. vergl. Physiol.* 47:596–655.

Matsuka, M., N. Watabe, and K. Takeuchi. 1973. Analysis of the food of larval drone honeybees. *J. Apicult. Res.* 12:3–7.

Maurizio, A. 1959. Factors influencing the lifespan of bees. *Ciba Foundation Symposium, The Lifespan of Animals,* pp. 231–243.

———. 1961. Lebensdauer und Altern bei der Honigbiene (*Apis mellifica* L.). *Gerontologia* 5:110–128.

Mautz, D. 1971. Der Kommunikationseffekt der Schwänzeltänze bei *Apis mellifica carnica* (Pollm.). *Zeitschr. vergl. Physiol.* 72:197–220.

———, R. Boch, and R. A. Morse. 1972. Queen finding by swarming honey bees. *Ann. Entomol. Soc. Amer.* 65:440–443.

May, D. G. K. 1970a. Chemical nature, origin and biological significance of the brood cell lining of a sweat bee, *Augochlora pura.* Ph.D. Thesis, Univ. Kansas, pp. 1–76.

———. 1970b. Brood care in halictid bees. *Science* 170:651–652.

———. 1972. Water uptake during larval development of a sweat bee, *Augochlora pura. J. Kansas Entomol. Soc.* 45:439–449.

May, J. 1959. *Čmeláci v ČSR.* Praha: Ceskoslovenská Akad. Zemědělských Věd.

Meder, E. 1958. Über die Einberechnung der Sonnenwanderung bei der Orientierung der Honigbiene. *Zeitschr. vergl. Physiol.* 40:610–641.

Medler, J. T. 1965. Variation in size in the worker caste of *Bombus fervidus* (Fab.). *Proc. XII Internat. Congr. Entomol.* [London], pp. 388–389.

Meidell, O. 1934. Fra Dagliglivet i et homlebol. [The daily life in a bumblebee nest.] *Naturen* [Bergen] 58:85–95, 108–116.

———. 1968. *Bombus jonellus* (Kirby) has two generations in a season. *Norsk Entomol. Tidsskr.* 14:31–32.

Menzel, R. 1967, 68, 69. Das Gedächtnis der Honigbiene für Spektralfarben. *Zeitschr. vergl. Physiol.* 56:22–62; 60:82–102; 63:290–309.

———, and J. Erber. 1972. The influence of the quantity of reward on the learning performance in honeybees. *Behaviour* 41:27–42.

Michener, C. D. 1944. Comparative external morphology, phylogeny, and a classification of the bees (Hymenoptera). *Bull. Amer. Mus. Natur. Hist.* 82:151–326.

———. 1946. Notes on the habits of some Panamanian stingless bees. *J. New York Entomol. Soc.* 54:179–197.

———. 1953a. The biology of a leafcutter bee (*Megachile brevis*) and its associates. *Univ. Kansas Sci. Bull.* 35:1659–1748.

———. 1953b. Comparative morphological and systematic studies of bee larvae with a key to families of hymenopterous larvae. *Univ. Kansas Sci. Bull.* 35:987–1102.

———. 1954. Bees of Panamá. *Bull. Amer. Mus. Natur. Hist.* 104:1–175.

———. 1958. The evolution of social behavior in bees. *Proc. X Internat. Congr. Entomol.* [Montreal] 2:441–448.

———. 1960a. Notes on the behavior of Australian colletid bees. *J. Kansas Entomol. Soc.* 33:22–31.

———. 1960b. Observations on the behavior of a burrowing bee (*Amegilla*) near Brisbane, Queensland. *Queensland Natur.* 16:63–68.

———. 1961a. Observations on the nests and behavior of *Trigona* in Australia and New Guinea. *Amer. Mus. Novitates* 2026:1–46.

———. 1961b. Social polymorphism in Hymenoptera. In *Insect Polymorphism, Symposium no. 1,* ed. J. S. Kennedy, pp. 43–56. London: Royal Entomological Society.

———. 1961c. Probable parasitism among Australian bees of the genus *Allodapula. Ann. Entomol. Soc. Amer.* 54:532–534.

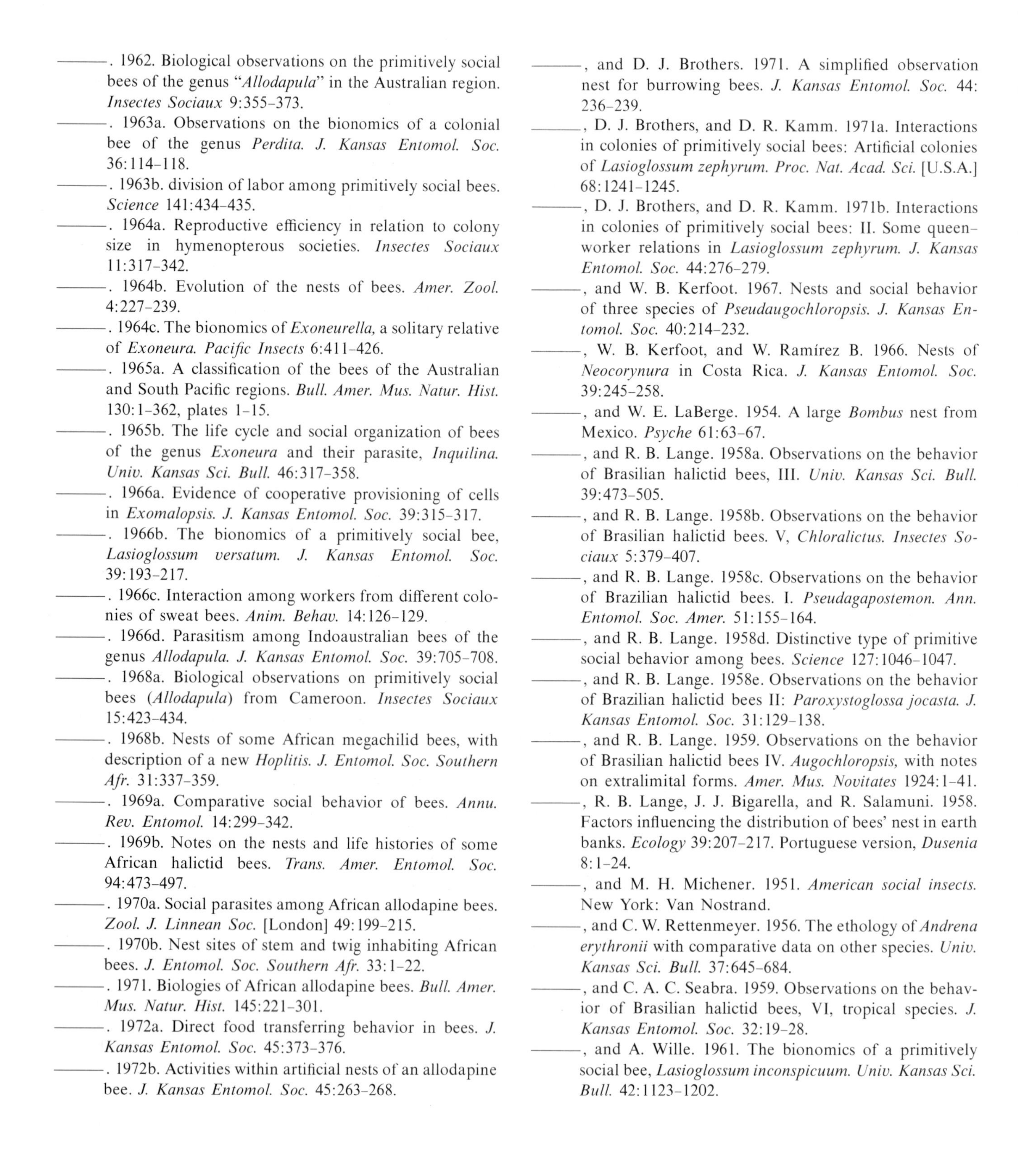

———. 1962. Biological observations on the primitively social bees of the genus "*Allodapula*" in the Australian region. *Insectes Sociaux* 9:355–373.

———. 1963a. Observations on the bionomics of a colonial bee of the genus *Perdita*. *J. Kansas Entomol. Soc.* 36:114–118.

———. 1963b. division of labor among primitively social bees. *Science* 141:434–435.

———. 1964a. Reproductive efficiency in relation to colony size in hymenopterous societies. *Insectes Sociaux* 11:317–342.

———. 1964b. Evolution of the nests of bees. *Amer. Zool.* 4:227–239.

———. 1964c. The bionomics of *Exoneurella*, a solitary relative of *Exoneura*. *Pacific Insects* 6:411–426.

———. 1965a. A classification of the bees of the Australian and South Pacific regions. *Bull. Amer. Mus. Natur. Hist.* 130:1–362, plates 1–15.

———. 1965b. The life cycle and social organization of bees of the genus *Exoneura* and their parasite, *Inquilina*. *Univ. Kansas Sci. Bull.* 46:317–358.

———. 1966a. Evidence of cooperative provisioning of cells in *Exomalopsis*. *J. Kansas Entomol. Soc.* 39:315–317.

———. 1966b. The bionomics of a primitively social bee, *Lasioglossum versatum*. *J. Kansas Entomol. Soc.* 39:193–217.

———. 1966c. Interaction among workers from different colonies of sweat bees. *Anim. Behav.* 14:126–129.

———. 1966d. Parasitism among Indoaustralian bees of the genus *Allodapula*. *J. Kansas Entomol. Soc.* 39:705–708.

———. 1968a. Biological observations on primitively social bees (*Allodapula*) from Cameroon. *Insectes Sociaux* 15:423–434.

———. 1968b. Nests of some African megachilid bees, with description of a new *Hoplitis*. *J. Entomol. Soc. Southern Afr.* 31:337–359.

———. 1969a. Comparative social behavior of bees. *Annu. Rev. Entomol.* 14:299–342.

———. 1969b. Notes on the nests and life histories of some African halictid bees. *Trans. Amer. Entomol. Soc.* 94:473–497.

———. 1970a. Social parasites among African allodapine bees. *Zool. J. Linnean Soc.* [London] 49:199–215.

———. 1970b. Nest sites of stem and twig inhabiting African bees. *J. Entomol. Soc. Southern Afr.* 33:1–22.

———. 1971. Biologies of African allodapine bees. *Bull. Amer. Mus. Natur. Hist.* 145:221–301.

———. 1972a. Direct food transferring behavior in bees. *J. Kansas Entomol. Soc.* 45:373–376.

———. 1972b. Activities within artificial nests of an allodapine bee. *J. Kansas Entomol. Soc.* 45:263–268.

———, and D. J. Brothers. 1971. A simplified observation nest for burrowing bees. *J. Kansas Entomol. Soc.* 44: 236–239.

———, D. J. Brothers, and D. R. Kamm. 1971a. Interactions in colonies of primitively social bees: Artificial colonies of *Lasioglossum zephyrum*. *Proc. Nat. Acad. Sci.* [U.S.A.] 68:1241–1245.

———, D. J. Brothers, and D. R. Kamm. 1971b. Interactions in colonies of primitively social bees: II. Some queen–worker relations in *Lasioglossum zephyrum*. *J. Kansas Entomol. Soc.* 44:276–279.

———, and W. B. Kerfoot. 1967. Nests and social behavior of three species of *Pseudaugochloropsis*. *J. Kansas Entomol. Soc.* 40:214–232.

———, W. B. Kerfoot, and W. Ramírez B. 1966. Nests of *Neocorynura* in Costa Rica. *J. Kansas Entomol. Soc.* 39:245–258.

———, and W. E. LaBerge. 1954. A large *Bombus* nest from Mexico. *Psyche* 61:63–67.

———, and R. B. Lange. 1958a. Observations on the behavior of Brasilian halictid bees, III. *Univ. Kansas Sci. Bull.* 39:473–505.

———, and R. B. Lange. 1958b. Observations on the behavior of Brasilian halictid bees. V, *Chloralictus*. *Insectes Sociaux* 5:379–407.

———, and R. B. Lange. 1958c. Observations on the behavior of Brazilian halictid bees. I. *Pseudagapostemon*. *Ann. Entomol. Soc. Amer.* 51:155–164.

———, and R. B. Lange. 1958d. Distinctive type of primitive social behavior among bees. *Science* 127:1046–1047.

———, and R. B. Lange. 1958e. Observations on the behavior of Brazilian halictid bees II: *Paroxystoglossa jocasta*. *J. Kansas Entomol. Soc.* 31:129–138.

———, and R. B. Lange. 1959. Observations on the behavior of Brasilian halictid bees IV. *Augochloropsis*, with notes on extralimital forms. *Amer. Mus. Novitates* 1924:1–41.

———, R. B. Lange, J. J. Bigarella, and R. Salamuni. 1958. Factors influencing the distribution of bees' nest in earth banks. *Ecology* 39:207–217. Portuguese version, *Dusenia* 8:1–24.

———, and M. H. Michener. 1951. *American social insects*. New York: Van Nostrand.

———, and C. W. Rettenmeyer. 1956. The ethology of *Andrena erythronii* with comparative data on other species. *Univ. Kansas Sci. Bull.* 37:645–684.

———, and C. A. C. Seabra. 1959. Observations on the behavior of Brasilian halictid bees, VI, tropical species. *J. Kansas Entomol. Soc.* 32:19–28.

———, and A. Wille. 1961. The bionomics of a primitively social bee, *Lasioglossum inconspicuum*. *Univ. Kansas Sci. Bull.* 42:1123–1202.

Milliron, H. E. 1967. A successful method for artificially hibernating *Megabombus f. fervidus,* and notes on a related species. *Can. Entomol.* 99:1321–1332.

———. 1971. A monograph of the western hemisphere bumblebees I. *Mem. Entomol. Soc. Can.* 82:1–80.

———, and D. R. Oliver. 1966. Bumblebees from northern Ellesmere Island, with observations on usurpation by *Megabombus hyperboreus* (Schönh.). *Can. Entomol.* 98:207–213.

Mitsui, T., T. Sagawa, and H. Sano. 1964. Studies on rearing honeybee larvae in the laboratory. I. The effect of royal jelly taken from different ages of queen cells on queen differentiation. *J. Econ. Entomol.* 57:518–521.

Miyamoto, S. 1963. Biology of *Bombus ignitus* Smith. *Kontyu* 31:91–98.

Morse, R. A. 1966. Honeybee colony defense at low temperatures. *J. Econ. Entomol.* 59:1091–1093.

———. 1970. Annotated bibliography on *Apis dorsata;* same on *Apis florea.* Bibliographies 9 and 10, pp. 1–25 and 1–16, Bee Research Association [mimeographed].

———. 1972. Environmental control in the bee hive. *Sci. Amer.* 226:93–98.

———, and R. Boch. 1971. Pheromone concert in swarming honey bees. *Ann. Entomol. Soc. Amer.* 64:1414–1417.

———, and F. M. Laigo. 1969. *Apis dorsata* in the Philippines. *Monogr. Philippine Assoc. Entomol.* 1:1–96.

———, and J. L. McDonald. 1965. The treatment of capped queen cells by honeybees. *J. Apicult. Res.* 4:31–34.

———, D. A. Shearer, R. Boch, and A. W. Benton. 1967. Observations on alarm substances in the genus *Apis. J. Apicult. Res.* 6:113–118.

Moure, J. S., P. Nogueira-Neto, and W. E. Kerr. 1958. Evolutionary problems among meliponinae. *Proc. X Internat. Congr. Entomol.* [Montreal] 2:481–493.

———, and S. F. Sakagami. 1962. As mamangabas sociais do Brasil (*Bombus* Latreille). *Studia Entomol.* 5:65–194.

Munakata, M., and S. F. Sakagami. 1958. Zum Verfliegen der Hummeln unter den künstlich zusammengesetzten Völkchen. *Kontyu* 26:15–19.

Nedel, J. O. 1960. Morphologie und Physiologie der Mandebeldrüse einiger Bienen-Arten. *Zeitschr. Morph. Ökologie Tiere* 49:139–183.

Nelson, J. A. 1915. *The embryology of the honeybee.* Princeton, New Jersey: Princeton Univ. Press.

New, D. A. T., and J. K. New. 1962. The dances of honeybees at small zenith distances of the sun. *J. Exp. Biol.* 39:271–291.

Newton, D. C. 1968. Behavioural response of honeybees to colony disturbance by smoke. I. Engorging behavior. *J. Apicult. Res.* 7:3–9.

Nielsen, J. C. 1902. Biologiske Studier over danske enlige Bier og deres Snyltere. *Videnskabelige Meddelelser fra den Naturhist. Forening i Kjøbenhavn,* pp. 75–106.

Nixon, H. L., and C. R. Ribbands. 1952. Food transmission within the honeybee community. *Proc. Roy. Soc.* [London] (B)140:43–50.

Nogueira-Neto, P. 1948. Notas bionômicas sôbre meliponíneos. I. Sôbre a ventilação dos ninhos e as construções com ela relacionadas. *Rev. Brasileira Biol.* 8:465–488.

———. 1949. Notas bionomicas sôbre meliponíneos. II Sôbre a pilhagem. *Papéis Avulsos Dept. Zool.* [São Paulo] 9:13–31.

———. 1950. Notas bionomicas sôbre meliponíneos. IV Colonias mistas e questões relacionadas. *Rev. Entomol.* [Rio de Janeiro] 21:305–367.

———. 1954. Notas bionômicas sôbre meliponíneos, III-Sôbre a enxameagem. *Arq. Mus. Nacional* [Rio de Janeiro] 42:219–452.

———. 1970a. *A criação de abelhas indígenas sem ferrão.* [1st ed., 1953] São Paulo: Chacaras e Quintais.

———. 1970b. Behavior problems related to the pillages made by some parasitic stingless bees. In *Development and evolution of behavior, essays in memory of T. C. Schneirla,* pp. 416–434. San Francisco: Freeman and Co.

Noll, J. 1931. Untersuchungen über die Zeugnung und Staatenbildung des *Halictus malachurus. Zeitschr. Morphol. Ökologie Tiere* 23:285–368.

Núñez, J. A. 1966. Quantitative Beziehungen zwischen den Eigenschaften von Futterquellen und den Verhalten von Sammelbienen. *Zeitschr. vergl. Physiol.* 53:142–164.

———. 1970. The relationship between sugar flow and foraging and recruiting behavior of honey bees (*Apis mellifera* L.). *Anim. Behav.* 18:527–538.

———. 1971. Beobachtungen an sozialbezogenen Verhaltensweisen von Sammelbienen. *Zeitschr. Tierpsychologie* 28:1–18.

Nye, W. P., and O. Mackensen. 1968, 1970. Selective breeding of honeybees for alfalfa pollen collection. *J. Apicult. Res.* 7:21–27; 9:61–64.

Oettingen-Spielberg, T. 1949. Über das Wesen der Suchbiene. *Zeitschr. vergl. Physiol.* 15:431–487.

Oliveira, B. L. de. 1960. Mudas ontogenéticas em larvas de *Melipona nigra schencki* Gribodo. *Bol. Univ. Paraná, Zool.* no. 2:1–16.

Oliveira, M. A. C. de, and V. L. I. Fonseca. 1973. Observações sobre o comportamento de uma colônia mista de *Plebeia saiqui–Plebeia droryana. Ciencia e Cultura* [São Paulo] 25:460–462.

Opfinger, E. 1931. Ueber die Orientierung der Biene an der Futterquelle. *Zeitschr. vergl. Physiol.* 15:431–487.

———. 1949. Zur Psychologie der Duftdressuren bei Bienen. *Zeitschr. vergl. Physiol.* 31:441–453.

Ordetx, G. S., and D. Espina Pérez. 1966. *La apicultura en los tropicos.* Mexico City: B. Trucco.

Ordway, E. 1964. *Sphecodes pimpinellae* and other enemies of *Augochlorella. J. Kansas Entomol. Soc.* 37:139–152.

———. 1965. Caste differentiation in *Augochlorella* (Hymenoptera, Halictidae). *Insectes Sociaux* 12:291–308.

———. 1966. The bionomics of *Augochlorella striata* and *A. persimilis* in eastern Kansas. *J. Kansas Entomol. Soc.* 39:270–313.

Pain, J. 1954. sur l'ectophormone des reines d'abeilles. *Compt. Rend. Acad. Sci.* [Paris] 239:1869–1870.

———. 1955. Influence des raines mortes dur le développement ovarien de jeunes ouvrières d'abeilles (*Apis mellifica*). *Insectes Sociaux* 2:35–43.

———. 1961. Sur la phéromone des reines d'abeilles et ses effets physiologiques. *Ann. Abeille* 4:73–153.

———, and M. Barbier. 1963. Structures chimiques et propriétés biologiques de quelques substances identifiées chez l'abeille. *Insectes Sociaux* 10:129–142.

———, and F. Ruttner. 1963. Les extraits de glandes mandibulaires des raines d'abeilles attirent les mâles, lors du vol nuptial. *Compt. Rend. Acad. Sci.* [Paris] 256:512–515.

Perkins, R. C. L. 1917. *Andrena bucephala* Steph. and *Nomada bucephalae* Perk. in Devonshire, and notes on their habits. *Entomol. Monthly Mag.* 3:198–199.

Pflumm, W. 1969a. Beziehungen zwischen Putzverhalten und Sammelbereitschaft bei der Honigbiene. *Zeitschr. vergl. Physiol.* 64:1–36.

———. 1969b. Stimmungsänderungen der Biene während des Aufenthalts an der Futterquelle. *Zeitschr. vergl. Physiol.* 65:299–323.

Pijl, L. van der, and C. H. Dodson. 1966. *Orchid flowers: their pollination and evolution.* Miami, Florida: Univ. Miami Press.

Plateaux-Quénu, C. 1959. Un noveau type de société d'insectes: *Halictus marginatus* Brullé. *Ann. Biol.* 35:235–444, plates I–IX.

———. 1960a. Utilisation d'un nid de *Halictus marginatus* par une fondatrice de *Halictus malachurus. Insectes Sociaux* 7:349–352.

———. 1960b. Nouvelle preuve d'un déterminisme imaginal des castes chez *Halictus marginatus* Brullé. *Compt. Rend. Acad. Sci.* [Paris] 250:4465–4466.

———. 1961. Les sexués de remplacement chez les insectes sociaux. *Ann. Biol.* 37:177–216.

———. 1962. Biology of *Halictus marginatus* Brullé. *J. Apicult. Res.* 1:41–51.

———. 1963. Sur les femelles d'été de *"Halictus calceatus"* Scopoli. *Compt. Rend. Acad. Sci.* [Paris] 256:2247–2248.

———. 1964. Sur quelques traits de la biologie de *Halictus calceatus* Scopoli. *Insectes Sociaux* 11:91–96.

———. 1965a. Sur le cycle biologique de *Halictus nigripes* Lep. *Compt. Rend. Acad. Sci.* [Paris] 260:2331–2333.

———. 1965b. Sur la succession des activités de la fondatrice de *Halictus nigripes* Lep. *Compt. Rend. Acad. Sci.* [Paris] 260:2609–2612.

———. 1967. Tendances évolutives et degré de socialisation chez les Halictinae. *Ann. Soc. Entomol. France* (NS)3:859–866.

———. 1972. *La biologie des abeilles primitives.* Paris: Masson et Cie.

———, and G. Knerer. 1968. Régulation sociale dans les sociétés orphelines d'*Evylaeus nigripes* (Lep.). *Insectes Sociaux* 15:31–35.

Plath, O. E. 1923. Notes on the egg-eating habit of bumblebees. *Psyche* 30:193–202.

———. 1934. *Bumblebees and their ways.* New York: Macmillan Co.

Plowright, R. C., and S. C. Jay. 1968. Caste differentiation in bumblebees: the determination of female size. *Insectes Sociaux* 15:171–192.

Pooley, A. C., and C. D. Michener. 1969. Observations on nests of stingless bees in Natal. *J. Entomol. Soc. Southern Africa* 32:423–430.

Portugal-Araújo, V. de. 1955. Colmeias para "abelhas sem ferrão." *Bol. Inst. Angola* no. 7:5–34, figs. 1–29.

———. 1958. A contribution to the bionomics of *Lestrimelitta cubiceps. J. Kansas Entomol. Soc.* 31:203–211.

———. 1963. Subterranean nests of two African stingless bees. *J. New York Entomol. Soc.* 71:130–141.

Pouvreau, A. 1971. Sur le déterminisme des castes chez les bourdons. *Ann. Zool.—Écol. Anim.* 3:501–507.

Qayyum, H. A., and N. Ahmad. 1967. Biology of *Apis dorsata. Pakistan J. Sci.* 19:109–113.

Quénu, C. 1957. Sur les femelles d'été de *Halictus scabiosae* (Rossi). *Compt. Rend. Acad. Sci.* [Paris] 244:1073–1076.

Rau, P. 1933. *Jungle bees and wasps of Barro Colorado Island.* Kirkwood, Missouri: published by author.

Rayment, T. 1951. Biology of the reed bees, with descriptions of three new species and two allotypes of *Exoneura. Australian Zool.* 11:285–313, plates xxvii–xxxii.

Reinhardt, E. 1960. Kernverhaltnisse, Eisystem und Entwicklungsweise von Drohnen und Arbeiterinneneiern der Honigbiene (*Apis mellifera*). *Zool. Jahrbucher* (*Anat.*) 78:167–234.

Reinhardt, J. F. 1952. Some responses of honeybees to alfalfa flowers. *Amer. Natur.* 86:257–275.

Renner, M. 1957. Neue Versuche über den Zeitsinn der Honigbiene. *Zeitschr. vergl. Physiol.* 40:85–118.

———. 1960. Das duftorgan der Honigbiene und die physiologische Bedeutung ihres Lockstoffes. *Zeitschr. vergl.*

Physiol. 43:411–468.
Rezende, J. A. 1967. Uma colônia mista de duas espécies de *Plebeia* (Hymenoptera, Apoidea). *Papéis Avulsos Dept. Zool.* [S. Paulo] 20:9–12.
Ribbands, C. R. 1953. *The behaviour and social life of honeybees.* London: Bee Res. Assn. Ltd. [Republished, 1964, by Dover Publ., Inc., New York.]
———. 1954. The defense of the honeybee community. *Proc. Roy. Soc.* [London] (B)142:514–524.
———. 1955. The scent perception of the honeybee. *Proc. Roy. Soc.* [London] (B)143:367–379.
———. 1956. The scent language of honeybees. *Smithsonian Report for 1955,* pp. 369–377.
Richards, O. W. 1927. The specific characters of the British humble-bees. *Trans. Entomol. Soc. London* 75:233–268.
———. 1946. Observations on *Bombus agrorum* (Fabricius). *Proc. Roy. Entomol. Soc. London* 21:66–71.
———. 1953. *The social insects.* London: Macdonald.
———. 1968. The subgeneric divisions of the genus *Bombus* Latreille. *Bull. Brit. Mus. (Natur. Hist.) Entomol.* 22:209–276.
Roberts, R. B., and C. H. Dodson. 1967. Nesting biology of two communal bees, *Euglossa imperialis* and *Euglossa ignita* including description of larvae. *Ann. Entomol. Soc. Amer.* 60:1007–1014.
Rockstein, M. 1950. The relation of cholinesterase activity to change in cell number with age in the brain of the adult worker honeybee. *J. Cell. Comp. Physiol.* 35:11–23.
Rohlf, F. J. 1968. Stereograms in numerical taxonomy. *Syst. Zool.* 17:246–255.
Rösch, G. A. 1925, 1930. Untersuchungen über die Arbeitsteilung im Bienenstaat, I and II. *Zeitschr. vergl. Physiol.* 2:571–631; 12:1–71.
———. 1927. Über die Bautätigkeit im Bienenvolk und das Alter der Baubienen. *Zeitschr. vergl. Physiol.* 6:264–298.
Röseler, P. F. 1965. Beobachtungen uber die Verhaltensweisen in künstlich erzielten polygynen Hummelvölkern. *Insectes Sociaux* 12:105–116.
———. 1967a. Arbeitsteilung und Drüsenzustände in Hummelvölkern. *Naturwissenschaften* 54:146–147.
———. 1967b. Untersuchungen über das Auftreten der 3 Formen im Hummelstaat. *Zool. Jahrbucher (Physiol.)* 74:178–197.
———. 1970. Unterschiede in der Kastendetermination zwichen den Hummelarten *Bombus hypnorum* und *Bombus terrestris. Zeitschr Naturforsch.* 25:543–548.
Roth, M. 1965. La production de chaleur chez *Apis mellifica. Ann. Abeille* 8:5–77.
Rothenbuhler, W. C. 1964. Behaviour genetics of nest cleaning in honeybees I. *Anim. Behav.* 12:578–583.
———. 1967. American foulbrood and bee biology. *Proc. 21st Internat. Beekeeping Congr.* pp. 179–188.
———, J. M. Kulinčević, and W. E. Kerr. 1968. Bee genetics. *Ann. Rev. Genet.* 2:413–438.
Rozen, J. G., Jr. 1965. The biology and immature stages of *Melitturga clavicornis* (Latreille) and of *Sphecodes albilabris* (Kirby) and the recognition of the Oxaeidae at the family level. *Amer. Mus. Novitates* 2224:1–18.
———. 1968. Biology and immature stages of the aberrant bee genus *Meliturgula. Amer. Mus. Novitates* 2331:1–18.
———. 1971. Biology and immature stages of Moroccan panurgine bees. *Amer. Mus. Novitates* 2457:1–37.
Ruttner, F. 1956. The mating of the honeybee. *Bee World* 37:2–15, 23–24.
———. 1966. The life and flight activity of drones. *Bee World* 47:93–100.
———, and H. Ruttner. 1966, 1968. Untersuchungen über die Flugaktivität und das Paarungsverhalten der Drohnen. *Zeitschr. Bienenforschung* 8:332–354; 9:259–265.
Sakagami, S. F. 1953a. Untersuchungen über die Arbeitsteilung in einem Zwergvolk der Honigbiene. Beiträge zur Biologie des Bienenvolkes, *Apis mellifera* L., I. *Jap. J. Zool.* 11:117–185.
———. 1953b. Arbeitsteilung der Arbeiterinnen in einem Zwergvolk, bestehend aus gleichaltrigen Volksgenossen. *J. Fac. Sci. Hokkaido Univ.* (6, Zool.) 11:343–400.
———. 1954. Occurrence of an aggressive behavior in queenless hives, with considerations on the social organization of honeybees. *Insectes Sociaux* 1:331–343.
———. 1958a. The false queen: fourth adjustive response in dequeened honeybee colonies. *Behaviour* 8:280–296.
———. 1958b. An attempt to rear the Japanese bee in a framed hive (Studies on the Japanese honeybee, *Apis cerana cerana* Fabr. IV). *J. Agr. Hokkaido Univ.* (VI, Zool.) 14:1–8.
———. 1959. Some interspecific relations between Japanese and European honeybees. *J. Anim. Ecol.* 28:51–68.
———. 1960a. Ethological peculiarities of the primitive social bees, *Allodape* Lepeltier (*sic*) and allied genera. *Insectes Sociaux* 7:231–249.
———. 1960b. Preliminary report on the specific difference of behavior and other ecological characters between European and Japanese honeybees. *Acta Hymenopterologica* 1:171–198.
———. 1965. Über den Bau der mannlichen Hinterschiene von *Eulaema nigrita* Lepeletier. *Zool. Anzeiger* 175:347–354.
———. 1966. Techniques for the observation of behavior and social organization of stingless bees by using a special hive. *Papéis Avulsos Dept. Zool.* [São Paulo] 19: 151–162.
———. 1968. Nesting habits and other notes on an Indomalayan halictine bee, *Lasioglossum albescens* with description of *L. a. iwatai* ssp. nov. *Malayan Natur. J.* 21:85–99.

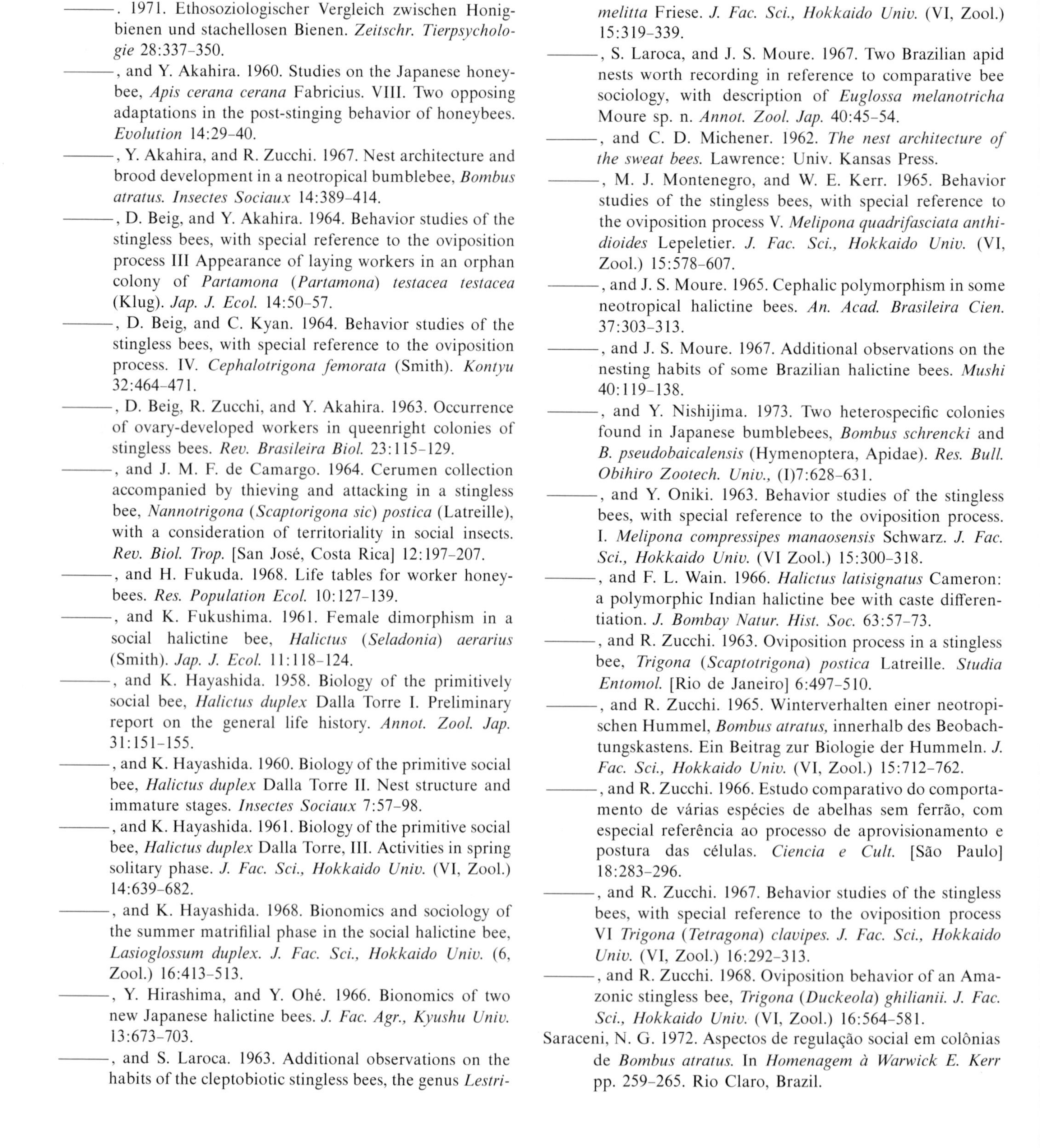

———. 1971. Ethosoziologischer Vergleich zwischen Honigbienen und stachellosen Bienen. *Zeitschr. Tierpsychologie* 28:337–350.

———, and Y. Akahira. 1960. Studies on the Japanese honeybee, *Apis cerana cerana* Fabricius. VIII. Two opposing adaptations in the post-stinging behavior of honeybees. *Evolution* 14:29–40.

———, Y. Akahira, and R. Zucchi. 1967. Nest architecture and brood development in a neotropical bumblebee, *Bombus atratus*. *Insectes Sociaux* 14:389–414.

———, D. Beig, and Y. Akahira. 1964. Behavior studies of the stingless bees, with special reference to the oviposition process III Appearance of laying workers in an orphan colony of *Partamona* (*Partamona*) *testacea testacea* (Klug). *Jap. J. Ecol.* 14:50–57.

———, D. Beig, and C. Kyan. 1964. Behavior studies of the stingless bees, with special reference to the oviposition process. IV. *Cephalotrigona femorata* (Smith). *Kontyu* 32:464–471.

———, D. Beig, R. Zucchi, and Y. Akahira. 1963. Occurrence of ovary-developed workers in queenright colonies of stingless bees. *Rev. Brasileira Biol.* 23:115–129.

———, and J. M. F. de Camargo. 1964. Cerumen collection accompanied by thieving and attacking in a stingless bee, *Nannotrigona* (*Scaptorigona sic*) *postica* (Latreille), with a consideration of territoriality in social insects. *Rev. Biol. Trop.* [San José, Costa Rica] 12:197–207.

———, and H. Fukuda. 1968. Life tables for worker honeybees. *Res. Population Ecol.* 10:127–139.

———, and K. Fukushima. 1961. Female dimorphism in a social halictine bee, *Halictus* (*Seladonia*) *aerarius* (Smith). *Jap. J. Ecol.* 11:118–124.

———, and K. Hayashida. 1958. Biology of the primitively social bee, *Halictus duplex* Dalla Torre I. Preliminary report on the general life history. *Annot. Zool. Jap.* 31:151–155.

———, and K. Hayashida. 1960. Biology of the primitive social bee, *Halictus duplex* Dalla Torre II. Nest structure and immature stages. *Insectes Sociaux* 7:57–98.

———, and K. Hayashida. 1961. Biology of the primitive social bee, *Halictus duplex* Dalla Torre, III. Activities in spring solitary phase. *J. Fac. Sci., Hokkaido Univ.* (VI, Zool.) 14:639–682.

———, and K. Hayashida. 1968. Bionomics and sociology of the summer matrifilial phase in the social halictine bee, *Lasioglossum duplex*. *J. Fac. Sci., Hokkaido Univ.* (6, Zool.) 16:413–513.

———, Y. Hirashima, and Y. Ohé. 1966. Bionomics of two new Japanese halictine bees. *J. Fac. Agr., Kyushu Univ.* 13:673–703.

———, and S. Laroca. 1963. Additional observations on the habits of the cleptobiotic stingless bees, the genus *Lestrimelitta* Friese. *J. Fac. Sci., Hokkaido Univ.* (VI, Zool.) 15:319–339.

———, S. Laroca, and J. S. Moure. 1967. Two Brazilian apid nests worth recording in reference to comparative bee sociology, with description of *Euglossa melanotricha* Moure sp. n. *Annot. Zool. Jap.* 40:45–54.

———, and C. D. Michener. 1962. *The nest architecture of the sweat bees.* Lawrence: Univ. Kansas Press.

———, M. J. Montenegro, and W. E. Kerr. 1965. Behavior studies of the stingless bees, with special reference to the oviposition process V. *Melipona quadrifasciata anthidioides* Lepeletier. *J. Fac. Sci., Hokkaido Univ.* (VI, Zool.) 15:578–607.

———, and J. S. Moure. 1965. Cephalic polymorphism in some neotropical halictine bees. *An. Acad. Brasileira Cien.* 37:303–313.

———, and J. S. Moure. 1967. Additional observations on the nesting habits of some Brazilian halictine bees. *Mushi* 40:119–138.

———, and Y. Nishijima. 1973. Two heterospecific colonies found in Japanese bumblebees, *Bombus schrencki* and *B. pseudobaicalensis* (Hymenoptera, Apidae). *Res. Bull. Obihiro Zootech. Univ.*, (I)7:628–631.

———, and Y. Oniki. 1963. Behavior studies of the stingless bees, with special reference to the oviposition process. I. *Melipona compressipes manaosensis* Schwarz. *J. Fac. Sci., Hokkaido Univ.* (VI Zool.) 15:300–318.

———, and F. L. Wain. 1966. *Halictus latisignatus* Cameron: a polymorphic Indian halictine bee with caste differentiation. *J. Bombay Natur. Hist. Soc.* 63:57–73.

———, and R. Zucchi. 1963. Oviposition process in a stingless bee, *Trigona* (*Scaptotrigona*) *postica* Latreille. *Studia Entomol.* [Rio de Janeiro] 6:497–510.

———, and R. Zucchi. 1965. Winterverhalten einer neotropischen Hummel, *Bombus atratus,* innerhalb des Beobachtungskastens. Ein Beitrag zur Biologie der Hummeln. *J. Fac. Sci., Hokkaido Univ.* (VI, Zool.) 15:712–762.

———, and R. Zucchi. 1966. Estudo comparativo do comportamento de várias espécies de abelhas sem ferrão, com especial referência ao processo de aprovisionamento e postura das células. *Ciencia e Cult.* [São Paulo] 18:283–296.

———, and R. Zucchi. 1967. Behavior studies of the stingless bees, with special reference to the oviposition process VI *Trigona* (*Tetragona*) *clavipes*. *J. Fac. Sci., Hokkaido Univ.* (VI, Zool.) 16:292–313.

———, and R. Zucchi. 1968. Oviposition behavior of an Amazonic stingless bee, *Trigona* (*Duckeola*) *ghilianii*. *J. Fac. Sci., Hokkaido Univ.* (VI, Zool.) 16:564–581.

Saraceni, N. G. 1972. Aspectos de regulação social em colônias de *Bombus atratus.* In *Homenagem à Warwick E. Kerr* pp. 259–265. Rio Claro, Brazil.

Schmid, J. 1964. Zur Frage der Störung des Bienengedächtnisses durch Narkosemittel, zugleich ein Beitrag zur Störung der sozialen Bindung durch Narkose. *Zeitschr. vergl. Physiol.* 47:559–595.

Schneider, P. 1972. Akustische Signale bei Hummeln. *Naturwissenschaften* 59:168.

Schnetter, B. 1968. Visuelle Formunterscheidung der Honigbiene im Bereich von Vier- und Sechsstrahlsternen. *Zeitschr. vergl. Physiol* 59:90–109.

Schremmer, F. 1972a. Beobachtungen zum Paarungsverhalten der Männchen von *Bombus confusus* Schenck. *Zeitschr. Tierpsychol.* 31:503–512.

———. 1972b. Der Stechsaugrüssel, der Nektarraub, das Pollensammeln und der Blütenbesuch der Holzbienen (*Xylocopa*) (Hymenoptera, Apidae). *Zeitschr. Morph. Tiere* 72:263–294.

Schricker, B., and W. P. Stephen. 1970. The effect of sublethal doses of parathion on honeybee behavior. I. Oral administration and the communication dance. *J. Apicult. Res.* 9:141–153.

Schwarz, H. F. 1948. Stingless bees of the western Hemisphere. *Bull. Amer. Mus. Natur. Hist.* 90:i–xviii + 1–546.

Seibt, U. 1967. Der Einfluss der Temperatur auf die Dunkeladaptation von *Apis mellifica. Zeitschr. vergl. Physiol.* 57:77–162.

Sekiguchi, K., and S. F. Sakagami. 1966. Structure of foraging population and related problems in the honeybee, with considerations on the division of labor in bee colonies. *Hokkaido Nat. Agr. Expt. Sta. Rep.* no. 69:1–65.

Shearer, D. A., R. Boch, R. A. Morse, and F. M. Laigo. 1970. Occurrence of 9-oxodec-*trans*-2-enoic acid in queens of *Apis dorsata, Apis cerana,* and *Apis mellifera. J. Insect. Physiol.* 16:1437–1441.

Shiokawa, M. 1966. Comparative studies of two closely allied sympatric *Ceratina* bees, *C. flavipes* and *C. japonica* II. Nest structure. *Kontyu* 34:44–51.

Shuel, R. W., and S. E. Dixon. 1959. Studies in the mode of action of royal jelly in honeybee development, II. *Can. J. Zool.* 37:803–813.

———, and S. E. Dixon. 1968. Respiration in developing honeybee larvae. *J. Apicult. Res.* 7:11–19.

Silva, D. L. N. da. 1972. Considerações em torno de um caso de substituição de rainha em *Plebeia* (*Plebeia*) *droryana* (Friese, 1900). In *Homenagem à Warwick Kerr* pp. 267–273. Rio Claro, Brazil.

———, R. Zucchi, and W. E. Kerr. 1972. Biological and behavioral aspects of the reproduction in some species of *Melipona. Anim. Behav.* 20:123–132.

Simpson, J. 1955. The significance of the presence of pollen in the food of worker larvae of the honeybee. *Quart. J. Microscop. Sci.* 96:117–120.

———. 1960. The functions of the salivary glands of *Apis mellifera. J. Insect Physiol.* 4:107–121.

———. 1963. The factor that causes swarming by honeybee colonies in small hives. *J. Apicult. Res.* 2:50–54.

———. 1964. The mechanism of honeybee queen piping. *Zeitschr. vergl. Physiol.* 48:277–282.

———, and S. M. Cherry. 1969. Queen confinement, queen piping and swarming in *Apis mellifera* colonies. *Anim. Behav.* 17:271–278.

———, I. B. M. Riedel, and N. Wilding. 1968. Invertase in the hypopharyngeal glands of the honeybee. *J. Apicult. Res.* 7:29–36.

Singh, S. 1962. *Bee keeping in India.* New Delhi: India Council of Agr. Research.

Skaife, S. H. 1952. The yellow-banded carpenter bee, *Mesotrichia caffra* Linn, and its symbiotic mite, *Dinogamasus Braunsi* Vitzthun (*sic*). *J. Entomol. Soc. Southern Africa* 15:63–76.

———. 1953. Subsocial bees of the genus *Allodape* Lep. and Serv. *J. Entomol. Soc. Southern Africa* 16:3–16.

Sladen, F. W. L. 1912. The humblebee, its life-history and how to domesticate it. London: Macmillan and Company.

Smith, F. G. 1960. *Beekeeping in the tropics.* London: Longmans.

Smith, J. B. 1901. Notes on some digger bees—II. *J. New York Entomol. Soc.* 9:52–72.

Smith, M. V. 1959. Queen differentiation and the biological testing of royal jelly. *Cornell Univ. Agr. Exp. Sta. Mem.* 356:1–56.

Snodgrass, R. E. 1910. *The anatomy of the honeybee.* U.S. Dept. Agr., Bur. Entomol. Tech. Ser., no. 18:1–162, 57 figs.

———. 1925. *Anatomy and physiology of the honeybee.* New York: McGraw Hill Co.

———. 1942. The skeleto-muscular mechanisms of the honey bee. *Smithsonian Misc. Collections,* 103(2):1–120.

———. 1956. *Anatomy of the honeybee.* Ithaca, New York: Cornell Univ. Press.

Steche, W. 1957. Soziale Steuerung des Alarmiwertes der Bienentänze. *Naturwissenschaften* 44:597–598.

Stein, G. 1956. Beiträge zur Biologie der Hummel (*B. terrestris* L., *B. lapidarius* L. u.a.). *Zool. Jahrbucher* (*Syst.*) 84:439–462.

———. 1963. Uber den Sexuallockstoff von Hummelmännchen. *Naturwissenschaften* 50:305.

Steiner, A. 1947. Der Wärmehaushalt der einheimischen sozialen Hautflügler. *Beih. Schweiz. Bienen-Zeitg.* 2:139–256.

Stejskal, M. 1962. Duft als 'Sprache' der tropischen Bienen. *Südwestdeut. Imker* 49:271.

Stephen, W. P., and B. Schricker. 1970. The effect of sublethal doses of parathion, II. Site of parathion activity, and signal integration. *J. Apicult. Res.* 9:155–164.

Stockhammer, K. A. 1966. Nesting habits and life cycle of a sweat bee, *Augochlora pura*. *J. Kansas Entomol. Soc.* 39:157–192.

———. 1967. Some notes on the biology of the blue sweat bee, *Lasioglossum coeruleum*. *J. Kansas Entomol. Soc.* 40:177–189.

Stöckhert, E. 1923. Über Entwicklung und Lebensweise der Bienengattung *Halictus* Latr. und ihrer Schmarotzer. *Konowia* 2:48–64, 146–165, 216–247.

Stoeckhert, F. K. 1933. Die Bienen Frankens. *Beiheft Deutschen Entomol. Zeitschr.* (1932) viii + 294 pp.

Stussi, T. 1967. Thermogenese de l'abeille et ses rapports avec le niveau thermique de la ruche. Thesis Fac. Sci. Univ. Lyon.

Syed, I. H. 1963. Comparative studies of larvae of Australian ceratinine bees. *Univ. Kansas Sci. Bull.* 44:263–280.

Taber, S., III, and C. D. Owens. 1970. Colony founding and initial nest design of honeybees, *Apis mellifera*. *Anim. Behav.* 18:625–632.

———, and J. Wendel. 1958. Concerning the number of times queen bees mate. *J. Econ. Entomol.* 51:786–789.

Taranov, G. F. 1955. On the biology of swarming in bees [in Russian]. *Pchelovodstvo* 32:32–35.

Terada, Y. 1972. Enxameagem em *Frieseomelitta varia* Lepeletier, In *Homenagem à Warwick E. Kerr,* pp. 293–299. Rio Claro, Brazil.

Thakar, C. V., and K. V. Tonapi. 1961. Nesting behavior of Indian honeybees I. Differentiation of worker, queen and drone cells in the combs of *Apis dorsata* Fab. *Bee World* 42:61–62.

———, and K. V. Tonapi. 1962. Nesting behaviour of Indian honeybees II. Nesting habits and cell differentiation in *Apis florea* Fab. *Indian Bee J.* 24:27–31.

Thompson, V. C., and W. C. Rothenbuhler. 1957. Resistance to American Foulbrood in Honeybees. II. *J. Econ. Entomol.* 50:731–737.

Thorp, R. W. 1969. Ecology and behavior of *Anthophora edwardsii*. *Amer. Midland Natur.* 82:321–337.

Tkalců, B. 1966. *Megabombus* (*Fervidobombus*) *abditus* sp. n. aus Aequatorial-Afrika. *Acta Mus. Morav.* 51:271–274.

Torchio, P. F., and N. N. Youssef. 1968. The biology of *Anthophora* (*Micranthophora*) *flexipes* and its cleptoparasite, *Zacosmia maculata,* including a description of the immature stages of the parasite. *J. Kansas Entomol. Soc.* 41:289–302.

Townley, R. 1970. The queen in relation to comb building [summary of studies of R. Darchen]. *Bee World* 51:74–78.

Tucker, K. W. 1958. Automictic parthenogenesis in the honeybee. *Genetics* 43:299–316.

Urban, D. 1967. As espécies do gênero *Thygater* Holmberg. *Bol. Univ. Federal Paraná, Zool.* 2:177–309.

Vareschi, E. 1971. Duftunterscheidung bei der Honigbiene—Einzelzell-Ableitungen und Verhaltensreaktionen. *Zeitschr. vergl. Physiol.* 75:143–173.

Vasić, Z. 1966. Nekoliko momenata iz života vrste *Halictus marginatus* Brullé [Einige Momente aus dem Leben von *Halictus marginatus* Brullé]. *Bull. Mus. Hist. Natur.* [Beograd] (B)21:103–118.

———. 1967. *Halictus quadricinctus* F. et le probleme de la polygynie. *Bull. Mus. Hist. Natur.* [Beograd] (B)22:181–187.

Velthuis, H. H. W. 1970. Queen substances from the abdomen of the honeybee queen. *Zeitschr. vergl. Physiol.* 70:210–222.

———. 1972. Observations on the transmission of queen substance in the honey bee colony by the attendants of the queen. *Behaviour* 41:103–129.

———, F. J. Verheijen, and A. J. Gottenbos. 1965. Laying worker honeybee: similarities to the queen. *Nature* 207: 1314.

Verheijen-Voogd, C. 1959. How worker bees perceive the presence of their queen. *Zeitschr. vergl. Physiol.* 41:527–582.

Verhoeff, C. 1897. Zur Lebensgeschichte der Gattung *Halictus* (Anthophila), insbesondere einer Übergangsform zu socialen Bienen. *Zool. Anzeiger* 20:369–393.

Viswanathan, H. 1950a. Temperature readings on *Apis dorsata* combs. *Indian Bee J.* 12:72.

———. 1950b. Note on *Apis dorsata* queen cells. *Indian Bee J.* 12:55.

Vleugel, D. A. 1961. Uber die Zahl der Individuen in den Nestern der halbsozialen Biene *Halictus calceatus* Scop. und verwandten in der ersten Generation. *XI Internat. Kongr. Entomol.* [Wien] 1:586-588.

Vogel, S. 1966. Scent organs of orchid flowers and their relation to insect pollination. *Proc. V World Orchid Conf.* p. 253–259.

Voogd, S. 1955. Inhibition of ovary development in worker bees by extraction fluid of the queen. *Experimentia* 11:181–182.

Weaver, N. 1956. The foraging behavior of honeybees on hairy vetch—foraging methods and learning to forage. *Insectes Sociaux* 3:537–549.

———. 1957. Effects of larval age on dimorphic differentiation of the female honeybee. *Ann. Entomol. Soc. Amer.* 50:283–294.

———. 1966. Physiology of caste determination. *Annu. Rev. Entomol.* 11:79–102.

Webb, M. C. 1961. The biology of the bumblebees of a limited area in eastern Nebraska. Ph.D. Thesis, Univ. Nebraska, vi + 337 pp.

Wehner, R., and M. Lindauer. 1966. Zur Physiologie des Formensehens bei der Honigbiene, I Winkelunterscheidung am vertikal orientierten Streifenmustern. *Zeitschr. vergl.*

Physiol. 52:290–324.
Weiss, K. 1962. Untersuchungen über die Drohnerzeugung im Bienenvolk. *Arch. Bienenkunde* 39:1–7.
Wells, P. H., and A. M. Wenner. 1971. The influence of food scent on behavior of foraging honeybees. *Physiol. Zool.* 44:191–209.
Wenner, A. M. 1967. Honeybees: Do they use the distance information contained in their dance maneuver? *Science* 155:847–849.
———, and D. L. Johnson. 1967. [Reply to von Frisch.] *Science* 158:1076–1077.
———, P. H. Wells, and D. L. Johnson. 1969. Honeybee recruitment to food sources: olfaction or language? *Science* 164:84–86.
———, P. H. Wells, and F. J. Rohlf. 1967. An analysis of the waggle dance and recruitment in honeybees. *Physiol. Zool.* 40:317–344.
Wheeler, W. M. 1923. *Social life among the insects.* New York: Harcourt, Brace and Company.
———. 1928. *The social insects.* London: Paul, Trench, Trubner and Company.
White, J. W., Jr., M. H. Subers, and A. A. Schepartz. 1963. The identification of inhibine, the antibacterial factor in honey, as hydrogen peroxide and its origin in a honey glucose-oxidase system. *Biochim. Biophy. Acta* 73: 57–70.
Wiese, H. 1972. Abelhas africanas, suas characteristicas e technologia de manejo. *Proc. 1st Congr. Brasileiro Apicult.* [Florianopolis] pp. 95–110.
Wille, A. 1959. A new fossil stingless bee from the amber of Chiapas, Mexico. *J. Paleo.* 33:849–852.
———, and L. Chandler. 1964. A new stingless bee from the Tertiary amber of the Dominican Republic (Hymenoptera; Meliponini). *Rev. Biol. Trop.* [San José, Costa Rica] 12:187–195.
———, and C. D. Michener. 1971. Observations on the nests of Costa Rican *Halictus* with taxonomic notes on Neotropical species (Hymenoptera: Halictidae). *Rev. Biol. Trop.* [San José, Costa Rica] 18:17–31.
———, and C. D. Michener. In press. The nest architecture of stingless bees with special reference to those of Costa Rica. *Rev. Biol. Trop.* [San José, Costa Rica].
———, and E. Orozco. 1970. The life cycle and behavior of the social bee *Lasioglossum* (*Dialictus*) *umbripenne. Rev. Biol. Trop.* [San José, Costa Rica] 17:199–245.
Williams, G. C. 1966. *Adaptation and natural selection.* Princeton, New Jersey: Princeton Univ. Press.
Wilson, E. O. 1966. Behaviour of social insects. In *Insect Behaviour, Symposium no. 3,* ed. P. T. Haskell, pp. 81–96. Royal Entomol. Soc. London.
———. 1971. *The insect societies.* Cambridge, Massachusetts: Harvard Univ. Press.
Witcomb, W., Jr., and H. F. Wilson. 1929. Mechanics of digestion of pollen by the adult hive bee and the relation of undigested parts to dysentery of bees. *Agr. Exp. Sta., Univ. Wisconsin, Res. Bull.* 92:1–27.
Witherell, P. C. 1971. Duration of flight and of interflight time of drone honeybees, *Apis mellifera. Ann. Entomol. Soc. Amer.* 64:609–612.
Wittekindt, W. 1960. Schwänzelbewegungen als Ausdruck gesteigerter Erregung innerhalb des Tanzverhaltens der Honigbiene. *Naturwissenschaften* 47:335–336.
———. 1966. Das Durchwinden, eine Tanz- und Alarmierungsform der Honigbiene. *Bienenzucht* 19:14–24.
Witthöft, W. 1967. Absolute Anzahl und Verteilung der Zellen im Hirn der Honigbiene. *Zeitschr. Morphol. Ökologie Tiere* 61:160–184.
Wohlgemuth, R. 1957. Die Temperaturregulation des Bienenvolkes unter regeltheoretischen Gesichtspunkten. *Zeitschr. vergl. Physiol.* 40:119–161.
Wójtowski, F. 1963. Studies on heat and water economy in bumble-bees nests. *Zool. Poloniae* 13:19–36.
Wolf, E. 1927. Ueber das Heimkehrvermögen der Bienen, II. *Zeitschr. vergl. Physiol.* 6:227–254.
Woyke, J. 1958. Przebieg kopulacji u pszczól. [The process of mating in the honeybee.] *Pszczelnicze Zeszyty Naukowe* 2:1–42.
———. 1963. Rearing and viability of diploid drone larvae. *J. Apicult. Res.* 2:77–84.
———. 1964. Causes of repeated mating flights in queen honeybees. *J. Apicult. Res.* 3:17–23.
———. 1965. Study on diploid drone honeybees. *Compt. Rend. V Congr. Internat. Étude Insectes Sociaux,* pp. 257–262.
———. 1967. Diploid drone substance–cannibalism substance. *Proc. XXIst Internat. Apicult. Congr.,* pp. 57–58.
———. 1971. Correlations between the age at which honeybee brood was grafted, characteristics of the resultant queens, and results of insemination. *J. Apicult. Res.* 10:45–55.
———. 1973. Reproductive organs of haploid and diploid drone honeybees. *J. Apicult. Res.* 12:35–51.
———, and Z. Adamska. 1972. The biparental origin of adult honeybee drones proved by mutant genes. *J. Apicult. Res.* 11:41–49.
———, and A. Knytel. 1966. The chromosome number as proof that drones can arise from fertilized eggs of the honeybee. *J. Apicult. Res.* 5:149–154.
———, and F. Ruttner. 1958. An anatomical study of the mating process in the honeybee. *Bee World* 39:3–18.
Yadava, R. P. S., and M. V. Smith. 1971. Aggressive behaviour of *Apis mellifera* L. workers towards introduced queens I. Behavioural mechanisms involved in the release of worker aggression. *Behaviour* 39:212–236.

Yarrow, I. H. H. 1970. Is *Bombus inexspectatus* (Tkalcu) a workerless obligate parasite? *Insectes Sociaux* 17:95–112.

———, and K. M. Guichard. 1941. Some rare Hymenoptera Aculeata, with two species new to Britain. *Entomol. Monthly Mag.* 77:2–13.

Yoshikawa, K., R. Ohgushi, and S. F. Sakagami. 1969. Preliminary report on entomology of the Osaka City University 5th scientific expedition to southeast Asia 1966. *Nature and Life in Southeast Asia* [Tokyo] 6:153–181, 17 plates.

Zander, E. 1951. *Der Bau der Biene.* Stuttgart: Eugen Ulmer.

———. 1964. *Das Leben der Biene* (*Handbook der Bienenkunde,* Vol. IV). 6th ed. Stuttgart: K. Weiss.

Zmarlicki, C., and R. A. Morse. 1963. The mating of aged virgin queen honeybees. *J. Apicult. Res.* 2:62–63.

———, and R. A. Morse. 1964. The effect of mandibular gland extirpation on the longevity and attractiveness to workers of queen honeybees, *Apis mellifera. Ann. Entomol. Soc. Amer.* 57:73–74.

Zucchi, R. 1966. Aspectos evolutivas do comportamento social entre abelhas. *Reunião Anual de Soc. Brasil. Genetica, Programa e Resumos* [Piracicaba], pp. 6–9.

———, and S. F. Sakagami. 1972. Capacidade termo-reguladora em *Trigona spinipes* e em algumas outras espécies de abelhas sem ferrão. In *Homenagem à Warwick E. Kerr,* pp. 301–309. Rio Claro, Brazil.

———, S. F. Sakagami, and J. M. F. de Camargo. 1969. Biological observations on a neotropical parasocial bee, *Eulaema nigrita,* with a review of the biology of Euglossinae. A comparative study. *J. Fac. Sci., Hokkaido Univ.* (VI, Zool.) 17:271–380.

Index